# Civil Engineering and Urban Planning

# Civil Engineering and Urban Planning

Edited by Seth Royal

CLANRYE INTERNATIONAL
www.clanryeinternational.com

Clanrye International,
750 Third Avenue, 9th Floor,
New York, NY 10017, USA



ISBN: 978-1-63240-779-5

**Cataloging-in-Publication Data**

Civil engineering and urban planning / edited by Seth Royal.
p. cm.
Includes bibliographical references and index.
ISBN 978-1-63240-779-5
1. Civil engineering. 2. City planning. 3. Urban policy. 4. Urban renewal. I. Royal, Seth.
TA145 .C58 2019
624--dc23

For information on all Clanrye International publications
visit our website at www.clanryeinternational.com

CLANRYE INTERNATIONAL

Printed in China.

# Contents

# Preface

This book aims to highlight the current researches and provides a platform to further the scope of innovations in this area. This book is a product of the combined efforts of many researchers and scientists, after going through thorough studies and analysis from different parts of the world. The objective of this book is to provide the readers with the latest information of the field.

The field of civil engineering is concerned with the design, construction and maintenance of infrastructure projects like roads, bridges, railways, etc. Urban Planning is a technical study of the optimum utilization of space. It is a combination of societal, economic and environmental welfare considerations to create a sustainable planning model for an urban, semi-urban or rural area. Many factors such as topography, agricultural land, natural drainage, presence or absence of erodible land are duly considered while planning an area. This book provides significant information to help develop a good understanding of civil engineering and urban planning. This book explores modern approaches and methodologies relevant in the present day scenario. With state-of-the-art inputs by acclaimed experts of this field, this book targets students and professionals. It will provide comprehensive knowledge to the readers.

I would like to express my sincere thanks to the authors for their dedicated efforts in the completion of this book. I acknowledge the efforts of the publisher for providing constant support. Lastly, I would like to thank my family for their support in all academic endeavors.

**Editor**

# Numerical Methodology for Predicting the Nonlinear Elastic Deflection of Curved Beams and Plates Using Nonlinear Integral Equations

**Jung Kwan Seo, So Young Bae, Jeom Kee Paik and Taek Soo Jang***

*The Ship and Offshore Research Institute, Pusan National University, Busan, Korea*

**Abstract**

Ships, ship-shaped offshore structures, land-based structures and aerospace structures typically consist of various curved beam and plate components. An important issue is how to manufacture and construct systems correctly and accurately when their variously shaped curved beams and plates display structural nonlinear behaviour. Multi-point forming is a general process that can be used to form variously shaped curved plates and beams. It has been difficult to apply the integrated systems to thin and/or thick metal plates and beams that have the characteristic of non-linear structural mechanics, such as nonlinear behaviour when loading is applied to the plates. The aim of this study was to derive a simplified numerical calculation based on the formulation of a nonlinear integral equation and a mathematical solution, to predict the nonlinear elastic deflection, load profile and elastic foundation of curved plates or/and beams during the early stages of the advanced manufacturing process of plate and beam formation. The validity of the method is demonstrated through simulations of the proposed numerical approach with experimental results.

**Keywords:** Nonlinear structural behavior; Nonlinear integral equation; Curved plate/beam; Elastic foundation; Cold forming process

## Nomenclature

$a$ Length of beam

$b_{eff}$ Effective breadth

$E$ Young's modulus of elasticity

$f[u(x)]$ Nonlinear restoration

$G$ Green's function

$I$ Mass moment of inertia

$k$ Spring force

$L(x)$ Applied loading

$p(x)$ Net loading

$t$ Plate thickness

$u(x)$ Vertical deflection

$x, y$ Principal axes

$\beta$ Convergence of parameter

$\delta_1$ Maximum magnitude of curved beam edges

$\delta_2$ Maximum magnitude of curved plate from original plate

$\lambda(\mathrm{u})$ Nonlinear part of the beam deflection

$\pi(\mathrm{x})$ Linear part of the beam deflection

## Introduction

The general forming process for curved metal plates can be roughly classified into two types, hot forming and cold forming, according to the thermal condition of the metal during the forming process. A typical method of hot forming is line heating, a popular technique used in shipyards to form sections of ship hulls using the properties of metal plates: heating for expansion and rapid cooling for contraction. Even today, skilled workers still perform the forming process using their empirical intuition, as shown in Figure 1, and no systematic method for ensuring effective formation has been developed.

**Figure 1:** Workers using the line heating method.

Cold forming can be categorised as either conventional solid die forming or multi-point forming (MPF) for sheet metal, as shown in Figure 2. MPF is a cold forming manufacturing technique that combines three-dimensional sheet metal with a pair of multi-point dies consisting of upper and lower matrices of square punches with spherical ends in an MPF press [1]. By controlling each punch, a sheet part can be manufactured along a specific predefined forming path. In this mode, the shape of the multi-point die (MPD) varies continuously during the forming process and all of the punches are kept in contact with the sheet at all times. Accordingly, the deforming load applied by the die is distributed throughout the sheet.

The flexibility of MPD and the variability of the forming path using

***Corresponding author:** Taek Soo Jang, The Ship and Offshore Research Institute, Pusan National University, Busan, Korea, E-mail: taek@pusan.ac.kr

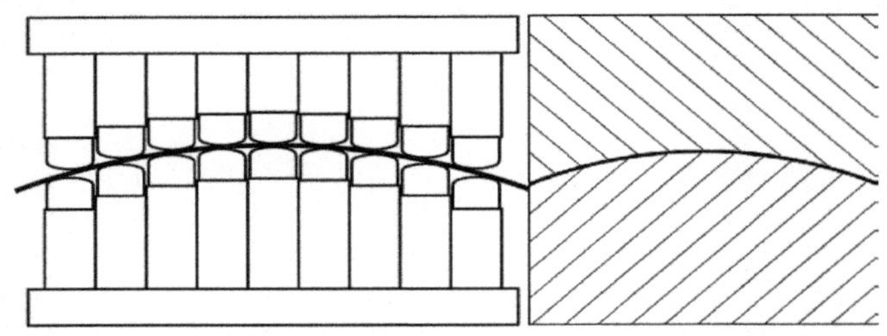

**Figure 2:** Press-type and MPF-type of solid die forming system.

MPF make it possible to form curved metal plate into different shapes in an MPF press. MPF has many other advantages, including a better work environment and independent use of the press device.

A number of researchers have made important contributions in this area by developing numerical, experimental and analytical approaches. The idea of forming sheets of variable shape has always been attractive as a means of reducing costs, as it enables rapid and almost cost-free design iterations. The method of "multi-point forming (MPF) for sheet metal" was first described by one of the authors in [2]. A series of improvements in MPF technology have been made in recent years [1]. In MPF, two matrices of punches replace the conventional stamping dies; the relative positions of the punches can be varied to suit specific requirements. The method is most suitable for forming large shell parts, such as the shells of ships and other large vessels.

Although the MPF system has several technical advantages, as previously described, it inevitably comes with some undesirable consequences. When the forming force is released from the cold-worked metal plate in the MPF forming process, the material has a tendency to return partially to its original shape because of the elastic recovery of the material. A critical problem is how to predict nonlinear structural behaviour that varies with the position, especially with three-dimensional curved plate. The problem with predicting nonlinear structural behaviour using a numerical approach is that it always includes nonlinearities, which renders the problem complicated and difficult to analyse. To overcome this, workers spend a great deal of time and effort repeatedly carrying out the same work and checking whether the MPF is behaving correctly in the forming process.

Paik et al. [3] developed an advanced changeable die system that applies the characteristics of nonlinear structural mechanics to the process of forming three-dimensionally curved metal plates. A commercial nonlinear finite element analysis (FEA) code was used to predict nonlinear structural behaviour based on the elastic-plastic large deformation theory with automation program, which provides information about the relative position of each punch. It is a simple operation because it only supplies the geometry and material information relevant to the desired part. Their changeable die system with FEA techniques considered nonlinear structural behaviour and it was successfully implemented in their integrated MPD system. However, predicting the quantitative loads and restraints at various positions on the plate with different plate thicknesses requires additional work to the numerical approach and a simplified system for the early stages of the forming process.

Recently, Jang et al. [4] proposed a transformed non-linear integral equation that governs nonlinear beam deflection behaviour, which was formulated to develop a new method for nonlinear solutions. Their results provide many examples of the solution's application and quick convergence, whereas traditional nonlinear solution procedures are generally quite complex to apply. However, as the practical application of this approach has not yet been demonstrated, it is still considered an 'applicable numerical method'.

In the present study, we propose a simplified numerical approach to overcome the problem of the nonlinear structural deflection of an infinite beam on a non-linear elastic foundation of a curved plate, considered as a one-dimensional beam. When the non-linear effects of the elastic foundation are taken into account, numerical solution procedures are usually employed.

This study obtained reasonable nonlinear structural behaviour and an ideal forming force and path by comparing the proposed numerical method with an experimental study to shape the desired part. It can also provide useful information for developing CAD/CAE/CAM systems to automate the MPF system to operate effectively with the results of previous research.

## The Proposed Numerical Methodology

The steps in the proposed numerical methodology are shown in Figure 3. First, a representation of the target plate and beam is developed by obtaining the desired curved plate and beam deflection data. Then, a governing differential equation is expressed for an infinite beam with a non-linear elastic foundation with simplified loading and boundary conditions, and its corresponding integral equation and the iterative solution are employed to obtain a numerical model of the shape of the required piece of plate.

A mathematical algorithm is used to conduct an analysis of convergence and obtain a unique solution. The nonlinear integral equations and mathematical solution [4] are then employed by including the elastic foundation and loading condition terms for the nonlinear deflection of the beam. The proposed numerical methodologies that demonstrate and validate the feasibility of the proposed method with curved plate models are presented in Section 3. The final section presents the discussion and conclusions. In the following section, an ideal one-dimensional beam on a nonlinear elastic foundation and integral equation formulation are described.

### Representation of a beam on a nonlinear elastic foundation

The system to be considered is an infinitely long beam resting on

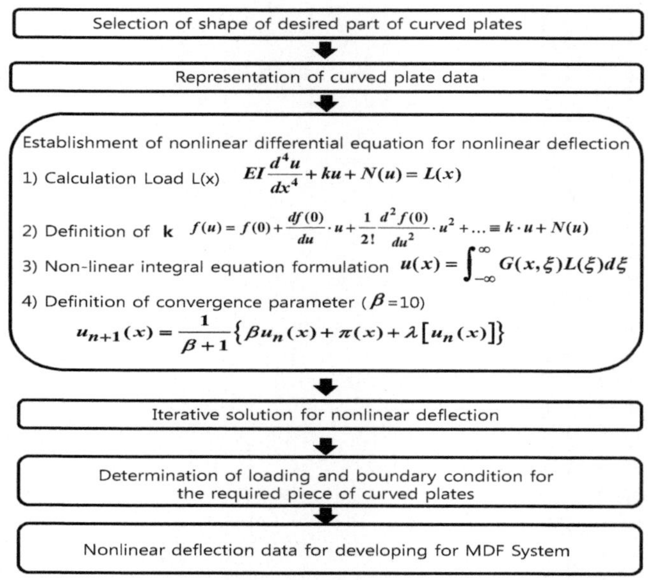


**Figure 3:** Flowchart of the proposed methodology.

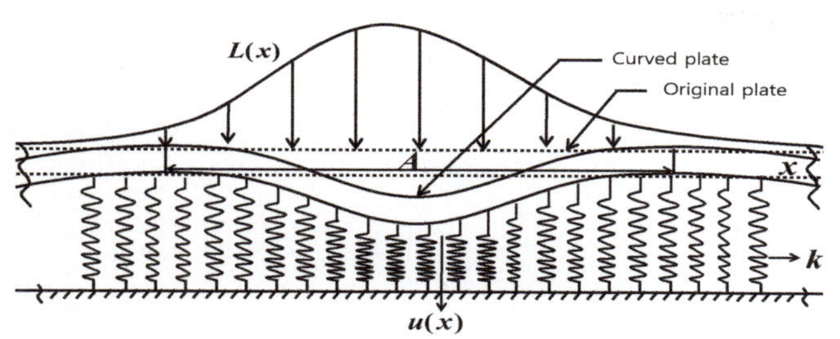


**Figure 4:** Infinite beam on a nonlinear elastic foundation.

a nonlinear elastic foundation, as shown in Figure 4. For the beam shown, A represents the cross-section of the beam, I is the moment of inertia of the beam cross-section, and $A$ is the projected length of the beam, which is the effective longitudinal length of the desired part of the curved plate. According to the classical Euler beam theory, the vertical deflection $u(x)$ that results from the load distribution $p(x)$ satisfies the fourth-order ordinary differential equation, written as

$$EI\frac{d^4u}{dx^4} = p(x) \tag{1}$$

As $p(x)$ is the net loading, consisting of the applied downward loading $L(x)$ and the nonlinear upward spring force $f[u(x)]$, the net loading can be written as

$$p(x) = L(x) - f\left[u(x)\right] \tag{2}$$

For simplicity, we ignore the weight of the beam. Substitution of Eq. (2) into Eq. (1) gives

$$EI\frac{d^4u}{dx^4} + f(u) = L(x) \tag{3}$$

Assume that the nonlinear restoration $f(u)$ is odd and analytic, such that we have the Taylor series of $f(u)$ around the equilibrium of $u = 0$:

$$f(u) = f(0) + \frac{df(0)}{du}\cdot u + \frac{1}{2!}\frac{d^2f(0)}{du^2}\cdot u^2 + \ldots \equiv k\cdot u + N(u)\,, \tag{4}$$

where $df(0)/du$ is defined as $k$, and $f(0)$ is set to zero because $f$ is an odd function. Therefore, a nonlinear ordinary differential equation for the nonlinear deflection can be expressed as

$$EI\frac{d^4u}{dx^4} + ku + N(u) = L(x) \tag{5}$$

## Nonlinear integral equation and iterative solution

It is considered advantageous to employ an integral operator in nonlinear analysis. The nonlinear differential equation, Eq. (6), was transformed into an equivalent integral equation. It can be viewed as the linear deflection solution when the nonlinear spring force term $N(u)$ is negligible.

Timoshenko, Kenney, Saito & Murakami and Fryba [5-8] derived the general linear solution of a nonlinear ordinary differential equation using the Fourier and Laplace transforms and expressed in Green's function as follows:

$$G(x,\xi) = \frac{\alpha}{2k}e^{-\alpha|\xi-x|/\sqrt{2}}\sin\left(\frac{\alpha|\xi-x|}{\sqrt{2}} + \frac{\pi}{4}\right),\ \text{where } a = \sqrt[4]{k/EI} \tag{6}$$

Green's function $G$ is calculated by the complex contour integration as follows:

$$u(x) = \int_{-\infty}^{\infty} G(x,\xi)L(\xi)d\xi \tag{7}$$

In the derivation of Eq. (7), it is assumed that the loading $L(x)$ is sufficiently localised for $u$, and $\frac{du}{dx}$, $\frac{d^2u}{dx^2}$ and $\frac{d^3u}{dx^3}$ all tend towards zero as $|x| \to \infty$. If we set $F = L(x) - N(u)$ then the nonlinear ordinary differential equation can be rewritten as Eq. (8). Eq. (7) expresses that (8) must be equivalent to the following relation in Eq. (9).

$$EI\frac{d^4u}{dx^4} + ku = L(x) - N(u) \tag{8}$$

$$u(x) = \int_{-\infty}^{\infty} G(x,\xi)\Phi(\xi)d\xi \tag{9}$$

$$u(x) = \int_{-\infty}^{\infty} G(x,\xi)L(\xi)d\xi - \int_{-\infty}^{\infty} G(x,\xi)N[u(\xi)]d\xi \tag{10}$$

After the substitution of $F = L(x) - N(u)$ into Eq. (9), the nonlinear relation for the nonlinear deflection u is defined Eq. (10), which is classified as a nonlinear Fredholm integral equation of the second kind for the nonlinear beam deflection.

The nonlinear integral equation was successfully derived for the nonlinear beam deflection [4]. It is necessary to develop a mathematical or analytical solution based on the derived nonlinear integral equation for the beam deflection. It can be defined by two terms, $\pi(x)$ for a linear beam deflection and $\lambda(u)$ for a nonlinear beam deflection. Therefore, Eq. (10) can be written as an operator equation, such as $u = p(x) + l(u)$. The operator equation is *our starting point* for calculating the nonlinear beam deflection. It can be expressed as an iterative solution for the nonlinear beam deflection of the desired part of the plate, as follows:

$$u_{n+1}(x) = \frac{1}{b+1}\left\{b\,u_n(x) + p(x) + l\left[u_n(x)\right]\right\} \tag{11}$$

# Validation of the Proposed Numerical Methodology

## Target curved-metal plates

To demonstrate the accuracy and application of the proposed numerical method to the manufacturing of curved plates, the mathematical explicit solution (integral equation) and test results of the curved plates were computed. Figure 5 presents the geometric properties of the curved plates considered in this study.

The target curved plate is from the side section of a typical container ship. Here, we consider two types of surface-curved plates (Type A and B), defined and manufactured by the MPF process, as described in Figure 5. Paik and Kim [9] tested and measured the material and geometric characteristics of their changeable die system and tested various types of curved plate using their advanced system. Type A and B plates are the most commonly used shape for the relevant part of ships' structures. Tensile coupon tests were carried out to determine the material characteristics of the metal plate using a 1000kN

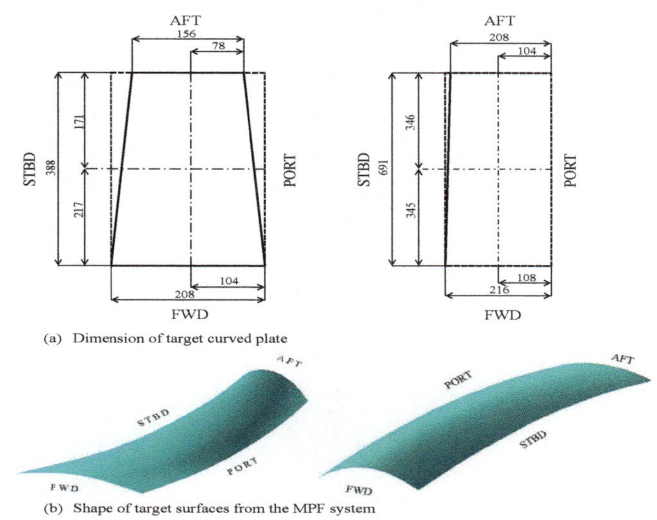


**Figure 5:** Two different target surfaces from conventional solid die forming and MPF.

universal testing machine. For this, a tensile coupon specimen was made according to the ASTM E8M standard. The elastic modulus and Poisson's ratio were 205.8 GPa and 0.3, respectively. The magnitudes of the plates' deformed shapes were measured for each test specimen.

## Numerical modelling

**Development of the numerical model:** Various analytical solutions are proposed to provide the most accurate and simplified solutions based on theoretical and mathematical analysis. However, it is important to realise that the modelling technique applied must be capable of representing the actual structural behaviour associated with geometrical nonlinearity, material nonlinearity, type boundary conditions, loading conditions and so on.

To determine the numerical model to assess the accuracy of the proposed iterative solution, it is necessary to define the one-dimensional numerical model of a nonlinear integral equation for the nonlinear beam deflection, as described in Figure 6. This model is suitable for the geometric properties and nonlinear structural behaviour of the plate considered here, because its plate deflection behaviour can deal with beam deflection.

To determine the effective breadth of the plate using a numerical model, an approximate effective breadth for a one-dimensional model is employed. Normally, the problem of determining the effective breadth and width is due to the in-plane compression of the steel plating [10]. In this study, the effective breadths ($b_{eff}$) are employed to determine the punch spacing for MPF under out-of-plane loading. In such a situation, and according to the assumptions of simplified modelling techniques, it is normal for the total load from the forming punch to be represented as carrying the maximum stress uniformly, rather than the actual stress distribution. Figure 7a, 7b show a representation of the target measurement deflection data and fitted deflection curves for the iterative solution.

Table 1 presents the geometric properties of the target plate surface as a one-dimensional numerical model. Each displacement ($\delta_1, \delta_2$) indicates the maximum magnitude of difference between the curved plate and the original plate.

Note that this study validates the reasonable nonlinear large deflection behaviour with elastic foundation and ideal forming force and path, by comparing the results of the proposed numerical method with the experimental target deflection data for the shape of the part.

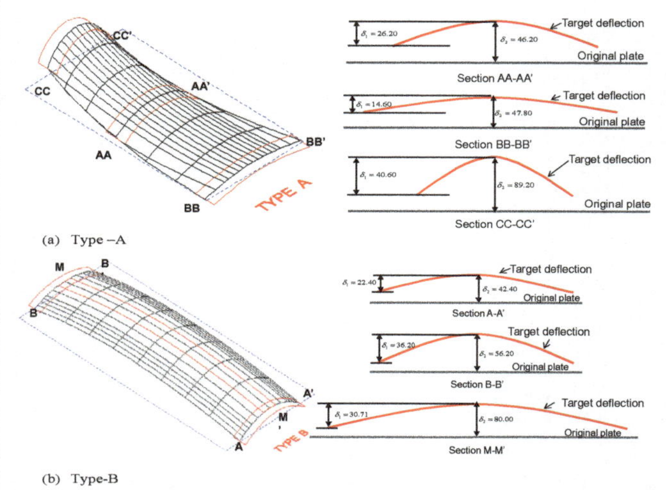


**Figure 6:** Schematic representation of the present numerical model.

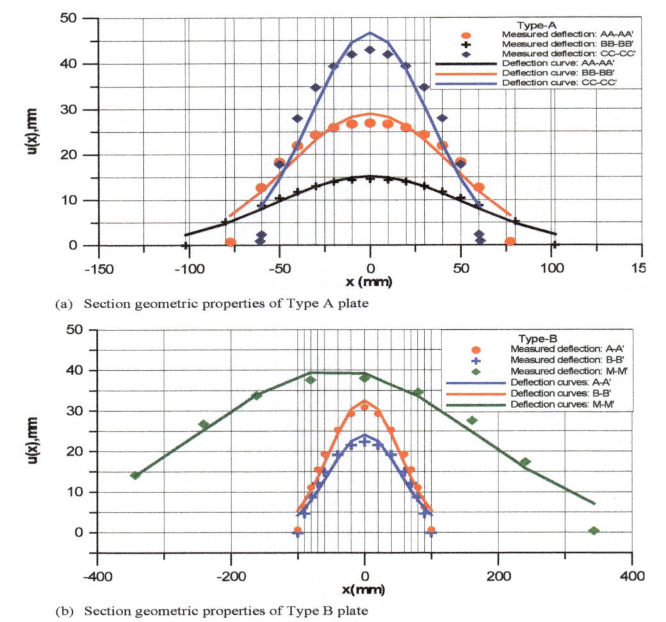


**Figure 7:** Measured deflection data and curves for the target surface (plate A and B).

Therefore, the numerical models are considered only as the most suitable nonlinear deflection and deformation effects for validation of the proposed method.

**Load and Boundary Conditions:** A one-dimensional beam model was applied in the length (x) direction to accurately account for the effects of the elastic foundation constraints, as described in Figure 4. The experimental results show that the shape of the MPD varied continuously during the forming process and all punches remained in contact with the sheet at all times. Therefore, we assume that the numerical MPD model represents a nonlinear elastic foundation.

To refine the model, the external loads corresponding to the target deflection obtained from the governing Eq. 3 were employed, by inputting the actual target deflection curves into $u(x) = Ae^{-((x-b)/c)^2}$. The external loading can be expressed by an exponential function, as in the following equation and as shown in Figure 8.

$$L(x) = 12e^{-x^2} - 48x^2e^{-x^2} + 16x^4e^{-x^2} + ke^{-x^2} + ge^{-3x^2} \tag{12}$$

It is necessary to consider an infinitely long beam resting on a nonlinear elastic foundation, as shown in Figure 4. According to the classical Euler beam theory, the target surface $u(x)$ that results from load distribution $p(x)$ satisfies the fourth-order ordinary differential equation. As $p(x)$ is the net loading, consisting of the applied downward loading L($x$) and the nonlinear upward spring force $f[u(x)]$, it can be assumed that the nonlinear restoration $f(u)$ is odd and analytic, such that it can be expressed by the Taylor series of $f(u)$ around the equilibrium of $u = 0$, as shown in Eq. (4).

## Results of the Proposed Method and Discussion

The one-dimensional numerical model was developed to simulate the loading and boundary conditions considered in the experiments as closely as possible. The desired target deflections of an infinite beam on a nonlinear elastic foundation are obtained using an iterative solution, representing a mathematical analysis of the proposed method.

Figure 9 compares the nonlinear deflection of the target surface and

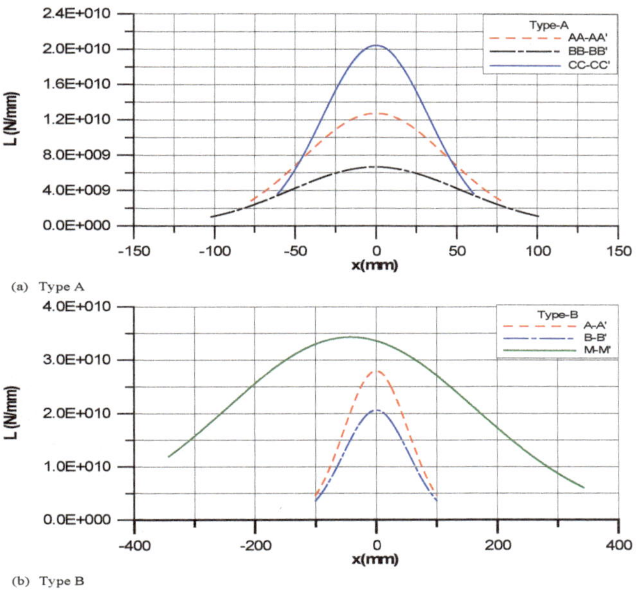


**Figure 8:** Applied loading conditions.

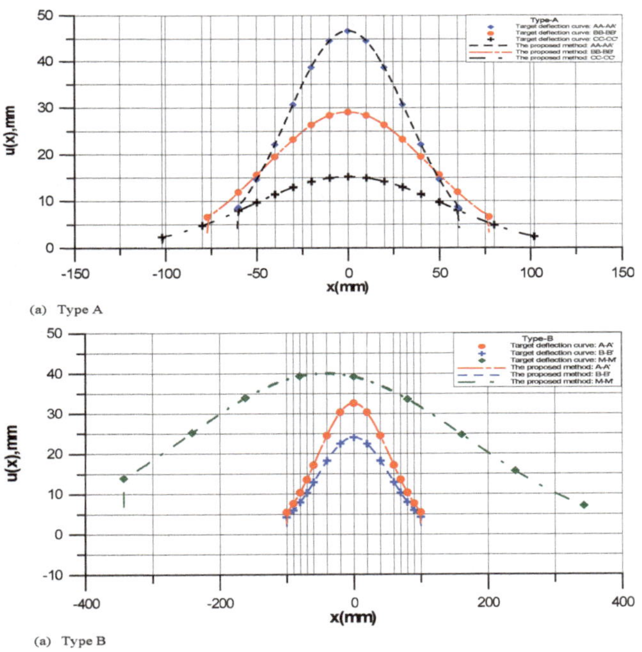


**Figure 9:** Comparison of the experimental data with the numerical solutions.

the proposed iterative method under ideal loads with elastic foundations for both models. The z-direction deformed values of the punch locations from the experimental data are calculated for the locations AA–AA', BB–BB' and CC–CC' in Figure 9a and the locations A–A', B–B' and M–M' in Figure 9b. It can be seen that the mathematically derived loads can be implemented successfully as the actual applied punch loads (displacement control method). The accuracy of the proposed numerical methods using the k values of nonlinear elastic foundation was determined by the methods described in Jang et al. [4].

Figure 9 presents the calculated non-linear deflections u(x), with zero initial guess $u_0 = 0$, for the model with 100 iterations, which shows good agreement with the target surface deflection.

Figure 10 illustrates the convergence behaviour of the iterative solution for the two models that have the most deflected shape compared with the original plate. As can be seen in the figure, the iterative solutions are close to the target surface deflection after only the first iteration

The errors of the approximate solutions [4] at the n[th] iteration are defined as follows.

$$Error(n) \equiv \frac{\left\| u_{\exp}(x) - u_{num}(x) \right\|_2}{\left\| u_{\exp}(x) \right\|_2}, \text{ where} \left\| z \right\|_2 \equiv \left( \sum_{i=1}^{N} \left| z_i \right|^2 \right)^{1/2}. \quad (13)$$

The errors of the proposed method are plotted against the iteration number in Figure 11. The errors decrease monotonically in most cases, and for different values of $\beta$ reach a steady state after 50-200 iterations. It is important to check the convergence behaviour of the proposed solutions at different convergence constant $\beta$s. The effect of constant $\beta$ in Eq. (11) on the number of iterations is shown in Figure 10. The analysis in Section 2 shows that the number of iterations is expected to increase as constant $\beta$ becomes larger.

The results demonstrate that the proposed method based on nonlinear integral equation is able to predict with accuracy the nonlinear deflection of curved plate, as shown in Figure 9. However, to simulate the actual structural behaviour of curved plates or beams during the cold-forming process, several issues need to be taken into consideration. When the forming force is released from the cold-worked metal plate in the MPF forming process, the material has a tendency to return partially to its original shape because of the elastic recovery of the material. This is called spring-back, and is influenced not only by the tensile and yield strengths, but also by the thickness, bend radius and bend angle. The model must be capable of modelling the structural

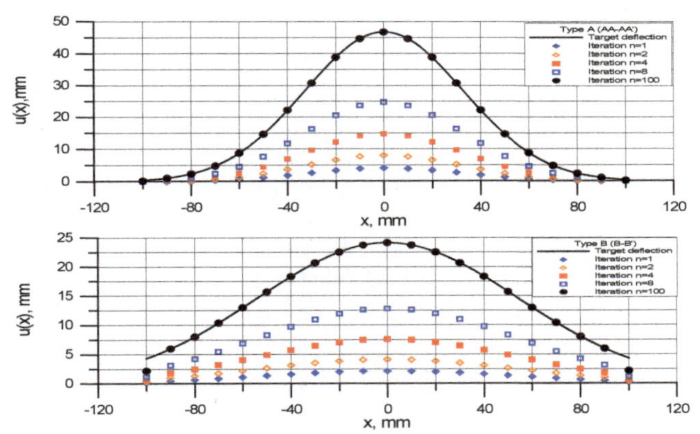


**Figure 10:** Convergence behaviour of the iterative method.

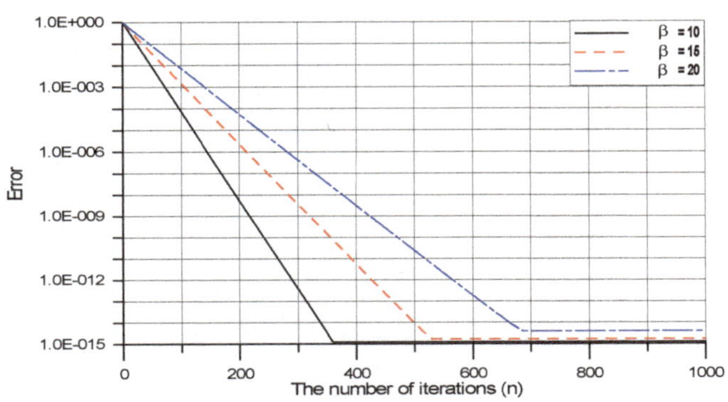


**Figure 11:** Error of the proposed iterative solutions vs. the number of iterations with different convergence $\beta$ coefficients.

| Type | Section | $a$ (mm) | $b_{eff}$ (mm) | $t$ (mm) | $\delta_1$ (mm) | $\delta_2$ (mm) |
|---|---|---|---|---|---|---|
| A | AA-AA' | 208 | 20 | 4 | 26.20 | 46.20 |
| | BB-BB' | 208 | 20 | 4 | 14.60 | 47.80 |
| | CC-CC' | 208 | 20 | 4 | 40.60 | 89.20 |
| B | A-A' | 216 | 20 | 5 | 22.40 | 42.40 |
| | B-B' | 216 | 20 | 5 | 36.20 | 56.20 |
| | M-M' | 691 | 20 | 5 | 30.71 | 80.00 |

**Table 1:** Geometric properties of the target beams for numerical modeling.

behaviour in both the linear and non-linear regions, including large displacements, elasto-plastic deformations and associated plasticity effects. This issue needs to be addressed using a numerical approach, to provide accurate predictions of the nonlinear structural behaviour of the plate and beam and the associated plasticity effects that vary according to position, especially with three-dimensional curved plate.

The boundary condition in the numerical model is considered to represent the elastic foundation with a spring force that is the basis of the flexibility of MPD. This also needs to be determined and clarified at the numerical modelling stage. However, the numerical elastic foundation model will give valuable information on the flexibility of the MPD to enable variability in the forming path in MPF. This technique makes it possible to form different shapes of curved metal plate in an advanced MPF press system based on the numerical approach. Furthermore, the proposed numerical method has many advantages, including a better work environment and independent use of the MPF press device.

Future research will need to modify the proposed numerical method to handle such problems and modelling issues. To overcome them, workers currently spend a great deal of effort and time repeating the same work and checking the accuracy of the predictions of nonlinear structural behaviour during the forming process of the plates and beams.

## Concluding Remarks

Multi-point forming (MPF) is a general forming process for three-dimensionally curved plates, which can form a variety of part shapes. It has been difficult to apply the MPF integrated system to thin and/or thick metal plate that displays the characteristics of non-linear structural mechanics, such as nonlinear behaviour when a loading is applied to the plates.

This paper proposes a simplified numerical approach based on the formulation of a nonlinear integral equation to predict the nonlinear structural deflection, load profile and elastic foundation of curved plates in advance of the manufacturing forming process of the required plates and beams during the early stages of the systems. The validity of the method is demonstrated through simulations of the proposed numerical approach with experimental results. The study obtained reasonable nonlinear structural behaviour and an ideal forming force and path by comparing the proposed numerical method with an experimental study to shape the required part. The MPF method has always been attractive as a means of reducing costs because it enables design iterations to be rapid and virtually cost free. It also enables workers, even unknowledgeable and unskilled workers, to form curved metal plates using an entirely automated computer-based CAE system.


## Acknowledgment

The first and the third author were supported by Leading Foreign Research Institute Recruitment Program through the National Research Foundation of Korea(NRF) funded by the Ministry of Science, Ict & future Planning(MSIP) (Grant no.: 2013044761). The corresponding author was supported by Basic Science Research Program through the National Research Foundation of Korea (NRF) funded by the Ministry of Education, Science and Technology (Grant no.: 2011-0010090).


## References


1. Li MZ, Cai ZY, Sui Z, Yan QG (2002) Multi-point forming technology for sheet metal. Journal of Materials Processing Technology 129: 333–338.
2. Li MZ, Nakamura K, Watanabe S (1992) Study of the basic principles (first report: research on multi-point forming for sheet metal). In: Proceedings of the Japanese Spring Conference for Technology of Plasticity 519–522.
3. Paik JK, Kim JH, Kim BJ, Tak CH (2010) Analysis of Spring-Back Behavior in the Cold-Forming Process of Three-Dimensionally Curved Metal Plates. The 29th Int. Conference on Ocean, Offshore and Arctic Engineering, Shanghi, China.
4. Jang TS, Baek HS, Paik JK (2011) A new method for the nonlinear deflection analysis of an infinite beam resting on a nonlinear elastic foundation. Int J of Non-Linear Mech 46: 339-346.
5. Timoshenko SP (1926) Statistical and dynamical stress in rails. Proceedings of the International Congress on Applied Mechanics, Zurich 407–418.
6. Kenney JT (1954) Steady-state vibrations of beam on elastic foundation for moving load. Journal of Applied Mechanics 21: 359–364.
7. Saito H, Murakami T (1969) Vibrations of an infinite beam on an elastic foundation with consideration of mass of a foundation. Japanese Society of Mechanical Engineering 12: 200-205.
8. Fryba L (1957) Infinite beam on an elastic foundation subjected to a moving load. Aplikace Matematiky 2: 105-132.
9. Paik JK, Kim JH (2009) Development of the changeable die system for the cold-forming of three-dimensionally curved metal plates. The Lloyd's Register Educational Trust (LRET) Research Centre of Excellence, Pusan National University, Korea.
10. Paik JK, Thayamballi AK (2003) Ultimate Limit State Design of Steel-plated Structures. Wiley, Chichester, UK.

# Sulfur Dioxide ($SO_2$) Monitoring Over Kirkuk City Using Remote Sensing Data

**Ghada Hasan*, Mohammed Ahmed Al Kubaisy, Faten Hamed Nahhas, Ali Adil Ali, Nibras Othman and Mahir Mahmod Hason**
*Civil Engineering Department, Universiti Putra Malaysia, Malaysia*

**Abstract**

Air pollution mapping is now being an important issue to manage and enhance the environment of a city. The major problems of air pollution mapping is the data acquisition due to the high cost of instruments and the high spatial distribution requirements. This study aimed to monitor Sulfur Dioxide over Kirkuk city using Landsat-8 thermal bands to provide Department of Environment Kirkuk with low-cost Sulfur Dioxide concentration maps to better manage the city. The study used correlation analysis to find a relationship between Sulfur Dioxide ground-based measurements and satellite data. The ground-based measurements were collected from (17) stations distributed in Kirkuk city in January, 2014 using NOVA device to measure $SO_2$ concentrations. The research showed a good correlation between ground-based measurements and satellite data with ($R^2$=0.48 for band 11 and $R^2$= 0.52 for band 10). Therefore, the study resulted that with band 10 of Landsat-8 data, better $SO_2$ can be monitored than using band 11. It is recommended to other researchers to investigate the ability of free remote sensing data to monitor all elements that specify the air quality of a city.

**Keywords:** Sulfur dioxide; Air pollution; Remote sensing; GIS; Environment; Kirkuk city; TIRS; $SO_2$; LST

## Introduction

Air pollution is a major concern in developing countries. The main sources of air pollution in developed cities are: emissions from vehicles (60-70%), coal based thermal power plants (10-15%), industrial units (10-15%) and domestic (5-10%) [1]. It causes a number of health problems and it has been linked with illnesses and deaths from heart or lung diseases [2]. Due to the high cost and limited number of air pollutant stations, it is difficult to provide a good spatial distribution of air pollutant measurements over a city. Although satellite measurements are a total column, not always correlated with surface air quality, satellite observations can help us in air pollution mapping with a high spatial distribution.

The objective of this study is to monitor Sulfur Dioxide pollutant over Kirkuk city, Iraq using free remote sensing data and in-situ measurements. The research uses Landsat 8 (TIRS) data to analyze the derived surface temperature and in-situ measurements of (17) ground stations around Kirkuk city. The relationship between satellite data and ground-based measurements will be investigated by correlation and regression analysis in GIS environment.

## Related Works

The monitoring of aerosol concentrations becomes a high environmental priority particularly in urban areas [3]. Several local and international studies have shown the possible relationships between satellite data and air pollution.

A study by [4] investigated the applicability of measuring $PM_{2.5}$ concentrations using satellite data in a semi-empirical approach. The study compared daily $PM_{2.5}$ sampled on the ground surface over six stations and satellite data (MODIS). It showed good agreement ($R^2$=0.68) [5,6]. Therefore, the research confirmed the potentiality of using satellite remote sensing observations for AQ monitoring.

Rohayu Haron Narashid [7] used images from Landsat (ETM+) and (8) continuous air quality monitoring stations to determine relationships between satellite thermal band data and air pollutant concentrations measured on the ground surface. The research resulted that the generated models by regression analysis can be used by environment staffs to monitor air quality in urban area.

Another research by Qihao Weng [8] investigated the relationship of air pollution pattern with urban thermal landscape using a GIS approach. The relationship between the spatial pattern of air pollution and land surface temperature were sought through GIS and correlation analysis.

The study resulted that there is a good relationship between satellite derived surface temperature and air pollutant concentrations. It has been concluded that thermal imagery could play a significant role in monitoring air quality.

Locally, a research by Al-Abdraba addressed monitoring and assessing of air pollution at several sites in North Oil Company in Kirkuk city. The study also tried to construct a mathematical equation for concentrations that discarded from sources close to ground surface. $SO_2$ concentration exceeded the international standards limitation in many studied sites. While Ali, studied Kirkuk Oil Refinery to assess the air, soil and plants contamination regarding the Kirkuk refinery system. The average concentration of $SO_2$ is recorded at 0.075 (ppm) in October, slightly higher than the allowable Iraqi National standards and (WHO), rather than lower concentration (0.006 ppm) is observed in March of that year.

Furthermore, Salah Saleh [1] investigated the relationship between Land Surface Temperature (LST) derived from Landsat TM6 (thermal infrared) satellite data and ground-based air quality parameters measurements over Baghdad city, Iraq. A linear regression analysis was performed using least square method to model air quality of the city. The results showed significant agreements between ground-based measurements and Landsat satellite data. Also, the study indicated that satellite imagery is capable for air pollution mapping.

***Corresponding author:** Ghada Hasan, Civil Engineering Department, Universiti Putra Malaysia, South City Academia A11-11, Selangor, Malaysia
E-mail: maherrsgis@gmail.com

## Sulfur Dioxide

Sulfur Dioxide ($SO_2$) is a significant indicator for volcanic movements [9]. Is produced mainly by combustion of fossil fuels or smelting of minerals, and is an intermediate product from organic sources in the ocean.

Global monitoring of sulfur dioxide gives information related to air quality, climate change, volcanology and volcanic hazards. Sulfate in air pollution is well known for production of acid rain and environmental degradation, as well as to reduce visibility and health effects.

It causes acidic pollution of lakes and streams and forms an aerosol that is important in climate change [10]. Other studies indicate that $SO_2$ causes nerve stimulation in the lining of the nose and throat. This can cause irritation, coughing and a feeling of chest tightness, which may cause the airways to narrow. People suffering from asthma are considered to be particularly sensitive to $SO_2$ concentrations.

## Study Area

The study area of this research is Kirkuk city located in the north part of Iraq and it is a vital place as shown in Figure 1. Generally, this area consists of industrial, residential, oil fields places. The area of the region is approximately equal to (93.16 $km^2$) and the population of the city center currently is estimated to (784444) (Statistical Center of Kirkuk, 2011). Oil is the main source of its economic activities, it has the largest oil field in Iraq, natural gas and sulfur sources.

Geographically, it is located between the two great circles of (35 31 20 N) and (35 20 20 S), and two longitudinal circles of (44 26 10 E) and (44 16 30 W). It is 236 kilometers north of the capital Baghdad. Furthermore, the climate of Kirkuk city is generally continental, with hot and dry summers, and cool rainy winters.

The wind directions play an important role in the distribution of pollutants in air. The wind of the study area prevail north-western (Iraqi Meteorological Organization, 2011) and follows the climate of the Mediterranean Basin.

## Data Sources

### Ground-based measurements

$SO_2$ pollutant concentrations were collected at (17) measuring stations around Kirkuk city in 9th of January, 2014. The used device was NOVA instrument which designed to monitor different types of ambient air pollutants such as CO, $CO_2$, $NH_3$, $NO_2$, $SO_2$ and $H_2S$ as shown in Figure 2.

**Figure 1:** Study Area.

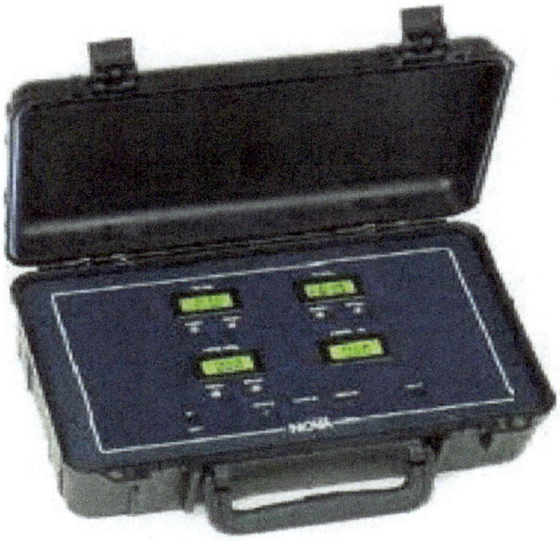


**Figure 2:** NOVA Device.

Table 1 below shows the average measurements of $SO_2$ pollutants concentration measurements of 9th of January, 2014 were higher than Iraqi and WHO standards air pollutants. So, air of Kirkuk city is polluted due to the traffic, personal and private electrical generators, industrial activities and poor vegetation cover in the city.

## Satellite data

Two bands of thermal infrared region (10.4-12.5 μm) of Landsat8 image acquired on 9th of January, 2014 for Kirkuk city was used for land surface temperature extraction. The Digital Number (DN) value for the thermal infrared of the corresponding ground base measurements are converted into radiance and apparent surface temperatures using NASA model [11-13].

## Methodology

**DN conversion to surface temperature:** In order to convert the raw data of Landsat thermal bands DN into surface temperature brightness [14], equation (1) provided by Chander and Markham and the Landsat 7 Science Data Users Handbook was used first to convert the DNs to spectral radiance.

$$TOAr = M * DN + B \quad (1)$$

Where: M is the Radiance Multiplier

B is the Radiance Add

The M, B values are in the metadata file of Landsat8 data. Table 2 shows the used parameters.

Where: K1 and K2 are parameters of band-specific thermal conversion constant.

After we have got the TOAr values, then brightness temperature is calculated using equation (2) below [15].

$$T\ (Kelvin) = K2/\ln(K1/TOAr+1) \quad (2)$$

After calculating temperature brightness in (Kelvin) unit, it is converted to Celsius and mapped in ArcGIS software as shown in Figure 3 [16].

**Correlation analysis:** Before investigating correlation analysis for ground-based and satellite data, the descriptive statistics are analysed for further understanding the nature of the collected data. Table 3 below shows important statistical parameters for two types of used data.

Relationship between Land Surface Temperatures (LST) [17] extracted from Landsat 8-TIRS satellite data with ground-based measurements of $SO_2$ concentrations around Kirkuk city were analyzed by correlation analysis. Table 4 summarizes the linear correlation coefficient analysis between surface temperature of thermal bands and ground-based air pollution concentrations of $SO_2$ pollutant.

The correlation analysis indicates to there are significant correlations

| ID | Station | X | Y | $SO_{2\ (ppm)}$ | Iraqi Standard | Who Standard |
|---|---|---|---|---|---|---|
| 1 | Chorao control | 443138 | 3929753 | 0.696 | | |
| 2 | Eternity Fire | 441111 | 3937665 | 1.105 | | |
| 3 | Compration gas station $AB_3$ | 440030 | 3932208 | 1.645 | | |
| 4 | Baba residential | 438943 | 3933565 | 0.871 | | |
| 5 | Araffa | 443362 | 3926437 | 0.501 | | |
| 6 | Alamal Alshaby | 439376 | 3929069 | 0.733 | | |
| 7 | Sharow hospital | 441948 | 3921311 | 0.276 | | |
| 8 | Rahimawa | 445035 | 3928489 | 0.517 | | |
| 9 | Sahat ihtifalat | 441948 | 3921311 | 0.276 | 0.14 | 0.01 |
| 10 | Sharow compound | 445386 | 3930344 | 0.434 | | |
| 11 | Baglar health centre | 443337 | 3925749 | 0.33 | | |
| 12 | Iraqi drilling company | 442724 | 3928192 | 0.461 | | |
| 13 | Almas | 444792 | 3926332 | 0.301 | | |
| 14 | Rzgary | 446694 | 3927582 | 0.595 | | |
| 15 | Imam qasim | 445321 | 3926091 | 0.484 | | |
| 16 | Khassa 2 | 444241 | 3923971 | 0.923 | | |
| 17 | Hay mualmeen | 444058 | 3920336 | 0.601 | | |

**Table 1:** Ground-based Measurements.

| | Band 10 | Band 11 |
|---|---|---|
| Radiance Multiplier (M) | 0.0003342 | 0.0003342 |
| Radiance Add (B) | 0.1 | 0.1 |
| K1 | 777.89 | 480.89 |
| K2 | 1321.08 | 1201.14 |

**Table 2:** Ground-based Measurements.

| Variable | Mean | SD | Minimum | Median | Maximum |
|---|---|---|---|---|---|
| **Temp Band 10** | 11.587 | 2.258 | 3.880 | 11.856 | 14.367 |
| **Temp Band 11** | 9.801 | 2.112 | 2.712 | 10.249 | 12.273 |
| **$SO_2$** | 0.6463 | 0.3416 | 0.276 | 0.5170 | 1.645 |

**Table 3:** Descriptive Statistical Parameters.

| r | Temp Band 10 | Temp Band 11 | $SO_2$ |
|---|---|---|---|
| **Temp Band 10** | | 0.99 | 0.465 |
| **Temp Band 11** | 0.99 | | 0.417 |
| **$SO_2$** | 0.465 | 0.147 | |

**Table 4:** Correlation between in-situ measurements and satellite data.

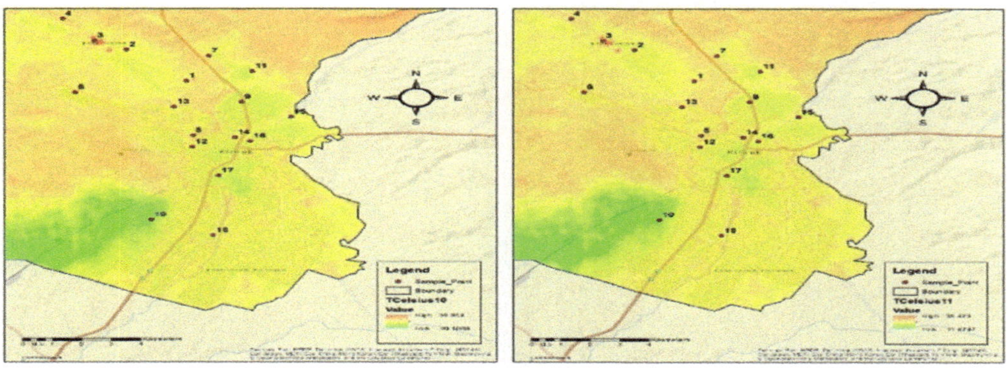

**Figure 3**: Surface Temperature Over Study Area.

between $SO_2$ pollutant concentrations and surface temperatures derived from satellite thermal bands. It also shows that the correlation between $SO_2$ ground-based concentrations and band 10 is stronger than band 11. The resulted correlation is accepted in terms of accuracy for monitoring purposes over large spatial distribution.

**Regression analysis:** A non-linear regression analysis was performed to establish relationship between surface temperatures from Landsat data as independent variables against the in-situ $SO_2$ measurements concentrations data.

A third-order polynomial expression between $SO_2$ pollutant parameter and surface temperatures extracted from thermal bands of Landsat data were determined using regression analysis in Excel Software. Figures 4 and 5 shows the scatter plot of in-situ and satellite data and the fitting line.

## Results

The regression analysis resulted an aerosol reflectance model for estimating $SO_2$ concentrations from raw DNs of thermal bands of Landsat 8 satellite data using the equations below:

$SO_2$ (ppm) = $0.0117T^3 - 0.3282T^2 + 2.837T - 6.4733$…. *For band 10*

$SO_2$ (ppm) = $0.0157T^3 - 0.362T2 + 2.524^2T - 4.219$....*For band 11*

Then, the two expressions above are used to calculate the $SO_2$ derived concentrations and it results as Table 5 below shows. The last

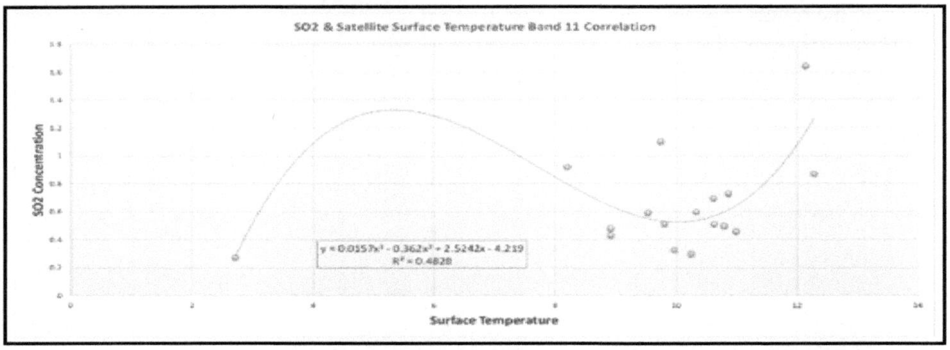


**Figure 4:** In-situ $SO_2$ & satellite data regression, Band 11.

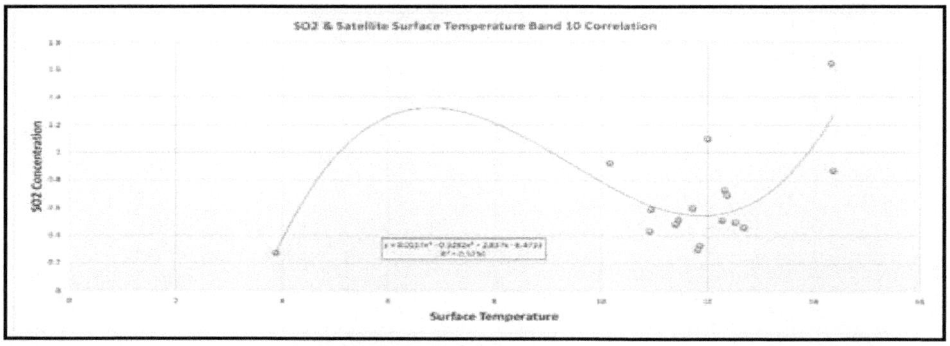


**Figure 5:** In-situ $SO_2$ & satellite data regression, Band 10.

| ID | Station | X | Y | $SO_{2\,(ppm)}$ | Temp Band 10 | Temp Band 11 | Derived $SO_2$-10 | Derived $SO_2$-10 |
|---|---|---|---|---|---|---|---|---|
| 1 | Chorao control | 443138 | 3929753 | 0.696 | 12.3572146 | 10.60898601 | 0.545 | 0.563 |
| 2 | Eternity Fire | 441111 | 3937665 | 1.105 | 12.00282186 | 9.741107607 | 0.528 | 0.532 |
| 3 | Compration gas station $AB_3$ | 440030 | 3932208 | 1.645 | 14.32778856 | 12.13584507 | 1.213 | 1.161 |
| 4 | Baba residential | 438943 | 3933565 | 0.871 | 14.36668769 | 12.27267818 | 1.238 | 1.257 |
| 5 | Araffa | 443362 | 3926437 | 0.501 | 12.51904401 | 10.78968171 | 0.562 | 0.594 |
| 6 | Alamal Alshaby | 439376 | 3929069 | 0.733 | 12.32087008 | 10.85511311 | 0.542 | 0.608 |
| 7 | Sharow hospital | 441948 | 3921311 | 0.276 | 12.27251023 | 10.62541247 | 0.539 | 0.566 |
| 8 | Rahimawa | 445035 | 3928489 | 0.517 | 11.45078024 | 9.803459561 | 0.546 | 0.528 |
| 9 | Sahat ihtifalat | 441948 | 3921311 | 0.276 | 3.880497196 | 2.711614217 | 0.277 | 0.277 |
| 10 | Sharow compound | 445386 | 3930344 | 0.434 | 10.91460935 | 8.911683558 | 0.606 | 0.638 |
| 11 | Baglar health centre | 443337 | 3925749 | 0.33 | 11.85561683 | 9.963183437 | 0.527 | 0.523 |
| 12 | Iraqi drilling company | 442724 | 3928192 | 0.461 | 12.68085875 | 10.98347322 | 0.584 | 0.638 |
| 13 | Almas | 444792 | 3926332 | 0.301 | 11.81584764 | 10.24863329 | 0.528 | 0.529 |
| 14 | Rzgary | 446694 | 3927582 | 0.595 | 10.93845311 | 9.536156362 | 0.603 | 0.548 |
| 15 | Imam qasim | 445321 | 3926091 | 0.484 | 11.39168989 | 8.914164909 | 0.550 | 0.638 |
| 16 | Khassa 2 | 444241 | 3923971 | 0.923 | 10.16220027 | 8.1970426 | 0.742 | 0.796 |
| 17 | Hay mualmeen | 444058 | 3920336 | 0.601 | 11.72208342 | 10.32468207 | 0.530 | 0.533 |

**Table 5:** $SO_2$ Derived Concentrations.

two columns are $SO_2$ concentrations in (ppm) unit that derived from the model which generated by regression analysis.

After that, the information that we have got from the generated model are used to map $SO_2$ pollutant dispersion over Kirkuk city using GIS tools. The below maps (Figure 6) are produced using ArcGIS 10.2 software.

First of all, the $SO_2$ ground-based measurements are mapped to be compared with the derived $SO_2$ concentrations maps later and to assess the model applicability and accuracy for different purposes.

The highest concentration of $SO_2$ is found in the north of Kirkuk city and this is expected due to the oil fields and industries in the north part of the study area.

As we see from the Figure 7, the results of generated models using satellite thermal and in-situ data are reliable for large spatial distribution area. It can be seen that the highest concentration of $SO_2$ pollutant is emphasized in the West-North of the city in three maps. In addition, the resulted models will provide the Directory of Environment Kirkuk with low-cost and quickly produced maps for spatial decision making.

## Air pollution Patterns & Landscape Temperature

Urban areas are associated with sources of a variety of air pollutants. In addition, cities are major contributors to global air pollution related to ozone depletion and carbon dioxide ($CO_2$) warming. Understanding air pollution patterns and thermal landscape patterns are important to better manage the cities [18].

Air pollution patterns in cities are mainly relate to the distribution of different land use and land cover categories. Land use of Kirkuk city are consists mainly of industrial, built-up, oil refinery and fields, bare areas with poor vegetated zones. It can be seen from Figure 7 that the high-concentrated of $SO_2$ is in the north and middle parts of the city which mostly are oil fields and built-up areas.

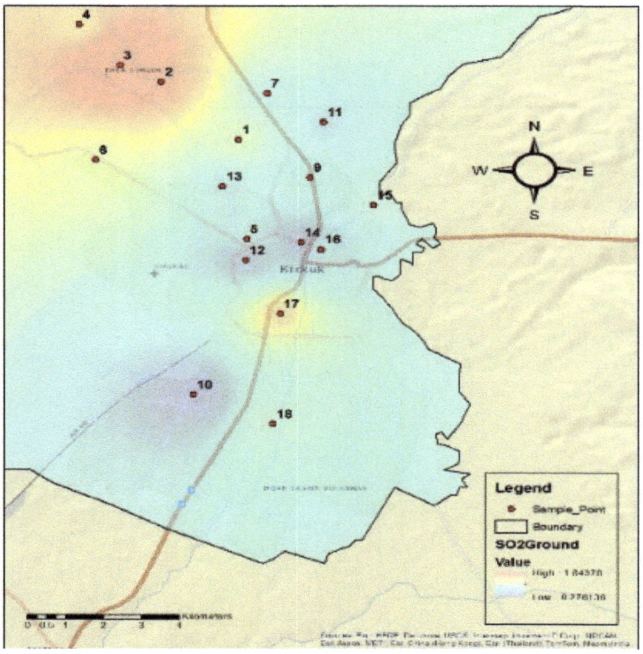


**Figure 6:** $SO_2$ Ground-based Dispersion.

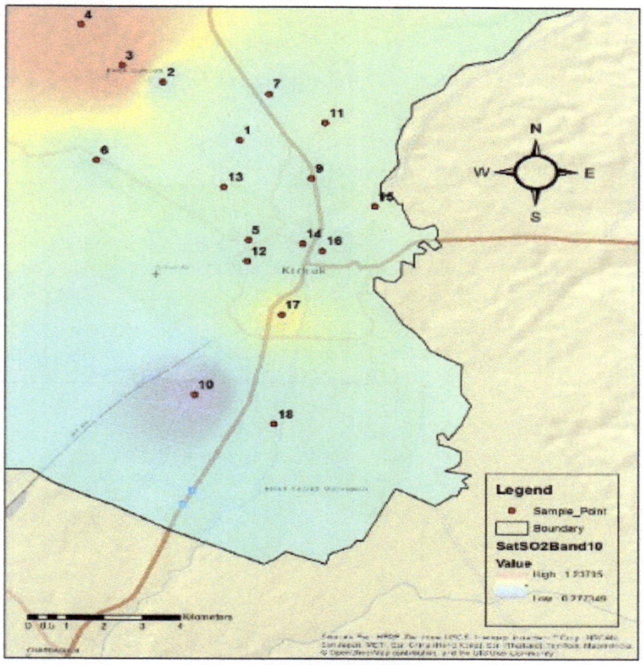


**Figure 7:** $SO_2$ Satellite Data Dispersion.

Investigating thermal landscape patterns that derived from satellite data of the study area gave us an imagination on how air pollutants are distributed in the city. Spatial pattern of air $SO_2$ pollutant is positively correlated with urban built-up density, oil refineries and with satellite derived land surface temperature in the study area.

## Conclusion

This research showed good agreements in the relationship between Landsat-8 (TIRS) [19,20] data and in-situ $SO_2$ pollutant measurements in Kirkuk city. The correlation and regression analysis performed using "Analysis it" plug-in in Excel software and it showed the ability to analysis multi-variant data and also in plotting scatter diagrams, PCA analysis and curve fitting as well. While GIS technique is used in the study to map the collected and derived data. The IDW interpolation used in GIS environment to produce maps for $SO_2$ concentrations dispersion over Kirkuk city. The technique indicates to reliable results over the large spatial distribution when it compared with other researcher results.

The research resulted two models for $SO_2$ concentration over Kirkuk city, one for band 10 of Landsat TIRS data and the second is for band 11. However, the band 10 model is more accurate than band 11. Also the study generated three maps for $SO_2$ pollutant concentration over the study area using ground-based, values of the model of band 10 and 11. It showed a high correlation between three maps and by this we could conclude that remote sensing is a powerful tool for air quality mapping [21,22].

The results of this research indicate that air pollution can be mapped using satellite data to provide a larger area of coverage. Further research can be done to enhance the models using numerical and statistical analysis.

### Acknowledgment

This research is supported by Mr. Ahmed Qusay providing the ArcGIS software and Analysis it plug-in. Also we thank the ministry of higher education of Iraq granting us for study in Malaysia to complete MSc degree.

## References


1. Salah AH (2011) Air Quality Over Baghdad City Using Earth Observation and Landsat Thermal Data. Journal of Asian Scientific Research 1: 291-298.

2. Lim HS, Jafri MZM, Abdullah K (2010). Algorithm for Air Quality Mapping Using Satellite Images. Air Quality.

3. Ung A, Wald L, Ranchin T, Weber C, Hirsch J, et al. (2001b) Satellite data for Air Pollution Mapping Over A City- Virtual Stations, Proceeding of the 21th EARSeL Symposium, Observing Our Environment From Space: New Solutions For A New Millenium, Paris, France, Abingdon, Exton (PA) Tokyo.

4. Nicolantonio WD, Cacciari A, Tomasi C (2009) Particulate matter at surface: Northern Italy monitoring based on satellite remote sensing, meteorological fields, and in-situ samplings. Selected Topics in Applied Earth Observations and Remote Sensing 2: 284-292.

5. Wang J, Christopher SA (2003) Intercomparison between satellite-derived aerosol optical thickness and $PM_{2.5}$ mass: Implications for air quality studies. Geophysics Research Letters.

6. Nicolantonio WD, Cacciari A, Bolzacchini F, Ferrero L, Volta L, et al. (2007) Modis aerosol optical properties over North Italy for estimating surface-level $PM_{2.5}$. In Proc. Envisat Symp, Montreux, Switzerland, ESA SP-636.

7. Narashid, Haron R, Mohd WMNW (2010) Air quality monitoring using remote sensing and GIS technologies. Science and Social Research (CSSR), 2010 International Conference.

8. Qihao W, Yang S (2006) Urban air pollution patterns, land use, and thermal landscape: an examination of the linkage using GIS. Environmental monitoring and assessment 117: 463-489.

9. Rix M, Valks P, HaoN, Geffen JV (2009) Satellite monitoring of volcanic sulfur dioxide emissions for early warning of volcanic hazards. Selected Topics in Applied Earth Observations and Remote Sensing 2: 196-206.

10. Krueger AJ, Krotkov NA, Yang K, Carn S (2009) Applications of satellite-based sulfur dioxide monitoring. Selected Topics in Applied Earth Observations and Remote Sensing. IEEE 2: 293-298.

11. Liu D, Zhu X (2012) An enhanced physical method for downscaling thermal infrared radiance. Geoscience and Remote Sensing Letters 9: 690-694.

12. Jhabvala M, Reuter D, Choi K, Jhabvala C, Sundaram M (2009) QWIP-based thermal infrared sensor for the Landsat Data Continuity Mission. Infrared Physics & Technology 52: 424-429.

13. Harvey NR, Brumby SP, Perkins S, Theiler J, Szymanski JJ, et al. (2001) Image feature extraction: GENIE vs conventional supervised classification techniques. IEEE Transactions on Geoscience and Remote Sensing 40: 393-404.

14. Thome K, Reuter D, Lunsford A, Montanaro M, Smith R, et al. (2011) Calibration of the Thermal Infrared Sensor on the Landsat Data Continuity Mission. In Geoscience and Remote Sensing Symposium (IGARSS), 2011 IEEE International, IEEE.

15. Balçik FB (2014) Determining the impact of urban components on land surface temperature of Istanbul by using remote sensing indices. Environmental monitoring and assessment 186: 859-872.

16. Sameen MI, Kubaisy MAA (2014) Automatic Surface Temperature Mapping in ArcGIS using Landsat-8 TIRS and ENVI Tools, Case Study: Al Habbaniyah Lake. Journal of Environment and Earth Science 412: 12-17.

17. Walawender JP, Szymanowski M, Hajto MJ, Bokwa A (2014) Land Surface Temperature Patterns in the Urban Agglomeration of Krakow (Poland) Derived from Landsat-7/ETM+ Data. Pure and Applied Geophysics 171: 913-940.

18. David BJ, Collins S, Elliott P, Fischer P, Kingham S, et al. (1997) Mapping urban air pollution using GIS: a regression-based approach. International Journal of Geographical Information Science 11: 699-718.

19. Jiménez-Muñoz JC, Sobrino JA, Skokovic D, Mattar C, Cristóbal J (2014) Land Surface Temperature Retrieval Methods from Landsat-8 Thermal Infrared Sensor Data, IEEE.

20. Rozenstein O, Qin Z, Derimian Y, Karnieli A (2014) Derivation of Land Surface Temperature for Landsat-8 TIRS Using a Split Window Algorithm. Sensors 14: 5768-5780.

21. Chitrini M, Reddy KV, Pratap D (2013) Air Pollution Modeling from Remotely Sensed Data Using Regression Techniques. Journal of the Indian Society of Remote Sensing 41: 269-277.

22. Richards JA, Jia X (1999) Remote sensing digital image analysis. Springer.

# Concept and Design of Environmentally Friendly Automobile Mechanic Village

**Nwachukwu MA[1]*, Umunna NA[1], Ntesat B[1] and Umunna CP[2]**
[1]*Department of Environmental Technology, Federal University of Technology Owerri, Nigeria*
[2]*Enugu State University of Science and Technology, Park Lane, Enugu, Nigeria*

**Abstract**

Global population increase has caused increased demand in automobile use and the shipment of used automobiles (Second hand) from the industrialized to the developing regions of the world. As a result, many developing countries now opt for auto mechanic village as against the citywide auto mechanic workshop practice because of the numerous advantages. The very poor standard and unsustainable setting of these mechanic villages cause environmental degradation leading to global change. This paper presents a total of 24 features and criteria that enabled AutoCAD design of an environmentally friendly mechanic village. Environmentally friendly mechanic village will improve environmental quality across developing countries of the world this century.

**Keywords:** Automobile repair; Waste and pollution; Global change; Sustainable engineering

## Introduction

A mechanic village represents several acres of land mapped out for automobile mechanics, where automobile owners must go for repairs and services to their motor vehicles. In some parts of Nigeria and many other developing countries, mechanic village (MV) is adopted against the city-wide automobile workshop practice. It is found that citywide auto workshop practice in these countries causes wider environmental pollution. The increasing world population and technological growth (Figure 1a) has continuously increased the use of automobile worldwide (Figure 1b), increasing the use of mechanic villages and global change.

World population was about 2.5 billion in 1950, rose to 6 billion in 2000, while World automobile was about 60 million in 1950 and rose to 700 m in 2000. According to Sousanis [1], which looked at government-reported registrations and historical vehicle-population trends, global registrations jumped from 980 million units in 2009 to 1.015 billion in 2010. The 3.6% rise in vehicles in 2010 was the largest percentage increase since 2000, while the 35.6 million year-to-year unit increase was the second-biggest gain in overall volume ever. The overall vehicle population growth since 2010 has seen China pushing ahead of Japan as the world second largest after United States. The trend has also shown tremendous increase in India vehicle registration, followed by Brazil. The registered figures may double before the end of the first quarter of this century, and similarly increasing the number of mechanic villages.

Mechanic village concept has been described as a twenty first century strategy to improve urban environmental quality in developing countries [2]. While holding on this, Nwachukwu et al. [3], Udebuani et al. [4], and Iwegbue [5] reported that these mechanic villages are poorly developed with no waste management plans. They described the soils as having severe to excessive heavy metal pollution, causing ecological and public health hazards. Nwachukwu et al. [6] found that there is no enforcement of any rule guiding automobile workshops and mechanic practice in the villages. They found the need for mechanic villages to be environmentally friendly. If automobile repair works in cities are confined to well planned and built mechanic villages; with collection and recycling of spent oil, and proper disposal of spent electrolyte, environmental quality will improve. This will reduce soil pollution, bioavailability of toxic heavy metal and poor nutrients availability to crops and vegetables within and around mechanic villages. Heavy metals such as Cadmium, Chromium, Copper, Lead and Nickel are poisonous even at less concentration within an environmental setting.

## Activities of Mechanics and their Environmental Implications

It became very necessary to investigate the various activities that go on in mechanic villages, assess their environmental implications to support the design criteria. Table 1 is a breakdown of the various activities of mechanics and other allied works and services that are not environmentally friendly. These activities are duly integrated in the design process.

## Background and Justification

Environmentally friendly mechanic village will serve as a base for the transfer of emission testing technology in developing countries, and provide emission testing services to motorists. This will make the enforcement and supervision of automobile emission testing possible in many developing countries, thereby providing the bases for developing countries to integrate climate change responses. It provides temporary storage for the final end of life vehicles. Waste automobiles moved from developed to developing nations at their primary end of life may have temporary storage in mechanic villages at their final end of life following lack of recycling facilities. Have suitable storm water best management practice (MV storm water BMP). This will allow the removal of metal contaminants in the MV storm water, thereby preventing such contaminants from entering the urban waterway, and reducing soil pollution in the MVs [6]. There may be need for extended producer responsibility for used automobile oil. This will end the ugly habit of spilling waste oil on the ground and reduce soil pollution. Lindquist [7] recommended EPR for materials that are recyclable and continuously marketable. Used motor oil is recyclable and also marketable, and the administrative, the economic and the informative tools are affordable and available at this time.

Provision of toilet facilities by government owners of the MVs,

***Corresponding author:** Nwachukwu MA, Department of Environmental Technology, Federal University of Technology Owerri, Nigeria
E-mail: futo.essg@hotmail.com

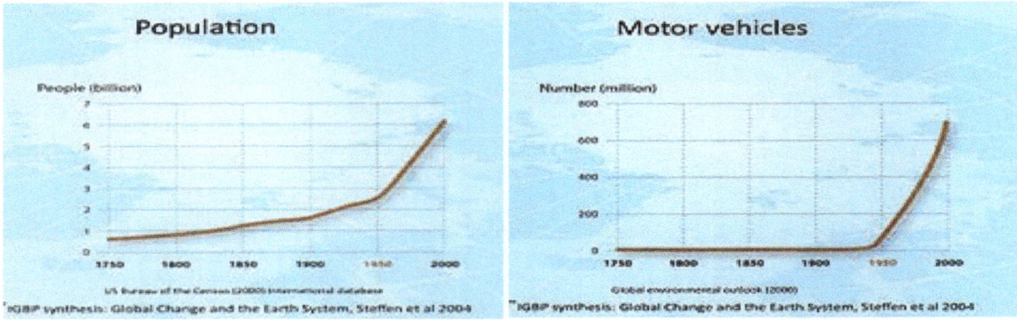


**Figure 1:** (a) Rapid global population growth (b) Rapid increase in global automobile needs.

| S. No. | Activities | Environmental impacts |
|---|---|---|
| 1 | Servicing of vehicles; change oil. | Spill spent engine oil on the ground. |
| 2 | Repair of transmission systems. | Spill spent transmission oil on the ground. |
| 3 | Repair of fuel tanks. | Spill petrol and diesel on bare ground. |
| 4 | Repair or charging of batteries. | Spill used electrolyte on the ground, discard battery casing, and lead plates on bare ground. |
| 5 | Repair of braking systems. | Spilling of brake fluid on bare ground. |
| 6 | Repair of clutch systems. | Spilling of hydraulic fluid on the ground. |
| 7 | Overhauling of vehicle engines. | Discharge used oil, metal particles on ground. |
| 8 | Panel beating of vehicle bodies Scraping of old body coats. | Metal bits, metallic coats and dusts on ground. |
| 9 | Grinding, threading, wiring etc. | Metal bits are filed onto bare ground |
| 10 | Greasing and oiling of parts. | Waste wires, metals and solders are dropped. |
| 11 | Welding, soldering vehicle parts. | Grease and oils spill on the ground. |
| 12 | Spray painting of vehicle bodies' | Litter carbide, solder, and electrode on ground. |
| 13 | Wash potty fill of vehicle body. | Air pollution, pills paints on the ground. |
| 14 | Rainfall; Runoff, Storm water | Washery on ground, waste filler in dumpsite. |
| 15 | Washing of vehicles and parts. | Roof dirt, Corrosion, Trace metals; Pollutants |
| 16 | Human waste | Petrol, Kerosene, Detergent on the ground. |
| | | Deposited to the ground, no toilet facilities |

**Table 1:** Activities of mechanics and their environmental implications.

which may be supported by the mechanics directly or indirectly will reduce incidents of high coliform counts in shallow wells around MVs. Nwachukwu et al. [8] observed that about 60% of shallow wells within 500 m in the vicinity of Orji and Nekede mechanic villages in the Imo River basin have between 40–70 CF/100 ml of coliform, exceeding the 10 CF/100 ml of water allowable by the world health organization. Be isolated from residential areas in a 50 to 100 year development plan, with farming or vegetable gardens not allowed within and at least 350 m to the MV. Not sited at relatively high elevated areas or at slope up to 10% to reduce transport distance of contaminants. Not sited in areas with shallow water table (water table <45 m in the basin). This will prevent easy transport of pollutants to groundwater and shallow water wells. Have tarred roads and drainage linked to storm water BMPs, as well as concrete floor workshops, to reduce infiltration of contaminants to soil. Have groundwater monitoring wells. This will enable regular study of groundwater chemistry in the vicinity of mechanic villages to prevent dynamic exposure of citizens to water of poor quality. Be provided with health center and recreational facilities, for the occupational health, and mental and physical wellbeing of mechanics. Have continued education program facility for mechanics, to improve awareness of occupational waste management and environmental health and safety in mechanic villages.

Environmentally friendly mechanic villages will stop the neglect of automobile wastes in developing countries, thus strengthening the ability of developing countries to integrate climate change responses. Presently, the worldwide motor vehicle greenhouse gas carbon dioxide emission is said to be over 900 million metric tons each year, predominantly from developed countries according to the World Resources Institute. "These emissions account for more than 15 percent of global fossil fuel $CO_2$ releases." This obviously has devastating consequences on the environment. Of this volume, the United States that has five percent of the world's population and 30 percent of the world's automobiles contributes 45 percent according to a 2006 report of the advocacy group environmental defense. This quantity may double before the end of the first quarter of this century due to the ever increasing demands of used vehicles by developing countries, and the matching supplies of old fossil fuel guzzling vehicles by the industrialized nations. As the industrialized nations transit to hybrid green electric cars, the need for environmentally friendly automobile mechanic village in developing countries becomes more compelling. To this effect therefore, automobile waste management including final end of life vehicles in developing countries stand as a major global environmental issue of this century.

Lack of storm water treatment facilities in the mechanic villages indicates poor environmental awareness, because infiltration and detention basins are relatively cheap to build and do not involve complex hydrological designs. Galli [9], Hidding [10], Pitt and Clark [11], Gobel et al. [12], Obropta and Kados [9-13], and others have discussed the use of storm water management facilities to improve urban environmental quality. Local phyto-remediation plants will be planted in the central basin to remove the toxic heavy metal contaminants.

Gilbert and Alberto [14] and Remon et al. [15] and others have discussed the principles of phyto-remediation for an effective site restoration. According to Remon et al. [15], indigenous plants species predominant in an area is usually most preferred. This is because indigenous species can easily adapt to relocations and cope with local conditions in a phyto-restoration program. Fakayode and Onianwa [16] obtained strong correlation between heavy metal concentration in soil and in local guinea grass (Panicum maximum) around Ikeja industrial estate Nigeria. Their result shows significant correlation with respect to Cd (0.83), Mn (0.94), Ni (0.90), and Pb (0.73). This result suggests that Guinea grass could be used for phyto-remediation in mechanic villages.

Another essential component of the waste management plan for environmentally friendly mechanic village is the installation of observation wells near the mechanic villages. Shallow wells (≤ 70 m deep) shall be placed near mechanic villages suspected to be at risk of pollution due to poor waste management. Such wells must be continuously monitored by analyzing the water. Observation wells may be installed ≥ 500 m away from another in a mechanic village, in the direction of drainage. The selected distance of each observation well from its respective mechanic village will depend on the proximity of domestic water wells and human residence._

## Methodology and Design Criteria

In this study, a number of Mechanic villages in Nigeria have been visited in order to design a model mechanic village that will be environmentally friendly and stand the test of the 21st century. The field study confirms that the mechanic villages are poorly developed. They are rather brown fields, and obvious source of environmental degradation. Waste management is not at all remembered or considered as an issue, which implies that all waste generated in the mechanic villages are continuously disposed on the ground. Notwithstanding, storm water management is never practiced. There is no environmental regulation guiding the activities of the mechanics neither is there any regulatory agency enforcing any rule. The mechanics are not properly trained, and neither conscious of any regulatory requirement. The workshop structures are predominantly make-shift and ill equipped. Safety standards are low to zero, and quality of work or auto maintenance is poor. The roads and streets are not built, and drainages not provided. Majority of the mechanic villages visited are wrongly sited; often within residential areas and at relatively high elevation areas.

Field investigation of the present state of mechanic villages is recorded in photographs. Auto card was used in the design, making the drawings easy with enough details. Details of the village are contained in the layout design divided into four zones, which describes a mechanic village as a community. Twenty four design features were considered in order to achieve the aim and objectives of an environmentally friendly mechanic village.

## Interview with Mechanics

Mechanics interviewed during the field study lamented the poor environmental state of the mechanic villages affecting their health. They informed that on average, they spend twelve hours in the mechanic village Monday through Saturday, while many of their apprentices live in the workshops. They continued that most of their apprentices are on their own, no guardian, no assistance from government or organization. Because of the poor hazardous environment, apprentices are no longer forth coming. Young school leavers and school drop-outs who would have considered automobile repair trade now prefer commercial motor cycle riding obtainable at hire purchase. Our members include mechanics, technicians, artisans, apprentices and other support workers like spare parts dealers and restaurants operators etc. Many of our members are not as educated. Majority are primary and secondary school drop-outs. They are ignorant of environmental health requirements and proper automobile waste management system. We are in search of daily bread but found ourselves in hazardous neglected mechanic villages. We are not happy with the ugly situation but look up to God for rescue they concluded.

## Field Assessment of Mechanic Village Infrastructure

Important field observations were captured in photographs

**Figure 2:** Picture frame view of selected mechanic villages.

representing the physical appearance of mechanic villages (Figure 2). Each photograph in the frame is of a different mechanic village in Nigeria. Greater of the mechanic village activities go on in the open air and on natural ground surface. The entire area is highly congested with unserviceable vehicles and vehicles undergoing repairs.

The earth roads are in deplorable condition contains stagnant pools of water, almost impassable in rainy seasons. Field Records of the present structural situation at selected mechanic villages show evidence of poor development as contained in the photo frame (Figure 2). The environment is hostile to the mechanics and other allied workers including families dwelling within and around the mechanic villages. Most of the mechanic villages lack electricity, water supply, clinic and recreational facilities. There is virtually no engineering input in the overall development, making the environment not conducive for human operation 13 hours per day, 7 days per week.

## Project Designs

The concept and design of environmentally friendly mechanic village is hereby presented in Auto Cad with Figure 3 representing the conceptual plan of a medium size mechanic village accommodating 300 - 600 workshops. This size of mechanic village will provide work space for 1000-2000 auto repair workers with similar number of apprentices. The village will also accommodate several allied workers and associate business. In the master plan are four distinguished zones numbered 1 to 4. Zone 1 (Figure 4) is for Toyota, Honda, Hyundai and KIA. This zone has a free area where toilet, bank, power station, police station, gas station, storm water treatment facility and used oil and acid storage tanks are located. This zone deserves to be at most lower elevation within the mechanic village, so that storm water collection and drainage could be supported by natural slope. There is provision for emission testing workshop and a monitoring well which must be located in the direction of drainage. Open spaces and freeways exist in this zone.

Zone 2 (Figure 5) is divided into four quadrants allocated to Volkswagen, Mercedes and Volvo, BMW and Ford, and Nissan. Space is reserved for restaurants, toilets, sorted wastes and used acid and oil

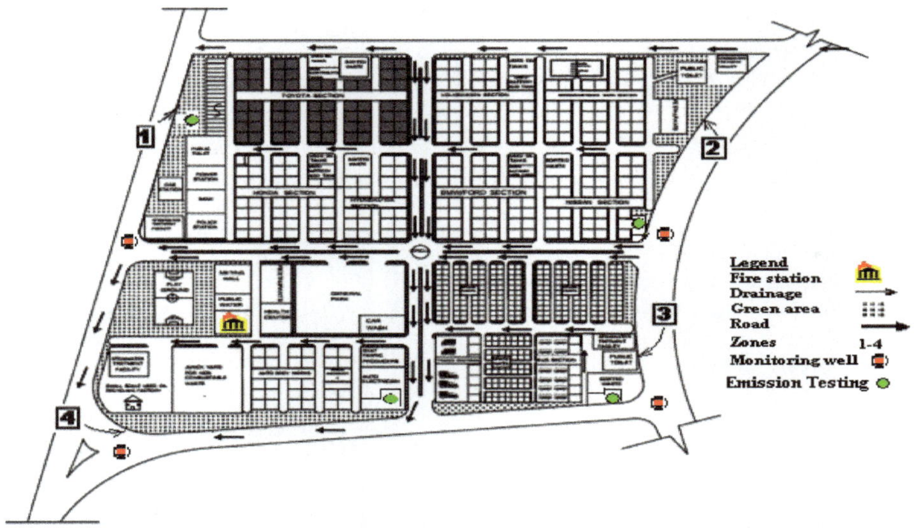


**Figure 3:** Conceptual plan of a medium size environmentally friendly mechanic village with four zones.

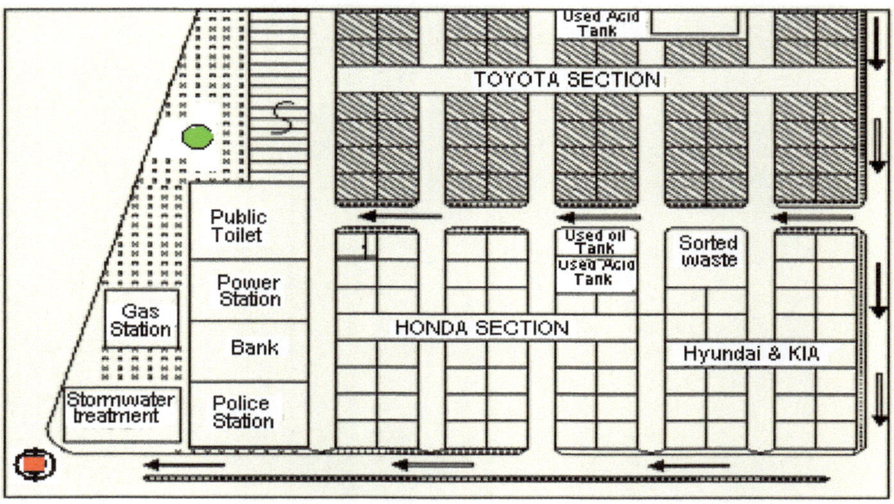


**Figure 4:** Layout of zone one.

storage tanks. There is provision for emission testing workshop and a monitoring well which must be located in the direction of drainage. Open spaces and freeways all exist in the zone. Zone 3 (Figure 6) is a mixed zone with abundant free way. The zone accommodates a play ground, meeting hall, public water supply scheme, restaurants, health center, central park and a fire service station. Others are: car wash, leather work section, auto electrical section, spray painting section and used acid storage facility. Other major activities located in this section are auto body works, used oil recycling business, junk yard where old parts dismantling and sales take place with storm water treatment. Emission testing workshop and a monitoring well are also provided, with the storm water treatment and monitoring well following the direction of drainage in line with zone 1.

Zone 4 (Figure 7) is the fourth quadrant in the conceptual plan and the most elaborate. This zone accommodates spare parts shops, the truck and bus sections, as well as sorted waste, storm water treatment and toilet facilities. Emission testing facility and a monitoring well are not left out in this zone. Figure 8 is showing a front sectional view of the workshops, with free way in between the workshop sections. Top of the workshops is concrete deck lined with treated soil to provide a green roof suitable for vegetable garden. Figure 9 is also showing a front view of body work/spray painting workshops equipped with gas control chambers to reduce air pollution in mechanic villages.

## Explanation and discussion of design

Table 2 comprises a detailed discussion and explanation of the design features and criteria. These features and criteria are conservatively articulated to project the concept of environmentally friendly mechanic village. These design features and criteria are grouped under headings:

- Standardization features and criteria
- Structural features and criteria

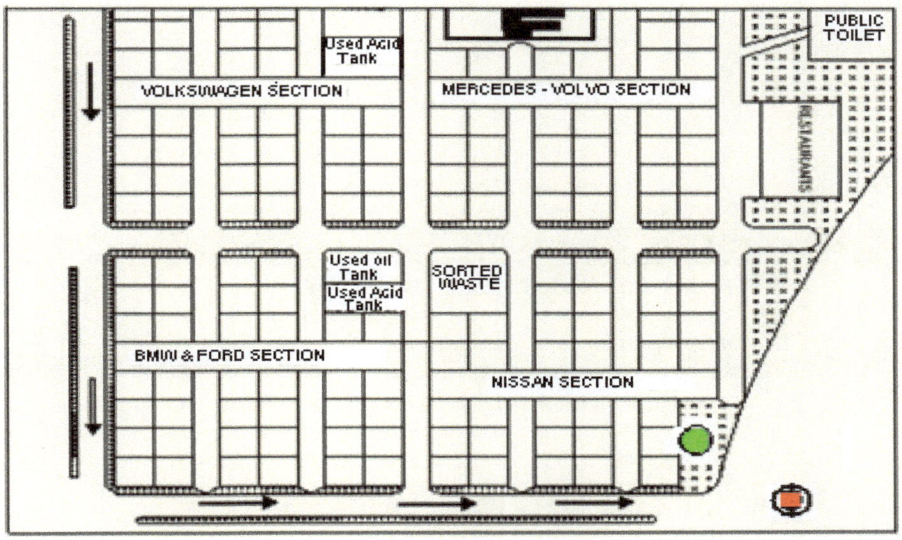


**Figure 5:** Layout of zone 2.

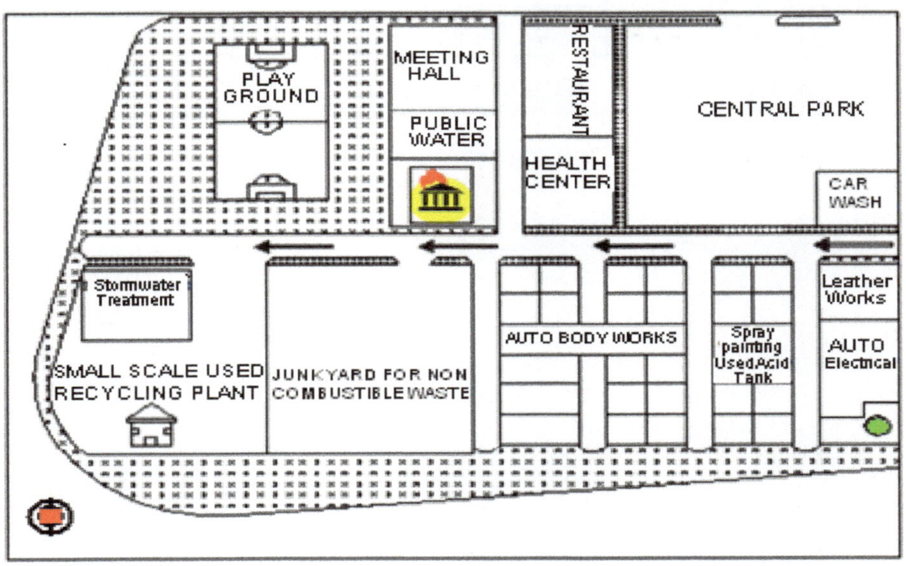


**Figure 6:** layout of zone 3.

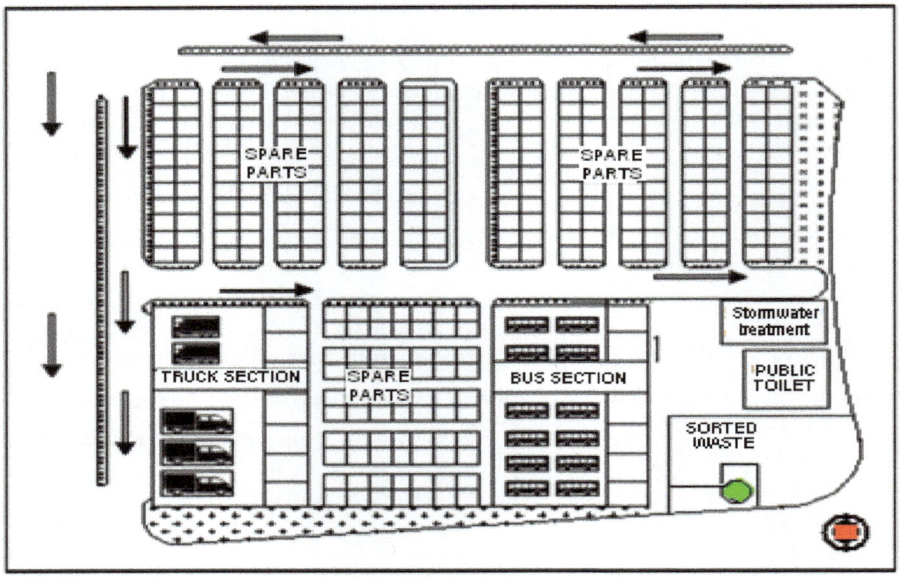


**Figure 7:** Layout of zone 4.

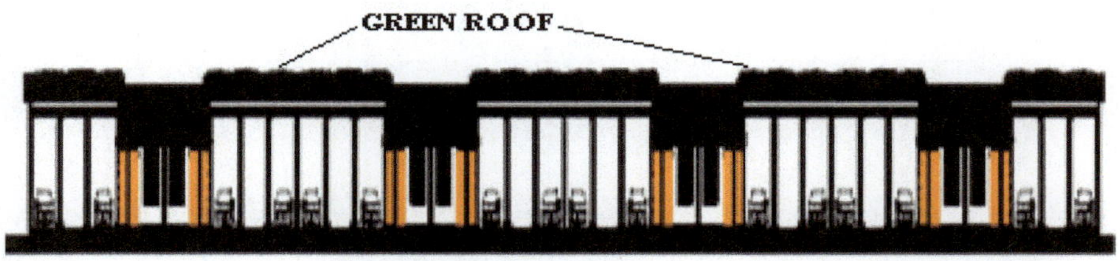


**Figure 8:** Sectional view of the mechanic village workshop.

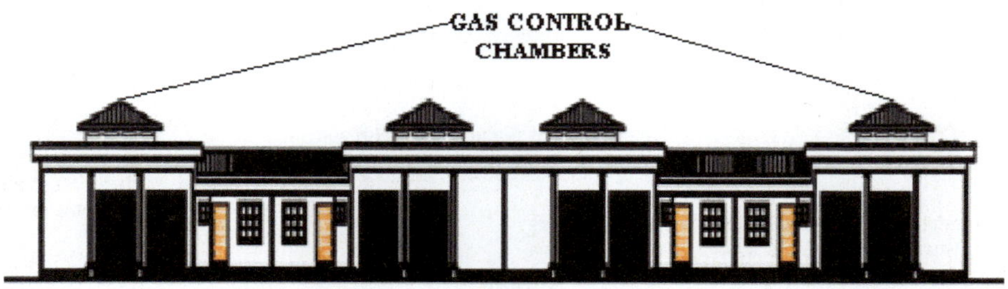


**Figure 9:** Body work spray painting workshops equipped with gas control chambers.

- Green facilities and criteria
- Support services and criteria

## Conclusions

Environmentally friendly mechanic village is a 21st century strategy to improve global environmental quality. The concept appear consistent with the United Nations framework convention on climate change; control of harmful substances and hazardous waste, as well as sustainable development. The concept will rather create an enabling environment for better investments that would attract highly qualified and educated individuals into the business of automobile maintenance. This will create more employment and will endeavor in poverty alleviation as mechanic villages may be described as a training center reachable to the ordinary citizen. The less privileged, which have no support for school and the school drop-outs have position in MVs. It is anticipated that environmentally friendly mechanic village concept may be supported by the United Nations Environmental Program, World Bank, United Nations Development Program, United Nations Industrial Development Organization, and the African Development Bank, etc. Models of MVs may be established in developing countries by governments, the United Nations and NGOs, using the design presented here as a guide. Modifications of the conceptual plan is welcome for design and development processes, but the major features of pollution control should be included to upgrade any existing mechanic villages.

| S. No. | Design Features | Specific criteria |
|---|---|---|
| **Standardization** | | |
| 1 | Size of mechanic village | Small: < 100 acres; Medium: 100-200 acres<br>Large: > 200 acres; based on number of workshops and available land. |
| 2 | Criteria for sitting a mechanical village | Elevation; Water table; urban development plan (50-100 years); Accessibility; Population |
| 3 | Workshops and Environment standards | As approved by government regulation, and duly enforced by the appropriate agency. |
| 4 | Personnel per workshop applicable to all sections | Number of personnel varies from workshop to workshop. This is a function of specialization, expertise, credibility and overall performance. |
| **Structures** | | |
| 5 | Workshop layout and design | Workshops are built in zones with access roads according to different operational activities and specialties. Spare parts dealers and other allied services are properly accommodated in the design. |
| 6 | Storm water best management practice | All surface runoff must be directed to treatment facility (storm water BMPs) before it joins the urban waterway. Choice of BMPs, e.g. infiltration or detention basins go with subsurface geology. |
| 7 | Accessible road network | Good tarred road network must exist with properly covered concrete drainage. |
| 8 | Junk yard | Segregated waste dump area: Junk vehicles; Scrap metals; Auto Plastics for temporary storage. |
| **Green Facilities** | | |
| 9 | Used oil storage | Storage facility per workshop specially designed and supplied by government agency. Stored oil regularly retrieved for recycling or taken care of through extended producer responsibility (EPR). Reduce soil and water pollution. |
| 10 | Used electrolyte storage | Storage facility in every auto-electrician workshop. Proper disposal of used electrolyte a condition. Reduce soil and water acidity. |
| 11 | Fume and gas collection | Gas collector on top all sprays painting workshops. Reduce air pollution. |
| 12 | Green Roof | All workshop buildings concrete tops overlain with good soil for planting food crops and vegetables. Improve air quality. |
| 13 | Emission testing facility | Provided by government under regulation to improve air quality. |
| 14 | Pollution monitoring wells | A number of observation wells at predetermined locations to monitor groundwater quality. |
| **Support Services** | | |
| 15 | Government owned Fire station | Service truck; Functional water well must be available. |
| 16 | Restaurants | Private low, Middle and high class restaurants for native and continental foods will function in MVs. |
| 17 | Conveniences | Toilet and shower in every workshop |
| 18 | Car wash | One or more, according to size of village |
| 19 | Public water supply | Two standard boreholes fully reticulated |
| 20 | Car park | General car park and car hire services |
| 21 | Transformer station | The MV shall exclusively have its own transformer of high capacity, to enable constant and regulated power supply. |
| 22 | Bank | The village union shall allow a bank of their choice to serve them and their customers. |
| 23 | Gas station | One or two gas stations are required in a mechanic village to facilitate repairs and run automobiles. |
| 24 | Police post | A police post is required to keep law and order. |

**Table 2:** Design features and criteria.

**Acknowledgements**

The authors remain very grateful to GEOPROBE International Consultants Limited Nigeria for design instrumentation and logistics support that enabled this research. Special thanks go to all persons who helped in the field assessment photographs and processing.

## References

1. Sousanis J (2011) World Vehicle Population Tops 1 Billion Units. Ward's Auto, Information Center.
2. Nwachukwu MA, Feng H, Alinnor J (2011) Trace Metal Dispersion in soil from Auto-Mechanic Village to Urban Residential Areas in Owerri, Nigeria. Procedia Environmental Sciences 4: 310–322.
3. Nwachukwu MA, Feng H, Alinnor J (2010a) Assessment of heavy metal pollution in soil and their implications within and around mechanic villages. Int J Environ Sci 7: 347-358.
4. Udebuani AC, Okoli CI, Nwaigwe H, Ozoh P (2011) Effect of Spent Engine Oil Pollution on Arable Soil of Nekede Mechanic Village Owerri, Imo State Nigeria, International Journal of Natural and Applied Science 7: 257-260.
5. Iwegbue CM (2007) Metal fractionation in soil profiles at automobile mechanic waste dumps. Waste Manag Res 25: 585-593.
6. Nwachukwu MA, Huan F, Achilike K (2010b) Integrated studies for automobile wastes management in developing countries; in the concept of environmentally friendly mechanic village. Environ Mon & Assess 178: 581-593.
7. Lindquist T (2000) Extended Producer Responsibility in Cleaner Production. Policy Principle to Promote Environmental Improvements of Product Systems. Doctoral Dissertation, Lund University.
8. Nwachukwu MA, Huan F, Amadi MI, Umunna FU (2010c) The Causes and the Control of Selective Pollution of Shallow Wells by Coliform Bacteria, Imo River Basin Nigeria. Water Qual Expo Health 2: 75-84.
9. Galli J (1992) Analysis of Urban BMP Performance and Longevity in Prince George's County Maryland: Metropolitan Washington Council of Governments, Wash., DC.
10. Hilding K (1996) Longevity of infiltration basins assessed in Puget Sound. Watershed Protect Tech 1: 124-125.
11. Pitt R, Clark SE (2008) Integrated Storm-Water Management for Watershed Sustainability. J Irrig Drain E-ASCE 134: 548-555
12. Göbel P, Stubbe H, Weinert M, Zimmermann J, Fach S, et al., (2004) Near-natural stormwater management and its effects on the water budget and groundwater surface in urban areas taking account of the hydrogeological conditions. Journal of Hydrology 299: 267-283.
13. Obropta CC, Kardos SJ (2007) Review of Urban Storm water Quality Models: Deterministic, Stochastic, and Hybrid Approaches[1] J. Am. Water Resour Assoc 43: 1508-1523.
14. Gilbert S, Alberto P (2008) Phytoremediation potentials of lead contaminated

soil, using Tropical grasses; Florida Scientist 71(1): 1-6, USDA- Agric. Res. Station.

15. Remon E, Bouchardon JL, Cornier B, Guy B, Leclerc JC, et al., (2005) Soil characteristics, heavy metal availability and vegetation recovery at a former metallurgical landfill: Implications in risk assessment and site restoration. Environ Poll 137: 316-323.

16. Fakayode SO, Onianwa PC (2002) Heavy metal contamination of soil, and bioaccumulation in Guinea grass (Panicum maximum) around Ikeja Industrial Estate. Environ Geol 43: 145-150.

# Effect of Single Fiber Pull Out Test Result on Flexural Performance of ECC

**Rathod JD** *

*Applied Mechanics Department, Faculty of Technology and Engineering, The M.S. University of Baroda, Vadodara, India*

**Abstract**

Study of effect of single fiber pull out test result on flexural performance of Engineered Cementitious Composite (ECC) is carried out in the present work. Cement: sand ratios of 1:0.5, 1:1, 1:1.5, and 1:2 are used in composition of ECC. Beams of 20 and 30 mm depth are tested under displacement control and four point loading arrangement. First crack strength, ultimate strength, toughness index, deflection hardening and reserved flexural strength are evaluated. Single fiber pull out test of 6 mm embedded length is carried out to evaluate frictional bond strength. An equation is suggested to evaluate ultimate flexural strength of ECC considering effect of micromechanical parameters in general and frictional bond strength in particular.

**Keywords:** Engineered cementitious composite; Slippage regime; Flexural performance; Micromechanical; Polypropylene fibers

## Introduction

When fracture failure occurs, a high strength material need not necessarily offer higher structural strength. Rather, a tougher material and in the extreme a ductile material can lead to a higher structural strength which, in turn, can play a significant role in the seismic response of the structure. Efforts have been made in recent past to make overall performance of the structure ductile by the introduction of ductile detailing. However, ductile detailing makes the member heavily congested by the reinforcement due to which there are practical difficulties of placing and compacting of concrete. Moreover, the concrete which consumes about 85% volume in a structure still remains inherently brittle. Concrete and steel should contribute simultaneously towards ductility to maintain compatibility and integrity of the composite to enhance the overall performance of the structure. A new innovative material, which can replace ordinary concrete at some key places, can improve overall performance of the structure dramatically is Engineered Cementitious Composites (ECC) [1]. Strain hardening through multiple cracking is the key property of this smart material. Micromechanics concept is used to develop ECC which relates macroscopic properties to the microstructures of a composite and form the backbone of materials design theory. Specifically it allows systematic deliberate tailoring of fiber, matrix, and interface as well as materials optimization. Most of the investigations on ECC performance all over the world have been carried out using PVA and PE fibers of circular cross section with a typical composition employing w/c ratio and c/s ratio of 0.5 with optimized fiber volume.

Cement: sand ratio and cross section shape of the fiber are two micromechanical parameters intended in this investigation for microstructure tailoring of matrix and fiber. Polyester fibers under trade name of Recron 3S in which aster is a functional polar group, responsible for good bond with cement. Its silicon coating helps in better dispersion in matrix which is one of the essential requirements in any fiber reinforced composite. Most of the synthetic fibers available are of circular cross section whereas Recron fibers are of substantial triangular in cross section. Specific gravity and tensile strength of Recron fibers is 1.36 g/cc$^3$ and 1000 N/mm$^2$ as per manufacturer's data. Performance of any fiber reinforced composite depends upon fiber/matrix interaction, which is largely influenced by the effectiveness of bond between fibers/matrix and fiber pull-out mechanism. Relative Fiber Intrinsic Efficiency Ratio (FIER) of Recron fibers is 2.2 because of its cross section shape.

## Single Fiber Pull Out Test

A method to directly assess fiber/mortar interfacial bond properties is to pull a single fiber out of its surrounding matrix. Fiber-matrix interfacial properties are important in controlling macroscopic properties of composite materials. Generally speaking, better the bond better is the composite. Increasing the bond leads to increase in both composite strength and toughness [2,3]. However, beyond certain bond strength, toughness begins to decrease due to fiber rupture. Therefore, fiber pull out rather than rupture confers a larger ductility to the fiber reinforced composite [4-6]. In addition, the phenomenon of pseudo strain hardening is strongly influenced by the maximum bridging stress, which is directly proportional to the interface bond strength [7,8].

The general profile of a single fiber pull out curve can be decomposed into three major regimes as depicted in (Figure 1) [9]. Initially a stable fiber debonding process occurs along the fiber/matrix interface. The load resisted by the fiber is then increasing up to $P_a$. The fiber embedded end, $l=l_e$, does not move. The debond length, $l_d$, increases towards $l_d=l_e$. The displacement corresponds only to elastic stretching of the debonded fiber segment and of the fiber free length. Then, the load decreases from $P_a$ to $P_b$. If the load drop is significant, it reveals that the chemical bond between the fiber and the matrix has broken. This means that debonding at the interface is governed by a fracture criterion [10]. In the case of nonchemically bonded fibers, such as steel, polyester, or polypropylene fibers, $P_a$ would be closed or equal to $P_b$. At point $P_b$, the embedded fiber end is just debonded. Finally, in the slippage regime, the fiber load is resisted by frictional forces. The

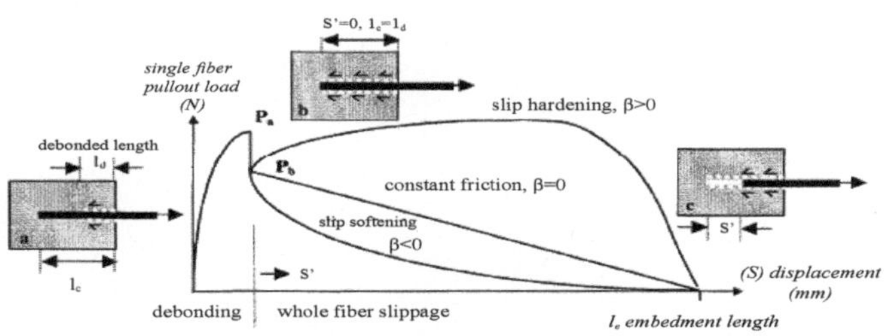


**Figure 1:** General profile of single fiber pull out curve.

***Corresponding author:** Rathod JD, Applied Mechanics Department, Faculty of Technology and Engineering, The M.S. University of Baroda, Vadodara, India, E-mail: jdrathodmsu@yahoo.com

fiber can undergo sliding with either slip hardening, constant friction or slip softening effect, characterized by the coefficient β, which is, respectively, positive, zero, or negative [4]. Slip hardening may occur with polymer fibers, if they are less hard than surrounding matrix. They would get damaged and a jamming effect can take place inside the matrix. This leads to an increasing load resisting fiber pull out. This phenomenon can be very beneficial as long as fiber tensile strength is not exceeded. Conversely, constant friction or slip softening is often observed when the fiber hardness is higher than that of the surrounding matrix. Frictional bond strength (τ) can be computed as $p_b/\pi d_f l_e$ when full debonding of fiber takes place with a constant friction.

## Preparation of Specimens and Test Set Up

### Flexure test

Kamal brand 53 grade OPC, 300μ passing silica sand, W/C ratio of 0.35, 4% fiber volume fraction of Recron and 2% dose of high performance concrete super plasticizer of Glenium 51 brand based on modified polycarboxylic ether were used for the present study. Ingredients of ECC were mixed in 5 liter capacity Hobart type mixer machine. Flow table test was performed to evaluate workability of ECC in fresh state. Beams of length 500 mm, width 100 mm having depth of 20 and 30 mm were cast. Three specimens each of C: S ratio 1:0.5, 1:1, 1:1.5 and 1:2 were cast. Beams were tested with four point loading arrangement on MTS machine under displacement control at a rate of 0.005 mm per second as shown in (Figure 2). Load and displacement at the first crack and at ultimate load were recorded during the test. Load-displacement readings were automatically recorded at an interval of 1 second.

### Single fiber pull out test

Moulds for pull out test were prepared for 6 mm thickness of mortar as shown in (Figure 3). Six fibers were embedded in each matrix specimen at a spacing of 2 cm. Long fibers were taken and tied to the nails provided on both sides of the mould. Cementitious matrices of C: S ratio 1: 0.5, 1:1, 1:1.5 and 1: 2 with w/c ratio of 0.35 and 0.5% Glenium 51 brand high performance super plasticizer were poured into mould and then covered with plastic for 24 hours. All the specimens were the taken out by untying the fibers and then specimens were put in the curing tank for 28 days. The extended length of fiber on one side was then cut while keeping other side as it is to facilitate proper grip in the machine. INSTRON 1121 machine of 10 KN capacity was used for testing as shown in (Figure 4).

Pull out specimens were gripped by mechanical grips and fibers were griped by pneumatic grips as shown in (Figure 5). Care was taken to ensure zero fiber free length to avoid elastic stretching of free fiber. Pull out force was applied and load was recorded with the help of 2 Kg load cell under displacement control at a rate of 3 mm/min. Load-pull out displacement graphs were plotted on plotter available with the machine as shown in (Figure 6). Results were then processed in excel software to have representative graphs. Six single fiber pull out tests were performed for each matrix out of which 3 most consistent results having variation upto 15% from the average are being considered in the present discussion.

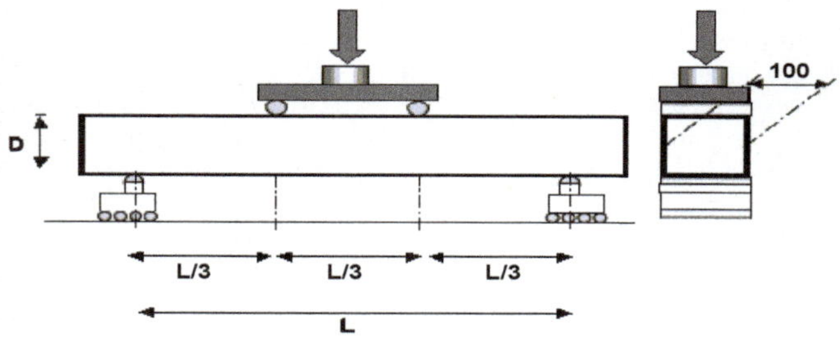

**Figure 2:** Four point loading set-up.

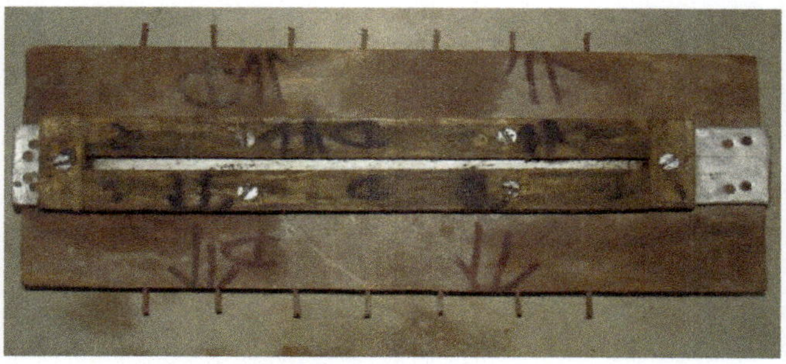
**Figure 3:** Pull out mould for 6 mm thickness.

**Figure 4:** Test set-up for pull-out test.

**Figure 5:** Grips for fiber and specimen.

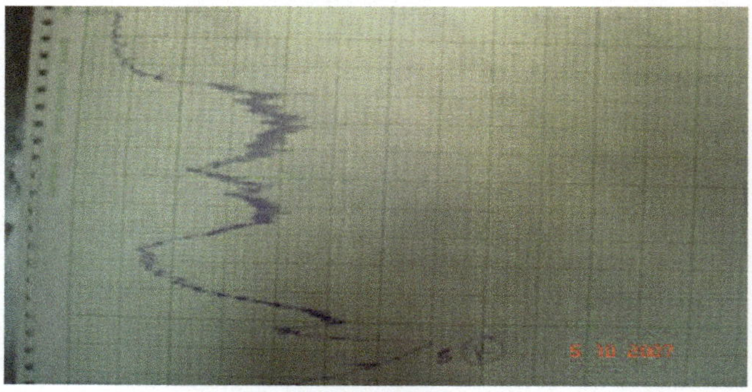
**Figure 6:** Graph produced by INSTRON.

## Discussion of Test Results

1. Strain hardening performance via multiple cracking in axial tension ensures strain hardening performance in flexure also but not vice versa. Experimental investigation carried out by Rathod et al. indicated strain hardening performance in the ECC in axial tension with C:S ratio of 1:0.5 and 4% fiber volume fraction [11]. In this experimental investigation, ECC with all C:S ratios indicated strain hardening performance in flexure. Average first crack strength and average ultimate strength values are indicated in Table 1. Strain hardening in flexural deformation is observed in all the specimens due to the contribution of fibers. Increase in strength after first crack is referred as reserved strength. First crack strength is considered to be the ultimate strength for ordinary concrete or normal FRC and that is taken as a reference. Any increase in strength after first crack will be as a result of subsequent multiple cracking by the fiber bridging action due to which load is getting transferred to other sections instead of being localized at a single crack unlike in normal FRC. Reserved strength shows additional load carrying capacity which represents damage tolerance of ECC material. All the beams of 20 and 30 mm depth with 1:0.5 and 1:1 C: S ratio showed multiple cracking. This validates strain hardening through multiple cracking, a unique characteristics of an ECC material. Percentage reserved strength decreases with increase in sand content in all matrices for 20 and 30 mm depth beams. C: S ratio of 1:0.5 gives the best performance. The well known size effect fact i.e. as depth of beam increases strength decreases is well validated here in all the matrices.
2. Shah and Rangan proposed the following general equation for predicting the ultimate flexural strength of the fiber reinforced composite [12]. Where $f_m$ is the maximum strength of the plain matrix, A and B are constants which can be determined experimentally. For plain concrete, A=1 and B=0. The constant B accounts for the bond strength of the fiber and randomness of fiber distribution. Swamy et al. (1974) established values for constant A and B as 0.97 and 4.94 for ultimate flexural strength of steel fiber reinforced concretes and 0.843 and 4.25 for its first cracking strength [13]. Because of linear dependence of the ultimate flexural strength of FRC on volume fraction of fibers and their aspect ratio, it could be stated that the ultimate flexural strength generally increases with the fiber reinforcing index, defined as the product of fiber volume fraction and aspect ratio ($V_f L/d_f$).

All the micromechanical parameters including the size effect are considered in the following equation. Addition of fiber in cementitious matrix, multiply the size effect due to random orientation and distribution of fibers in the composite. To avoid size effect of plain cementitious matrix, it is preferred to substitute the strength of the same size of specimen in the equation of flexural strength of ECC which simplify the governing equation. Constants A and B are proposed in the equation for strain hardening fiber reinforced composite.

$$F_{ue} = [A\, f_m\, (1-V_f^2)] + [B\, V_f^2\, (L_f / D_f)]$$

Where,

$F_{ue}$=Ultimate flexural strength of ECC matrix

$f_m$=Ultimate flexural strength of a plain cementitious matrix of same material composition as ECC except fibers; this will elimate size effect of plain matrix.

A=1.25. A constant considered for 2 dimensional specimens i.e. where two dimensions of specimen are more than atleast 3 times the length of fiber.

$V_f$=Fiber volume fraction

B=g x τ where g=Snubbing coefficient (frictional pull out effect when fibers are aligned in other than the stress direction) and τ=Frictional bond strength in MPa.

$L_f$, $D_f$=Length and diameter of fiber respectively.

Snubbing coefficient is assumed to be 2 for the calculation. $L_f$ and $D_f$ are considered as 12.5 mm and 30 μm respectively. Actually shape of cross section of Recron 3S fiber is substantial triangular for which equivalent diameter of fiber is taken as 30 μm. Frictional bond strength values are taken from Table 2 for various matrices. Little scattering is observed between calculated and experimental values indicated in Table 3 because snubbing coefficient is not actually evaluated and is assumed to be constant for all matrices. Closer values can be achieved by evaluating g. The required flexural strength can be achieved by tailoring micromechanical parameters such as τ, g, $L_f$, $D_f$ and $V_f$. Frictional bond strength values evaluated in this investigation helps to predict realistic flexural performance.

3. Deflection of the beam is measured under flexure, therefore, deflection hardening term may be the most appropriate to indicate enhancement in deformation after first crack up to ultimate load. It undergoes large inelastic deformation after formation of first crack. Percentage deflection hardening showed by 20 mm thick beam in a matrix having C: S ratio as 1:0.5 is 442.25%. It validates ECC material in a category of smart bendable concrete. Most of the specimens showed deflection hardening more than 100%.
4. Toughness index is used to evaluate post peak performance of the material (Table 4). Toughness index $I_5$ is calculated as area under load-displacement curve for 3 times first crack displacement divided by area under load-displacement curve for first crack displacement. Toughness index for plain cementitious matrix could not be represented because of

| C:S Ratio | Depth of Beam (mm) | Avg. First Crack Strength (N/mm²) | Avg. Ultimate Strength (N/mm²) | Displ. At First Crack Load (mm) | Displ. At Ultimate Load (mm) | Strain Hardening |
|---|---|---|---|---|---|---|
| 1:0.5 | 20 | 4.095 | 5.177 | 1.8541 | 10.0537 | Yes |
| | 30 | 3.553 | 4.277 | 2.0928 | 5.5386 | Yes |
| 1:1 | 20 | 2.554 | 2.834 | 2.3143 | 6.1545 | Yes |
| | 30 | 1.022 | 1.225 | 1.7719 | 4.2539 | Yes |
| 1:1.5 | 20 | 0.683 | 0.767 | 1.4041 | 2.2587 | Yes |
| | 30 | 0.563 | 0.604 | 1.4802 | 1.5695 | No |
| 1:2 | 20 | 0.625 | 0.687 | 1.4301 | 1.8057 | Yes |
| | 30 | 0.576 | 0.804 | 1.4473 | 3.1240 | No |

**Table 1:** Test results of beams in flexure.

| C: S Ratio | $P_{max}$ (N) | Bond Strength τ (N/mm²) | Average Bond Strength |
|---|---|---|---|
| 1: 0.5 | 0.451 | 0.798 | 0.792 |
| | 0.451 | 0.798 | |
| | 0.441 | 0.780 | |
| 1: 1.0 | 0.441 | 0.780 | 0.746 |
| | 0.422 | 0.747 | |
| | 0.402 | 0.711 | |
| 1: 1.5 | 0.402 | 0.711 | 0.723 |
| | 0.392 | 0.694 | |
| | 0.432 | 0.764 | |
| 1: 2.0 | 0.387 | 0.683 | 0.685 |
| | 0.358 | 0.633 | |
| | 0.417 | 0.738 | |

**Table 2:** Frictional bond strength.

| C: S Ratio | Depth of Beam (mm) | $f_m$ | Ultimate Flexural Strength [Experimental] (N/mm²) | Ultimate Flexural Strength [Theoretical] (N/mm²) |
|---|---|---|---|---|
| 1: 0.5 | 20 | 3.302 | 5.177 | 5.174 |
| | 30 | 2.581 | 4.277 | 4.301 |
| 1: 1 | 20 | 1.474 | 2.834 | 2.867 |
| | 30 | 0.184 | 1.225 | 1.232 |
| 1: 1.5 | 20 | 0.158 | 0.767 | 1.164 |
| | 30 | 0.288 | 0.604 | 1.326 |
| 1: 2 | 20 | 0.181 | 0.687 | 1.139 |
| | 30 | 0.087 | 0.804 | 1.022 |

**Table 3:** Comparison of experimental and theoretical values.

| C:S Ratio | Depth of Beam (mm) | Percentage Deflection Hardening | Percentage Reserved Strength | Toughness index $I_5$ |
|---|---|---|---|---|
| 1:0.5 | 20 | 442.253 | 26.43 | 7.155 |
| | 30 | 164.652 | 20.38 | 6.937 |
| 1:1 | 20 | 165.934 | 10.96 | 5.769 |
| | 30 | 140.069 | 19.93 | 7.766 |
| 1:1.5 | 20 | 60.866 | 12.37 | 5.095 |
| | 30 | 6.033 | 7.22 | 4.195 |
| 1:2 | 20 | 26.265 | 9.92 | 4.671 |
| | 30 | 115.857 | 39.74 | 5.932 |

**Table 4:** Flexural performance of ECC.

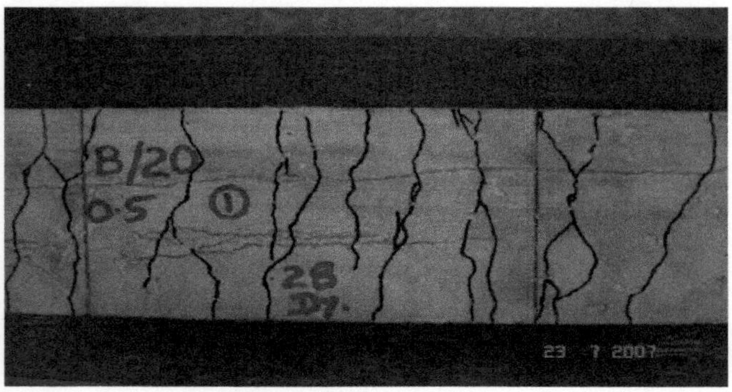

**Figure 7:** Multiple cracks in 20 mm beam.

catastrophic failure after first crack formation. Toughness index of any high strength concrete ranges from 1 to 2 whereas it ranges from 6 to 8 for ECC material. ECC with C: S ratio as 1:0.5 exhibited highest toughness index value.

5. Almost all the beams having 20 mm thickness exhibited multiple cracking as reflected in Figure 7. 30 mm thick beams having C: S ratio as 1:1.5, and 1:2 could not show multiple cracking due to size effect and less bond of fiber with cementitious matrix. Reduction in flexural strength is found with the increase in sand content as shown in Figure 8. It has direct relation with frictional bond strength.

6. Load-displacement curves for ECC matrix in 20 mm thick beam are shown in Figure 9. The performance showed by ECC matrices without steel reinforcement is exciting and is comparable to performance of steel reinforced concrete in which large displacement ability without unloading is exhibited due to yielding of reinforcement. This performance reveals deformation compatibility of ductile ECC material with the reinforcement without loosing integrity of the material. Such large displacement is accompanied by large spreaded damage area which gives sufficient time. After formation of first crack, the member doesn't unload immediately but exhibits inelastic deformation absorbing large energy associated with formation of other closely spaced cracks. It should be noted that this performance is only for 20 mm thick beams, which reduces with increase in beam depth which is largely reflected in load-displacement curves of all the specimens having 30 mm depth.

Significant load drop in the curves indicating initiation of micro-crack is observed in axial tension [11], while, sudden load drop for each initiation of micro crack may not be observed because of flexural deformation mechanism.

7. Single fiber pull out tests were conducted for different matrices with 6 mm embedded length. Bridging and pull out action of 12 mm length fiber through the crack in the composite is executed to its maximum possible extent on either side of crack by 6 mm. Therefore 6 mm pull out length can be considered as an effective length in single fiber pull out test for the composite with 12 mm fiber length. Fibers were gripped in pneumatic grip and attempt was made to get zero fiber free length.

8. Load-Pull out displacement graphs were automatically plotted by INSTRON machine. The results were then processed in Microsoft Excel and graphs are plotted as shown in Figure 10. These results are taken as average of best 3 results out of 6 fibers embedded in the particular composite. Performing the pull out test is difficult task as diameter of fiber is 30 μm. As explained earlier if there is significant load drop from the maximum, it is considered to be the chemical bond. Recron 3S fibers are polyester type and hydrophobic in nature. It has no chemical bond which is confirmed from all the graphs of pull out test.

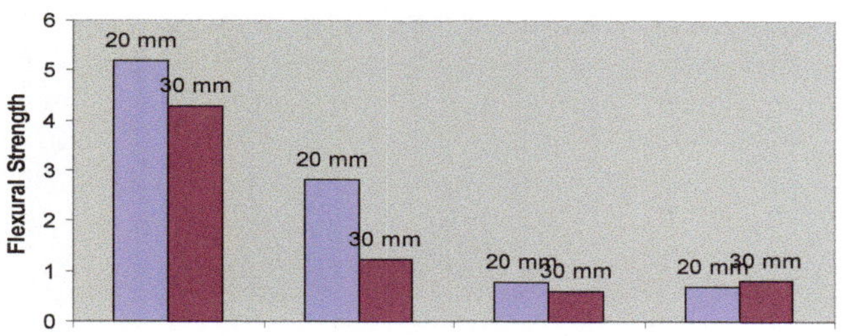


**Figure 8:** Flexural strength v/s C: S ratio.

9. Complete debonding portion of load-pull out curve is linear and then it reaches to maximum pull out force which is considered for the bond strength calculation. Frictional bond strength values are given in Table 2. Bond strength increases as the C: S ratio increases. Previous study revealed that as sand content decreases, compressive strength and indirectly, the packing density increases. This effect results in more sound and stable interface around the fiber [14]. The bond strength achieved with C: S ratio as 1:0.5 is better than PE (0.5 MPa) and PP (0.1-0.3 MPa) fibers [5,15]. Calculated values of bond strength are used in the proposed equation. The results of which conform the reliability of the experiment.

10. Tensile strength of Recron 3S fibers is 1000 N/mm$^2$. Rupture of the some of the fibers takes place at 644 N/mm$^2$ strength. The strength at rupture in composite material is known as apparent tensile strength of fiber, which is always less than its actual tensile strength. As the surrounding matrix is harder than fiber, abrasion damage occurs by delaminating the layers

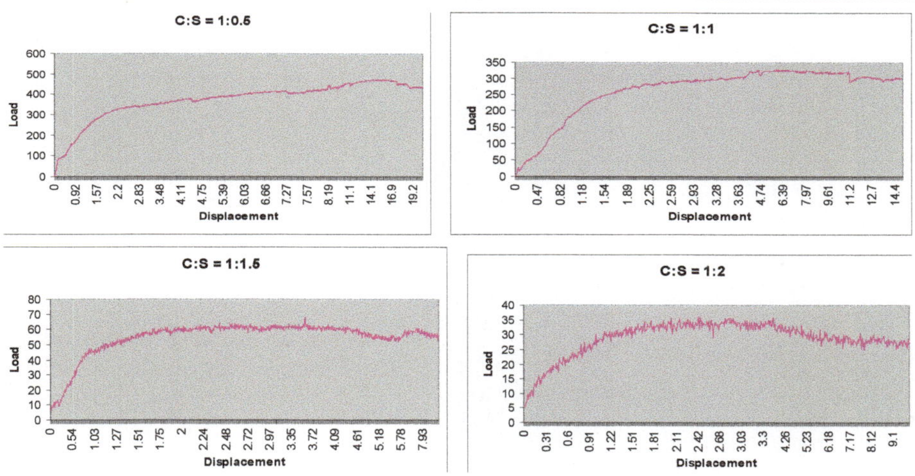


**Figure 9:** Load-displacement curves for 20 mm thick beam.

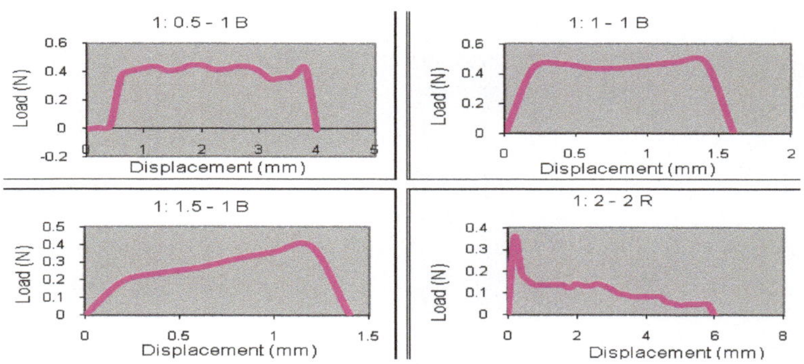


**Figure 10:** Load-pull out displacement curves in single fiber pull out.

through the fiber thickness. This mechanism is confirmed from the load-pullout curve very clearly. Complete pull out of fibers did not take place in the matrix with C: S ratio of 1:0.5, 1:1, 1:1.5 because after fiber gets its debonding process complete, it undergoes pull out with damage due to delamination of outer layer with the reduction in cross section which eventually resulted into rupture. This phenomenon is beneficial as it doesn't allow to unload the composite and the load resistance mechanism is enhanced. Fiber ruptures due to reduction in cross section hence before it pulls out completely the apparent tensile strength governs the mechanism. If the debonded end is viewed under microscopic lens, fiber tunnel is seen. Fiber did not rupture in the matrix with C: S ratio 1:2 due to lower bond strength. Fiber did not damage due to softer surroundings and hence resistance dropped suddenly. Frictional bond strength is calculated ignoring the $\beta$ effect which is more realistic. Fibers pull out of the matrix with constant friction as seen from all the graphs, hence $\beta$ value can be considered as zero.

11. Percentage deflection hardening is the phenomenon that takes place after the first crack when fibers get debonded and enhances load resistance mechanism up to the maximum load. Single fiber pull out test gives best prediction of deformation mechanism. C: S ratio of 1:0.5 gives highest bond strength hence first crack flexural strength is highest and then after debonding it pulls out about 3 mm maintaining load carrying capacity. It reflects maximum deformation by pull-out action without unloading which indicates maximum deflection hardening. Fibers do not rupture because when such large volume fraction of 4% is used, it gets fewer bonds with cementitious matrix due to interaction with each other. Deflection hardening and first crack strength in the composite with C: S ratio 1:1 is also validated by the single fiber pull out curve. Matrix with C: S ratio 1: 1.5 and 1: 2 could not undergo much deformation because fibers did not show much resistance after debonding and sudden drop could be seen in single fiber pull out curve.

## Conclusions

- Although ECC is characterized by its performance in axial tension, flexural behavior should be evaluated properly as beam members behave in different manner and are used as energy absorbing element by their large inelastic deformation.
- ECC accepts challenging seismic performance and its promising deformation mechanism makes it most suitable material in key places of a structure where ductile performance is expected. Experimental results show absolutely unique and stunning performance of ECC in flexure, not seen before in any cementitious composites. The flexural performance of ECC, however, reduces with the increase in sand content.
- ECC absorbs large energy by inelastic deformation mechanism. Cement: sand ratio of 1: 0.5 is suggested for optimum performance.
- Single fiber pull out test though difficult to perform provides best means to evaluate flexural strength and understand deformation mechanism.
- Proposed equation for ultimate flexural strength holds good to predict strength. However, predication can be further improved by evaluation snubbing coefficient.

### References


1. Naaman (2003) Engineered Steel Fibers with Optimal Properties for Reinforcement of Cement Composites, Journal of Advanced Concrete Technology 1: 241-252.
2. Bentur A, Mindess S (1990) Fiber reinforced cemenetitious composites, Elsevier Science Publishers, Amsterdam, Netherland.
3. Li VC, Wang Y, Backer S (1991) A micromechanical model of tension softening and bridging toughening of short random fiber reinforced brittle matrix composites, J. of Mech. and Physics of Solids 39: 607-625.
4. Cooper GA, Kelly A (1970) The contribution to the work of fracture of a composite material of pull-out fibers, Mechanics of Composite Materials Wendt FW, Liebowitz H, Perrone L, eds., Pergamon, Oxford, U.K. 653-661.
5. Lin Z, Li VC (1997) Crack bridging in fiber reinforced cementitious composites with slip-hardening interfaces, J. of Mech. Phys. Solids, Vol. 45(5): 763-787.
6. Li VC, Kanda T, Lin Z (1997) The influence of fiber/matrix interface properties on complementary energy and composite damage tolerance Proc., 3rd Conf. on Fracture and Strength of Solids, Hong Kong.
7. Li VC, Leung CKY (1992)Steady-state and multiple cracking of short random fiber composite,J. of Engg. Mech., ASCE 118: 2246-2264.
8. Li VC (1993) From micromechanics to structural engineering-the design of cementitious composites for civil engineering applications,J. of Struct. Mech. and Earthquake Engg., 10:37-48.
9. Li VC, Stang H (1996) Interface property characterization and strengthening mechanism in fiber reinforced cement based composites, J. of Advanced Cement Based Materials, Vol. 6(2), pp. 1-20.
10. Leung CKY, Li VC (1990)Strength-based and fracture based approaches in the analysis of fiber debonding, J. of Mat. Sci. Letter 9: 1140-1142.
11. Rathod JD, Patodi SC (2010)Interface tailoring of polyester type fiber in Engineered Cementitious Composite matrix against pull out, J. American Concrete Institute (Material).
12. Shah SP, Rangan BV (1971) Fiber reinforced concrete properties, J. of American Concrete Institute, 68: 126-135.
13. Swamy RN, Mangat PS, Rao, CVSK (1974) The mechanics of fiber reinforcement of cement matrices, Fiber-Reinforced concrete, J. of American Concrete Institute, Detroit.
14. Rathod JD, Patodi SC (2008) Effect of cement-sand ratio on shear strength of ECC in 1D and 2D fiber orientation,J. Civil Engineering and Construction Review 21:86-96.
15. Li VC, Hwai-Chung W, Chan YW (1996) Effect of plasma treatment of polyethelene fibers on interface and cementitious composite properties, J. of American Ceramic Society 79: 700-704.

# Estimation of the Residual Stress in Clay Sampled from the Field

**Owayo AA* and Ou CY**

*Department of Civil and Construction Engineering, National Taiwan University of Science and Technology, ROC Taipei-Taiwan*

**Abstract**

The objective of this study is to establish a reliable way to determine the residual stress in clay sampled from the field. During saturation in triaxial test, using residual stress as back pressure as opposed to the traditional saturation methods leads to better sample quality. In this research the residual stress within the sample is measured and compared against the one that can be predicted with the help of the equation proposed by Skempton based on the account of change in total stress during sampling. The results do not exhibit any predictable trend, in fact sometimes the prediction is quite good while the rest of the time not good. Thus in the absence of a perfect sample, the use of Skempton's equation is more reliable.

**Keywords:** Residual stress; Triaxial test; Skempton's equation; Clay; Perfect sample; Sampling; Saturation

## Introduction

During triaxial testing we require a sample with as minimal disturbance as possible. The in-situ conditions needs to be preserved as best as possible. Residual effective stress also called the initial mean effective stress, stored effective stress or effective stress after sampling is the effective stress remaining in the soil sample after sampling, handling and storage. This is a concept rooted in perfect sampling concept where there would be no disturbance except that from stress relief. The difference between sampling effective stress and the residual effective stress is caused by disturbance rather than stress relief. This implies that the residual stress can be a qualitative measure of sample disturbance [1]. The sample swelling thus disturbance during saturation, can be minimized by applying an effective stress equal to the measured residual stress before adding water to the specimen [1]. This implies that the sample quality may be improved further if the correct or 'true' residual stress is used. This informed by the fact that due to poor storage, and handling the sample may undergo pore pressure dissipation and hence the measured residual stress is not the 'true' value. The measured residual stress is applied as back pressure during saturation. In the conventional saturation process, usually very small back pressure is used; this could lead to sample swelling thus disturbance.

Skempton proposed an equation that may be used to estimate residual effective stress within a sample based on the account of total stress changes, equation 1. Equation 2 may be used to compute the pore water pressure change on the account of total stress change. This research seeks to measure the remaining residual stress in the in situ sample and compares it against the prediction from equation 2. Certain prediction matches the experimental values so well as elucidated in Figure 3 while others do not. Practically speaking, it is difficult to eliminate the sources of disturbance.

Inevitable disturbance may be encountered during sample extrusion, trimming etc. Non homogeneity of the soil i.e. presence of foreign material like silt or sand may lead to pore water pressure dissipation thus the measured residual stress is not the 'true' residual stress or the expected residual stress immediately after sampling. Since the residual stresses experimentally determined from the laboratory depends on uniformity of the sample and human factor such as sample handling, storage, trimming etc. The effects of disturbance due to these factors are difficult to quantify.

## Method

### Soil description

Tube samples of Taipei silty clay sampled from within the environs of Taipei 101 were used. Taipei Silty clay has its water table at 2 meters below the surface. The basic soil properties are as summarized below in Table 1.

- Unified soil classification system (USCS)
- Natural water content, $W_n$
- Liquid limit, LL
- Plastic, limit, PL
- Plastic index, PI
- Shrinkage Limit SL
- Bulk density $\gamma_t$
- Coefficient of swelling, $C_s$
- Coefficient of recompression, $C_r$
- $\sigma_1$ is total vertical stress
- $\sigma_3$ is horizontal total stress

| | ProTable 1perties |
|---|---|
| **USCS** | |
| $W_n$ (%) | 33.0~40.0 |
| LL (%) | 38.0~38.5 |
| PL (%) | 26.0~26.3 |
| PI (%) | 12.0~12.2 |
| SL (%) | 7.5 |
| $\gamma_t$ (kN/m³) | 18.0~20 |
| **Others** | |
| $C_c$ | 0.32 |
| $C_r$ | 0.038 |
| $G_s$ | 2.7 |
| $e_o$ | 0.89~1.10 |

**Table 1:** Summary of the basic soil properties.

***Corresponding author:** Owayo AA, Department of civil and construction engineering, National Taiwan University of science and Technology, ROC Taipei-Taiwan, E-mail: owayoachiando@gmail.com

- $\gamma_{sat}$ is saturated unit weight of the soil
- $\gamma_w$ is the unit weight of water
- z is the sample depth.

## Idealized concept of stress relief during sampling and the consequent pore water pressure changes

$$\Delta u = B\Delta\sigma_3 + A(\Delta\sigma_1 - \Delta\sigma_3) \qquad (1)$$

$$u_0 + \Delta u = p'_r \qquad (2)$$

A and B are Skemptons pore pressure parameters

The equation 1 as proposed by Skempton in 1954 can be used to estimate the changes due to stress relief. $k_0$ for saturated soil is taken as 0.5, B=1 and A=2/3 for axial extension synonymous to unloading during sampling [2].

Before sampling as shown in Figure 1, the soil sample experiences a typical triaxial compression test condition [3], an overburden total stress, $\sigma_1$ and a horizontal total stress, $\sigma_3$ the soil also experiences a hydrostatic water pressure, $u_0$.

After sampling, the soil sample is under no confinement but is still standing this implicitly means that there is a negative pore water pressure, $-u_r$ in the sample preventing it from collapsing. This is theoretically equals the residual stress in the sample. The resultant change in pore water pressure Δu, may be estimated by $u_0$-$u_r$. Δu may also be estimated from equation 1. In the laboratory we are able to measure $-u_r$, ideally this value is the residual stress immediately after sampling: but due to disturbance during sampling handling, pore water dissipation and storage this value may be less than the expected value. Theoretically when the estimation from equation 2 equals the measured residual stress in the laboratory, this implies that the sample has not experienced substantial disturbance hence is close to a perfect sample. The residual stress is the intercept of Y axis (negative value) when the cell pressure equal zero.

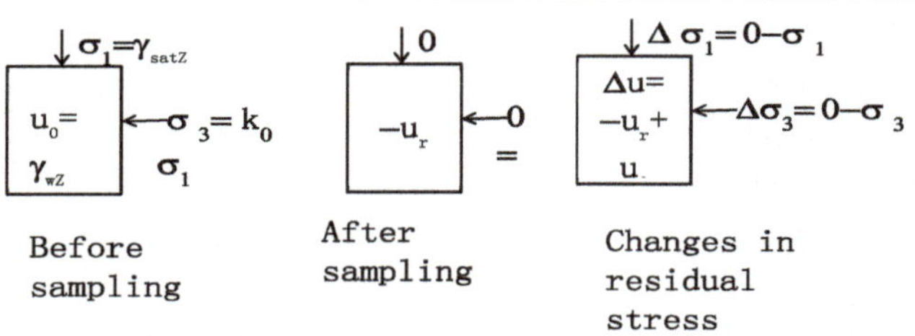


**Figure 1:** Illustration of the stress release concept.

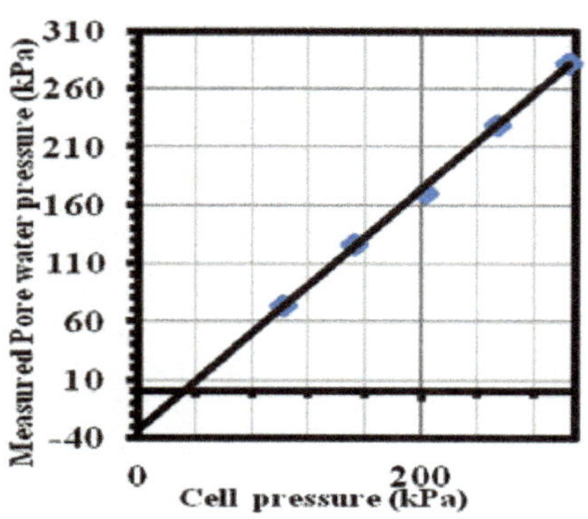


**Figure 2:** Graphical quantification of change in void ratio during traditional saturation and saturation using residual stress.

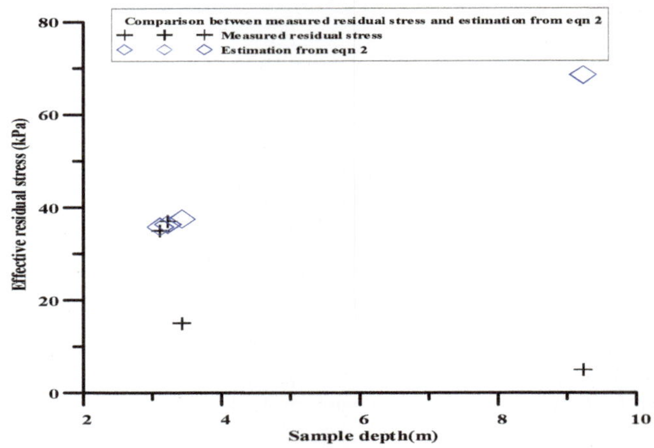


**Figure 3:** Estimated $p'_r$ versus estimated from equation 2.

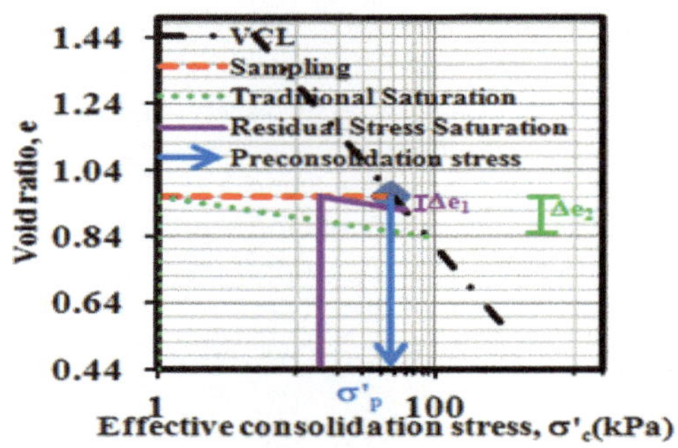


**Figure 4:** Graphical quantification of change in void ratio during traditional saturation and saturation using residual stress.

## Residual stress measurement

The effective stress, $p'_r$, was evaluated prior to saturation through the response of excess pore-water pressure as isotropic stress increaments applied with drainage lines closed. The total confining stress, $\sigma_c$ was increased in 50 kPa from 100 kPa to 300 kPa and reduced in 100 kPa from 300 kPa to 100 kpa. In each load increament step the pore water, u, was allowed to equillibrate within 30-60 min. The same procedure was followed in reducing $\sigma_c$. Expected response of u to $\sigma_c$ is as shown in (Figure 2). Linear regression applied to this data, the matric suction within the sample can be taken as th e value of u at $\sigma_c$=0. Following the general concept of effective stress,the residual effective stress, $p'_r$ may be defined:

$$p'_r = -u_r \qquad (3)$$

## Results

VCL, is the virgin consolidation line

## Quantifying the levels of disturbance

Figure 3 is established from the fact that, we know the in-situ pre-consolidation stress and the original void ratio can be estimated from the in situ water content for each sample. This point on the VCL can thus be established. The gradient of VCL is the value of $C_c$. Recompression of the sample follows the value of $C_r$ (Figure 4).

## Measured residual effective stress against the estimation from equation

Summary of measured residual stress against the estimation from

| Test | Sample depth, m | Water content, w (%) | Measured residual stress (kPa) | Calculated residual stress (kPa) equation 2 | Effective vertical stress (kPa) |
|---|---|---|---|---|---|
| Sample 1 | 3.23 | 36.3 | 37.05 | 36.4 | 46.26 |
| Sample 2 | 3.43 | 35.74 | 15 | 37.61 | 47.92 |
| Sample 3 | 9.23 | 33.07 | 5 | 68.6 | 96.55 |
| Sample 4 | 3.12 | 40.01 | 35 | 35.94 | 45.34 |

**Table 2:** Summary of measured residual stress against the estimation from Skempton's formula.

| Test | $e_0$ | $\Delta_e$ | $\Delta e_2/e_0$ | $\Delta e_1$ | $\Delta e_1/e_0$ |
|---|---|---|---|---|---|
| Sample1 | 0.98 | 0.03 | 0.03 | 0.24 | 0.24 |
| Sample 2 | 0.96 | 0.05 | 0.05 | 0.11 | 0.11 |
| Sample 3 | 0.89 | 0.08 | 0.09 | 0.12 | 0.13 |
| Sample 4 | 1.11 | 0.04 | 0.04 | 0.16 | 0.14 |

**Table 3:** Comparing the void ratio changes when traditional saturation is used and when residual stress is used as back pressure.

| Sample quality | Volume change De/e 0 |
|---|---|
| Very Good | <0.04 |
| Good to Fair | 0.04~0.07 |
| Fair to Poor | 0.07~0.14 |
| Very Poor | >0.14 |

**Table 4:** Classification of sample quality based on the void ratio change during recompression.

Skempton's formula is shown in Table 2.

## Discussions

Sampling relieves the sample of its overburden pressure effectively reducing the in-situ pre-consolidation stress to zero as shown in Figure 3. The sample is not allowed to swell since it is confined by the sample tube. If there is no disturbance, we expect no volumetric changes thus constant void ratio, equal to the field void ratio. A negative pore water pressure equal to the residual stress is developed in the sample to prevent it from collapsing and help it maintain its void ratio. However if any disturbance is caused to the sample this may result in volumetric changes thus a change in the void ratio. Traditional saturation method uses very small back pressure, normally close to zero kPa as shown in Figure 3. When a very small back pressure is used during saturation to recompress the sample back to the in-situ condition before triaxial test is done on the sample. A path similar to the one shown in dotted green line is followed. On the other hand, if measured residual stress (usually higher than zero) is used during saturation to recompress the sample back to the in-situ state. A path similar to the solid pink line is followed. Recompressing the soil sample back to the original in situ state leads to volumetric changes in the sample thus the changes in void ratio as in Figure 3. The subsequent changes in void ratio when traditional saturation is employed $\Delta e_{2,}$ is much higher than the changes in void ratio $\Delta e_1$ when measured residual stress is used as back pressure. Figure 3 also clearly indicates that when the measured residual stress is used as back pressure: when the consolidation stress equal to or higher than the in-situ pre-consolidation stress, if residual stress is used for saturation then the sample is much closer to the virgin consolidation line compared to when traditional saturation technique is applied, leading to better sample quality. For all the samples used in this study we established (Figure 3) for each sample, from which we derived (Table 3).

When the consolidation stress is higher or equal to the in-situ pre-consolidation stress the soil sample is at the virgin consolidation line (VCL), where it is normally consolidated. The prediction from equation 2 matches the measured residual stress in samples 1 and 4 very well. While for samples 2 and 3 there is a substantial difference between the prediction of equation 2 and the measured value i.e. the measured value is much smaller. This difference may be attributed to the possible accumulated effects of disturbance in the course of soil handling from the sampling stage, storage and performing the test. Presence of foreign material in the clay such as silt or sand may also make it easier for the dissipation of pore water pressure during sample extrusion and trimming so that the measured residual stress is way below the expected 'true value' [4].

Thus the measured residual stress may not be the actual, correct or the 'true' residual.

When the measured residual stress is used the volumetric changes are far much less as compared to when traditional saturation process is employed [5]. The classification of sample quality ranges from 'Fair to Poor' to Very good when residual stress is used as back pressure (Table 4). When conventional saturation is used as back pressure during saturation the classification is from 'Very Poor' to 'Fair to Poor'. It is expected thus, that if the 'true' residual stress estimated from equation 2 is used all the samples would fall within the 'Very Good' quality classification. It is clear thus that when the residual stress is used as back pressure during the saturation stage the sample quality is much more improved (Figure 5).

### CIU_AC-Isotropic consolidated undrained axial compression test

These results demonstrate that the use of Skemptons equation to predict residual stress agrees very well with the experimental data when the sample in question has the insitu conditions preserved as best as possible. While when the prediction does not agree with the experimental value; this is possibly due to disturbance in the course of soil handling,storage trimming, extrusion etc. Since all the samples used here in were stored within the same environment, and all the experiments carried out by the same individual. It is expected that the accumulated effects of disturbance that may be attributed to the environment, human judgement,method of sample preparation etc. are fairly controlled and are the same for all the samples. Employing the same procedure of sample preparation and experimental execution some samples turn out to give estimates close to equation 2 while others

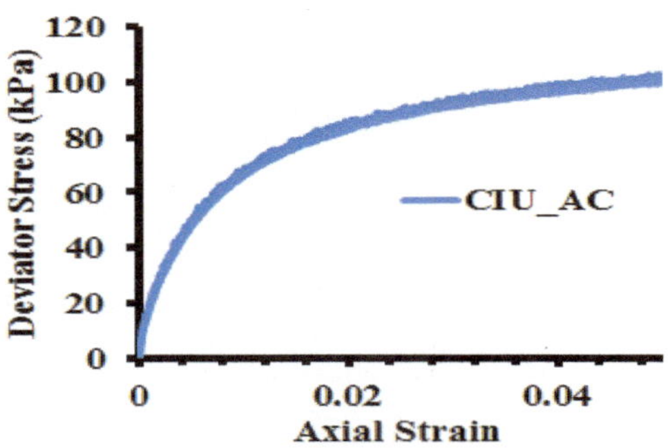


**Figure 5:** Isotropic Consolidated test with the sample quality classified as 'very good'

do not. We therefore deduce that this may be due to disturbance, and it's effects to the remaining residual stress in the sample. These effect appear to be random and are therefore difficult to quantify. This further implies that the samples whose prediction from equation 2 and measured values from the experiment are close, are very good quality samples with minimal disturbance. The samples whose prediction and measurement do not agree implies disturbance to the samples and thus are samples of lesser quality.

## Conclusions

When measured residual stress is used as back pressure during saturation in the triaxial test the sample quality is quite improved. If the sample is disturbed, pore water pressure dissipates in the course of sample handling thus, the measured residual stress is not the actual or the expected 'true' residual value. While when the true residual stress is used the sample quality is expected to be much more improved. The use of equation 2 to estimate the residual stress would be more reliable in cases where a sample is relatively disturbed.

### References

1. Cho W, Terence P, Holman, Jung YH, Finno RJ (2006) Effects of Swelling During Saturation in Triaxial Test in Clays. Geotechnical testing Journal.
2. Holtz RD, Kovacs WD (1991) An introduction to Geotechnical Engineering: Prentice-Hall.
3. Standard Test Method for Consolidated Undrained Triaxial Compression Test for Cohesive Soils. ASTM D4767 – 11.
4. Das BM (1978) Advanced Soil Mechanics. Taylor and Francis.
5. Teng FC, Ou CY (2011) Application of a Suction Control System in the Method of Specimen Saturation in Triaxial Tests. Geotechnical testing Journal.

# Investigation of Heat Impact Behavior on Exterior Wall Surface of Building Material at Urban City Area

**Mohd Fadhil Md Din[1], Hazlini Dzinun[1*], Ponraj Mohanadoss[1], Shreshivadasan Chelliapan[1], Zainura Zainun Noor[1], Dilshan Remaz Ossen[2] and Kenzo Iwao[3]**

*[1]Institute of Environmental Water Resources and Management (IPASA), Faculty of Civil Engineering, Universiti Teknologi Malaysia, 81310 UTM Skudai, Johor, Malaysia*
*[2]Faculty of Built Environment, Universiti Teknologi Malaysia, 81310 UTM Skudai, Johor, Malaysia*
*[3]Nagoya Institute of Technology, Gokiso-cho, Showa-ku, Nagoya, Aichi-Pref., 466-8555, Japan*

**Abstract**

This paper represents a comparative study aiming to investigate the heat impact on the vertical surfaces of buildings temperature based on their thermal behaviors. This study was carried out based on four building materials commonly used in Malaysia namely brick, concrete, granite and white concrete tiles. In order to investigate the thermal performance on the building materials, surface temperature sensors, data logging system as well as infrared thermography procedures were used, respectively. As the thermal impact to the materials was measured using infrared thermometric and thermographic, a field work of thermal value can be simply observed as a "seeing heat" effect. The results indicate that bricks had the capability to absorb and store heat greater than other materials during peak daytime event. The normalized (total heat/solar radiation) of brick materials was 0.093, which was the highest value compared to others. A brick material shows the highest impact of heat in 51% than white granite material and it releases a substantial amount of heat into the atmosphere through radiation and convection factors. The types of material used at exterior wall buildings have significant impact to the surrounding environment. The use of suitable materials contributes to the reduction of the air temperature due to heat transfer phenomena.

**Keywords:** Thermograph; Building materials; Exterior wall surfaces; Surface temperature; Urban Heat Island (UHI)

## Introduction

The high concentration of hard surfaces actually triggered many environmental issues. The primary root of heat island in cities is due to the absorption of solar radiation by mass building structures, roads, and other hard surfaces during daytime. The absorbed heat is subsequently re-radiated to the surroundings and increases ambient temperatures at night [1]. The use of appropriate materials to reduce heat island and to improve the thermal characteristics of the urban environment has gained increasing interest during recent years. Many research works have been carried out to evaluate the possible energy and environmental benefits, when light colored surfaces are used [2]. Furthermore, some other researchers tried to investigate the impact of the optical materials and thermal characteristics on the urban temperatures as well as the possible energy conservation from heat stroke during the midday.

The heat-retaining properties of building materials, roads, and other surfaces (e.g. asphalt, concrete) contribute to the formation of a phenomenon known as "urban heat island (UHI)" effect. The urban heat island effect is a phenomenon where air temperature in densely built cities is higher than the suburban rural areas. Some of the most important factors which may influence the UHI effect include canyon geometry, thermal properties of materials, anthropogenic heat, the urban greenhouse effect and evaporation surfaces [3]. The maximum UHI intensity usually occurs 3 to 4 h after sunset in the urban area [4]. Some of the radiation reflected from these surfaces leaves the urban environment much of it is intercepted and partially absorbed by other urban surfaces. As a result cities tend to have lower effective albedo than its surroundings.

In this study, the effect of construction materials that contribute for the UHI and the thermal behavior of different materials used in the vertical buildings of surfaces was investigated. Infrared thermograph was used as a technique for the detection of the thermal performance of the selected buildings. The surface temperature in various types of building materials in Johor Bahru, Malaysia were analyzed during peak sunshine and over brighthess condition. No such study has been conducted in this area previously. This study mainly focused on a comparison between different thermal behavior of four building materials which are commonly applied in Johor Bahru city, such as brick, granite, white concrete tiles and concrete (Figure 1) and thermal properties in (Table 1). The following factors were analyzed including type of building material, different height of vertical buildings and impact of materials used towards outdoor environment. This study can be used to increase the awareness of urban planners, designers and decision makers on the importance of the construction material choices, which is not only for their esthetical aspect but their effect on local climate and indirectly on energy consumption of buildings.

## Principles of the Thermal Heat

In a typical observation, building wall is affected by all three heat transfer mechanisms; conduction, convection, and radiation. The incoming of solar radiation into the outer wall surface will be converted to heat by absorption and transmitted into the building by conduction. In this research, the study focused on the outer surface of building. Therefore, heat of radiation and convection were considered. Total heat can be described as submission of heat of radiation ($q''_{rad}$) and heat of convection ($q''_{conv}$) as stated below:

$$\text{Total heat} = q''_{rad} + q''_{conv} \qquad (1)$$

where total heat, $q_{rad}$ and $q_{conv}$ in W/m$^2$. The convection and

***Corresponding author:** Hazlini Dzinun, Institute of Environmental Water Resources and Management (IPASA), Faculty of Civil Engineering, Universiti Teknologi Malaysia, 81310 UTM Skudai, Johor, Malaysia
E-mail: hazlinidzinun@gmail.com

| Location | Type of material in vertical surfaces | Density (kg m$^{-3}$) | Specific heat (J kg$^{-1}$ K$^{-1}$) | Thermal conductivity (W m$^{-1}$ K$^{-1}$) | Solar reflectance* |
|---|---|---|---|---|---|
| (a) Hospital Sultanah Aminah | Brick | 1700 | 800 | 0.84 | 0.15 |
| (b) Tabung Haji | Granite | 2662 | 816 | 2.81 | 0.31 |
| (c) Foh Chong Building | White concrete tiles | 1900 | 800 | 0.84 | 0.83 |
| (d) TJB Building | Concrete (Grey color) | 2300 | 1800 | 1.63 | 0.20 |

Note: * measurement by spectrophotometer

**Table 1:** Properties of external surface building materials used.

radiation heat transfer were calculated based on the equations [5]. The outgoing radiation $q''_{rad}$ is described by:

$$q''_{rad} = A\varepsilon\sigma(T_s^4 - T_a^4) \quad (2)$$

where $A$ = Area of the surface material (m$^2$); ε = Infrared emissivity of the surface; σ = Stefan-Boltzmann constant (=5.67x10$^{-8}$ $Wm^{-2}K^{-4}$); $T_s$ = Surface temperatures (°C); $T_a$ = Ambient temperatures (°C). The convection heat transfer was calculated using the formula:

$$q''_{conv} = K_{conv}A(T_s - T_a) \quad (3)$$

where $A$ = Area of the surface material (m$^2$); $K_{conv}$ = convection heat transfer coefficient (W/ m$^2$$K$); $T_s$ = Surface temperatures °C); $T_a$ = Ambient temperatures (°C) [6].

Considering heat loss by natural convection, this formula was implemented to account the heat loss due to convection. This formula only considered natural convection under the assumption that the heat loss is caused only by the temperature difference between the surface and its surroundings.

$$N_{GR} = (L^3\rho^2 g\beta\Delta T)/\mu^2 \quad (4)$$

$$N_{PR} = (Cp\mu)/k \quad (5)$$

Where $N_{GR}$ is the Grashof Number, L = length in vertical planes (m), ρ = density (kg/m$^3$), g = 9.80665 m/s$^2$, β = volumetric coefficient of expansion of fluid = $1/(T_{film}) = 1/(T_s+T_a)$ in (K), ΔT = |positive temperature different between the wall and bulk fluid| (K), μ = viscosity (kg/m.s), $N_{PR}$ = Prandtl Number, Cp = heat capacity (J/kg.K), k = thermal conductivity (W/m.K) [6].

## Methods and Procedure

### Selection of buildings for in-situ measurement

Four locations have been selected in Johor Bahru area as shown in (Figure 2). The choices of the buildings were determined based on types of material used for construction. As seen in the figure the buildings on the south-orientated of exterior wall building for different materials were investigated.

Hospital Sultanah Aminah (HSA) is geographically located at latitude 1°27' 31.64". North of the Equator and longitude 103°44' 46.16". East of the Prime Meridian where as latitude 1°27' 28.98". North of the Equator and longitude 103°45' 33.03". East of the Prime Meridian is location of Tabung Haji. Foh Chong Building is geographically located at latitude 1°27' 22.89". North of the Equator and longitude 103° 45' 52.61". East of the Prime Meridian where as latitude (1.4605 degrees) 1° 27' 37". North of the Equator and longitude (103.7594 degrees) 103° 45' 33". East of the Prime Meridian is location of Menara TJB.

### Data logger and solar intensity measurement

A simple data measurement was taken using thermostats with an accuracy of ±0.5°C, connected to data logger of non-volatile memory (GRAPHTEC data logger) for the registration of the surface temperature on the references samples. A data logger with temperature and relative humidity sensors were recorded with accuracy of ±0.2°C for temperature and 3% for relative humidity. Solar radiation was measured by a pyranometer types (EKO, MS 602).

### Thermal graphic measurement

The infrared camera was used to capture the 'picture of heat' at building surfaces; in a case of thermal images were recorded at every 2 min from 9:30 AM until 9:00 PM during this study. The type of infrared camera used was ThermaCAM from FLIR Systems with 640 x 480 pixels, full colour and working in the wavelength band of 7.5 to 13 μm and a specified accuracy of ±0.2°C of reading. These infrared cameras can detect the infrared radiation in the shorter wave length (3-5 μm) or in the longer wave length (8-14 μm). Shorter wave length cameras are more appropriate for high temperatures while longer wave length ones provide best results at low temperature (0-100°C). The infrared cameras commonly used in the thermal surveys of buildings are works in the long wave length spectrum because they are more sensitive to lower temperatures (ambient and above), able to detect small temperature differences and less affected by the reflection caused by solar rays.

### Data sampling

The study consists of *in-situ* temperature measurements on the building surfaces of selected buildings within 3km radius around Johor Bahru city in order to measure the UHI phenomena. The study was conducted in various influences of weather and climates; which recorded in a consistent date measurement for 3 month observation begun from June till August 2010. The periodic month is chosen because prolong clear and shiny condition compared to other month in a year. There are some limitations in this study which highly influence by winds that could reduce the effectiveness of outdoor surveys due to surface temperature shear effects. Similarly, rain may also lead to surface cooling, thus masking thermal effects from below the surface.

Information of thermal behaviors' of materials used was examined through the reflectivity, absorption and albedo aspects. A continuous measurement of surface temperature in various construction materials was plotted in graphical forms by using GRAPHTEC data logger.

The experimental equipment consists of temperature sensors which is thermocouple type T, connected to a Graphtec midi data logger. The ambient and surface temperature was recorded in Graphtec midi data logger in every 2min from 8:00 in the morning until 19:00 in the evening. The measurements were limited to this range of time, due to safety of the instrument. The temperature sensors were placed on the vertical surfaces of building with 0.5, 1.0, 1.5 and 2.0 m high from ground level.

For surface temperatures, the thermocouples were bonded with epoxy to the surface and covered with cloth construction tape. The thermocouple measuring ambient temperature was protected from solar radiation by a radiation shield. In comparison, the surface temperature readings were taken with an Infrared (IR) thermometer and camera thermal image at every 30 min from 9:30 AM until 9:00 PM throughout a day.

For ambient temperature, the temperature sensors were placed on the vertical surfaces of building with 1 m away from any building and 1.5 m high from ground level. For comparison, ambient temperature was also measured using Lutron 4 in 1 Environment Tester (Anemometer, Hygrometer, Light Meter & Thermometer). The measurement always taken directly exposed to sunlight, 1 m away from any building and 1.5 m high from ground level.

The measurements of IR thermometer to the vertical surface material are limited to a height of no more than 3 m, due to restrictions posed by the instrument. On each of the measured external walls, IR readings was measured at different heights, together with different parts of the exterior surfaces with the intention of correlation to the pedestrians' thermal comfort perception, in order to ensure that recorded surface temperature were representative. The ambient temperatures, relative humidity and air velocity were then determined using Lutron 4 in 1 Environment Tester (Anemometer, Hygrometer, Light Meter & Thermometer).

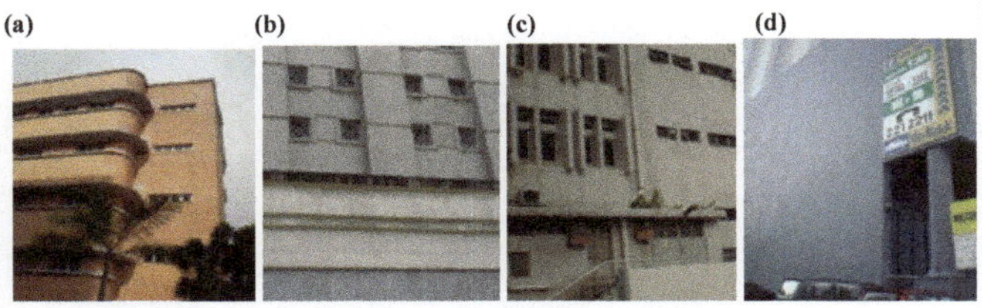


**Figure 1:** The selected facade buildings in Johor Bahru areas; (a) Hospital Sultanah Aminah, (b) Menara Tabung Haji (c) Foh Chong Building and (d) TJB Building.

**Figure 2:** Site selection in Johor Bahru, located in the southern Malaysia.

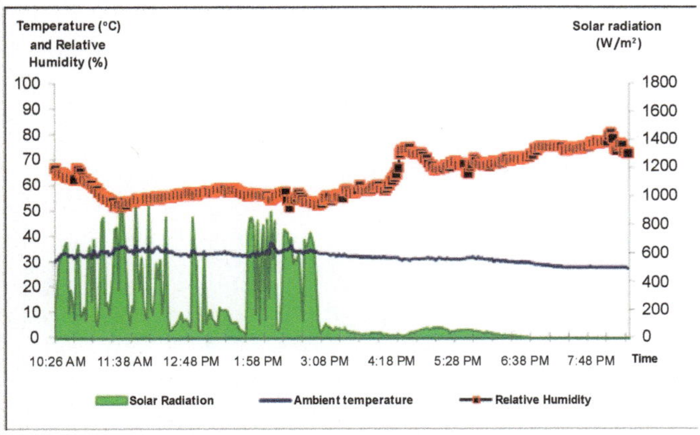


**Figure 3:** Ambient data record of solar radiation, temperature and relative humidity for Case 1.

## Results

### Analysis of existing exterior building based on different materials

This part discusses the four locations of the observed study area with a different types of exterior wall building material that most commonly used in Johor Bahru city, namely brick, granite, concrete and white concrete tiles. The impact of materials used and the solar radiation intensity were then analyzed to find out their influence on the outdoor surface temperature of the building.

**Case 1-Hospital sultanah aminah:** Figure 3 shows the ambient temperature, relative humidity and solar radiation between 10:30 AM until 9:00 pm. The mean wind speed values from data obtained during measurements is 0.63 m/s, which consider as moderate calm wind profile. The maximum ambient temperature was recorded up to 37.2°C at 2:12 PM and the minimum was depicted at 27.4°C at 9:00 pm, whereas the mean ambient temperature was recorded at 31.6°C. The maximum relative humidity was measured at 80.4%, whereas the minimum value and mean relative humidity were reported at 51.2% and 62.9% respectively. As shown in Figure 3, solar radiation was fluctuated started from 10.36 AM until 3 pm, then it suddenly drops until reached 0W/m$^2$ because of the cloudy weather.

The survey carried out during the day (presence of direct solar radiation) succeeded thermal images at different temperatures. Simultaneously, with the infrared camera observation, a sensitive contact thermometer (i.e. shot-gun temperature and thermo couple) was checked twice to ensure the accuracy of the online data logger. Once recorded, a correlation between infrared camera and contact thermometer is determined. From the images captured, a clear zone has been identified, which correspond to the colors' characterization. Lighter colors appeared at lower temperature than darker ones. As showed in the images, the temperature identification of each image is prepared according to the local time and weather.

Surface temperatures at brick wall had reached about 32.5°C at 9:30 AM and their maximum was 7°C higher than the ambient temperature. At 12:00 noon, surface temperature rose to 41.1°C higher than the ambient air temperature by 9°C different. It is due to the storage of solar heat absorbed by the buildings in the previous daytime, as the solar radiation was recorded highest up to 1005.73 W/m$^2$. Therefore, brick wall was absolutely absorbed more heat due to the heating direct solar radiation during daytime. As shown in the captured images, the brick wall appears in red dark colour.

At 4:30 pm, the solar transmission was lower than recorded at 12:00 noon. As a result, the surface temperatures decrease to 33.4°C and steadily reduce by 8°C lower than those monitored at 12:00 noon. Surface temperatures of exterior wall building varied during midday due to the heat capacity of brick wall. The temperature difference was caused by their different thermal inertia [7]. The bricks began to increase the temperature faster than the glass material, thus proves that brick had larger capacity to store heat. Furthermore, at midday hotter zones can be observed in the thermal images caused by the overheating due to the effect of solar rays on the building facades, the contrary is also true (see Figure 4).

**Case 2-Menara tabung haji:** Menara Tabung Haji is a single

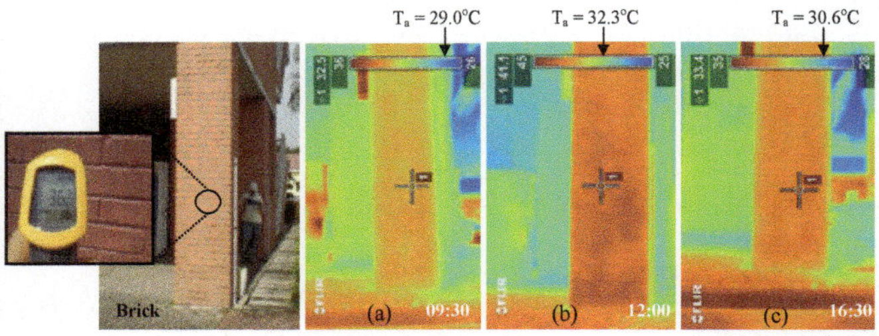


**Figure 4:** The visible and thermal image of brick wall at different time (a) morning (b) afternoon (c) evening.

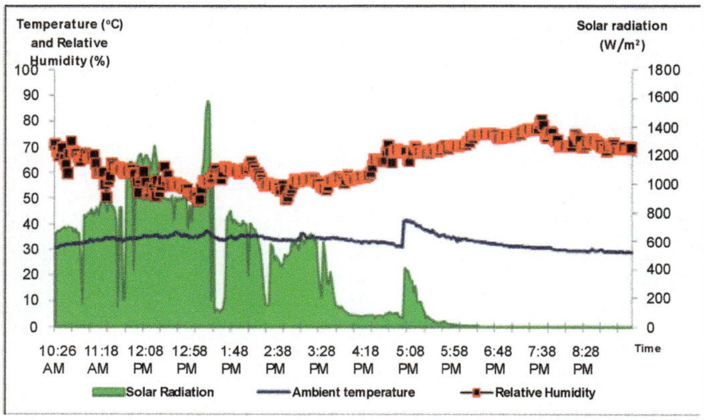


**Figure 5:** Ambient data record of solar radiation, temperature and relative humidity for Case 2.

detached-building located in Johor Bahru. Based on the observation, the east-side exterior wall of the building was reached the higher values of solar radiation. Figure 5 shows the ambient temperature, relative humidity and solar radiation at the continuous period from 10:52 AM until 10:00 pm.The highest value of solar radiation was depicted up to 1585.96 W/m² (Figure 5). However, since the orientation of exterior wall is not exposed consistently, the direct solar rays only fell upon the samples until 6:00 pm. The mean wind speed values from data obtained during measurements is measured at 1.8 m/s, which consider as calm condition. The maximum ambient temperature was depicted at 41.7°C at 5:39 pm, whereas the minimum value was reported at 29.0°C at 9:00 pm. Overall the ambient temperature was observed at 33.5°C. The maximum relative humidity recorded was 80.4% with minimum and relative humidity recorded at 48.9% and 64.3%, respectively.

Figure 6 shows that the colors in the afternoon are relatively intense, while in the evening and morning periods are relatively light. These phenomenons can most likely be attributed to the difference in the angle of the elevation and the intensity of direct solar radiation. As showed in the images, the temperature of each image is prepared according to the local time and weather. Surface temperatures for exterior wall building of Menara Tabung Haji had reached about 31.1°C at 11:00 AM and their maximum was 2°C higher than the ambient temperature. As evident, the exterior wall of buildings is just starting to absorb heat because of intensity of solar radiation. At 1:40 pm, surface temperature rose to 41.9°C higher than the ambient air temperature by 5°C different. It is due to the storage of solar heat absorbed by the buildings in the previous daytime and at this time, the solar radiation was recorded highest at 1585.96 W/m² which means that granite absorbs and stores heat from direct solar radiation during daytime. This is obviously shown in the images; in which the granite material appears in dark red colour.

**Case 3-Foh chong building:** Foh Chong Building is a commercial building located in Johor Bahru. Figure 7 shows the ambient measurement from 9:00 AM until 9:00 pm. The maximum ambient

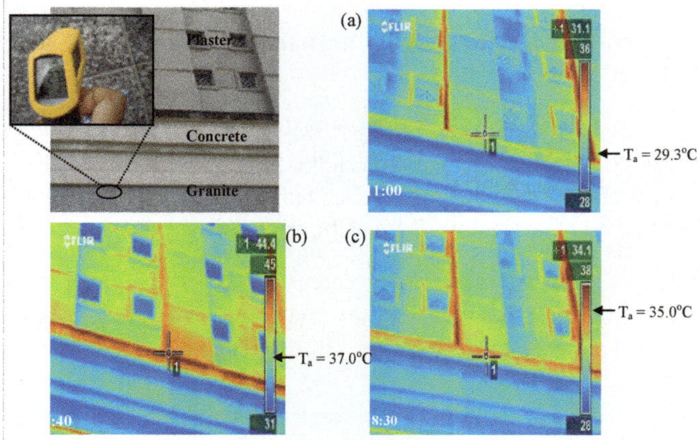


**Figure 6:** The visible and thermal image of granite at different time (a) morning (b) afternoon (c) evening.

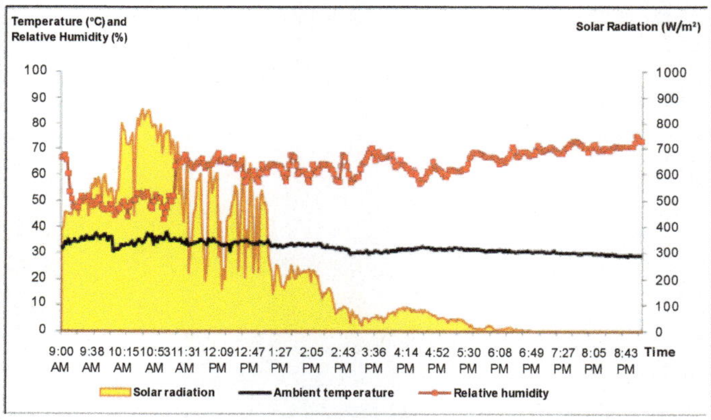


**Figure 7:** Ambient data record of solar radiation, temperature and relative humidity for Case 3.

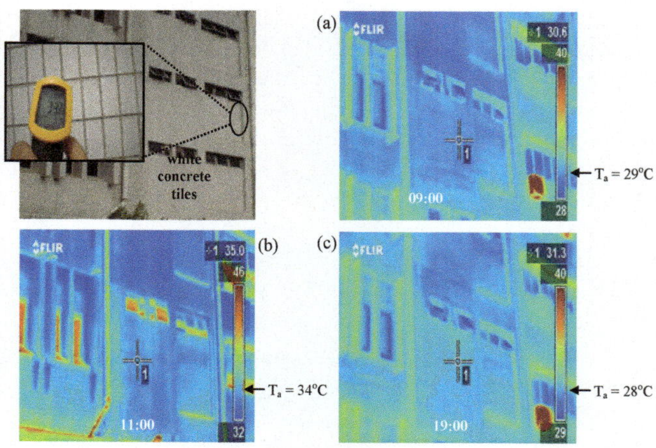


**Figure 8:** The visible and thermal image of wall tiles at different time (a) morning (b) afternoon and (c) evening.

temperature was recorded up to 38.1°C at 11:07 am, the minimum was determined at 28.8°C at 8:45 PM and the mean ambient temperature was depicted at 32.2°C. The maximum relative humidity was 74.9% with its minimum and mean value at 43.1% and 62.2%, respectively. The highest value of solar radiation was monitored at 859.6 W/m$^2$. Unfortunately, the direct solar rays fell upon the samples only recorded until 5:40 pm.

Thermography offers the possibility of obtaining a map of the surface temperatures of an object as shown in Figure 8. The scale of temperature had been changed according to the maximum and minimum values of temperature during the observation day. Surface temperatures for exterior wall building of Foh Chong Building had reached about 37.7°C at 11:00 AM and the maximum was up to 3°C higher than the ambient temperature. At 7:00 pm, the surface temperature was dropped and close to the air temperature. Parts of the area on the exterior surface are identified as 'blue' color. This showed that the exterior wall building is a "cool" material. From the temperature measurement, the wall tiles has least absorb any heat from the solar radiation showing the dynamic behavior of the surface material chosen. Based on material properties, the heat capacity of white concrete tiles is relatively low; heat from the daytime solar radiation can be prevented to store. Therefore, the incoming solar radiation has been reflected accordingly and it can cause energy balance at the surfaces.

**Case 4-TJB building:** TJB Building is a commercial building located in an urban region of Johor. The building is constructed of reinforced concrete and the east-side exterior surface of the building reached the higher values of solar radiation. Figure 9 shows the ambient temperature, relative humidity and solar radiation as monitored from 10:40 AM until 8:30 pm. The mean wind speed is recorded around 0.7 m/s. The highest value of solar radiation was recorded up to 1025.79 W/m$^2$, however the direct solar rays was only measurable until 6:19 pm. The maximum ambient temperature was recorded at 35.1°C at 2:54 PM with minimum and mean value at 30.0°C and 32.3°C, respectively. The maximum relative humidity was recorded at 73.9% with its minimum and mean value depicted at 56.1% and 63.5%, respectively.

From the obtained thermal images, it can be observed that the concrete structure was preferably storing heat. It showed in Figure 10, part of the area in the exterior surface material shows a red dark color, even the direct solar radiation is gone at 5.00 pm. The red areas in thermograph occur because the external surface of the building was exposed directly to the solar radiant without any sun-shading. The temperature was higher than the air temperature due to the direct solar radiation during the daytime. The surface temperatures for exterior wall building had reached about 31.5°C at 11:00 AM and their maximum was 1.5°C higher than the ambient temperature. At this time, the exterior walls of the buildings were just starting to absorb heat because the solar radiation had started falling into the buildings. This can be seen at 2:30 PM where part of the area in the facade surface showed a red dark color which indicated a high temperature. The red areas in the thermograph occurred on external surfaces of the building because there was no protection from the sun. The temperatures in those areas were higher than the ambient temperature due to the heat from direct solar radiation. The temperature of the concrete that was warmer than the surrounding concrete was shown in yellowish color. These characteristics showed that the wall has a capability to store heat. Although the surface temperature of these areas rose slightly in the late afternoon, the surface temperature fell and was close to the air temperature at sunset. At 5:00 pm, the surface temperature reduced to 33.9°C; higher than the ambient air temperature by 3°C.

## Analysis of sensible heat flux mechanism

The heat gain from within the exterior surface buildings is governed by their physical characteristics and of materials used. The heat exchanges that occur on exterior wall buildings are more intense between their external surfaces and their environment. As closer the surface temperature is to the air temperature, as lower is the transmission of heat to the environment by convection and by radiation [8]. In order to calculate the heat flux over the entire surface of the building, the ambient temperature and surface area of each element of the entire building were measured, respectively. The total heat, convection and radiation heat were calculated based on the equations stated in equation (1), (2) and (3) respectively.

It seems that the calculations for heat of radiation and convection are made based on the differences shown between surface and ambient temperature, accordingly for the current experiment the trend change in terms of heat flux intensity of convection and radiation for corresponding temperature changes measured are shown in Figure 11. It is obvious that if there is no difference between surface and ambient temperature, the heat of convection (q'conv) and heat of radiation (q'rad) the subsequent emission to the atmosphere will be significantly low. The values of Figure 11 (a) and (b) indicates that the values of heat of radiation are greater than the values of heat of convection and clearly shows that the heat flux of exterior surface changes smoothly due to the nature of thermal properties of wall materials.

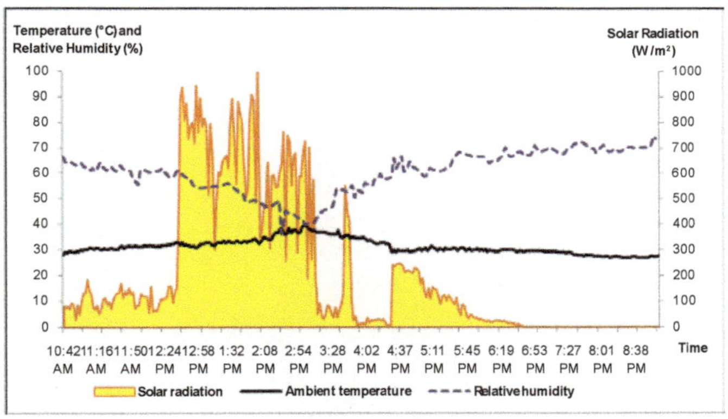


**Figure 9:** Weather data for Case 4.

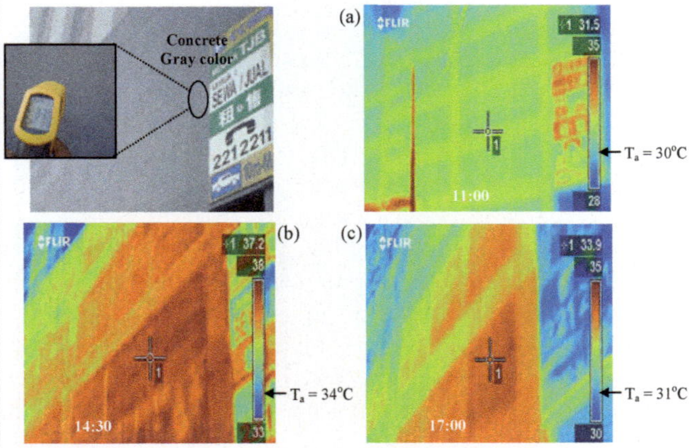


**Figure 10:** The visible and thermal image of concrete wall at different time (a) morning (b) afternoon and (c) evening.

The convection and radiation of heat transfer also depends on the intensity of solar radiation falling towards the exterior wall buildings. The comparison between these four materials cannot be made because of each buildings have different conditions of experimental measurement and it influenced via the values of convection and radiation heat transfer. Therefore, in order to determine the suitable materials at exterior wall building, the normalized of each material were calculated and analyzed as shown in Figure 12.

Figure 12 represents the normalized (total heat/Sr) which is defined as heat of convection plus heat of radiation and divided by average of solar radiation. Total heat indicated that the amount of heat would increase the ambient temperature as well as make contribution to the UHI impact. The normalized (total heat/Sr) of brick materials was 0.093, which was the highest value compared to others. Brick materials show the highest impact of heat in 51% than granite material and it releases a substantial amount of heat into the atmosphere through radiation and convection. The normalized for granite and white concrete tiles have relatively the same. This was expected due to the values of specific heat for both materials are relatively the same.

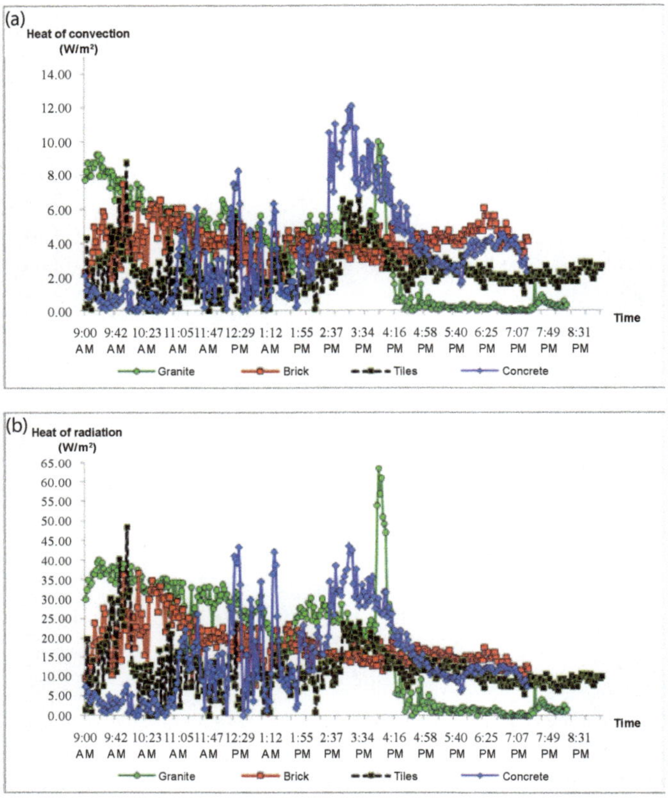


**Figure 11:** (a) Heat of convection and (b) Heat of radiation for different materials.

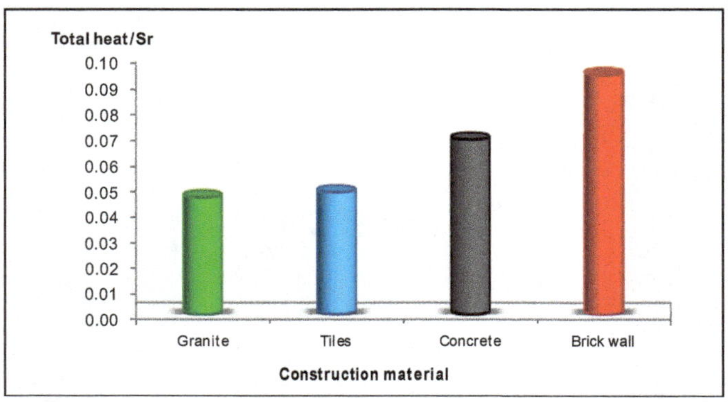


**Figure 12:** Normalized of (total heat/solar radiation) using average heat of convection (qconv), heat of radiation (qrad) and solar radiation (Sr).

It can be expected that the capacity of heat stored in different material was the main influence contribution to the UHI impact. For example, brick material is the evident of increment surrounding area because of its capacity of storing heat energy. The absorbed heat will be released back to the surrounding environment and warm up the ambient near the material surfaces throughout the day. This was because the energy accumulated within the exterior wall during midday and was released during the night time in the form of heat. This is actually the root of the urban heat island effect in the built environment [9].

## Conclusion

In this study, the applications of IR thermograph for the detection of different thermal behavior of two buildings were presented. The different types of material illustrated the different heat impact on the surrounding environment. Thus, by referring to the final result, the most sensitive of absorb and storing heat are recorded as follows; brick > concrete > granite and white concrete tiles. As a conclusion, granite and white concrete tiles are suitable material for the exterior wall building; providing a thermally comfortable for outdoor environment, while can contribute highly to improve the urban microclimate mainly for UHI phenomena. The results from this research could provide valuable information to the thermographer whose work is concerned with the making up of urban buildings.

### Acknowledgement

The authors wish to thank Dr Mohd Fadhil Md Din, Dr Remaz Dilshah Ossen and Dr Zainura Zainun Noor for their valuable help and support, Mr Salleh for his technical help in using infrared thermographic and also to Dr Ponraj for the final checking of the manuscript. The authors are also grateful to Professor Kenzo Iwao at Nagoya Institute Technology, Japan for his useful advice and instrumentation supplied to do this research. Our sincere gratitude goes to Universiti Teknologi Malaysia and Academic Science Malaysia for their financial supports under the Fundamental Research Grant Scheme (No. 4F041).

### References

1. Chow WTL, Roth M (2006) Temporal dynamics of the urban heat island of Singapore. Int J Clim 26: 2243-2260.
2. Doulos L, Santamouris M, Livada I (2004) Passive cooling of outdoor urban spaces. The role of materials. Solar Energy 7: 231-249.
3. Dzinun H, Md Din MF, Noor ZZ, Iwao K, Chellapan S (2010) Analysis of heat impacted behavior at vertical façade building based on heat flux mechanism. Proceedings of 2010. International Conference on Environment and Industrial Innovation (ICEII 2011) IACSIT Press, Singapore 7: 74-78.
4. Guerrero IC, Ocana SM, Requena IG (2005) Thermal-physical aspects of materials used for the construction of rural buildings in Soria (Spain). Construction and Building Materials 19: 197-211.
5. Ocana SM, Guerrero IC, Requena IG (2004) Thermographic survey of two rural buildings in Spain. Energy and Buildings 36: 515-523.
6. Prado RTA, Ferreira FL (2005) Measurement of albedo and analysis of its influence the surface temperature of building roof materials. Energy and Buildings 37: 295-300.
7. Santamouris M (2001) Energy and climate in the urban built environment. James & James Science Publishers, London.
8. Wong NH, Chen Y (2009) Tropical Urban Heat Islands: Climate, buildings and greenery. Taylor & Francis Group, London and New York.
9. Wong NH, Chena Y, Ongb CL, Siab A (2005) Investigation of thermal benefits of rooftop garden in the tropical environment. Building and Environment 38: 261–270.

# Experimental Study on Concrete Using Copper Slag as Replacement Material of Fine Aggregate

**Tamil Selvi P*, Lakshmi Narayani P and Ramya G**

*Department of Civil Engineering, Dr. Sivanthi Aditanar College of Engineering, Tiruchendur-628215, Tamilnadu, India*

**Abstract**

This paper reports the effect of concrete using copper slag as fine aggregate replacement. In this project work, the concrete grade M40 was selected and IS method was used for mix design. The properties of material for cement, fine aggregate, coarse aggregate and copper slag were studied for mix design. The various strength of concrete like compressive, flexural and split tensile were studied and non-destructive test such as rebound hammer test and ultrasonic pulse velocity measurement were studied for various replacements of fine aggregate using copper slag that are 0%, 20%, 40%, 60%, 80% and 100%. The maximum compressive strength of concrete attained at 40% replacement of fine aggregate at 7 and 28 days. The split tensile strength and the flexural strength were also obtained higher strength at 40% replacement level at 28 days. The rebound hammer test showed higher compressive strength at 40% fine aggregate replacement, this is due to uniformity of concrete. The pulse wave velocity is higher for the 40% fine aggregate replacement, it is understood that the density of the mix is high and free from pores.

**Keywords:** Copper slag; Waste material; Replacement of fine aggregate; Concrete; Properties of materials; Workability; Compressive strength; Split tensile strength; Flexural strength; Rebound hammer test; Ultrasonic pulse velocity measurement

## Introduction

Concrete is a composite material composed of water, coarse granular material (the fine and coarse aggregate or filler) embedded in a hard matrix of material (the cement or binder) that fills the space among the aggregate particles and glues them together. Concrete is widely used for making architectural structures, foundations, brick or block walls, pavements, bridges or overpasses, highways, runways, parking structures, dams, pools/reservoirs, pipes, footings for gates, fences and poles and even boats. Concrete is used in large quantities almost everywhere mankind has a need for infrastructure [1].

The amount of concrete used worldwide, ton for ton, is twice that of steel, wood, plastics, and aluminum combined. Concrete's use in the modern world is exceeded only by that of naturally occurring water. Concrete is also the basis of a large commercial industry. Globally, the ready-mix concrete industry, the largest segment of the concrete market, is projected to exceed $100 billion in revenue by 2015. In the United States alone, concrete production is a $30-billion-per-year industry, considering only the value of the ready-mixed concrete sold each year. Given the size of the concrete industry, and the fundamental way concrete is used to shape the infrastructure of the modern world, it is difficult to overstate the role this material plays today [2,3].

Copper slag is used in the concrete as one of the alternative materials. It is the waste product of copper from Sterlite Industries India Ltd, Tuticorin. The safe disposal of this waste is a lack, costly and causes environmental pollution. The construction industry is the only area where the safe use of waste material (copper slag) is possible. When it is introduced in concrete as a replacement material, it reduces the environmental pollution, space problem and also reduces the cost of concrete.

However, further additions of copper slag caused reduction in the strength due to an increase of the free water content in the mix. Mixes with 80% and 100% copper slag replacement gave the lowest compressive strength value of approximately 80 MPa, which is almost 16% lower than the strength of the control mix. The results also demonstrated that the surface water absorption decreased as copper slag quantity increases up to 40% replacement; beyond that level of replacement, the absorption rate increases rapidly. Therefore, it is recommended that 40 wt% of copper slag can used as replacement of sand in order to obtain HPC with good strength and durability properties [4,5].

## Materials and Methods

### Materials

There are many types of concrete available, created by varying the proportions of the main ingredients below. In this way or by substitution for the cementitious and aggregate phases, the finished product can be tailored to its application with varying strength, density, or chemical and thermal resistance properties.

**Aggregate:** Aggregate consists of large chunks of material in a concrete mix, generally a coarse gravel or crushed rocks such as limestone, or granite, along with finer materials such as sand. The fine aggregate used in this study is river sand conforming to grading zone II Table 1 of IS 383 [6]. The coarse aggregate used in this study is of angular in shape and the maximum nominal size of coarse aggregate is 20 mm and it is Conforming to Table 2 of IS 383 [6].

| Test | Cement |
|---|---|
| Specific gravity | 3.09 |
| Fineness modulus (%) | 1.4 |
| Standard consistency (%) | 31.5 |
| Initial setting time (minutes) | 100 |
| Final setting time (minutes) | 240 |

**Table 1:** Properties of cement.

***Corresponding author:** Tamil Selvi P, Department of Civil Engineering, Dr. Sivanthi Aditanar College of Engineering, Tiruchendur-628215, Tamilnadu, India
E-mail: tamse10@gmail.com

| Test | Fine aggregate | Copper slag | Coarse aggregate |
|---|---|---|---|
| Specific gravity | 2.47 | 3.45 | 2.68 |
| Fineness modulus | 2.89 | 3.84 | 7.09 |
| Moisture content (%) | 0.1 | 0.1 | 0.1 |
| Water absorption (%) | 1.2 | 0.4 | 1.0 |
| Density (loose state) (kg/m³) | 1700.67 | 2197 | 1513 |

**Table 2:** Properties of fine aggregate, copper slag and coarse aggregate.

**Figure 1:** Fine aggregate and copper slag used in the study.

**Cement:** Cement, commonly Portland cement, and other cementitious materials such as fly ash and slag cement, serve as a binder for the aggregate. The cement used in this study is of OPC 53 grade conforming to IS 12269 [7].

**Water:** Water is then mixed with this dry composite, which produces a semi-liquid that workers can shape (typically by pouring it into a form). The concrete solidifies and hardens to rock-hard strength through a chemical process called hydration. The water reacts with the cement, which bonds the other components together, creating a robust stone-like material. The good quality water is used in this study.

**Copper slag:** Copper slag is an irregular, black, glassy and granular in nature and its properties are similar to the river sand. In this project, Copper slag used is brought from Sterile Industries India Ltd, Tuticorin. Every ton of copper will generate approximately 2.2-3 tons of copper slag [8,9]. Sterlite Industries India Ltd produces 400,000t/year of copper and during the process, around 800,000t of copper slag is generated in a year. The chemical traces such as copper, sulphate and alumina present in the slag are not harmful (Figure 1).

## Objective of this study

Many researchers had already found, copper slag possible to use as a material in concrete. In this experimental study copper slag is used in concrete as replacement material of fine aggregate. For this study, M40 grade of concrete is used and the tests are conducted for various replacement of fine aggregate using copper slag as 0%, 20%, 40%, 60%, 80% and 100% in concrete. The obtained results are compared with those of control concrete made with fine aggregate [10].

## Properties of materials

The preliminary tests were carried out for finding the materials properties of cement, fine aggregate and coarse aggregate as per IS 12269 [7] and IS 383 [6]. The studied properties are given in Tables 1 and 2.

## Methodology

In fresh state; the workability parameters such as slump value and compacting factor were studied. In hardened state; the strength tests such as compressive strength, split tensile strength and flexural strength were studied and the Non-destructive tests such as rebound hammer test and ultrasonic pulse velocity test were conducted for each mix. The obtained results are tabulated. The conclusions were made from the results and discussions (Table 3).

## Mix Design and Experimental Work

### Mix design

An integral part of concrete mix proportioning is the preparation of trial mixes and effect adjustments to such trials to strike a balance between the requirements of placement, that is, workability and strength, concomitantly satisfying durability requirements 4 [11]. As per IS 10262 [12], the method is adopted for finding the proportion of M40 grade concrete and the arrived mix ratio was 1:1.36:2.52:0.41 (Tables 4 and 5).

### Experimental work

**Workability:** Workability is a measure of the ease with which a fresh mix of concrete or mortar can be handled and placed. For various mixes the concrete were prepared. In the fresh concrete, the slump cone test and compaction factor test were carried out (Figure 2).

- **Slump cone test:** The slump test result is a measure of the behaviour of a compacted inverted cone of concrete under the action of gravity as per IS 1199 [13]. It measures the consistency or the wetness of concrete.
- **Compacting factor test:** Compacting factor of fresh concrete is done to determine the workability of fresh concrete by compacting factor test as per IS 1199 [13]. The apparatus used is Compacting factor apparatus (Table 6).

| Strength test | Compression test | Split tensile strength | Flexural strength |
|---|---|---|---|
| **Sample type** | cube | cylinder | beam |
| **Sample size (mm)** | 150×150 × 150 | 150×300 | 100×100×500 |
| **Days of testing** | 7 & 28 | 28 | 28 |
| **Total no. of samples for one series** | 6 | 3 | 3 |

**Table 3:** Details of specimen used in the study.

| Cement | Fine aggregate | Coarse aggregate | w/c ratio |
|---|---|---|---|
| 450 | 610.49 | 1132.06 | 186 |
| 1 | 1.36 | 2.52 | 0.41 |

**Table 4:** Mix proportions (kg/m³) and mix ratio.

| S.No | % of replacement of fine aggregate | Cement (kg/m³) | Fine aggregate (kg/m³) | Copper Slag (kg/m³) | Coarse aggregate (kg/m³) | Water (kg/m³) |
|---|---|---|---|---|---|---|
| 1 | 0 | 450 | 610.49 | 0 | 1132.06 | 186 |
| 2 | 20 | 450 | 488.40 | 122.10 | 1132.06 | 186 |
| 3 | 40 | 450 | 366.30 | 244.20 | 1132.06 | 186 |
| 4 | 60 | 450 | 244.20 | 366.30 | 1132.06 | 186 |
| 5 | 80 | 450 | 122.10 | 488.40 | 1132.06 | 186 |
| 6 | 100 | 450 | 0 | 610.49 | 1132.06 | 186 |

**Table 5:** Various concrete mix details.

**Figure 2:** Slump cone and compacting factor test of fresh concrete.

| Replacement of fine aggregate (%) | Slump (mm) | Compacting factor | Degree of workability |
|---|---|---|---|
| 0 | 26 | 0.82 | Low |
| 20 | 27.5 | 0.86 | Low |
| 40 | 30 | 0.88 | Low |
| 60 | 32.5 | 0.90 | Low |
| 80 | 34.5 | 0.92 | Low |
| 100 | 36 | 0.96 | Low |

**Table 6:** Slump and compacting factor value.

**Preparation of sample:** The moulds were prepared as per IS 10086 [14] and the samples were prepared and cured as per IS 516 [15]. For compression test 6 cubes of size 150×150×150 mm were cast and carried out test at 7 and 28 days for each mix. For Split tensile test and Flexure test 3 cylinders of size 150×300 mm and 18 beams of size 100×100×500 mm were cast and carried out test at 28 days for each mix. The Rebound hammer test and the Ultrasonic Pulse Velocity test are carried out for concrete cubes before carrying out strength tests.

**Strength tests:** In the hardened concrete, the Semi-destructive tests like Compressive strength test, Split tensile strength test and flexural strength test were carried out as per IS 516 [15].

**Compressive strength:** Compressive strength is often measured on a universal testing machine. By definition, the ultimate compressive strength of a material is that value of uniaxial compressive stress reached when the material fails completely. The compressive strength (N/mm²) is calculated by using the formula,

$$\text{Compressive strength} = \frac{\text{Ultimate load (N)}}{\text{Area of cross section}\left(\text{mm}^2\right)}$$

**Split tensile strength:** Ultimate tensile strength (UTS), often shortened to tensile strength (TS) or ultimate strength, is the maximum stress that a material can withstand while being stretched or pulled before failing or breaking. Tensile strength is not the same as compressive strength and the values can be quite different [16]. The split tensile strength (N/mm²) is calculated by using the formula,

$$\text{Split tensile strength} = \frac{2P}{\pi LD}$$

Where, P=Ultimate load at failure (N), L=Length of specimen (mm) and D=Diameter of specimen (mm)

**Flexural strength:** Flexural strength, also known as modulus of rupture, bend strength, or fracture strength, a mechanical parameter for brittle material, is defined as a material's ability to resist deformation under load. The flexural strength represents the highest stress experienced within the material at its moment of rupture (Figure 3 and Tables 7-9). The flexural strength (N/mm²) of rectangular cross-section using a three point flexural test technique is obtained by the formula,

$$\text{Flexural strength} = \frac{PL}{BD^2}$$

Where, P=Ultimate load at failure (N), L=Length of specimen (mm), B=Breadth of specimen (mm) and D=Depth of specimen (mm)

**Non-destructive test:** In the hardened concrete, the Non-destructive tests like Rebound hammer test & Ultrasonic pulse velocity test were carried out as per IS 13311 [17,18].

**Rebound hammer test:** A Schmidt hammer, also known as a Swiss hammer or a rebound hammer, is a device to measure the elastic properties or strength of concrete or rock, mainly surface hardness and penetration resistance. The hammer measures the rebound of a spring-loaded mass impacting against the surface of the sample. By reference to the conversion chart, the rebound value can be used to determine the compressive strength.

**Ultrasonic pulse velocity test:** Ultrasonic Pulse Velocity (UPV) testing of concrete is based on the pulse velocity method to provide information on the uniformity of concrete, cavities, cracks and defects. The test equipment must provide a means of generating a pulse, transmitting this to the concrete, receiving and amplifying the

**Figure 3:** Compressive strength, split tensile strength and flexural strength test of specimen.

| Replacement of fine aggregate (%) | Weight of the cube (kg) | | |
|---|---|---|---|
| | Before curing | 7 days curing | 28 days curing |
| 0 | 8.037 | 8.309 | 8.494 |
| 20 | 8.407 | 8.797 | 8.907 |
| 40 | 8.699 | 8.852 | 9.059 |
| 60 | 9.107 | 9.163 | 9.437 |
| 80 | 9.326 | 9.443 | 9.726 |
| 100 | 9.631 | 9.525 | 9.819 |

**Table 7:** Average weight of the cube.

| Replacement of fine aggregate (%) | 7th day Compressive strength (N/mm²) | 28th day Compressive strength (N/mm²) |
|---|---|---|
| 0 | 29.50 | 38.80 |
| 20 | 33.72 | 40.70 |
| 40 | 38.37 | 42.95 |
| 60 | 30.88 | 34.44 |
| 80 | 29.65 | 31.39 |
| 100 | 26.87 | 27.66 |

**Table 8:** Average compressive strength of the specimen.

| Replacement of fine aggregate (%) | Average split tensile strength (N/mm²) | Average flexural strength (N/mm²) |
|---|---|---|
| 0 | 2.45 | 4.79 |
| 20 | 2.68 | 7.00 |
| 40 | 3.09 | 7.73 |
| 60 | 2.43 | 6.27 |
| 80 | 2.22 | 5.46 |
| 100 | 1.85 | 4.42 |

**Table 9:** Average split tensile strength and flexural strength of the specimen.

pulse and measuring and displaying the time taken (Figure 4, Tables 10 and 11).

## Results and Discussions

### Study on fresh concrete

Based on the observations from the experimental work, the following results and discussions are presented in this chapter. For workability study, the slump and compacting factor test were conducted on fresh concrete for different proportions as mentioned in Table 5.

**Slump value:** From Figure 5 it was noted that the slump value increased with the percentage of copper slag increases in concrete. The measured slump was 26 mm for the control mixture whereas 36 mm with 100% replacement of fine aggregate using copper slag. This considerable increase is due to the low water absorption characteristics of copper slag and its glassy surface compared with sand which caused surplus quantity of free water to remain after the absorption and hydration processes have completed. It should be noted that mixes with high contents of copper slag showed signs of bleeding and segregation which can have detrimental effects on concrete performance [19].

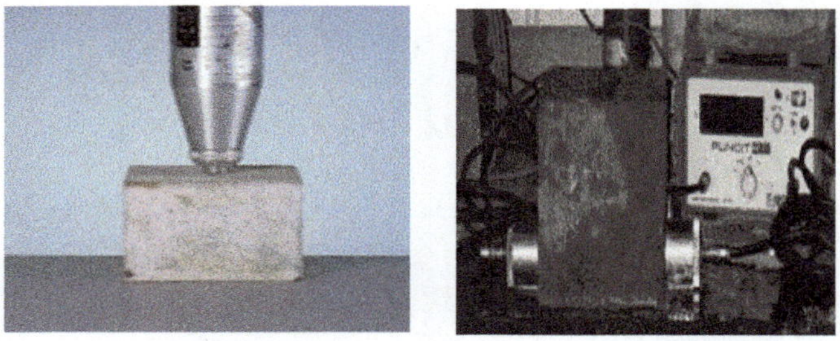

Figure 4: Rebound hammer and ultrasonic pulse velocity test for cube.

| Replacement of fine aggregate (%) | 7th day Compressive strength ($N/mm^2$) | 28th day Compressive strength ($N/mm^2$) |
|---|---|---|
| 0 | 9.81 | 10.00 |
| 20 | 16.19 | 25.75 |
| 40 | 26.49 | 35.64 |
| 60 | 22.72 | 30.43 |
| 80 | 19.98 | 24.32 |
| 100 | 14.73 | 20.08 |

Table 10: Average compressive strength of cube by rebound hammer test.

| Replacement of fine aggregate (%) | 7th day | | 28th day | |
|---|---|---|---|---|
| | velocity (km/sec) | Concrete quality | velocity (km/sec) | Concrete quality |
| 0 | 3.95 | Good | 4.26 | Good |
| 20 | 3.68 | Good | 4.54 | Excellent |
| 40 | 4.52 | Excellent | 5.21 | Excellent |
| 60 | 4.21 | Good | 4.83 | Excellent |
| 80 | 3.86 | Good | 4.54 | Excellent |
| 100 | 3.79 | Good | 4.23 | Good |

Table 11: Ultrasonic pulse velocity test of cube.

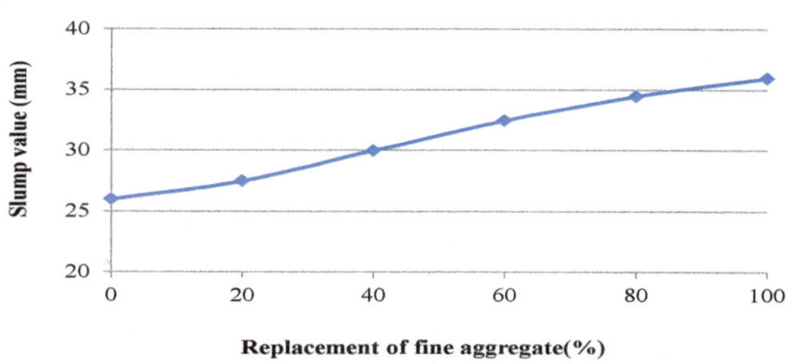


Figure 5: Relationship of slump value and percentage of replacement of fine aggregate.

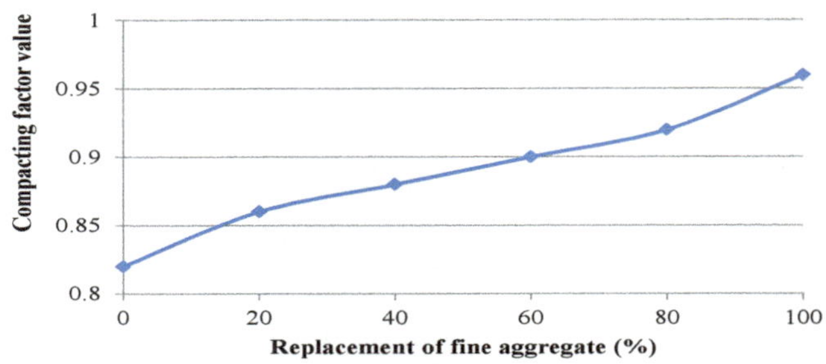


Figure 6: Relationship of compacting factor value and percentage of replacement of fine aggregate.

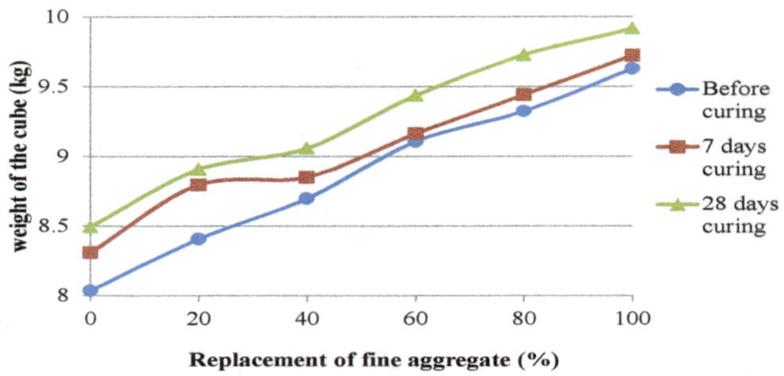


Figure 7: Relationship of weight of the cube and percentage replacement of fine aggregate.

**Compacting factor:** From Figure 6 it was noted that the value of compacting factor slightly increased with the replacement of copper slag in concrete. The calculated value of compacting factor was 0.82 for the control mixture whereas 0.96 with the 100% replacement of copper slag. This is mainly due to the higher specific gravity of copper slag which was 3.45 compared with sand which has a specific gravity of 2.47.

### Study on hardened concrete

In this section, based on the experimental results of the compressive strength, split tensile strength, flexural strength, rebound hammer test and ultrasonic pulse velocity measurement discussions are made as follows.

**Compressive strength:** From the Figure 7 it was found that the weight of the cube showed lowest value in control concrete at before curing, 7 days curing and 28 days curing were 8.037 kg, 8.309 kg and 8.494 kg respectively whereas in concrete with 100% fine aggregate replacement using copper slag, the weight of the concrete value found the highest at before curing, 7 days curing and 28 days curing were 9.631 kg, 9.725 kg and 9.918 kg. The weight of specimen increased with the addition of copper slag into concrete. This is because of higher density of copper slag which was 2197 $Kg/m^3$ compared with density of sand which was 1700.67 $Kg/m^3$.

Figure 8 showed that the compressive strength of cube was found to be 29.50 $N/mm^2$ at 0% fine aggregate replacement and of 38.88 $N/mm^2$ at 100% fine aggregate replacement of 7 days. And the compressive strength of cube at 28 days was found to be 26.87 $N/mm^2$ at 0% fine aggregate replacement and of 27.66 N/mm² at 100% fine aggregate replacement. The maximum compressive strength was found to be at 40% fine aggregate replacement of about 38.37 N/mm² at 7 days and of 42.97 N/mm² at 28 days. The compressive strength of concrete at 7 and 28 days increased gradually up to 40% fine aggregate replacement and then decreased with increase in percentage of replacement.

**Split tensile strength:** Figure 9 showed that the split tensile strength of cylinder was found to be 2.45 N/mm² at 0% fine aggregate replacement and of 1.85 N/mm² at 100% fine aggregate replacement. The maximum split tensile strength was found to be at 40% fine aggregate replacement of about 3.09 N/mm². The split tensile strength of copper slag added concrete was gradually increased up to 40% replacement and then decreased with further fine aggregate replacement. The split tensile strength of cylinder showed a similar behaviour to the compressive strength of the cube for all mixtures.

**Flexural strength:** Figure 10 showed that the flexural strength of beam was found to be 4.79 N/mm² at 0% fine aggregate replacement and of 4.42 N/mm² at 100% fine aggregate replacement. The maximum flexural strength of beam was found to be at 40% fine aggregate replacement of about 7.73 N/mm². The flexural strength of beam showed higher strength at 40% fine aggregate replacement. The flexural strength of all mixtures showed a similar behaviour to the compressive strength results.

**Rebound hammer test:** From the Figure 11 it is noted that the compressive strength was found to be 9.81 N/mm² at 0% fine aggregate replacement and of 14.73 N/mm² at 100% fine aggregate replacement of

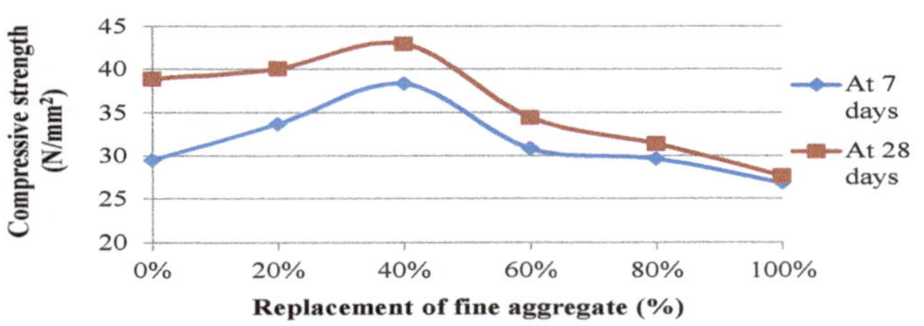


**Figure 8:** Relationship of compressive strength of cube and percentage of replacement of fine aggregate at 7 and 28 days.

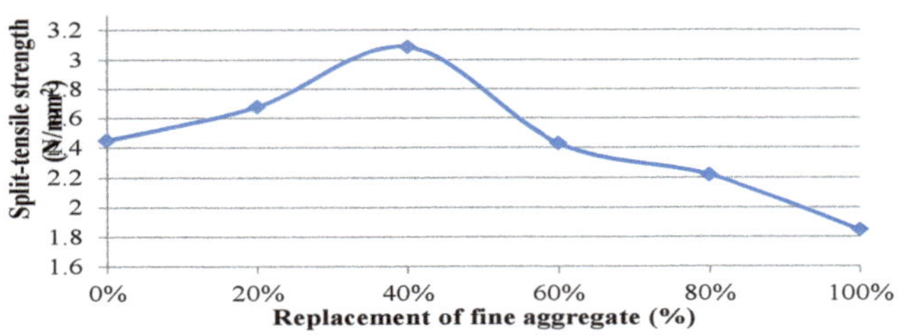


**Figure 9:** Relationship of split tensile strength and percentage of replacement of fine aggregate at 28 days.

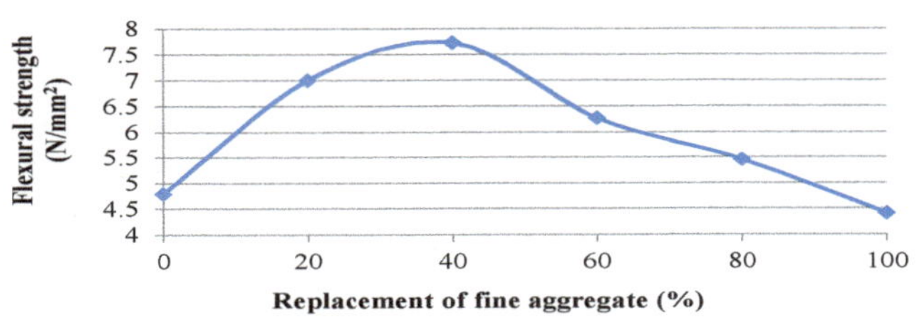


**Figure 10:** Relationship of flexural strength and percentage of replacement of fine aggregate at 28 days.

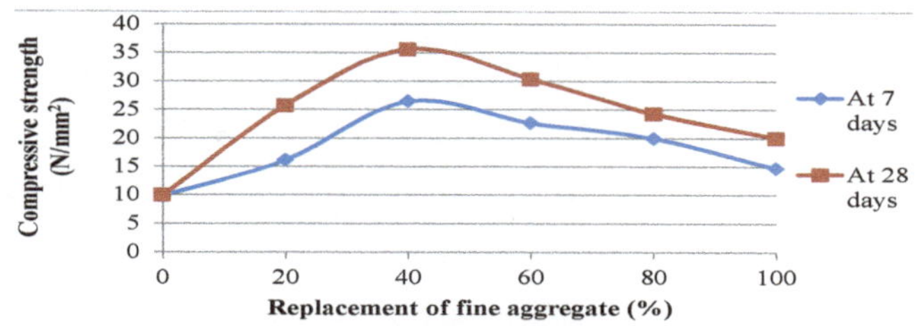


**Figure 11:** Relationship of compressive strength and percentage of replacement of fine aggregate at 7 and 28 days of rebound hammer test.

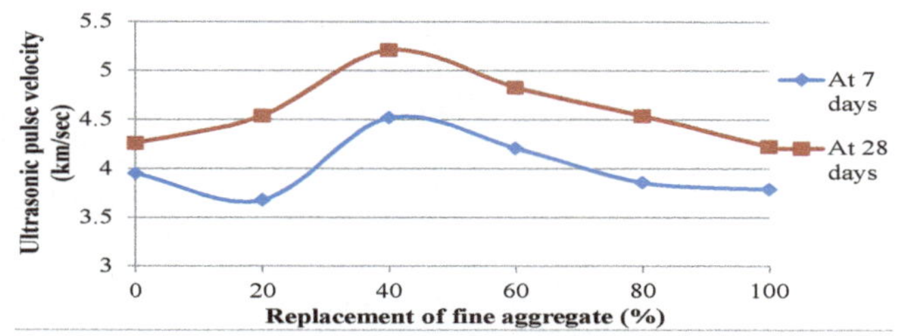


**Figure 12:** Relationship of pulse wave velocity and percentage of replacement of fine aggregate at 7 and 28 days.

**Figure 13:** Copper powder Slug.

7 days. And the compressive strength at 28 days was found to be 10 N/mm² at 0% fine aggregate replacement and of 20.08 N/mm² at 100% fine aggregate replacement. The maximum compressive strength was found to be at 40% fine aggregate replacement of about

26.49 N/mm² at 7 days and 35.64 N/mm² at 28 days. The compressive strength of concrete at 7 and 28 days obtained with the rebound number was increased gradually up to 40% fine aggregate replacement and then decreased with increase replacement. This is due to homogeneity of concrete and the strengths were lie in between ± 15% and ± 20% of compressive strength of compression test.

**Ultrasonic pulse velocity measurement:** The ultrasonic pulse velocity measurement is the measure of quality of concrete. It is mainly related to its density and modulus of elasticity which in turn, depends upon the materials and mix proportions used in making concrete as well as the method of placing, compaction and curing of concrete. From Figure 12 it was observed that the pulse wave velocity showed above 4.5 km/sec at 7th day measurement and above 5.2 km/sec at 28th day measurement. This is due to the density of the mix is high and free from pores. It is also observed that the concrete at 40% replacement level showed excellent quality and good quality for other percentages at 7th day measurement [20]. At 28th day measurement, the concrete showed excellent quality for 20%, 40%, 60% and 80% replacement of fine aggregate and showed good quality for 0% and 100% replacement of fine aggregate. The important observation was that the addition of copper slag definitely reduced the pores of concrete and made the concrete impermeable (Figure 13).

## Conclusion

From the results and discussions, the following conclusions were made

- The replacement of fine aggregate using copper slag in concrete increases the density of concrete thereby increases the self-weight of the concrete.
- The workability of concrete increased with the increase in copper slag content of fine aggregate replacements at same water-cement ratio.
- Form the results of compressive strength, split tensile strength and flexural strength, the concrete shown higher value at 40% replacement of fine aggregate using copper slag. So it is recommended that 40% of fine aggregate can be replaced by copper slag.
- The rebound hammer test revealed the uniformity of concrete and their compressive strength.
- The ultrasonic pulse velocity test indicated the excellent quality of concrete at 40% replacement level.
- The construction industry is the only area for safe use of waste materials, which reduces the environmental problems, space problems and cost of construction.

### References


1. Al-Jabri KS, Hisada M, Al-Saidy AH, Al-Oraimi SK (2009) Performance of High Strength Concrete made with Copper Slag as a Fine Aggregate. Construction and Building Materials 23: 2132-2140.
2. Santhakumar AR (2006) Concrete Technology. Oxford University Press, Engineering and Computer Science.
3. Shetty MS (2006) Concrete Technology Theory and Practice. S Chand & company ltd, India.
4. Brindha D, Baskaran T, Nagan S (2010) Assessment of Corrosion and Durability Characteristics of Copper Slag Admixed Concrete. International Journal of Civil and Structural Engineering 1: pp. 192-211.
5. Reddy SS, Kumar KM (2013) Utilization of Copper Slag as a Partial Replacement of Fine Aggregate in Concrete. International Journal of Mechanical Engineering and Computer Applications, 1: 140-144.
6. IS 383 (1970) Indian Standard Specifications for Coarse and Fine aggregates from Natural Sources for Concrete. CED 2: Cement and Concrete.
7. IS 12269 (1987) Indian Standard Ordinary Portland cement, 53 grade-specification. CED 2: Cement and Concrete.
8. Kharade AS, Kapadiya SV, Chavan R (2013) An Experimental Investigation of Properties of Concrete With Partial or Full Replacement of Fine Aggregates through Copper Slag. IJERT 2: 01-10.
9. Al-Jabri KS, Hisada M, Al-Oraimi SK, Al-Saidy AH (2009) Copper Slag as Sand Replacement for High Performance Concrete. Cement & Concrete Composites 31: 483-488.
10. Gupta RC, Thomas BS, Prachi Gupta (2012) Utilization of Copper Slag and Discarded Rubber Tyres in Construction. International Journal of Civil and Structural Engineering 3: 271-281.
11. Brindha D, Nagan S (2011) Durability Studies on Copper Slag Admixed Concrete. Asian Journal of Civil Engineering (Building and Housing) 12: 563-578.
12. IS 10262 (2009) Guidelines for Indian Standard Concrete Mix Design Proportioning. CED 2: Cement and Concrete.
13. IS 1199 (1959) Indian Standard Methods of sampling and analysis of concrete. CED 2: Cement and Concrete.
14. IS 10086 (1982) Indian Standard Specification for moulds for use in tests of cement and concrete. CED 2: Cement and Concrete.
15. IS 516 (1959) Indian Standard Methods of tests for strength of concrete. CED 2: Cement and Concrete.
16. IS 5816 (1999) Indian Standard Method of test for Splitting tensile strength of concrete. CED 2: Cement and Concrete.
17. IS 13311-1 (1992) Indian Standard Methods Non-destructive testing of concrete: Part 1 Ultrasonic pulse velocity. CED 2: Cement and Concrete.
18. IS 13311-2 (1992) Indian Standard Methods Non-destructive testing of concrete: Part 2 Rebound hammer. CED 2: Cement and Concrete.
19. Brindha D, Nagan S (2010) Utilization of Copper Slag as a Partial Replacement of Fine Aggregate in Concrete. International Journal of Earth Sciences and Engineering 3: 579-585.
20. IS 456 (2000) Indian Standard Code of practice for Plain and Reinforced Concrete. CED 2: Cement and Concrete.

# 8

# Evaluation of Combined Excavation Waste and Reclaimed Asphalt Pavement Aggregates for Use in Road Bases and Sub-bases

**Ramzi Taha[1]*, Okan Sirin[1] and Husam Sadek[1,2]**

*[1]Department of Civil & Architectural Engineering, Qatar University, Doha, Qatar*
*[2]Centre for Engineering Sustainability, University of Liverpool, Liverpool, UK*

**Abstract**

More than 15 million tons of aggregates are imported each year to Qatar from neighbouring countries. Large quantities of waste materials (around 20 million tons/year) from excavation waste, demolition rubble and reclaimed asphalt pavement (RAP) are being generated. The main objective of this paper is to present the results on the use of combined excavation waste (EW) and RAP aggregates in the construction of road bases and sub-bases. Physical and chemical properties were determined. Different combinations of both materials were subjected to compaction and California Bearing Ratio (CBR) testing in accordance with Qatar Construction Specifications. Results indicated that, for the materials tested, it is not feasible to use excavation waste, RAP aggregates or a combination of the two materials in road bases and sub-bases. The materials failed to meet some Qatari standards such as Los Angeles abrasion, liquid limit, plasticity index and CBR specified for road construction.

**Keywords:** Excavation; Waste; Reclaimed; Asphalt; Road; Recycling

## Introduction

Waste materials are commonly used in construction projects in order to save natural resources for future generations. Road construction is one of the main users of these natural resources. Utilizing these materials in unbound base/sub-base construction will provide sustainable development in a country by saving virgin materials, conserving energy and diverting materials from landfills [1,2]. Recycled materials used in asphalt pavements and unbound base/sub-base applications mainly come from construction and demolition wastes (CDW), solid wastes (SW) and by-products from industrial processes. Most commonly used ones include coarse and fine aggregates, tiles, bricks obtained from CDW and excavation wastes (EW), reclaimed asphalt pavement (RAP), incinerator bottom ash (IBA), crumb rubber, fly ash, plastics and glass [3]. Demolition waste is the material that results from the destruction of a structure, specifically reinforced concrete structures, or generally from a structure/object that contains sand and cement.

In some countries where aggregate and sand resources are limited, these wastes are usually recycled and harvested for other construction purposes. Another source of wastes that could be recycled for construction use is the materials that are generated from ground excavation. These materials are formed due to excavation work during the construction phase of a structure, underground infrastructure and metro stations. The material usually consists of natural soil of the excavated location; it also might contain other materials like wood and organic material.

Reclaimed asphalt pavement (RAP) is another waste material generated when asphalt pavements are removed for reconstruction and resurfacing. They are commonly used in hot-mix asphalt courses and unbound base/sub-base applications as well Kallas [4], Decker and Young [5].

Economic, industrial and population growth in Qatar will generate increasing amounts of waste materials that must be disposed of. As the volume of wastes continues to grow, the approval and availability of facilities for waste processing and proper disposal will become more difficult to obtain. In addition, Qatar suffers from the availability of quality aggregates and materials that could be utilized in road, parking and building construction. Most of the aggregate used in the country is imported from the United Arab Emirates, Oman and Saudi Arabia. It is estimated that approximately 15 million tons of aggregate are imported every year. At the same time, Qatar generates large quantities of building and construction wastes, including milled asphalt pavements and demolition rubble (close to 20 million tons a year), which could be recycled in road construction.

## Research Objective and Scope of Work

The main objective of this research project was to investigate the potential for constructing road bases and sub-bases from waste materials generated in the State of Qatar. To meet this objective, physical and chemical properties of excavation waste (EW) and reclaimed asphalt pavement (RAP) aggregates were determined. Then combinations of EW and RAP were subjected to compaction and California Bearing Ratio (CBR) tests. Results were compared with Qatar Construction Specifications [1] to establish the viability of using such materials in the base and sub-base structure.

## Materials

Three waste materials, namely Excavation Waste-1 (EW1), Excavation Waste-2 (EW2) and RAP were collected from several locations around Doha (Capital of Qatar). RAP materials consist of limestone aggregate and aged Pen 60/70 bitumen. No pre-treatment was done to the materials of this study. A total of 25 buckets for each type of waste material were sampled from stockpiles and brought to our construction materials laboratories for physical, chemical, and mechanical testing. The following tests were conducted on the three waste materials in accordance with Qatar Construction Specifications (QCS-2010).

- Sieve analysis (ASTM C 136)
- Density, bulk specific gravity and absorption of coarse and fine aggregates (ASTM C 127 and ASTM C 128)

***Corresponding author:** Ramzi Taha, Department of Civil and Architectural Engineering, Qatar University, P.O. Box 2713, Doha, Qatar
E-mail: ramzitaha@qu.edu.qa

- Los Angeles abrasion test (ASTM C 131)
- Sand equivalency test (ASTM D 2419)
- Liquid limit and plasticity index tests (BS 1377) [6]
- Chemical leachate analysis
- Compaction (ASTM D698)
- California Bearing Ratio (CBR) test (ASTM D 698 and ASTM D 1883)

| Property | Specific gravity | Absorption (%) | Bulk density (kg/m³) | Los Angeles abrasion (%) |
|---|---|---|---|---|
| EW1 | 2.22 | 8.74 | 1,501 | 50.1 |
| EW2 | 2.47 | 6.57 | 1,665 | 60.3 |
| RAP | 2.41 | 2.35 | 1,692 | 43.1 |
| QCS-2010 | N/A | N/A | N/A | ≤40 |

N/A: Typical values are not mentioned in QCS

**Table 1:** Physical properties of coarse aggregates.

| Property | Specific gravity | Absorption (%) | Bulk density (kg/m³) | Sand equivalent (%) | Liquid limit (%) | Plasticity index (%) |
|---|---|---|---|---|---|---|
| EW1 | 2.15 | 11.8 | 1,570 | 23 | 39.2 | 12 |
| EW2 | 2.28 | 12.4 | 1,225 | 29 | 48 | 20.1 |
| RAP | 2.01 | 1.2 | 1,487 | 96 | NP | NP |
| QCS 2010 | N/A | N/A | N/A | ≥25 | ≤25 | ≤6 |

N/A: Typical values are not mentioned in QCS
NP: non-plastic

**Table 2:** Physical properties of fine aggregates.

| Sieve size (mm) | Percent passing | | | | | |
|---|---|---|---|---|---|---|
| | EW1 | EW2 | RAP | Class A (QCS-2010) | Class B (QCS-2010) | Class C (QCS- 2010) |
| **63** | 100 | 100 | 100 | 100 | - | - |
| **50** | 100 | 100 | 100 | 90-100 | 100 | - |
| **37.5** | 93.9 | 85.2 | 100 | 60-90 | 70-100 | - |
| **25** | 78.7 | 77.8 | 98 | 42-77 | 55-85 | 100 |
| **20** | 69.1 | 70.5 | 95 | 35-70 | 50-80 | 90-100 |
| **10** | 52.4 | 53.0 | 65 | 25-60 | 40-70 | 50-85 |
| **5** | 41.2 | 40.2 | 22 | 15-40 | 30-60 | 35-65 |
| **2.36** | 33.6 | 32.5 | 7 | 10-26 | 20-50 | 25-50 |
| **0.425** | 19.8 | 19.0 | 0.6 | 5-15 | 10-30 | 15-30 |
| **0.075** | 5.5 | 5.3 | 0.1 | 2-9 | 5-15 | 5-15 |

**Table 3:** Sieve analysis results and gradation limits in QCS.

# Test Results and Discussion

## Physical tests

Physical properties of each of the three materials were determined in the laboratory. A summary of test results for coarse and fine aggregates are presented in Tables 1 and 2, respectively. Table 1 indicates that all three materials did not meet the maximum 40% LA Abrasion criteria in QCS. However, Los Angeles abrasion value for the RAP material (43.1%) was only slightly higher than that of the QCS requirement.

Table 2 indicates that liquid limit and plasticity index results obtained for EW1 and EW2 didn't meet QCS standards. However, the sand equivalent for EW2 and RAP materials were satisfactory. The sand equivalent value for EW1 was slightly less than that of the QCS requirements of a minimum of 25%.

In order to determine the grain size distribution of materials collected from the field, sample sizes were reduced using a splitter to collect enough material to run the sieve analysis test according to ASTM C 136 [7]. Figure 1 shows the sieve analysis results for EW1 material along with the upper and lower limits for Class A base course given in QCS. Figure 1 indicates that % passing amount for some sieve sizes are outside the upper and lower limits and thus do not meet the specification requirement.

All sieve analysis results for the three materials and % passing requirements for two types of bases (Classes A and B) and one type of sub-base (C) given in QCS are presented in Table 3. The results indicates that EW1 and EW2 materials can meet the requirements for a Class B type base course, but not those of Class A base and Class C sub-base. The materials failed to meet the percent passing requirements for many of the gradation sizes. Also, RAP material did not meet the % passing requirement for any type of unbound base and sub-base courses given in QCS.

## Chemical tests

Table 4 shows the concentrations of chemical constituents present in the leachate of the two excavation waste sources (EW1 and EW2) and the reclaimed asphalt pavement (RAP) aggregate. The QCS doesn't have standard for the concentrations of chemical constituents were results are reported in part per billion and they are extremely low to present. Such materials are safe to use and no water pollution should be anticipated from their use in road construction.

## Mechanical tests

Mechanical tests in this project focused on two main tests, the

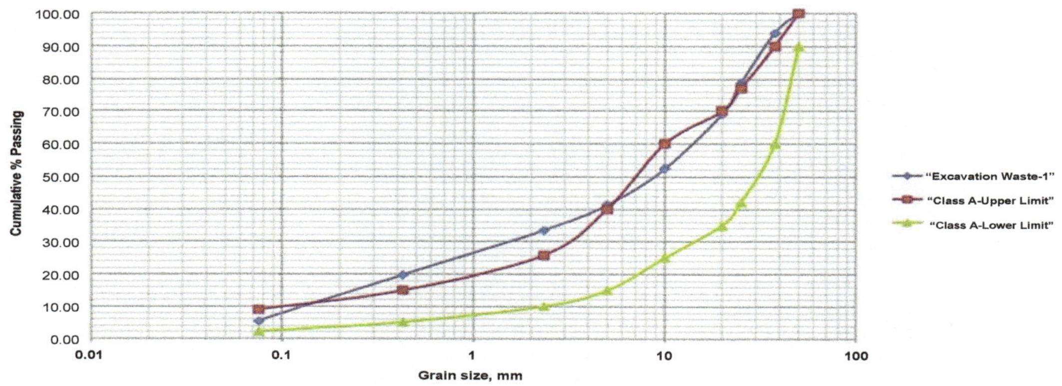


**Figure 1**: Sieve analysis results for excavation waste from site 1 (EW1).

Standard Proctor Compaction test (ASTM D698) and the California Bearing Ratio (CBR) test (ASTM D1883). Table 5 presents the different RAP/EW mixtures tested in our laboratories. The main reason for blending excavation waste and reclaimed asphalt pavement aggregate is to improve the shearing strength of excavation waste materials as they tend to be the weak part of the matrix.

| Element | EW1 | EW2 | RAP |
|---|---|---|---|
| Mg | 53,060 | 41,990 | 34,990 |
| Al | 8,943 | 10,870 | 41,060 |
| Ca | 191,800 | 207,400 | 152,500 |
| Ti | 410.8 | 924.7 | 2,973 |
| V | 20.92 | 29.46 | 164.4 |
| Cr | 30 | 53.41 | 197.2 |
| Mn | 111.3 | 158.8 | 629.3 |
| Fe | 4,243.5 | 6,596.5 | 34560 |
| Co | < 0.02 | 1.053 | 24.49 |
| Ni | 10.81 | 19.28 | 76.56 |
| Cu | 7.844 | 5.249 | 32.82 |
| Zn | 4.055 | 9.677 | 28.56 |
| As | 2.54 | 3.891 | 1.353 |
| Se | 0.462 | 0.5619 | 0.4113 |
| Ag | < 0.1 | < 0.1 | < 0.1 |
| Cd | 0.0729 | 0.1149 | 0.1426 |
| Ba | 470.6 | 313.9 | 54.1 |
| Hg | 0.1276 | 0.115 | 0.09333 |
| Pb | 3.018 | 3.401 | 1.931 |

**Table 4:** Concentrations of chemical constituents in leachate samples (ppb: part per billion).

**Optimum moisture content:** Laboratory compaction tests are usually used for determining the percent compaction and optimum water content needed for a dense mix. For this purpose, different amounts of water were added to sample mixes to determine the optimum moisture content (OMC) using the standard Proctor compaction test (ASTM D 698). Samples were compacted by 25 blows in 3 layers using a 24.5 N rammer dropped from a height of 305 mm. Corrected moisture contents were determined after drying samples at an oven temperature of 110 ± 5°C for 24 hours. A total of 21 sample mixes were prepared using different percentages of two materials (RAP+EW1 and RAP+EW2). The compaction curve for each mixture was established after a sufficient number of water contents were used.

Compaction graphs for all mixtures are shown in Figures 2-5. A summary of optimum moisture content (OMC) results are presented in

| Mix # | RAP (%) | EW1 (%) | Mix # | RAP (%) | EW2 (%) |
|---|---|---|---|---|---|
| 1 | 0 | 100 | 12 | 0 | 100 |
| 2 | 10 | 90 | 13 | 10 | 90 |
| 3 | 20 | 80 | 14 | 20 | 80 |
| 4 | 30 | 70 | 15 | 30 | 70 |
| 5 | 40 | 60 | 16 | 40 | 60 |
| 6 | 50 | 50 | 17 | 50 | 50 |
| 7 | 60 | 40 | 18 | 60 | 40 |
| 8 | 70 | 30 | 19 | 70 | 30 |
| 9 | 80 | 20 | 20 | 80 | 20 |
| 10 | 90 | 10 | 21 | 90 | 10 |
| 11 | 100 | 0 | | | |

**Table 5:** Sample mixtures combinations' matrix.

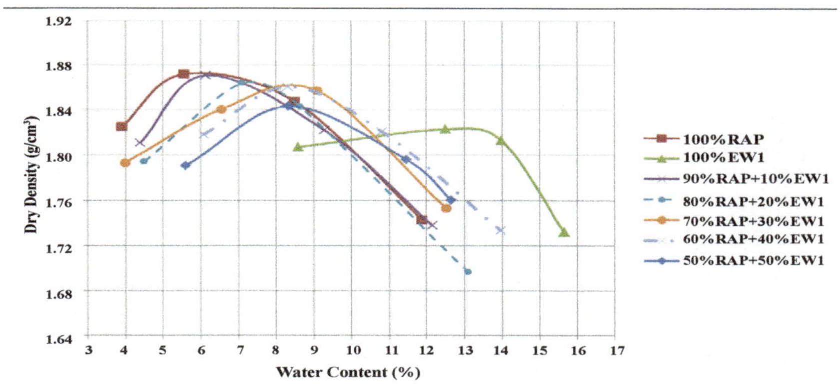


**Figure 2**: Compaction results for mix group 1.

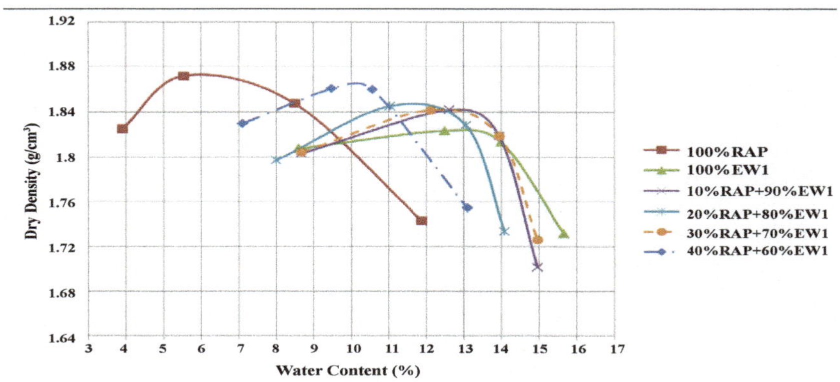


**Figure 3:** Compaction results for mix group 2.

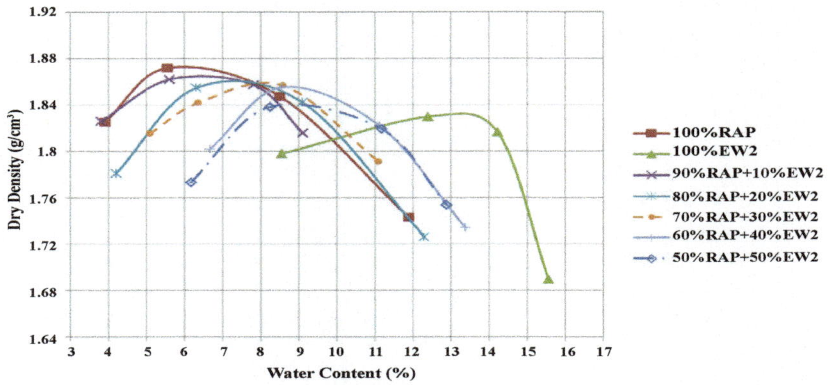


**Figure 4:** Compaction results for mix group 3.

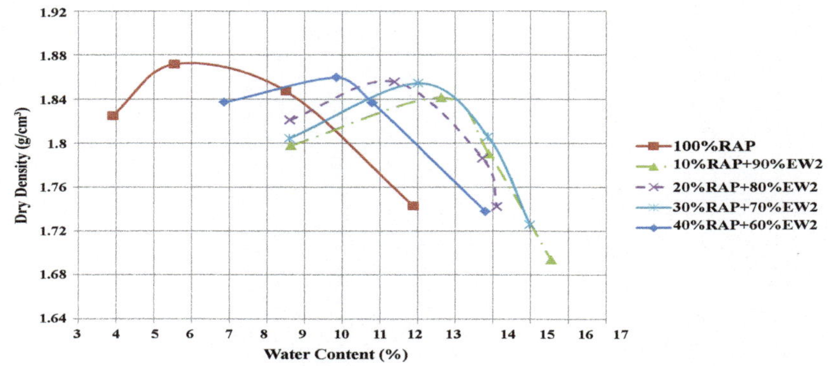


**Figure 5:** Compaction results for mix group 4.

| Mix # | RAP (%) | EW1 (%) | OMC (%) | Mix # | RAP (%) | EW2 (%) | OMC (%) |
|---|---|---|---|---|---|---|---|
| 1 | 0 | 100 | 13.9 | 12 | 0 | 100 | 13.7 |
| 2 | 10 | 90 | 13.6 | 13 | 10 | 90 | 13.4 |
| 3 | 20 | 80 | 12.5 | 14 | 20 | 80 | 12.9 |
| 4 | 30 | 70 | 12.7 | 15 | 30 | 70 | 13.1 |
| 5 | 40 | 60 | 10.1 | 16 | 40 | 60 | 10.1 |
| 6 | 50 | 50 | 9.6 | 17 | 50 | 50 | 9.8 |
| 7 | 60 | 40 | 9.1 | 18 | 60 | 40 | 9.1 |
| 8 | 70 | 30 | 9.0 | 19 | 70 | 30 | 8.3 |
| 9 | 80 | 20 | 7.9 | 20 | 80 | 20 | 7.9 |
| 10 | 90 | 10 | 6.4 | 21 | 90 | 10 | 6.8 |
| 11 | 100 | 0 | 5.8 | | | | |

**Table 6:** Summary of compaction results.

Table 6. The results indicate that as the percentage of RAP is increased in the mix there is a decrease in the OMC and an increase in dry density. This can be attributed to the better shear strength interlocking between the RAP particles and the fines from the EW.

**California bearing ratio (CBR) results:** Samples were compacted at the OMC in accordance with ASTM D 698 (standard Proctor compaction test). Then, compacted samples were soaked in water for 96 hours prior to conducting the California Bearing Ratio (CBR) test (ASTM D 1883). No swelling was observed in the soaked samples.

Load versus penetration curves for all mixtures were obtained for all mixtures. Figure 6 shows a typical load versus penetration graph for Mix Group 1. A summary of CBR results obtained at penetrations of 2.5 and 5 mm are presented in Table 7. A maximum CBR of 19.3% was obtained for 100% RAP. 100% excavation waste from Sites 1 and 2 generated low CBR values of 6.1 and 7.1%, respectively. As the percentage of RAP was increased in the mixture, there was an increase in the CBR value. However, none of the CBR values obtained for all the mixtures met the minimum QCS-2010 requirements of 60 and 80% for road sub-bases and bases, respectively

## Conclusion and Recommendations

### Conclusions

Two excavation waste and one reclaimed asphalt pavement (RAP) aggregate materials collected from different locations in Qatar were investigated for possible use in road base and sub-bases in the State of Qatar. Physical, chemical and mechanical tests were conducted on a

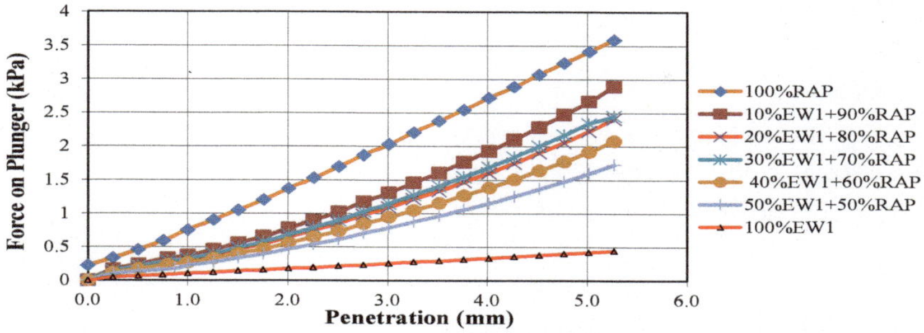


**Figure 6:** Load versus penetration for mix group 1.

| Sample | RAP (%) | EW1 (%) | OMC (%) | CBR (%) @ 2.5 mm | CBR (%) @ 5 mm | Sample | RAP (%) | EW2 (%) | OMC (%) | CBR (%) @ 2.5 mm | CBR (%) @ 5 mm |
|---|---|---|---|---|---|---|---|---|---|---|---|
| 1 | 0 | 100 | 13.9 | 3.4 | 6.1 | 12 | 0 | 100 | 13.7 | 4.1 | 7.1 |
| 2 | 10 | 90 | 13.6 | 3.8 | 7.2 | 13 | 10 | 90 | 13.4 | 4.3 | 7.4 |
| 3 | 20 | 80 | 12.5 | 4.3 | 7.4 | 14 | 20 | 80 | 12.9 | 5.1 | 8.6 |
| 4 | 30 | 70 | 12.7 | 4 | 7.6 | 15 | 30 | 70 | 13.1 | 6 | 10 |
| 5 | 40 | 60 | 10.1 | 5.1 | 8.11 | 16 | 40 | 60 | 10.1 | 6.8 | 11.6 |
| 6 | 50 | 50 | 9.6 | 6.3 | 9.7 | 17 | 50 | 50 | 9.8 | 7.1 | 13.6 |
| 7 | 60 | 40 | 9.1 | 6.4 | 11.6 | 18 | 60 | 40 | 9.1 | 7.7 | 15.2 |
| 8 | 70 | 30 | 9.0 | 8.2 | 15.4 | 19 | 70 | 30 | 8.3 | 8 | 16.4 |
| 9 | 80 | 20 | 7.9 | 8.7 | 17.6 | 20 | 80 | 20 | 7.9 | 8.1 | 17.5 |
| 10 | 90 | 10 | 6.4 | 9.6 | 18 | 21 | 90 | 10 | 6.8 | 9.8 | 18.1 |
| 11 | 100 | 0 | 5.8 | 11 | 19.3 | | | | | | |

**Table 7:** CBR and penetration results for mixes.

variety of mix proportions to determine their suitability according to QCS. Based on the data obtained from laboratory test, the following conclusions can be made:

- All three materials did not satisfy the Los Angeles Abrasion, liquid limit, plasticity index and CBR requirements for unbound materials specified in QCS.
- EW1, EW2 and RAP materials are not suitable in their current conditions for road bases and sub-bases construction in the State of Qatar.

## Recommendations

Recommendations for future work should focus on:

- Stabilization of such materials with Portland cement to improve their strength.
- Blending with other virgin aggregates such as gabbro, gravel or limestone.
- Mixing with discarded aggregate generated in Qatar such as steel slag, aluminum dross and others.

### Acknowledgments

This publication was made possible by a grant from the Qatar National Research Fund under its Undergraduate Research Experience Program (UREP), Award Number UREP 12-014-2-003. Its contents are solely the responsibility of the authors and do not necessarily represent the official views of the Qatar National Research Fund.

### References

1. Qatar Construction Specifications-QCS (2010) Qatar.
2. El-Assaly A, Ellis R (2001) Evaluation of recycling waste materials and by-products in highway construction. Int. Journal of Sustainable Development and World Ecology 8: 299-308.
3. Horvath A (2003) Life-Cycle Environmental and Economic Assessment of Using Recycled Materials for Asphalt Pavements. Technical Report, University of California-Berkeley.
4. Kallas BF (1984) Flexible pavement mixture design using reclaimed asphalt concrete. (FHWA/RD-84/088), Asphalt Institute, Lexington.
5. Decker DS, Young TJ (1996) Handling RAP in an HMA Facility. Proc. Canadian Technical Asphalt Association. Edmonton, Alberta, Canada.
6. BS 1377-1 (1990) Methods of test for soils for civil engineering purposes. General requirements and sample preparation. Standard Publication by BSI Group.
7. ASTM C136 – 06, Standard Test Method for Sieve Analysis of Fine and Coarse Aggregates.

# Specimen Size, Slenderness Ratio and Loading Rate Effect on Dynamic Simulation of HPC

**Hasan ASMZ*[1], Hamid R[2] and Ariffin AK[3]**

[1]*Department of Civil Engineering, Rajshahi University of Engineering and Technology, Bangladesh*

[2]*Department of Civil and Structural Engineering, Universiti Kebangsaan Malaysia, Malaysia*

[3]*Department of Mechanical and Materials Engineering, Universiti Kebangsaan Malaysia, Malaysia*

**Abstract**

The length to diameter ratio (L/D), loading rate, size and specimen geometry are significant factors that affect the compressive strength of high performance concrete (HPC) under dynamic loading conditions. A numerical study is carried out on selected conventional laboratory size prismatic specimen between the slenderness ratio 0.5 to 3.5 to evaluate the variation of stress-strain by performing finite element (FE) solving procedure. The results show that the maximum stress is found in specimen with slenderness ratio one. The stresses gradually decreased with the increase of the value of slenderness ratio (from one and above). The HPC dynamic strength dependency on strain rate and L/D is correlated using multiple regression equation. The accuracy of the predicted regression equation is verified with published experimental data and the result is 3% and 25% deviation under the applied strain rate 100 and 300 $s^{-1}$ respectively. The regression equation yields significant result within the limited stain rate 150 $s^{-1}$. The size effect analyses also satisfy existing published experimental data of concrete.

**Keywords:** Dynamic load; Finite element; High performance concrete; Slenderness ratio; Stress-strain; Size effect; Loading rate

## Introduction

Many experimental and theoretical investigations have showed that behavior of concrete loaded in tension, compression, shear or torsion is largely influenced by the specimen size and loading rate. Numerous researches were performed to understand the effect of fracture characteristics on material properties of concrete to evaluate the load carrying capacity. The results of these researches were inconsistence due to the large number of variables that affect the results such as specimen size, geometry, length to diameter ratio (L/D) as well as moisture content in the specimens, concrete quality, curing procedure and age [1-3]. The most effective experimental method is to determine the size effect. The term size effect means concrete material strength dependency on specimen size variation.

Concrete compressive strengths are mostly evaluated by tests performed on cylindrical specimens with slenderness ratio of two. Nevertheless, when it is necessary to evaluate in situ concrete strength (drilled cores); specimen with slenderness ratio lower than two can be used. Added to this fact, other type of specimens, that is prism or cube has been adopted in many countries, due to the greater amount of experimental work based on them.

Over the last 10 years, significant size effect studies on compressive failure of concrete were performed by many researchers in static loading conditions [4-9]. In their study Kim et al. [4], the numbers of specimens tested with height to diameter ratio L/D=2.0 and L/D ≠ 2 are 222 and 456, respectively. The failure stress of a series of concrete prism is expressed by the following equation [9]:

$$f_{pr} = (d) = \frac{1.02 f_c'}{\sqrt{1+\frac{d}{2.6}}} + 0.52 f_c' \quad , \qquad (1)$$

where compressive strength of prisms, $f_{pr}(d)$, compressive strength, $f'_c$, are in MPa and the diameter of cylinder or in case of prism, the smallest side of the cross section of prism, d, is in cm. These studies are based on the static loading conditions and the maximum size of coarse aggregate is 13 mm.

It is well known that the load carrying capacity increases substantially at higher strain rates. The way of loading rate and the size effect phenomena interact with each other, however, is not well understood. Elfahal et al. [10-12] have investigated the size effect phenomenon numerically and experimentally on both normal and high strength concrete cylinders subjected to static and dynamic compression within the limited strain rates of 0.111 $s^{-1}$ and below. Few or little information is available on effect of L/D to the compressive strength of concrete subjected to high rate of loading. There is still a lack of knowledge on the HPC response in dynamic loading conditions like blast or impact loading [11,12].

Analysis on behavior of concrete under dynamic loading conditions are quite complicated and the parameters that can affect the results include the inertia effects, specimen geometry, uniformity of stress and strain along the specimen length and wave propagation effects. The effect of L/D on the dynamic behavior becomes particularly important at very high rates of loading. In this study, dynamic strength dependency on different L/D specimens and loading rate are presented. The regression model between dynamic strength with different L/D specimens and loading rate developed can be used in predicting the dynamic behavior of concrete accurately. Also, this paper contributes by discussing the stress-strain behavior of HPC under the dynamic load regarding to the influence of size. Different modes of failure at the extreme loading such as greater than 1300 GPa/s are also described.

## Material and Concrete Mixes

The static properties of HPC incorporating silica fume used

***Corresponding author:** Hasan ASMZ, Department of Civil Engineering, Rajshahi University of Engineering and Technology, Rajshahi-6204, Bangladesh
E-mail: hasanzia2003@yahoo.co.uk

| Ingredients | Unit | Quantity |
|---|---|---|
| Cement (ordinary portland type I) | kg/m$^3$ | 500 |
| Silica fume | kg/m$^3$ | 55 |
| Fine aggregate | kg/m$^3$ | 667 |
| Coarse aggregate (max. size 19 mm) | kg/m$^3$ | 999 |
| Super plasticizer | Liter/m$^3$ | 22.9 |
| Water | kg/m$^3$ | 136 |
| Water-cement ratio | kg/m$^3$ | 0.25 |
| Density | kg/m$^3$ | 2400 |
| 28 days compressive strength, $f'_c$ | MPa | 87 |
| Poisson ratio, $\upsilon$ | - | 0.2 |
| Modulus of elasticity, $E$ | GPa | 36.5 |

**Table 1:** Mix proportion and static properties of the HPC [13].

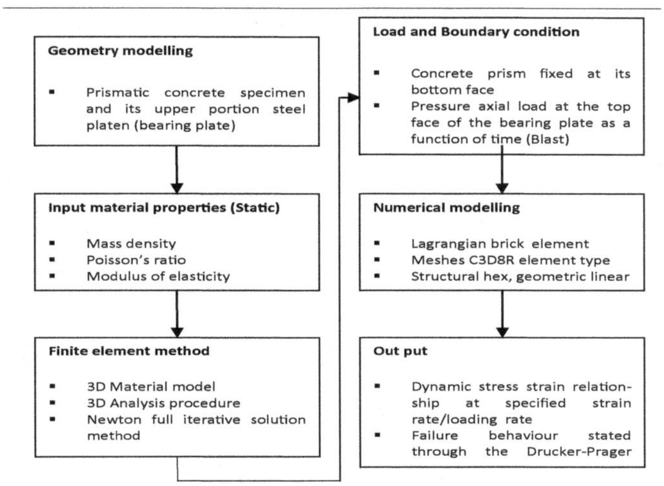


**Figure 1:** Block diagram of FE model.

| Prismatic specimens | | | | Cylindrical specimens | | |
|---|---|---|---|---|---|---|
| Specimens notification | Depth, D (mm) | Width, b (mm) | Length, L (mm) | Specimens ID | Diameter, D (mm) | Length, L (mm) |
| L/D 0.50 | 100 | 100 | 50 | Cy75 | 75 | 150 |
| L/D 0.75 | 100 | 100 | 75 | Cy150 | 150 | 300 |
| L/D 1.00 | 100 | 100 | 100 | Cy300 | 300 | 600 |
| L/D 1.50 | 100 | 100 | 150 | | | |
| L/D 2.00 | 100 | 100 | 200 | | | |
| L/D 3.50 | 100 | 100 | 350 | | | |

**Table 2:** Details of specimen geometry and dimensions.

for simulation in this research was determined by Ilham [13]. The ingredients, details of mixing proportions and properties are shown in Table 1.

## Finite Element Modelling

A non-associated plastic flow rule along with combination of isotropic properties of elasticity is used on the whole structure of concrete specimen. The Drucker-Prager linear function is used to define the plastic flow potential. The strength hardening used to trace the evaluation of the yield surface is the effective plastic strains in compression and the start of compressive plastic strains in these numerical analyses are considered to be zero. The physical model is formed with two parts namely bearing plate and concrete prism. Details descriptions of the modeling of concrete are illustrated in the subsections 3.1-3.4 and the block diagram of the model is shown in Figure 1.

### Geometry of the specimens

Numerical studies are performed on selected conventional laboratory size prismatic specimens of L/D 0.5 to 3.5 conforming to the experimental specimens used by Cotsovos and Pavlovic [2] and cylindrical specimens of L/D are equal to 2 following investigators del Viso et al. [14], Krauthammer et al. [12] and Elfahal et al. [11]. The diameter of a cylinder is equivalent to the least dimension of the cross section of a prism. In this study, square cross section is opted, so that D is equivalent to either depth or width of the square. The detailed dimensions of the prismatic and cylindrical specimens are summarized in Table 2.

### Rigid body and contact

A general contact algorithm is selected between the concrete prism and steel bearing plate (rigid body). The bearing platen bottom surface is defined as the master surface due to dynamic load applied on the top surface; whereas the concrete surface in contact is define as the slave surface. Tie constraint is used to interlock the interface of two contact surfaces for transmission of shear and normal forces across the interface properly. A mechanical interaction model excluding friction loss is defined to model the contact between the bearing plate and concrete surface. A hard contact is used in ABAQUS to formulate the tangential and normal forces in the present model. A central difference rule is used to integrate the equation of motion explicitly through time. The schematic diagram of the simulation setup is shown in Figure 2.

### Load and boundary condition

The bottom face of concrete prism is assumed to be fixed and high magnitude time dependent pressure load from 500 to 2200 GPa/s are imposed on the top face of the bearing plate. In this model, no separation is permitted between the bearing plate-concrete prism contact interfaces to avoid failure occurrence near the top face of concrete specimen. The load is allowed to all three degrees of freedom to avoid the wave deflection problems on the boundaries of the prism. In order to vary the rate of loading, the load increase is kept constant and the time step is varied.

### Numerical modeling

The simulations are conducted using the general purpose finite element code ABAQUS/Explicit version 6.7.2 [15]. The analysis is performed employing a 3D material model to account for the triaxial behavior of concrete. Both the concrete prism and bearing plate are

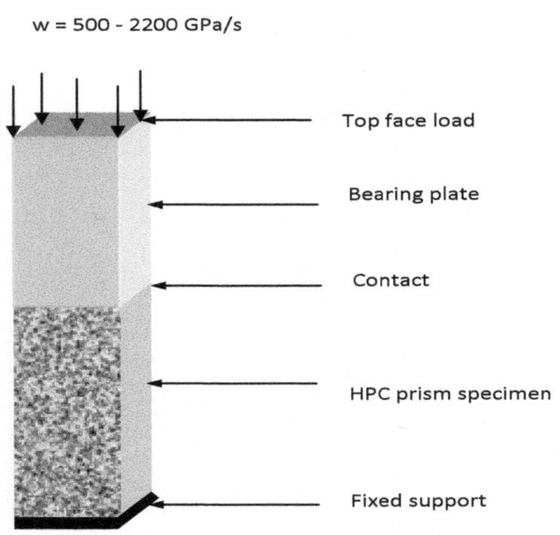


**Figure 2:** Schematic diagram of the simulation arrangement.

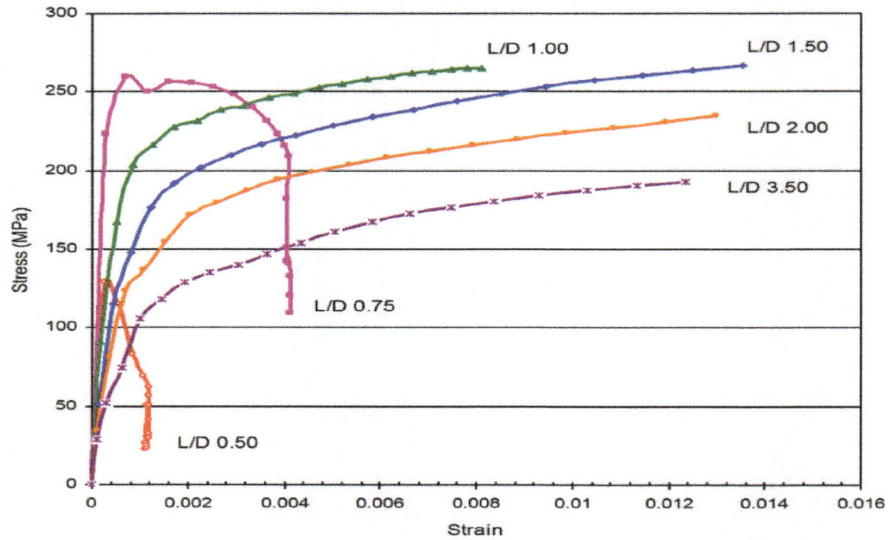


**Figure 3:** Stress-strain behavior of different L/D prismatic HPC specimen.

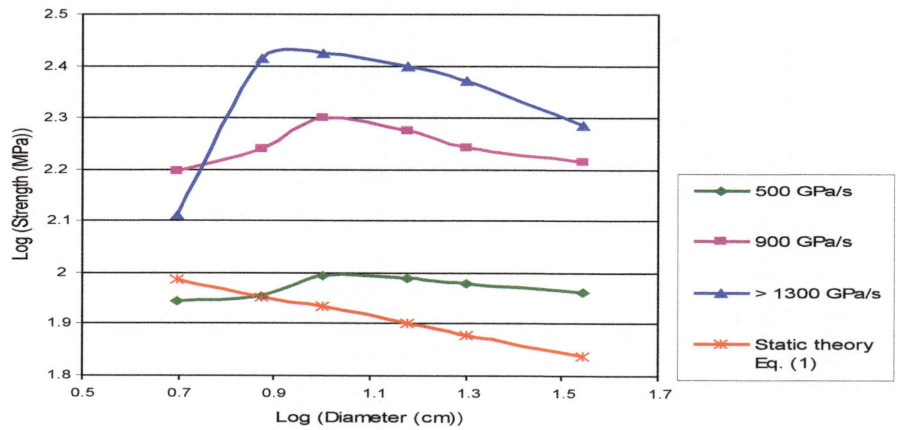


**Figure 4:** Smallest side vs strength with the variation of loading.

modeled using Lagrangian brick element. Meshes of both concrete prism and bearing plate are selected as structural hex shape, geometric linear and element type C3D8R (An 8-node linear brick, reduced integration, hourglass control). Finer meshes are preferred during the simulation procedure though Cotsovos and Pavlovic [2] in part I and II adopted coarse meshes in his material model. Usually, dense FE meshes are preferred in order to model the concrete specimen with several investigations adopting FE analysis as small as 2-3 mm [16,17]. Hence, finer meshes are adopted in the present study. In order to obtain more accurate results the concrete specimen is meshed into 3 mm size. Finer meshes are examined in several trail simulations to obtain the most accurate results.

## Numerical Results and Discussions

The results are presented on the basis of the affecting factors, i.e., L/D ratio, loading rate and size. The discussion will be on, firstly, the effect of the specimen length to diameter ratio at different loading rate (the loading rate that produce the highest load bearing capacity) to the stress-strain diagram, next, the effect of L/D at different loading rate, then, the correlation between dynamic strength, L/D and strain rate and finally, the failure patterns at the extreme loading, i.e. greater than 1300 GPa/s. Size effect is discussed through simulations of different sizes of cylindrical specimens at L/D equal to 2.

### Effect of L/D ratio at different loading rate on stress-strain diagram

The stress and strain relationship simulation of the HPC prismatic specimens is presented graphically in Figure 3. The curve for each L/D presented in Figure 3 are taken as the highest load carrying capacity it can take from simulation of each specimen with different L/D at different loading rate between 500-2200 GPa/s The results clearly show that axial stresses increase when the L/D is raised from 0.50 to 1.00 but further increase of the L/D (from 1.50 to 3.50) had resulted in the reduction of the ultimate strength capacity. In addition, Figure 3 also reveals that strains of HPC increase when L/D is raised from 0.50 to 1.5 but the reverse occurred when L/D is raised from 1.50 to 3.50. The failure of the specimens with small slenderness ratio of L/D 0.50 and L/D 0.75 occur at strain value 0.0003 and 0.0007 at maximum stress 129 and 260 MPa respectively. The important factor not shown in Figure 3 is that the extreme load carrying capacity for these two specimens (L/D 0.50 and L/D 0.75) are found to be 2200 GPa/s and 2000 GPa/s accordingly. The specimens with L/D 0.50 and L/D 0.75 are applicable as bearing seats of girder in bridge, thus explained the high load bearing capacity they can held. On the other hand, the load carrying capacities for the other specimens (L/D 1.0-L/D 3.5) are found to be at loading rate 1300 GPa/s.

The effect of different loading rate (500, 900 and 1300 GPa/s) are clearly found in all sets of results of different L/D. Figure 4 reveals that with the increase of loading rate, stresses increase i.e., the strength at the loading rate 900 GPa/s is nearly 85% higher than at 500 GPa/s in the each sample of the specimen. Furthermore, at 1300 GPa/s applied loading rate, the stresses are on average 138% and 28.5% higher than the stresses at applied loading rate of 500 GPa/s and 900 GPa/s respectively. The only exception is found in case of specimen L/D 0.50, which shows lower strength but it carries the maximum load (2200 GPa/s). The energies applied are high for short specimen (50 mm compared to 100 x 100 mm cross-section) showing that short specimens are better in withstanding high strain rate of loading.

The analogous data demonstrates that the maximum stresses are at the slenderness ratio equal to one under the application of different loading. The stresses decrease with increasing value of slenderness ratio (i.e. one and above). Whereas, in the theoretical static loading conditions, according to eq. (1), stresses decrease with the increasing value of slenderness. The divergent observed also can be attributed to difference in the maximum coarse aggregate size used in the concrete mix and the specimen size. In the present work, simulations are carried out based on the properties of concrete with maximum size of coarse aggregate 19 mm. On the contrary, the maximum size of coarse aggregate used is limited to 13 mm in the static analyses. Hence, the influence of maximum coarse aggregate size is reflected in numerical results of the smaller size of specimen L/D 0.50 and L/D 0.75 as shown in Figures 3 and 4. If the results of these two specimens are disregarded, the results of other specimens follow the existing static law as described in eq. (1) earlier only qualitatively due to loading rate is not considered separately. In this research, the effect of L/D and loading rate are considered separately by using a statistical package to formulate a more realistic correlation. Details of this correlation are illustrated as comparative study with the existing published experimental data in Table 3.

| Strain rate, $\varepsilon$ ($s^{-1}$) | L/D | Avg. dynamic strength, $\sigma_d$ (MPa) (simulation) | Predicted dynamic strength, $\sigma_d$ (MPa) according to eq. (3) |
|---|---|---|---|
| 14 | 1.0 | 98.80 | 103.82 |
| 14 | 1.5 | 97.30 | 100.52 |
| 14 | 2.0 | 95.10 | 97.22 |
| 14 | 3.5 | 91.50 | 87.31 |
| 25 | 1.0 | 126.30 | 114.89 |
| 25 | 1.5 | 114.25 | 111.59 |
| 25 | 2.0 | 108.30 | 108.29 |
| 25 | 3.5 | 104.30 | 98.38 |
| 36 | 1.0 | 132.38 | 125.67 |
| 36 | 1.5 | 118.05 | 122.37 |
| 36 | 2.0 | 109.25 | 119.07 |
| 36 | 3.5 | 106.15 | 109.16 |
| 18 | 2.0 | 100.37 | 101.02 |
| 10 | 2.0 | 89.16 | 93.89 |

**Table 3:** Loading rate effect on strength at different L/D of specimens.

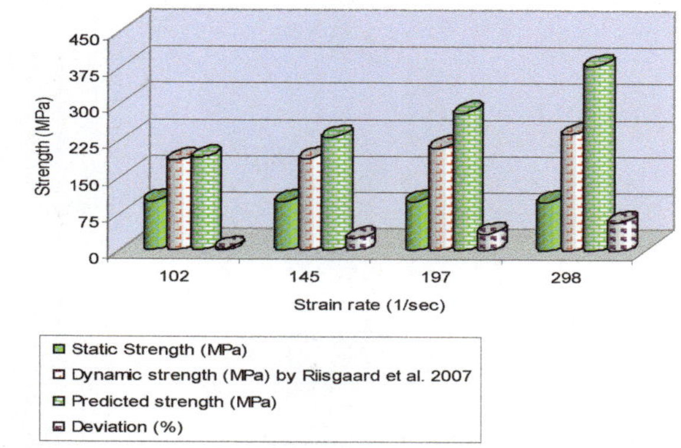


**Figure 5:** Comparative study with existing test results.

## Correlation between dynamic strength, L/D and strain rate

The obtained dynamic strength with L/D from 1 to 3.5 and strain rate range from 10 to 36 $s^{-1}$ are tabulated in Table 3. The relationships between dynamic strength as a dependent variable and two independent variables of slenderness ratio and strain rates is as expressed by the mathematical regression equation. The values of regression parameters are predicted by using statistical package of "Finite Mathematics & Applied Calculus Resources Page". The mathematical regression equation is as follows:

$$\sigma_d = a + b\dot{\varepsilon} + c(\text{L/D}), \quad (2)$$

where is the dynamic strength, *a*, *b* and *c* are regression parameters, and (L/D) are the strain rate and length to diameter ratio respectively.

The values of regression parameters are found to be *a*=-97.00585, *b*=0.97975 and *c*=6.60524. Hence, the dynamic strength of concrete can be obtained from the analogous data is expressed by the following equation:

$$\sigma_d = 97.00585 + 0.97975(\dot{\varepsilon}) - 6.60524(\text{L/D}). \quad (3)$$

Developed regression model eq. (3) reveals the dynamic strength dependency on loading rate at different length to diameter ratio. The results from the predicted regression eq. (3) are compared with experimental results conducted by Riisgaard et al. [18] with the static strength of concrete 100 MPa in order to verify the accuracy. The experimental test is conducted under the applied strain rate from 100 to 300 $s^{-1}$ using Split Hopkinson Pressure Bar with the value of L/D is equal to 1 for each specimen. The predicted regression model shows that the deviations between the predicted results and the experimental results are 3%, 24%, 34%, 59% with respect to strain rates 102, 145, 197 and 300 $s^{-1}$ accordingly (Figure 5). This comparative study demonstrates that the predicted model generates the accurate results with reasonable difference to be 25% under the application of strain rate 150 $s^{-1}$. For strain rate higher than 150 $s^{-1}$, the percentage of error is not acceptable.

## Analysis of failure pattern

Several modes of failure are detected during the simulations. The deformable contour shapes prior to failure are shown in Figure 6. The short specimen, that is, L/D 0.50 and L/D 0.75 are found to exhibit cone shaped shear failure as shown in Figure 6A and B. The prism fails at the side forming two cones at the top and bottom in the cone shaped shear failure modes. Compressive belly failures are found in specimens L/D 1.00, L/D 1.50 and L/D 2.00 with the inflation of the prism centre, forming a belly shape as shown in Figure 6C-E. In practical, this type of failure happens combined with the shell-bursting. Buckling failure

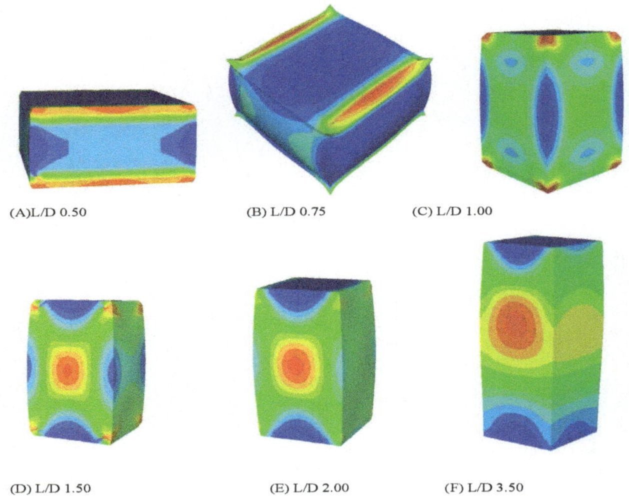


**Figure 6:** Predicted failure pattern.

| Specimen shape | Specimen ID | Maximum stress, $\sigma_{max}$ (MPa) | Time Δt (μs) | Strain at, $\sigma_{max} = \varepsilon_0$ (με) | Strain rate, $\bar{\varepsilon} = (\varepsilon_0)/\Delta t$ ($s^{-1}$) |
|---|---|---|---|---|---|
| Cylinder | Cy75 | 100.37 | 150 | 2637 | 17.58 |
| | Cy150 | 92.58 | 400 | 2400 | 6.00 |
| | Cy300 | 27.06 | 751 | 697 | 0.93 |

**Table 4:** Simulation results of cylindrical specimens.

| Cylindrical specimen size (mm) | Experimental and simulated results by Krauthammer et al. (2003) | | | | | Present study using Drucker-Prager model | |
|---|---|---|---|---|---|---|---|
| | Dynamic test | | Static strength $f_c'$ (MPa) | Simulated results | | $\bar{\varepsilon}$ ($s^{-1}$) | $\sigma_d$ (MPa) |
| | $\bar{\varepsilon}$ ($s^{-1}$) | $\sigma_d$ (MPa) | | Drucker-Prager model | Brittle fracture model | | |
| φ75 × 150 | 0.11 | 52.76 | 89.46 | 92.33 | 51.63 | 17.58 | 100.37 |
| φ150 × 300 | 0.10 | 68.06 | 82.86 | 70.47 | 31.78 | 6.00 | 92.58 |
| φ300 × 600 | 0.05 | 39.29 | 71.10 | 38.85 | 52.23 | 0.92 | 27.00 |

**Table 5:** Comparison of predicted results with existing data.

is distinguished of specimen L/D 3.5 with the inflation of the prism from the top and bulging of material as shown in Figure 6F. This type of failure is observed due to the greater value of slenderness ratio (L/D=3.5).

## Specimen size effect

Numerical studies are carried out on selected cylindrical specimens with the slenderness ratios of two and the obtained results are summarized in the Table 4.

The present numerical results are compared with the existing results with static strength 89 MPa as shown in Table 5. Comparative study demonstrates that differences of 9%, 31%, 39% for the specimens φ75 × 150, φ150 × 300, φ300 × 600 between the present study and the simulated results using Drucker-Prager model performed by Krauthammer et al. (2003). The variation of result occurs due to the loading rate and different method of loading and not satisfying eq. (3) for specimen Cy300. Table 5 shows that predicted simulation results also agree with the dynamic test results rather than the Brittle fracture model.

## Conclusion

Dynamic strength dependency on length to diameter ratio size and loading rate effect of prismatic specimen with slenderness ratios between 0.5 to 3.5 and cylindrical specimen with slenderness ratio of 2.0 of each specimen is studied. From the numerical results the following conclusion can be pointed out; (i) the maximum stresses are found in specimen with slenderness ratio one. (ii) The stresses gradually decreased with the increase of the value of slenderness ratio (from one and above), (iii) three different modes of failure are found: cone shape shear failure; compressive belly failure; and top buckling failure, (iv) The multiple model regression equation shows good agreements with the existing experimental results showing 25% deviation at strain rate 150 $s^{-1}$, (v) The comparative study with L/D equal to 2 for each cylindrical specimen satisfy the simulation results performed by the investigator Krauthammer et al. 2003. Therefore, it can be concluded that the proposed eq. (3) can described well the dependency of dynamic strength with the rate of loading, size and L/D of specimen.

## Acknowledgement

This research is sponsored by the Ministry of Higher Learning Malaysia under the grant UKM-KK-02-FRGS0014-2006.

## References

1. Bischoff PH, Perry SH (1991) Compressive behaviour of concrete at high strain rates. J Eng Mech 24: 425-450.
2. Cotsovos DM, Pavlovic MN (2008) Numerical investigation of concrete subjected to compressive impact loading. Part 1: A fundamental explanation for the apparent strength gain at high loading rates. Computers & Structures 86: 145-163.
3. Ross CA, Jerome DM, Tedesco JW, Hughes ML (1996) Moisture and strain rate effects on concrete strength. ACI Materials Journal 93: 293-300.
4. Kim JK, Yi ST, Park CK, Eo SH (1999) Size effect on compressive strength of plain and spirally reinforced concrete cylinders. ACI Structural Journal 96: 88-94.
5. Kim JK, Yi ST, Yang EI (2000) Size effect on flexural compressive strength of concrete specimens. ACI Struct J 97: 291-296.
6. Kim JK, Yi ST and Kim JHJ (2001) Effect of specimen sizes on flexural compressive strength of concrete. ACI Struct J 98: 416-424.
7. Kim JH, Yi ST and Kim JK (2004) Size effect of concrete members applied with flexural compressive stresses. Int J Fracture 126: 79-102.
8. Yi ST, Kim JHJ, Kim JK (2002) Effect of specimen sizes on ACI rectangular stress block for concrete flexural members. ACI Struct J 99: 701-708.
9. Yi ST, Yang E, Choi J (2006) Effect of specimen sizes, specimen shapes, and placement directions on compressive strength of concrete. Nuclear Engineering and Design 236: 115-127.
10. Elfahal MM, Krauthammer T (2005) Dynamic size effect in normal and high strength concrete cylinders. ACI Materials Journal 102: 77-85.
11. Elfahal MM, Krauthammer T, Ohno T, Beppu M, Mindess S (2005) Size effect for normal strength concrete cylinders subjected to axial impact. International Journal of Impact Engineering 31: 461-481.
12. Kruathammer T, Elfahal MM, Lim J, Ohno T, Beppu M et al. (2003) Size effect for high-strength concrete cylinders subjected to axial impact. International Journal of Impact Engineering 28: 1001-1016.
13. Ilham A (2003) Mix design method and properties of high performance concrete for compressive strength between 50 and 100 MPa. Universiti Kebangsaan Malaysia, Malaysia.
14. del Viso JR, Carmona JR, Ruiz G (2008) Shape and size effects on the compressive strength of high-strength concrete. Cement and Concrete Research 38: 386-395.
15. Hibbitt, Karlsson, Srensen (2007) ABAQUS/explicit user manual, USA.

16. Park SW, Xia Q, Zhou M (2001) Dynamic behavior of concrete at high strain-rates and pressures: II. Experimental characterization. Int J Impact engng 25: 887-910.

17. Li QM, Meng H (2003) About the dynamic strength enhancement of concrete-like materials in a split Hopkinson pressure bar test. Int J Solids Struct 40: 343-360.

18. Riisgaard B, Ngo N, Mendis P, Georgakis CT, Stang H (2007) Dynamic increase factors for high performance concrete in compression using Split Hopkinson Pressure Bar. International conference of Fracture Mechanics of Concrete Structures, Technical University of Denmark, Denmark.

# Investigation of Performance of Glass Fibre Impregnated with Bitumen Emulsion against Reflective Cracking

**Ogundipe Olumide Moses***
*Civil Engineering Department, Ekiti State University, Ado-Ekiti, Nigeria*

**Abstract**

The study investigates the use of glass fibre impregnated with bitumen emulsion with 6mm aggregates compacted on it as a Stress Absorbing Membrane Interlayer (SAMI) to retard reflective cracking. The causes and mechanism of reflective cracking in an overlay over a cracked pavement were highlighted. Tests were carried out in a wheel tracking facility to induce modes I and II cracking. The effects of temperature, load level, composition of SAMI and overlay thickness on the performance of the SAMIs were evaluated. It was found that the two SAMIs investigated in this study were both able to retard reflective cracking. Better relative performance was observed with low overlay thickness and at high load level. Also, the study showed that temperature influenced the performance of the SAMIs and indicated that difference in the viscosity of the bitumen emulsion used in the SAMIs led to difference in performance. It was observed that the interface shear strain of the overlay-SAMI interface influenced the performance of the SAMI.

**Keywords:** Cracks; Overlay; Interlayer; Temperature; Load level

## Introduction

The method commonly used to rehabilitate deteriorated pavements is by overlaying with a new surfacing material. Cleveland et al. [1] stated that for both flexible and composite pavements, a common technique used by many agencies for preventive maintenance and/or rehabilitation was simply to construct thin Hot Mix Asphalt (HMA) overlay, normally between 25 and 50 mm thick. Overlaying cracked pavements prevents water from infiltrating through the cracks into the pavement structure, thus preventing the deterioration of the pavement structure and increasing its structural capacity. Also, it reduces roughness, restores skid resistance, and improves the overall ride quality to the travelling public. However, it has been found that the cracks propagate through the overlay to its surface, a phenomenon called reflective cracking.

Reflective cracking can be defined simply as the propagation of existing cracks in an old pavement through the underside of the overlay to the surface. (Figure 1) shows a schematic of reflective cracking in HMA overlay on Portland Cement Concrete (PCC) slab or Asphalt Concrete (AC). Reflective cracking is often initiated at the bottom of the overlay material and grows until it appears at the surface. Also top-down cracking (cracks initiating at the surface and growing into the lower layers) occurs especially where there are large temperature variations in the pavement (thermal cracking).

A number of factors have been identified as the causes of reflective cracking. Palacios et al. [2] reported that cracks propagate to new overlays due to vertical movement of the underlying pavement layer which may be due to traffic loading, frost heave and consolidation of the sub grade soils and/or the horizontal movement of the pavement upper layers due to temperature changes. Quintus et al. [3] reported three causes of reflective cracking, the major being horizontal movements from the expansion and contraction of the base pavement that is caused by temperature changes; the differential vertical deflections between the approach and departure slabs or across transverse cracks, which create shear stresses; and the curling of PCC slabs during colder temperature when the HMA overlay is stiff and brittle.

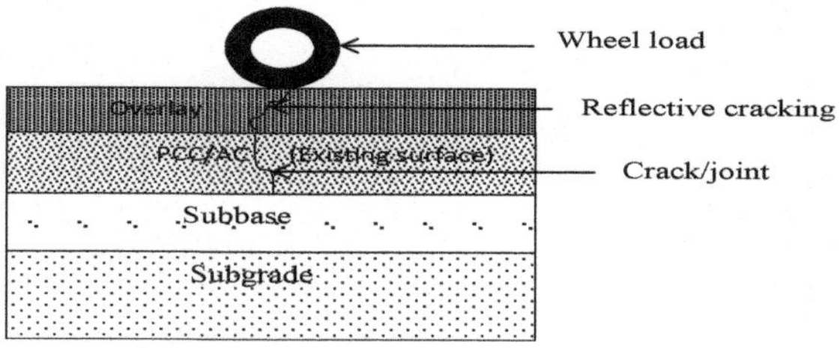


**Figure 1:** Schematic of reflective cracking in pavement.

Abe et al. [4] stated that cracking was caused on the surface of the overlay at an early stage by the movement of the pavement and traffic load. Smith [5] also mentioned the differential vertical movement at a crack or slab joint in the old pavement which induces a vertical shear stress in the overlay, horizontal movement associated with temperature or moisture changes in the old pavement which induces tensile stress in the overlay or live load flexural stress in the overlay, which tends to concentrate directly over discontinuities. Based on the causes identified by researchers, there seems to be an agreement on the causes of reflective cracking. While it is common knowledge that factors such as sub grade conditions, pavement material quality, workmanship etc affect the performance of a pavement; the principal factors responsible for reflective cracking are the action of traffic loading on an overlay on a cracked pavement and thermal stresses developed as a result of daily/seasonal temperature variation. This study has concentrated on reflective cracking caused by the action of traffic loading.

It is important that the mechanisms of reflective cracking are well understood to be able to provide a solution to the problem. This is illustrated in Figures 2 and 3. Figure 2a depicts a situation where the moving wheel on the edge of the pavement above the crack produces maximum shear stresses, A and C respectively as shown in Figure 3, while in Figure 2b, the wheel directly on the overlay above the cracked pavement generates flexural stresses producing maximum bending stress, B as indicated in Figure 3. This illustration indicates two of the three modes of reflective cracking - Modes I and II. Molenaar [6]

***Corresponding author:** Ogundipe Olumide Moses, Civil Engineering Department, Ekiti State University, Ado-Ekiti, Nigeria
E-mail: momide2002@yahoo.com

established the three modes: mode I, mode II and mode III cracking. Mode I cracking occur due to tensile stresses caused by a drop in temperature or flexure under traffic loading. Mode II cracking is caused by the effects of shear stress induced by a loaded wheel crossing from one side of a transverse crack or joint to the other. Mode III, referred to as the tearing mode, is less common. This occurs in pavements when the wheel load travels along (parallel to) a crack. Figure 4 shows the three modes of cracking.

A number of measures have been adopted by engineers with the aim of retarding reflective cracking in overlays and results achieved have been mixed. These measures include: thick overlays, overlay mixture modification, overlay reinforcement, stress or strain absorbing interlayer's and reinforcing interlayer's. Al-Qadi et al. [7] stated that various interlayer materials namely: geosynthetics, geocomposites, steel reinforcement netting, and modified HMA, have been used as interlayer systems for the purpose of reducing reflective cracking, and that their effectiveness is still not well quantified due to lack of understanding of the reflective cracking mechanism and because imperfect evaluation approaches are used. No current pavement rehabilitation techniques have been shown to solve completely the problem of reflective cracking. This has been attributed to the number of variables that are involved in reflective cracking [8].

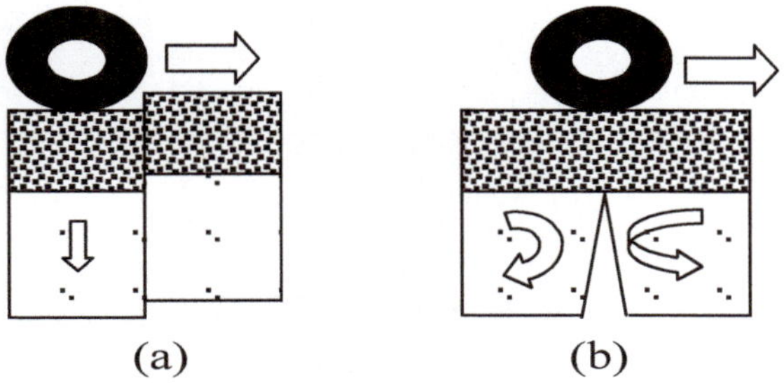

**Figure 2:** Movements in pavement joints and cracks.

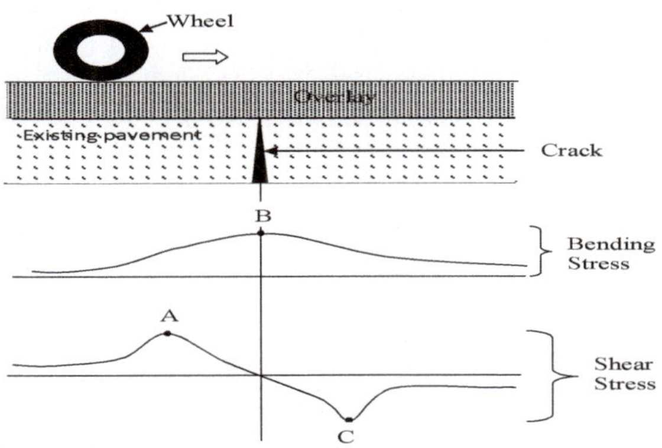

**Figure 3:** Stresses induced by wheel load (Adapted after Lytton, [7]).

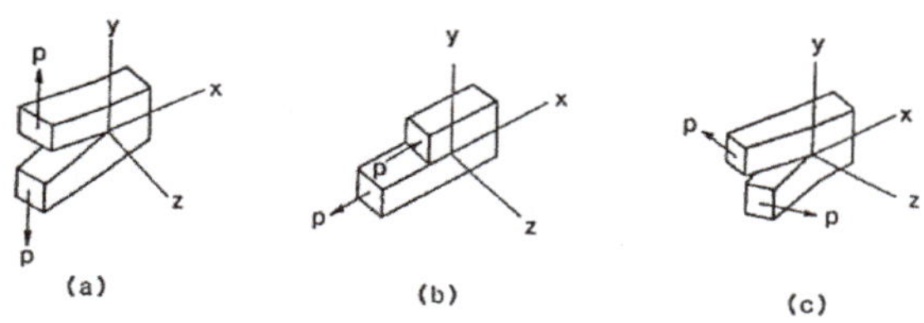

**Figure 4:** Modes of cracking (a) mode I (b) mode II and (c) mode III.

| Sample type | Percent by composition of aggregate |
|---|---|
| 10mm aggregate | 37% |
| 6 mm aggregate | 26% |
| Dust | 36% |
| Filler | 1% |
| **Binder type** | 10/20, 40/60 bitumen |
| **Binder content** | 5.3% by mass of total mix |
| **Target air void** | 5% |

**Table 1:** Mix composition for 10 mm asphalt concrete.

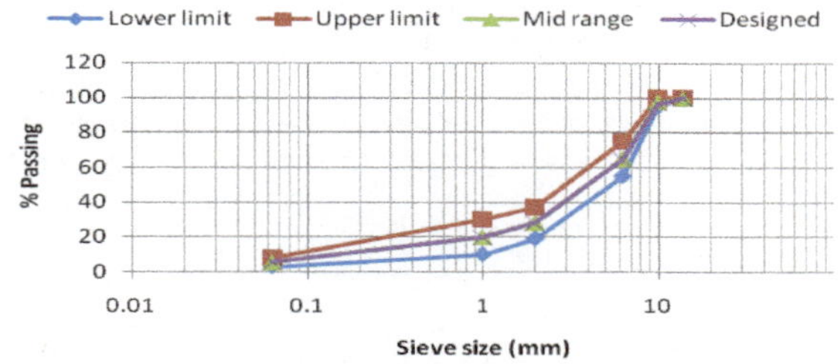

**Figure 5:** Particle size distribution curve for the blend of aggregates for 10 mm asphalt concrete.

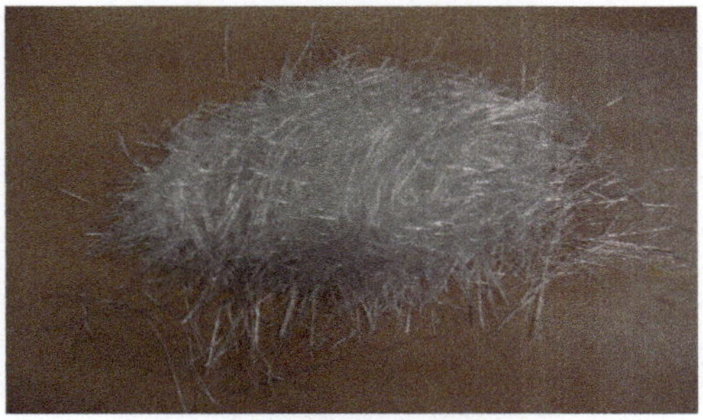
**Figure 6:** Chopped glass fibres.

This research work aims to evaluate the contribution of a particular type of stress absorbing membrane interlayer (SAMI) designed to deform horizontally or vertically, therefore allowing the movement (vertical/horizontal) of the underlying pavement layers without causing large tensile stresses in the asphalt overlay. The effect of the SAMI on overlaid pavement performance in delaying reflective cracking and the influence of certain variables such as temperature overlay thickness, and load levels have all been studied.

## Materials and Methods

### Materials

The materials used in this study were as follows:

- 10 mm asphaltic concrete with 10/20 or 40/60 penetration grade bitumen. The mix composition is as shown in Table 1 and the particle size distribution curve for the blend of aggregates is shown in Figure 5.
- Proprietary SAMIs designated as SAMI 1 and SAMI 2. SAMI 1 is prepared by sandwiching chopped glass fibre shown in Figure 6 at 120 g/m$^2$ between layers of ordinary bitumen emulsion at 0.9 L/m$^2$ with 6 mm aggregates spread and compacted on top at a rate of 8 kg/m$^2$, while SAMI 2 varies in its use of polymer modified bitumen emulsion.
- A 10 mm thick rubber mat of 6.45MPa stiffness was used to simulate the pavement foundation.

| Asphalt concrete (AC) | Stiffness (MPa) | | |
|---|---|---|---|
| | Temperature | | |
| | 10°C | 20°C | 30°C |
| AC (40/60) | 10000 | 3900 | 1100 |
| AC (10/20) | 15400 | 9600 | 5000 |

**Table 2:** Indirect tensile stiffness moduli.

| Bitumen emulsion | Viscosity (Pa.s) @ 25°C | Viscosity (Pa.s) @ 30°C | Viscosity (Pa.s) @ 40°C |
|---|---|---|---|
| Ordinary bitumen emulsion | 0.700 | 0.580 | 0.390 |
| polymer modified bitumen emulsion | 0.184 | 0.194 | 0.180 |

**Table 3:** Viscosity of bitumen emulsion.

**Figure 7:** A typical 500 mm by 500 mm slab.

- The indirect tensile stiffness moduli (ITSM) of the mixtures were determined using the Nottingham asphalt tester (NAT) in accordance with British standard [9]. In the ITSM a load pulse is applied to the vertical diameter of a cylindrical specimen (100mm in diameter and 40mm thick) positioned centrally between the upper and the lower platens and the resultant peak transient deformation along the horizontal diameter is measured. The ITSM results for the 10mm asphalt concrete with 10/20 and 40/60 penetration grade bitumen at 10°C, 20°C and 30°C are shown in Table 2. The viscosity of the bitumen emulsion is shown in Table 3.

## Methods

**Sample preparation:** The test specimens were made up of beams of length 404mm and width 50mm. The bottom (base) layer was a 30mm thick 10mm asphaltic concrete with 10/20 penetration grade bitumen, the middle layer (where present) was SAMI and the top layer (overlay) was a 10mm asphaltic concrete with 40/60 penetration grade bitumen.

The beams were obtained by manufacturing a slab of dimension 500 mm × 500 mm (Figure 7). For the base layer, the aggregates and binder were batched as shown in Table 1, heated at 185°C and compacted in a mould of dimension 500 mm × 500 mm × 205 mm with a roller compactor at a temperature of 180°C to a thickness of 30 mm. The SAMIs were prepared by sandwiching 60 mm glass fibre strands between layers of bitumen emulsion and 6 mm aggregates compacted over them. The aggregates for the top layer were batched as shown in Table 1, heated to 160°C and compacted to the required thickness at 150°C. All the mixtures were mixed in accordance with British standard [10] and compacted in accordance with British standard [11].

Ten beams of length 404 mm and width 50 mm were cut from each slab and a 10 mm wide notch was sawn at the centre of the beam through the 30 mm base layer to simulate the crack. Two aluminium brackets were glued to each side of the split base as LVDT measurement points. Typical beams with 10 mm notch in the base layer and aluminium brackets attached are shown in Figure 8.

**Test procedure:** The test was carried out by placing a 10mm thick rubber mat on the base of a steel mould on a reciprocating table. The beams were placed in a conditioning cabinet at test temperature for a minimum of five hours. The specimen sides were painted white to allow monitoring of crack growth during the test. The beams were placed in the steel mould on the wheel tracking facility and they were clamped at the top and sides to prevent them moving sideways. The Linear Variable Differential Transformers (LVDTs) were positioned on the aluminium brackets and connected to the data acquisition system. The specimens were trafficked and the numbers of wheel cycles for crack growth to the top of the overlay were recorded. The wheel tracker with a sample in place is shown in Figure 9. The test was carried out at 10°C, 20°C and 30°C. Two replicates were tested in each case. The test plan is shown in Table 4.

**Test results and discussion:** The results from the wheel tracking test were processed. Firstly, the number of wheel cycles for crack propagation to failure (crack appearing at the top of the overlay) was recorded for all specimens. The vertical displacements of the specimens during test were obtained from the LVDTs. The results presented are the averages of the replicates.

The specimen names and references used for the presentation of results are shown in Table 5. The results for the numbers of wheel cycles to failure at 10°C, 20°C and 30°C with a 2.4 kN (approximately 1.1 MPa) load applied are shown in Figures 10-12, respectively. Figure 13 shows the results at 30°C with a 1.35 kN (approximately 0.6 MPa) load applied. The figures show both SAMI 1 and SAMI 2 were able to increase the life of the overlay in most cases. Also, it can be seen from

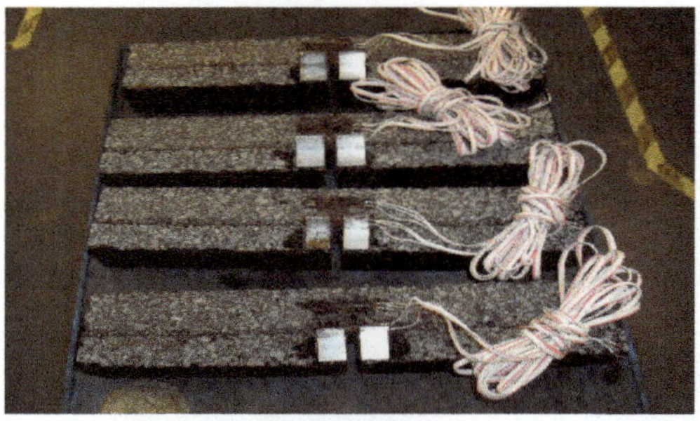

**Figure 8:** Typical beams with 10 mm notch and LVDT measurement points.

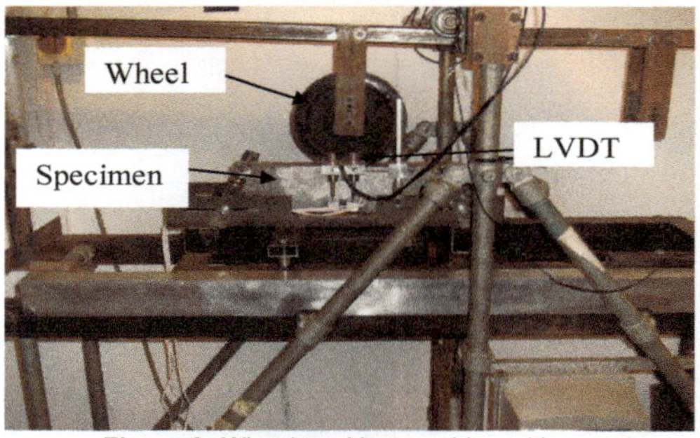


**Figure 9:** Wheel tracking machine with sample.

| Load (kN) | | 2.4 | | | 1.35 |
|---|---|---|---|---|---|
| Base thickness (mm) | Overlay thickness (mm) | Temperature (°C) | | | |
| | | 10 | 20 | 30 | 30 |
| 30 | 40 | √ | √ | √ | √ |
| 30 | 60 | √ | √ | √ | √ |

**Table 4:** Test plan.

| Specimen name | Specimen reference |
|---|---|
| 40 mm Overlay (Control) | O40 |
| SAMI 1 with 40 mm Overlay | SA1O40 |
| SAMI 2 with 40 mm Overlay | SA2O40 |
| 60 mm Overlay (Control) | O60 |
| SAMI 1 with 60 mm Overlay | SA1O60 |
| SAMI 2 with 60 mm Overlay | SA2O60 |

**Table 5:** Specimen names and references.

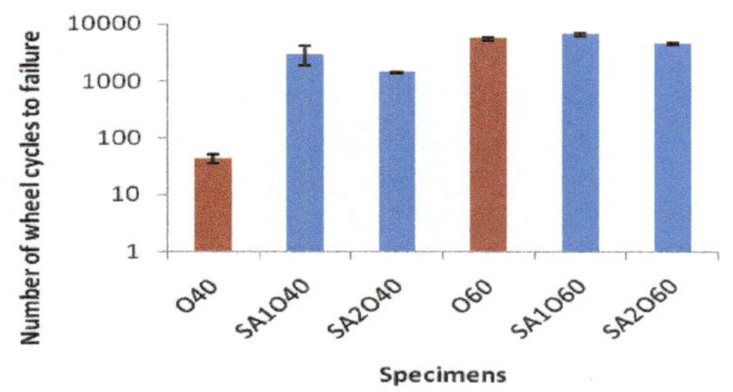


**Figure 10:** Number of wheel cycles to failure at 10°C with a 2.4kN load applied.

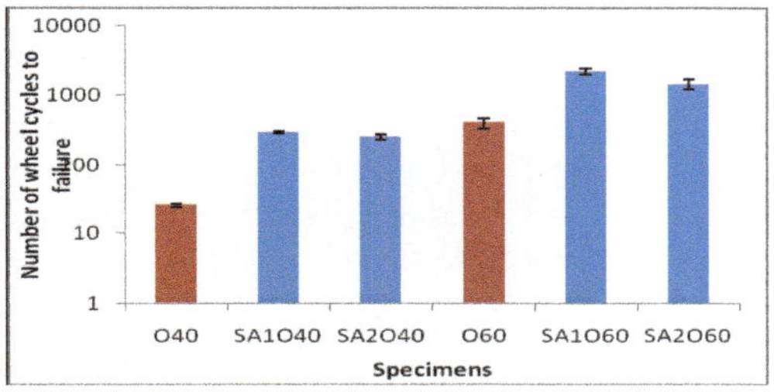


**Figure 11:** Number of wheel cycles to failure at 20°C with a 2.4kN load applied.

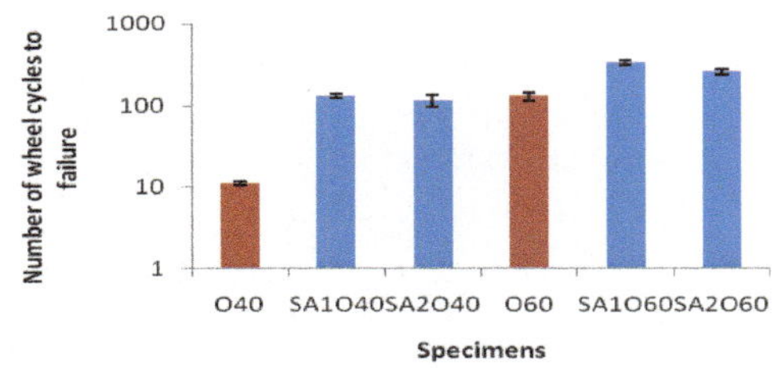


**Figure 12:** Number of wheel cycles to failure at 30°C with a 2.4kN load applied.

the figures that the specimens with SAMI 1 gave better performance than those with SAMI 2. This may be because the bitumen emulsion used in SAMI 2 was of insufficient viscosity, although the exact reason why this should give reduced performance is unclear.

**Effect of load levels:** Two load levels were used at 30°C: 2.4 kN (1.1 MPa) and 1.35 kN (0.6 MPa). The former is representative of tyre pressures on airfields and the latter representative of tyre pressures on highways. The numbers of wheel cycles to failure were normalized against the respective control specimens of equivalent thickness. The results as shown in (Figure 14) indicate that the contribution of the SAMIs in resisting reflective cracking was more pronounced in specimens tested under a load of 2.4 kN (1.1 MPa) than under 1.35 kN (0.6 MPa), in most cases.

The reason for the better relative performance under a load of 2.4 kN compared to 1.35 kN may be because of the permanent deformation associated with creep at high temperature. Further studies are to be carried out to evaluate the effect of load level at 10°C and 20°C.

**Effect of temperature:** The effect of temperature on the performance of SAMI 1 and SAMI 2 against reflective cracking was considered by carrying out tests at 10°C, 20°C and 30°C with a 2.4 kN (1.1 MPa) load applied. The results presented in Figure 15 show the number of wheel cycles to failure for the specimens with SAMIs as ratio of those without SAMIs (control). It can be seen from the figure that when an overlay of 40 mm thick was used, the life to failure as ratio of control decreases with increase in temperature, which may be related to change in stiffness of the overlay (Table 2) and SAMI with temperature and the reduction in the stiffness of the overlay-SAMI interface as temperature increases [12]. Also, the figure shows that when an overlay of 60mm was used, the best performance was recorded at 20°C, while the worst performance was observed at 10°C. The reason for this is probably because the overlay becomes brittle due to long exposure to cold temperature (10°C), while it becomes susceptible to permanent deformation at 30°C.

**Effect of overlay thickness:** The effect of overlay thickness on the performance of the SAMIs was considered by using two different overlay thicknesses of 40 mm and 60 mm for specimens with SAMI 1 and SAMI 2. The results were normalized by dividing the number of cycles to failure of the test specimens by the number of wheel cycles to failure for control specimens of equivalent overall thickness. The results are shown in Figures 16-19.

The figures show that specimens with SAMIs gave better performance against the offset of crack when a 40mm overlay was

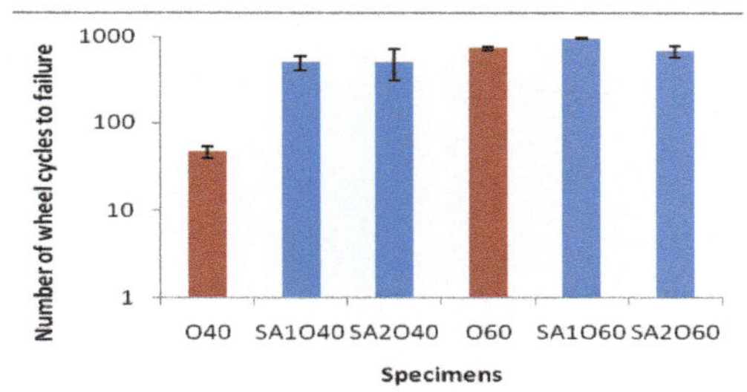


**Figure 13:** Number of wheel cycles to failure at 30°C with a 1.35kN load applied.

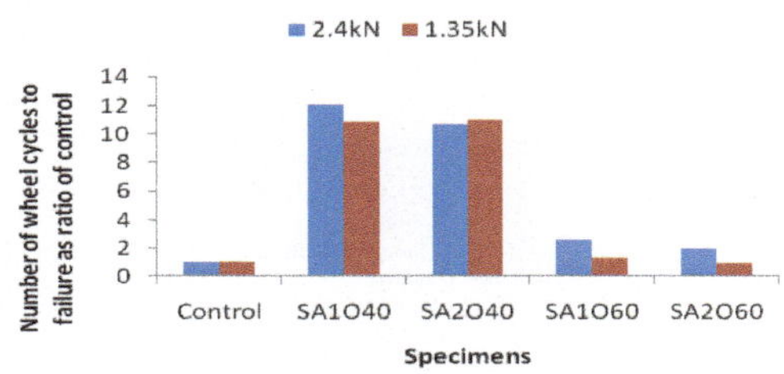


**Figure 14:** Number of wheel cycles to failure as a ratio of control with 2.4kN and 1.35kN loads applied at 30°C.

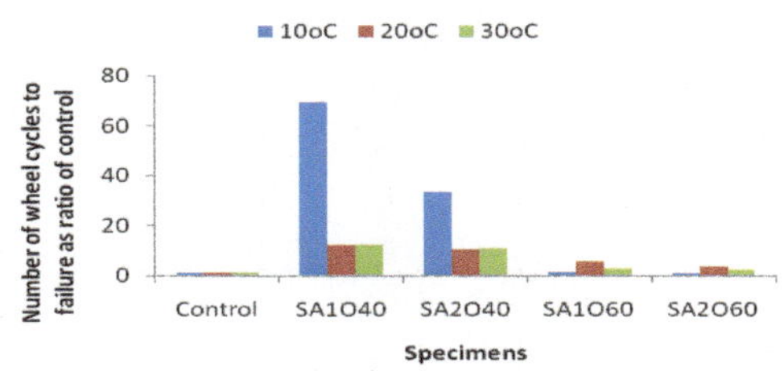


**Figure 15:** Number of wheel cycles to failure as a ratio of control with a 2.4kN load applied at 10°C, 20°C and 30°C.

used than a 60mm overlay. This indicates that the SAMIs provide better resistance to the offset of crack with a lower overlay thickness. Molenaar et al. [13] found that when SAMIs are used against reflective cracking, a thin overlay would have a longer life than a thick overlay. The reason for the better performance of thinner overlay over a SAMI on a cracked pavement is probably because the SAMI becomes ineffective once crack is initiated at the bottom of the overlay. While it may take greater number of wheel load cycles for the cracks to propagate to the surface when thicker overlay is used, the additional cost of the overlay may outweigh the life gained. Therefore, having thinner overlay over a SAMI provides a better relative benefit than thicker overlay.

**Displacement:** The initial absolute displacements of the cracked base layer of the specimens are shown in Figure 20. It can be seen that the vertical movement of the specimens increases with increasing temperature. Also the figure shows that lower absolute displacements were measured in the control specimens than in those with SAMIs. This

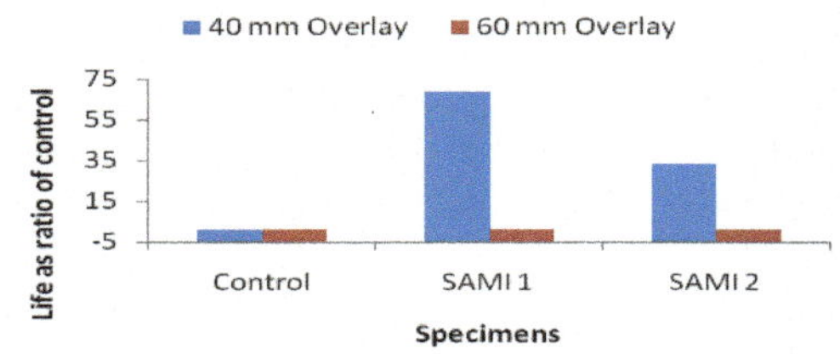


**Figure 16:** The number of wheel cycles to failure as a ratio of control for specimens with 40 mm and 60 mm overlays and a 2.4kN load applied at 10°C.

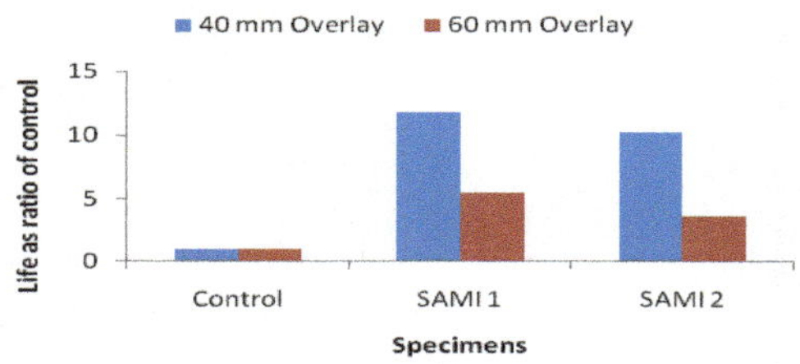


**Figure 17:** The number of wheel cycles to failure as ratio of control for specimens with 40 mm and 60 mm overlays and a 2.4kN load applied at 20°C.

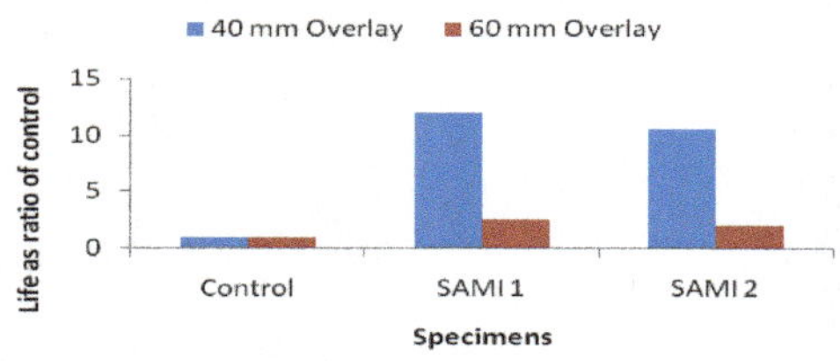


**Figure 18:** The number of wheel cycles to failure as ratio of control for specimens with 40 mm and 60 mm overlays and a 2.4kNload applied at 30°C.

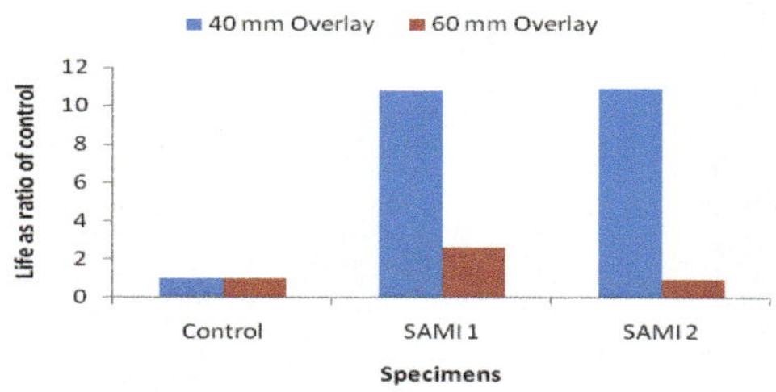


**Figure 19:** The number of wheel cycles to failure as ratio of control for specimens with 40 mm and 60 mm overlays and a 1.35kN load applied at 30°C.

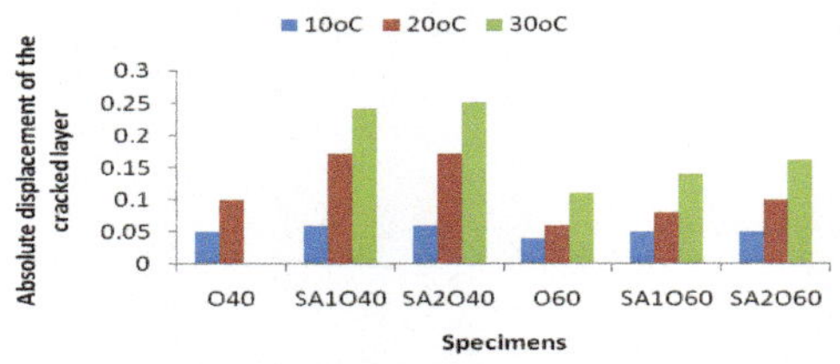


**Figure 20:** Initial absolute displacement of the specimens under a 2.4kN load at 10°C, 20°C and 30°C.

shows that the increased deflection of the specimen that accompanies the use of the SAMIs does not have a detrimental effect on their performance. The shear strain at the overlay-SAMI interface isolates the overlay from the strain concentration immediately above the base.

## Conclusion

This study has shown that the SAMIs evaluated (comprising chopped glass fibre, bitumen emulsion and single-sized aggregate) are able to delay the appearance of cracks in an overlay over a cracked pavement. The degree of improvement varied up to a factor of 69 for thin 40 mm overlay at 10°C. It was discovered that SAMI 1 gave better performance against reflective cracking than SAMI 2, which may be due to the viscosity of the bitumen emulsion used in the SAMIs.

In general, it was found that the SAMIs gave better relative performance when a higher load level was adopted for the trafficking of the specimens. It was also found that the relative benefit of the SAMIs decreased as the overlay thickness increased. This is possibly because the SAMIs become less effective once a crack has been initiated in the overlay. Furthermore, the test results showed that temperature influences the performance of the SAMIs, but in a complex way. The shear strain at the overlay-SAMI interface plays a role in isolating the overlay from strain concentration around the crack region, therefore increasing the life to failure of the overlay. Large scale testing is to be carried out to evaluate the practical/field significance of the study. Also, the use of different types of binder with the glass fibre and modelling of the test are to be considered in future study.

### References

1. Cleveland GS, Button JW, Lytton RL (2002) Geosynthetic in flexible and rigid pavement overlay systems to reduce reflection cracking. Texas Transport Institute, Texas A&M University System. Report 1777-1.
2. Palacios C, Chehab GR, Chaignon F, Thompson M (2008) Evaluation of fiber reinforced bituminous interlayer's for pavement preservation. Proc. 6th International RILEM Conference.
3. Von Quintus HL, Mallela J, Weiss W, Shen S (2009) Techniques for mitigation of reflective cracking. Applied Research Associates, Champaign, IL, Interim Report AAPTP 05-04.
4. Abe N, Maehara H, Maruyama T, Ooba K (2000) An examination of factor which affects reflective cracking, Proc. 4th International RILEM Conf.
5. Smith (1983) Laboratory testing of fabric interlayer's for asphalt concrete paving, Transportation Research Record 916, 6-17.
6. Moleenar AAA (1993) Evaluation of pavement structure with emphasis on reflective cracking, Proc. 2nd International RILEM Conf.
7. Al-Qadi IL, Baek J, Buttlar WG, (2008) Development of reflective cracking index to determine the effectiveness and service life of strip interlayer systems. Proc. 6th International RILEM Conf.
8. Bhosale SS and Mandal JN, (2008) Open graded asphalt concrete for mitigation of reflection cracking on asphalt concrete overlays, Proc. 12th International Conference of International Association for Computer Methods and Advances in Geomechanics (IACMAG).
9. BSI (1993) Method for determination of the indirect tensile stiffness modulus of bituminous mixture. British Standard Institution, London, DD 213.

10. BSI (2004) Bituminous mixtures-test method for hot mix asphalt: Laboratory mixing. British Standard Institution, London, BS EN 12697-35.

11. BSI (2003) Bituminous mixtures-test method for hot mix asphalt: Specimen prepared by roller compactor, British Standard Institution, London, BS EN 12697-33.

12. Ogundipe OM (2011) Mechanical Behaviour of Stress Absorbing Membrane Interlayers. University of Nottingham, Nottingham.

13. Molenaar AAA, Heerkens JCP, Verhoeven JMH, (1986). Effects of stress absorbing membrane interlayers, Association of Asphalt Paving Technologists.

# Utilization of the Alkaline White Mud as Cement-Based Materials for the Production of Cement

**Shaoqi Zhou[1,2,3]*, Yiming Lin[1,3], Jingjing Zhao[1,3], Songfeng Zeng[4] and Jianhua Zhou[4]**

[1]*College of Environmental Science and Engineering, South China University of Technology, Guangzhou 510640, China*
[2]*State Key Laboratory of Subtropical Building Science, South China University of Technology, Guangzhou 510641, China*
[3]*Key Laboratory of Environmental Protection and Eco-remediation of Guangdong Regular Higher Education Institutions, South China University of Technology, Guangzhou Higher Education Mega Center, Guangzhou 510006, PR China*
[4]*Guangdong South Solda Industry, Co Ltd, Nangang West 488, HuanPu District, Guangzhou 510760, PR China*

**Abstract**

The effects of using alkaline white mud as substitutes for cement property in the process of clinker burning were investigated in this paper. Samples were prepared with adding 0.5-5.0% of alkaline white mud as substitutes for unit raw meal and then these mixtures were burned at 1450°C for 2 hrs. The results showed that the compressive strength and flexibility strength of the pastes were a little better than the plain sample. The concentrations of MgO, $SO_3$ and chloride ions in clinkers were far below the limits in the standards when the addition of alkaline white mud was below the content of 5.0%. The Results also showed that the concentrations of free lime were in a growing tendency with the increasing of alkaline white mud.

Moreover, the addition of AWM in raw meal increased the content of $C_3S$ phase in the clinker, something leading the increase of the setting times and enhances the strengths of the cement pastes in the early curing ages.

**Keywords:** Alkaline white mud (AWM); Clinker; Compressive strength; Particle-size; Hydration

## Introduction

Alkaline white mud (AWM) is an inevitable waste material that is formed during the Solvay process. According to the published statistics [1], in 2006, 1.5 × $10^7$ tons soda ash was produced and there was a growthing demand by (1.4-1.5) × $10^6$ tons per year. The increasing requirement for the soda ash coupled had caused an over-proportional increase of huge quantities of waste products. In recent years, despite part of them being used in agriculture as organic fertilizer and soil management [1], the most common method to dispose AWM are sanitary landfills and in ocean [2,3]. However, after more stringent regulatory limits and with an increasing problem of changes in natural ecosystems, the ocean deposition has already been banned [1]. Moreover, the high cost of disposal and the increasingly difficult to find suitable landfill sites. In fact there is a need to find an alternative solution to dispose the AWM.

In the Solvay process, the alkaline white mud was produced from the limestone, which is used in the form of milk lime for the recovery of ammonia [3]. Interestingly, the limestone is also one of the main raw materials in the producing clinkers of ordinary Portland cement (OPC), so it is possible to concern that it can be used as substitutes for conventional cement and aggregates in concrete [4-6]. Moreover, the waste heat of kiln can be recycled to dry the AWM and their increasing emissions also can be controlled by the flue gas treatment system that has already been equipped in the cement plant. Furthermore, the method for using cement plant to dispose the AWM, which is expected greatly for protecting of environment by virtue of reducing the natural resources for cement and relieving the load to disposal sites.

In general, the reuse of a waste material in concrete production depends on several characteristics in it, several characteristics, for example chemical composition. Frequently, The main components in AWM include $CaCO_3$, $CaSO_4$, $Ca(OH)_2$, $CaCl_2$ and $Mg(OH)_2$ and so on [7,1]. However, the mineral formation of clinkers with magnesium, sulfur and chlorine are harmful materials for the clinker. Consequently, the minerals formed by sulfate, chlorine and magnesium in the concrete producing processes should be taken into account and their contents must be detected.

In this study, our aim is to concern about the characteristic of cement clinkers by comparing the normal cement clinker, which were added AWM as substitutes for raw meals during the clinker production process. The flexural and compressive strengths and setting times of cement pastes, scanning electron microscope (SEM) micrographs, and the hydration characters of curing ages at 3, 7 and 28 days of the clinkers were also investigated.

## Materials and Methods

### Materials

**Alkaline white mud:** Alkaline white mud used in the current study, is collected from a soda ash manufacturing plant located in the suburban area of Guangzhou city (South China). The plant has a capacity of producing 350,000 tons soda ash per year. The AWM was collected after it was concentrated by plate and frame filter press in the plant and then it was dried at 105°C for 48 hrs. Finally, the dried AWM was further pulverized in a ball mill (porcelain mill, XMCO-ф 280×290) until the particles could pass through a # 200 mesh (with 75 μm pore diameter) metallic sieve and its chemical composition was characterized.

**Raw materials:** The limestone, iron, shale, fly ash, sand powders were collected from a cement plant in Guangzhou (South China). The plant has a potential production of 9000 tons' cement clinker per day. All the raw materials were over dried at 105°C for 24 hrs and the chemical compositions were detected by using wavelength dispersion

***Corresponding author:** Shaoqi Zhou, College of Environmental Science and Engineering, South China University of Technology, Guangzhou 510640, China
E-mail: fesqzhou@scut.edu.cn

X-ray fluorescence spectrometer (XRF). The raw materials were further pulverized in a ball-mill until the particles could pass through a # 200 mesh (with 75 μm pore diameter) metallic sieve. And then all of them were well blended by according to the cement clinker ratios, which were calculated with the consideration of the Lime Saturation Coefficient:

KH= $(CaO-1.65Al_2O_3-0.34Fe_2O_3/2.80SiO_2)$, $0.902<KH<0.915$;

the Silica Modulus:

SM = $SiO_2/(Al_2O_3+Fe_2O_3)$, $2.25<SM<2.35$;

the Iron Modulus:

IM = $(Al_2O_3/Fe_2O_3)$, $1.30<IM<1.40$;

and the Lime Saturation:

L.S.F = $(CaO/2.8SiO_2+1.18Al_2O_3+0.65Fe_2O_3)$, $0.94<L.S.F<0.96$ [8-10]. The oxide compositions of all the raw materials, including the AWM, and the mixed raw meal were shown in Table 1 and the particle size distribution analysis of raw meal was showed in Figure 1(a).

**Cement clinkers:** The cement raw meals were prepared by well mixing with various ratios of AWM/raw meal (by weight), which were 0:100, 1:99, 2.5: 97.5 and 5:95, respectively. All the grounded raw mixtures were burned in a programmable electrical heated furnace to form cement clinkers. The high temperature furnace temperature was raised by 6°C $min^{-1}$ from the room temperature and maintained at 1,450°C for 2 hrs, and then the resulting products were cooled rapidly in air. The next step is that, all the cement clinkers were further pulverized and grounded in a ball-mill until, most of them could pass through a # 200 mesh (with 75 μm pore diameter) metallic sieve (retained on mesh sieve was less than 5% in weight). After this step each sample of cement clinker was co-grounded with 5% and 15% per weight of gypsum and fly ash to produce its corresponding cement.

## Approach

**Molding and curing of cement mortar specimens:** The mortar mixtures of the above four cements were prepared according to GB/T 17671-2007 (Standard test method for strength of hydraulic cement mortar of China). The ratio of sand/cement (by weight) was 3:1. The water to cement ratio was fixed at 0.50. Mortars were put in a (40×40×160) mm molds for obtaining specimens and were stored in a moisture room (20 ± 1)°C for 24 hrs. And then all the specimens were demoulded and cured by immersion in a (20 ± 1)°C water bath for 3-28 days. Finally, the flexural strength and compressive strength development of specimens were measured after curing ages of 3, 7 and 28 days.

## Analyses

Chemical and physical analyses of the cement clinkers, ordinary Portland cement and cement pastes were conducted as follows:

- Particle distribution test: Particle distribution test was carried out by a laser diffraction particle size analyzer (SALD-201, Japan) and the measure range is from 0.25 to 350 μm.
- Flexural strength and compressive strength test: GB/T 17671-2007.
- Setting time: The setting times of the cement mixes were measured according to GB/T1346-2001 (Test methods for water requirement of normal consistency, setting time and soundness of the Portland cements of China). This standard was similar to the ASTM C 191. The initial setting time was defined as when a Vicat needle (1.13 ± 0.05) mm would penetrate the sample to a point (4 ± 1) mm from the bottom of the mould and the final setting time was defined as when the Vicat needle would penetrate the sample to a point 0.5 mm [9].
- Mineralogy: The x-ray diffraction (XRD) analysis was conducted out by using a Bruker D8 Advance X-ray diffractometer with Cu Ka radiation and 2θ scanning ranging from 10° to 70°. The

| Chemical composition | Limestone | Shale | Ferrate | Fly ash | Soda residue | Raw meal |
|---|---|---|---|---|---|---|
| LOI[a] (%) | 38.42 | 3.53 | 8.62 | 4.06 | 36.47 | 34.71 |
| $SiO_2$[b] (%) | 7.19 | 75.43 | 37.05 | 45.30 | 2.78 | 8.35 |
| $Al_2O_3$[b] (%) | 1.68 | 12.04 | 16.13 | 36.99 | 2.02 | 2.69 |
| $Fe_2O_3$[b] (%) | 0.98 | 4.44 | 34.09 | 5.98 | 0.57 | 2.25 |
| CaO[b] (%) | 49.28 | 0.18 | 0.65 | 4.65 | 39.16 | 42.66 |
| MgO[b] (%) | 0.56 | 0.51 | 0.44 | 0.59 | 7.53 | 0.26 |
| $K_2O$[b] (%) | 0.12 | 2.23 | 0.70 | 0.56 | 0.05 | 0.32 |
| $Na_2O$[b] (%) | 0.08 | 0.32 | 0.09 | 0.26 | 1.15 | 0.11 |
| $SO_3$[b] (%) | 0.45 | 0.02 | 0.55 | 0.24 | 4.22 | 0.38 |
| Cl[-b] (%) | 0.003 | 0.006 | 0.005 | 0.012 | 4.420 | 0.010 |
| Total | 98.76 | 98.70 | 98.32 | 98.64 | 98.37 | 97.98 |

[a]Analyzed by muffle furnace at temperature (950±25)°C
[b]Analyzed by XRF

**Table 1:** Chemical compositions of the raw materials and raw meal.

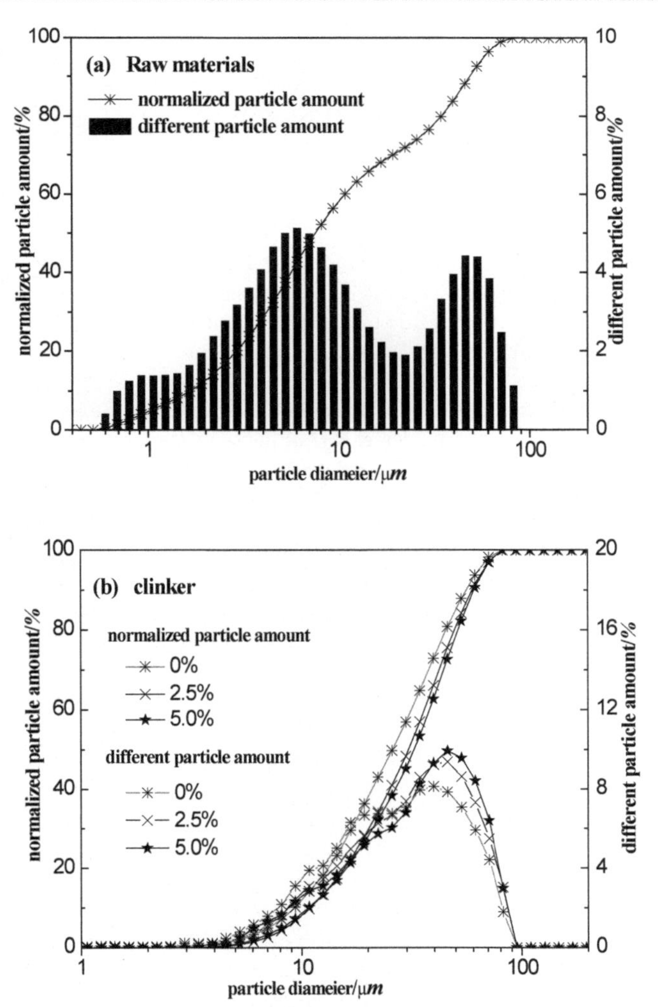


**Figure 1:** Particle size distribution of the raw meal (a) and cement clinkers (b).

XRD scans were ran in 0.02° steps, with a counting time of 17.7s.

- SEM: Several cement clinkers were investigated with an SEM (S-3700N, Hitachi Ltd of Japan).
- Chemical composition: The chemical compositions of the raw materials and raw meal were confirmed by using wavelength dispersion X-ray fluorescence spectrometer (XRF PANalytical Axios of Netherlands).
- Detection of chloride ions: A 2.50 ± 0.002 g sample was placed in 250 ml of beaker, and 25 ml of distilled water and 25 ml acid of $HNO_3$ solution ($HNO_3$:$H_2O$=1:2, by volume) was added to each beaker (included check sample). The samples were then treated by ultrasonic waves (300 W, 80 $KH_Z$) under the water temperature at 50°C with an ultrasonic reflux rotator (KQ-300VDE, Kun Shan of China). The slurries were filtered by using 0.45 μm Certified Glass Fiber Filter. The leachates were decomposed by acid ($HNO_3$:$H_2O$=1:100, by volume) and the specimen was analyzed by automatic potentiometric titration (CCQTC2006-4 chlorine ion detector, China).
- FTIR analysis: The hydration products and its changes with different AWM addition in raw clay of clinkers were analyzed with a Fourier Transform Infrared spectrometer (Nicolet 6700, USA). The samples were dehydrated by ethanol after curing ages at 3, 7 and 28 days, respectively, then it was dried at 80°C for 24 hrs and finally was measured [11].

## Results and Discussion

### Particle size distributions of raw meal and cement clinkers

The size distributions of raw meal and three cement clinkers (the AWM contents were 0%, 2.5% and 5.0%, respectively) were presented in Figure 1. As shown in Figure 1(a), the diameter of particles with percentages by volume less than 10%, 50% and 90% were 1.66, 7.57 and 49.3 μm, respectively. The maximum size of sample was 82.19 μm and the particle size of the raw meal mainly distributed in 0.34-71.11 μm, which attained to the requirements of production of burning cement clinker (retained on 80 mesh sieves was less than 10%). On the other hand, Figure 1(b) showed that the size distributions of three clinkers, the median diameter ($D_{50}$) of the three cement clinkers were 25.92, 30.72 and 32.54 μm, respectively. In addition, the diameters in 0-30 μm were 57.15, 48.66 and 45.47%; 10-40 μm were 61.24, 58.36 and 54.44%; 30-60 μm were 35.73, 41.48 and 43.63%, more than 60 μm were 7.12, 9.86 and 10.90%, respectively. Moreover, the diameter less than 80 μm ($D_{80}$) of the three cement clinkers were 45.42, 49.75 and 51.55%. Therefore, we drew a conclusion that the specific surface area of the clinkers were approximately from 350 to 600 m²/kg and the size distributions of raw meal and clinkers met the industrial requirements of production of cement.

### Cement clinkers composition

The compositions of the three clinkers, which were substituted the raw meal of 0, 2.5 and 5.0% by AWM, and the OPC clinker used in this study was analyzed by XRF. The theoretical phase compositions were predicted using a modified Bogue calculation [$C_3S$=3.80 $SiO_2$ (3KH-2); $C_2S$=8.61×$SiO_2$ (1-KH); $C_3A$=2.65$Al_2O_3$-1.69 $Fe_2O_3$; $C_4AF$=3.04 $Fe_2O_3$; KH= (CaO-1.65$Al_2O_3$-0.34$Fe_2O_3$/2.80$SiO_2$)] [1]. The results were shown in Table 2. It revealed that the main composition of $SiO_2$, CaO, $Al_2O_3$, and $Fe_3O_4$ were similar to those in the clinker of OPC. However, the amount of $SiO_2$ and $Fe_3O_4$ decreased with the AWM addition, while the compositions of CaO, MgO, $Al_2O_3$ and $SO_3$ increased. The contents of the MgO and $SO_3$ could had an influence on the property of cement, but Table 2 showed that the contents of those ranged from 1.05 to 1.18%, as well as 1.06 to 1.81%, which were much less than the limitation value (MgO≤5.0%; $SO_3$≤3.5%,) of the Standards of GB175-2007 (Common Portland Cement, China).

Table 2 also showed that the phase composition of $C_3S$ in the clinkers when the raw meal was substituted with 5.0% of AWM were somewhat higher than those in the clinker of OPC, while the $C_2S$ content was also lower than other clinkers. It indicated that this clinker could have a fast rate of hydration than other clinkers. However, it also could be seen that lime saturation coefficient (KH) of was higher than others, so it might had a higher free lime content that others in the clinkers.

### f-cao and chloride ions contents in the clinkers

Figure 2 presented the contents of the free lime and chloride ions in

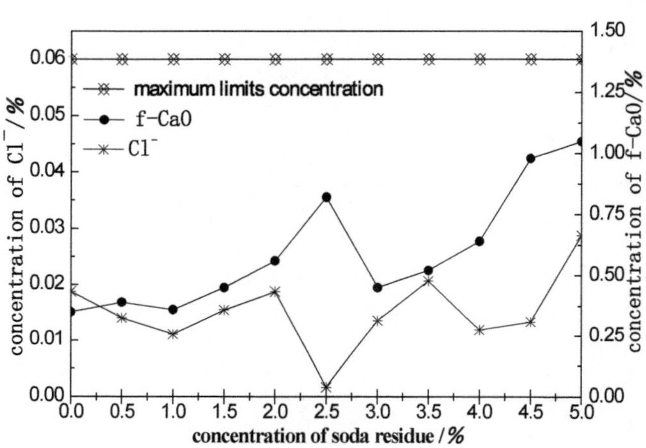


**Figure 2:** Changes of f-CaO and chloride ions contents in the clinker.

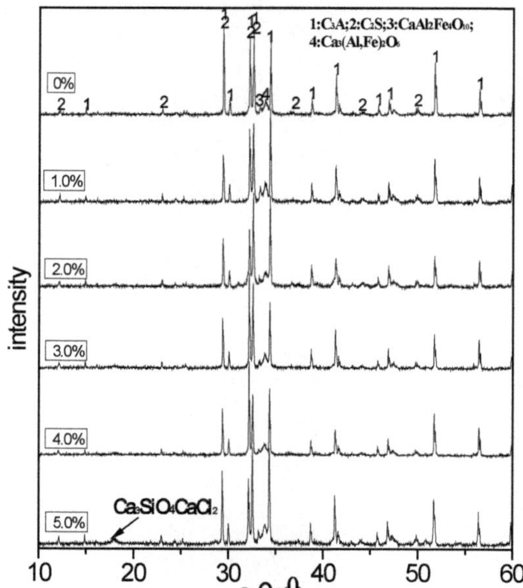


**Figure 3:** XRD patterns of clinkers.

| Composition | OPC | Sample 1 | Sample 2 | Sample 3 |
|---|---|---|---|---|
| $SiO_2$ (%) | 21.16 | 20.98 | 20.59 | 20.35 |
| $AL_2O_3$ (%) | 5.31 | 5.52 | 5.64 | 5.85 |
| $Fe_3O_4$ (%) | 3.85 | 3.93 | 3.84 | 3.77 |
| CaO (%) | 65.58 | 64.80 | 65.15 | 66.01 |
| MgO (%) | 1.09 | 1.05 | 1.15 | 1.18 |
| $K_2O$ (%) | 0.63 | 0.76 | 0.61 | 0.82 |
| $Na_2O$ (%) | 0.13 | 0.12 | 0.12 | 0.10 |
| $SO_3$ (%) | 0.097 | 1.06 | 1.70 | 1.81 |
| Constituents | | | | |
| $C_3S$ (%) | 58.71 | 57.17 | 56.63 | 59.99 |
| $C_2S$ (%) | 16.39 | 17.04 | 16.31 | 13.10 |
| $C_3A$ (%) | 7.54 | 7.99 | 8.45 | 9.14 |
| $C_4AF$ (%) | 11.73 | 11.94 | 11.67 | 11.47 |
| Moduli | | | | |
| LSF (%) | 0.953 | 0.943 | 0.956 | 0.974 |
| KH (%) | 0.910 | 0.906 | 0.908 | 0.925 |
| SIM (%) | 2.309 | 2.221 | 2.172 | 2.114 |
| ALM (%) | 1.376 | 1.405 | 1.469 | 1.551 |

**Table 2:** Chemical compositions of the four sample clinkers.

the clinkers with addition of AWM. It showed that all the contents of chloride ions in the clinkers were lower than the values (≤0.06%) of the maximum limits concentration of GB175-2007 (Standard of common portland cement of China). Thus, the results indicated that the chloride ions were vaporized during the melting process. In addition, Figure 2 also showed that the content of the free lime increased with the addition of AWM in the raw meal. From the previous discussion (3.2), the results could be attributed to the decreased contents of $SiO_2$ and the increased contents of CaO, because those changes could increase the melting temperature during the composing of clinkers. However, the results suggested that all the contents of free lime (Standard of China: less than 2.0%) and chloride ions (Standard of Japan: 0.35%, Standard of China: 0.06% and Standard of UK: 0.1) much lower than the maximum values of limits of the cement plants [12,13], so all the clinkers met the industrial requirements of production of cement.

## Characterization of the cement clinkers

Figure 3 showed the X-ray diffraction (XRD) pattern results for the plain clinker and the six cement clinkers. The results indicated that the major components in the cement clinkers were $C_3S$ [$Ca_3SiO_5$ and $Ca_3(SiO_4)O$], $C_2S$ [$Ca_2SiO_4$ and $Ca_2(SiO_4)$], $C_3A$ [$Ca_3Al_2O_6$ and $Ca_3(Al, Fe)_2O_6$], $C_4AF$ ($Ca_4Fe_2Al_2O_{10}$) and all were similar to those in ordinary Portland cement. Moreover, the phase formation of $Ca_{54}MgAl_2Si_{16}O_{90}$ was identified in all the clinkers. However, the $Ca_3SiO_4CaCl_2$ phase formation was detected in the clinkers, especially when the raw meal was replaced with 5.0% of AWM. Therefore, we confirmed that the addition of AWM as a part of raw meal could increase the content of chloride ions in the clinkers, whereas most of them were vaporized during the melting process.

## Microstructure of cement clinkers by SEM

To verify the X-ray diffraction (XRD) pattern results for the clinkers, SEM examination was performed. Figure 4 showed the microstructure of clinkers with AWM contents of 0% (a), 2.5% (b) and 5.0% (c). It was found that different amount of AWM replacement somewhat led to discrepancies in the microstructures of the clinkers. However, all the microstructures of the mixture were basically the same, so further study is needed in the next section.

## Setting time and strength development of OPC paste and cement pastes

The setting times for the three pastes and the OPC paste were given

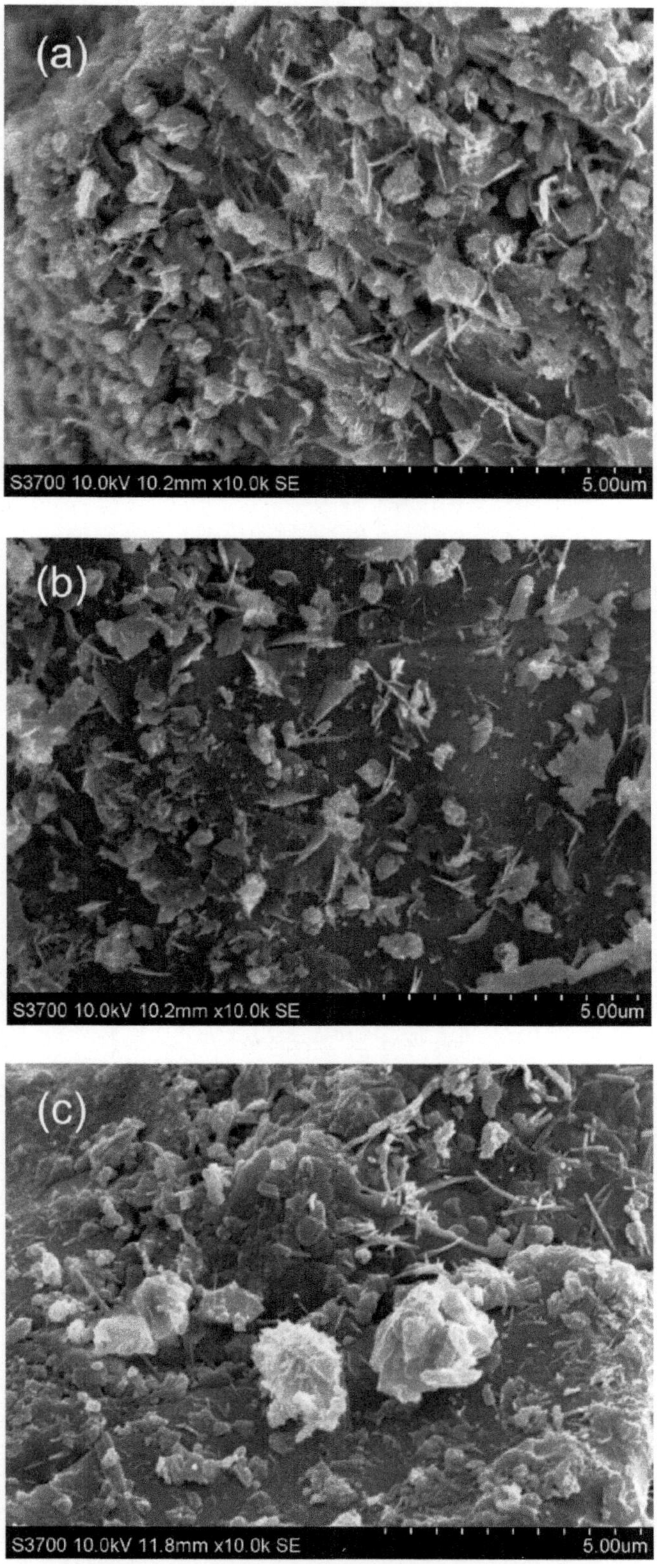


**Figure 4:** SEM micrographs of the cement clinkers with the replacement of different AWM in raw meals: (a) 0%, (b) 2.5%, (c) 5.0%.

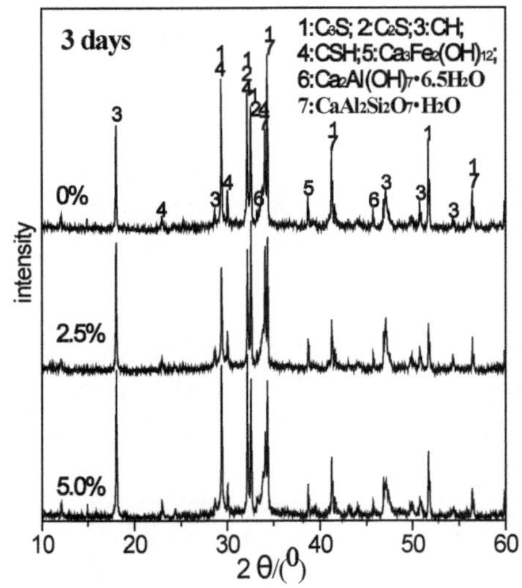


**Figure 5:** X-raw diffraction patterns for hydrated cement clinker pastes for 3 days.

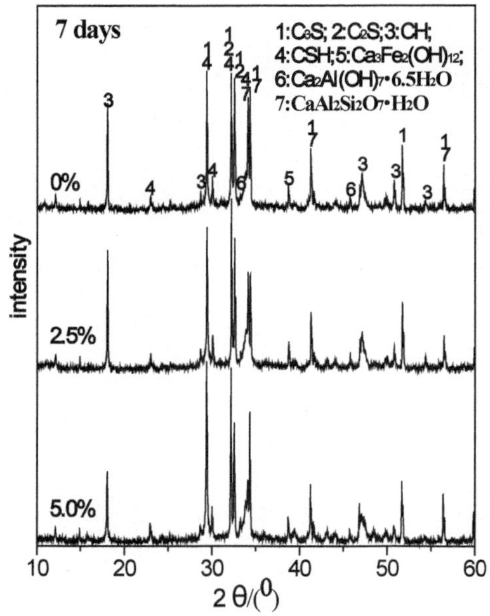


**Figure 6:** X-raw diffraction patterns for hydrated cement clinker pastes for 7 days.

in Table 3. The results showed that the initial setting time and final setting time of the OPC phase were 140 min and 196 min, respectively, while the other three cement mortars with AWM replacements at 0, 2.5 and 5.0% in raw meal were 145-202, 139-195 and 140-193 min, respectively. It indicated that the initial and final setting times of cement mortars, which had AWM replacements in raw meal, were slightly short than that both of the plain paste and the OPC paste. The reason for the shorter setting times might be attributed primarily to the change contents of $C_3S$ in the clinkers [9], because the $C_3S$ phase in OPC clinker and those clinkers with AWM replacements were higher than the plain phase.

Flexural and compressive strengths of the pastes and OPC paste after curing ages at 3-28 days also were shown in Table 2. It could be seen that the flexural and compressive strengths of the three pastes and OPC paste developed when the curing ages extended from 3 to 28 days. In comparison with the OPC and the plain paste, the results demonstrated that the flexural and compressive strengths of cement pastes were similar with the same conditions, although there had somewhat changes of the strengths. Those results might be responsible for the changes of the setting times in the process of hydration and the changes of compositions in the clinkers.

## Cement clinker paste hydrates

To confirm the results discussed above, tests of the hydration process for the clinker pastes were carried out. The hydration products of the three clinker pastes, at the various curing times (3-28 days), could be seen through the XRD patterns in figure 5, 6 and 7, respectively. It revealed that the higher peaks for $Ca(OH)_2$, calcium silicate hydrate (CSH) and calcium aluminum hydrates (CAH) could be identified in all the curing times, except for the peaks of $C_3S$ and $C_2S$. Moreover, it could be observed from these figures that the degrees of hydration for the three clinker pastes were increased as the amount of AWM replacement increased in raw meal, when curing age extended from 1 to 28 days. In addition, higher peaks of $Ca(OH)_2$ in figure 5 suggested a faster hydration process for the clinker pastes than the plain clinker paste. It was reported [14] that CSH was formed by consuming $Ca(OH)_2$, figure 6 and 7 showed that the peaks of CSH for sample clinker pastes were higher than the plain clinker paste, while the peaks

| Soda residue content (%) | Setting times (hrs : min) | | Flexural strengths (MPa) | | | Compressive strengths (MPa) | | |
|---|---|---|---|---|---|---|---|---|
| | Start | Final | 3 d | 7 d | 28 d | 3 d | 7 d | 28 d |
| OPC | 2:20 | 3:16 | 6.10 | 7.55 | 9.65 | 30.25 | 41.35 | 61.26 |
| 0 | 2:25 | 3:22 | 6.02 | 7.56 | 9.55 | 31.22 | 43.25 | 62.15 |
| 2.5 | 2:19 | 3:15 | 6.10 | 7.45 | 9.68 | 33.25 | 43.75 | 61.55 |
| 5.0 | 2:20 | 3:13 | 6.15 | 7.55 | 9.60 | 32.45 | 43.80 | 62.05 |

**Table 3:** Setting times, Flexural strengths and Compressive strengths of plain and cement mortars.

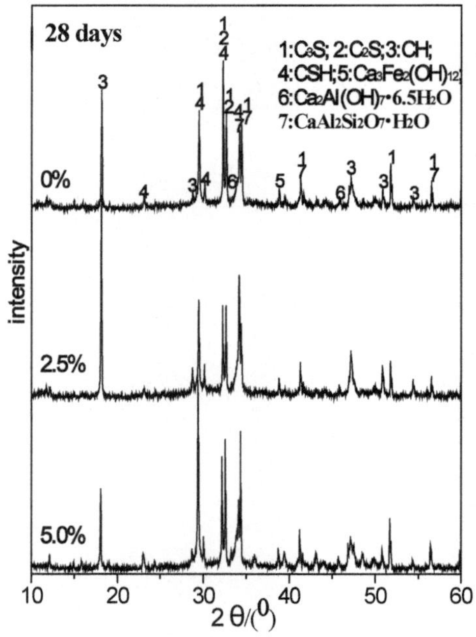


**Figure 7:** X-raw diffraction patterns for hydrated cement clinker pastes for 28 days.

of $Ca(OH)_2$ decreased. This was primary duo to the increased contents of $C_3S$ in the clinkers, so the amount of $Ca(OH)_2$ was formed at early curing ages, and then the hydration of the clinker pastes somewhat increased and the rate of the strength development was enhanced at the later curing ages.

## FTIR specimen analysis

FT-IR is one of the versatile tools in predicting the degree of hydration and monitoring the dynamics of changes during the hydration reaction in Portland cement [15-17]. In this study, in order to confirm the result of the hydrates, FT-IR was used to evaluate the clinker pastes when AWM was added as substitutes in the raw meal. The FTIR spectra of the cement clinker pastes were presented in figure 8, 9 and 10, respectively. It showed the appearance of bands at 3644, 3450, 1638, 1425, 975, 873 and 460 $cm^{-1}$. According to the spectra of hydrated products had been reported, the absorption band at 3644 $cm^{-1}$ was due to the stretching vibration of υ(OH) of $Ca(OH)_2$ which was formed as silicate phases in the cement dissolved [17]. Strong peak at 3450 $cm^{-1}$ may be due to –OH or $H_2O$ stretching vibration. The strong peaks at 973 $cm^{-1}$ indicated the shifting of anhydrous phase and /or (Si-O) absorption bands of $C_2SH_2$ and C-S-H phases stretching vibrations [18-20]. The peaks at 1637 $cm^{-1}$ were due to the presence of calcium sulphate in the form of ettringite (mono and tri-sulphate) in the cement [19,20]. The peaks at 1425 $cm^{-1}$ was due to the presence of C=O stretching vibration [21,20]. The peaks at 460 and 873 $cm^{-1}$ might be due to the Si-O and Al-O stretching vibration of silicate and aluminate groups [19].

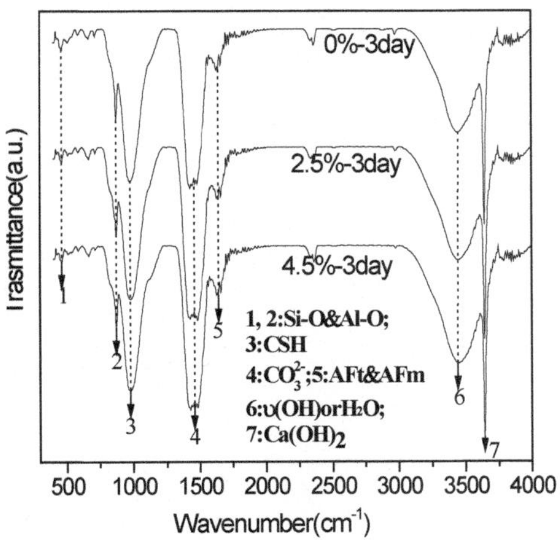


**Figure 8:** FT-IR spectra of cement clinkers specimens cured for 3 days.

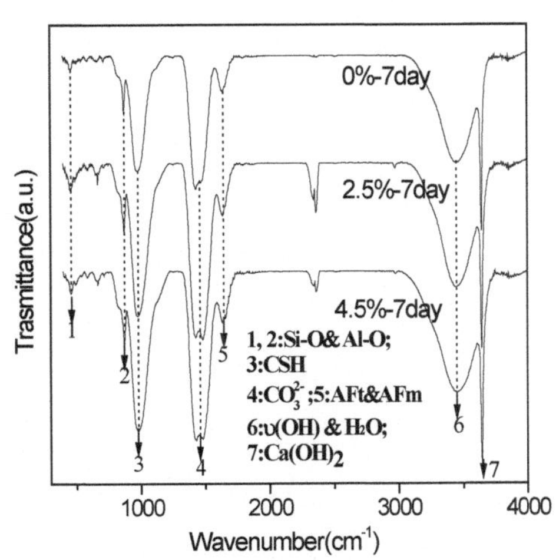


**Figure 9:** FT-IR spectra of cement clinkers specimens cured for 7 days.

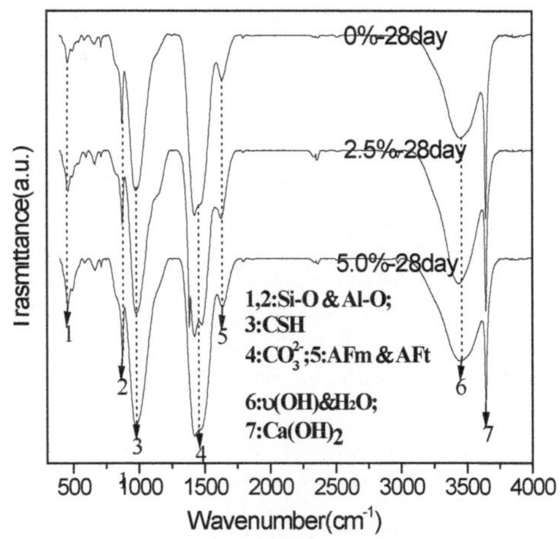


**Figure 10:** FT-IR spectra of cement clinkers specimens cured for 28 days.

As compared to the infrared absorption band of hydrated cement clinkers, it showed that all the hydrated products were similar in the curing ages of 1-28 days. However, it could be seen in the figures 8 and 9, the peaks for the hydrated products ($CO_3^{2-}$ and CSH) of the clinker pastes, which had been added AWM as substitutes in the raw meal, were somewhat higher than the plain clinker paste. In contrast, after 28 days (Figure 10) of curing, the differences disappeared. This indicated that the using the AWM as substitutes in the raw meal might increase the ratio of the hydration for the clinker pastes.

## Conclusions

The objective of this study was to investigate the possibility of disposing alkaline white mud by using the cement kiln of the cement plant and evaluating its effects on the properties of cement-based materials by adding different percentages of alkaline white mud as substitutes in raw meal. Based on the above results and discussion, it was found that there was feasibility of using AWM as substitutes in raw meal to produce OPC cement. The contents of the magnesium oxide, Sulphur trioxide and chloride ions, which might have an influence on the property of cement, were much less than the limitation value of the standards. However, the contents of free lime increased with the contents of AWM addition. Moreover, the addition of AWM in raw meal increased the content of $C_3S$ phase in the clinker, and it somewhat decreased the setting times and enhanced the strengths of the cement pastes in the early curing ages. Furthermore, this research only had been carried out in the laboratory, if it was applied to practical production, more research needed to be done.

### Acknowledgement

The authors wish to gratefully acknowledge the financial support from the Key Science and Technology Research Project from Peoples Republic of China (Grant No. 2008BAE64B05) and would like to acknowledge assistance from staffs who provided full support to this study. The kind suggestions from the anonymous reviewers are greatly acknowledged.

### References

1. Shi L, Luo H (2009) Preparation of soil nutrient amendment using white mud

produced in ammonia-soda process and its environmental assessment. T Nonferr Metal Soc 19: 1383-1388.

2. Kasikowski T, Buczkowski R, Lemanowska E (2004b) Cleaner production in the ammonia-soda industry: an ecological and economic study. J Environ Manage 73: 339-356.
3. Steinhauser G (2008) Cleaner production in the Solvay Process: general strategies and recent developments. J Clean Prod 16: 833-841.
4. Choy KKH, Ko DCK, Cheung WH, Fung JSC, Hui DCW, et al. (2004) Municipal Solid Waste Utilization for Integrated Cement Processing with Waste Minimization: A Pilot Scale Proposal Process. Saf Environ 82: 200-207.
5. Mijno V, Martin F, Bollinger JC, Catalan LJJ (2007) Stabilization–Solidification Process: Comparison of Synthetic Sludge and Volcanogenic Massive Sulphide Tailings. Process Saf Environ 85: 260-264.
6. Viguri J, Andrés A, Ruiz C, Irabien A, Castro F (2001) Cement-Waste and Clay-Waste Derived Products from Metal Wastes: Environmental Characterization. Process Saf Environ 79: 38-44.
7. Kasikowski T, Buczkowski R, Dejewska B, Peszynska-Bialczyk K, Lemanowska E, et al. (2004a) Utilization of distiller waste from ammonia-soda processing. J Clean Prod 12: 759-769.
8. Lin KL, Lin DF, Luo HL (2009) Influence of phosphate of the waste sludge on the hydration characteristics of eco-cement. J Hazard Mater 168: 1105-1110.
9. Lin KL, Chiang KY, Lin CY (2005) Hydration characteristics of waste sludge ash that is reused in eco-cement clinkers. Cement Concrete Res 35: 1074-1081.
10. Lin KL, Lin CY (2005) Hydration characteristics of waste sludge ash utilized as raw cement material. Cement Concrete Res 35: 1999-2007.
11. Lee TC, Li ZS (2010) Conditioned MSWI ash-slag-mix as a replacement for cement in cement mortar. Constr Build Mater 24: 970-979.
12. Gao XB, Wang W, Ye TM, Wang F, Lan YX (2008) Utilization of washed MSWI fly ash as partial cement substitute with the addition of dithiocarbamic chelate. J Environ Manage 88: 293-299.
13. Ito R, Dodbiba G, Fujita T, Ahn JW (2008) Removal of insoluble chloride from bottom ash for recycling. Waste Manage 28: 1317-1323.
14. Wang KS, Lin KL, Huang ZQ (2001) Hydraulic activity of municipal solid waste incinerator fly-ash-slag-blended eco-cement. Cement Concrete Res 31: 97-103.
15. Creighton AT, Zhengxian Y, Peterson MM, Xianming S (2009) Effect of styrene-butadiene rubber latex on the chloride permeability and microstructure of Portland cement mortar. Constr Build Mater 23: 2283-2290.
16. Mollah MYA, Yu WH, Schennach R, Cocke DL (2000) A Fourier transform infrared spectroscopic investigation of the early hydration of Portland cement and the influence of sodium lignosulfonate. Cement Concrete Res 30: 267-273.
17. Ylmen R, Jaglid U, Steenari BM, Panas I (2009) Early hydration and setting of Portland cement monitored by IR, SEM and Vicat techniques. Cement Concrete Res 39: 433-439.
18. Geng Ying L, Pei Ming W, Xiaohua Z (2005) Mechanical behavior and microstructure of cement composites incorporating surface-treated multi-walled carbon nanotubes. Carbon 43: 1239-1245.
19. Rai US, Singh RK (2005) Effect of polyacrylamide on the different properties of cement and mortar. Mat Sci Eng A-struct 392: 42-50.
20. Singh NK, Mishra PC, Singh VK, Narang KK (2003) Effects of hydroxyethyl cellulose and oxalic acid on the properties of cement. Cement Concrete Res 33: 1319-1329.
21. Farooque KN, Yeasmin Z, Halim ME, Mahmood AJ, Mollah MYA (2010) Effect of Carboxymethyl Cellulose on the Properties of Ordinary Portland Cement. Bangladesh J Sci Ind Res 45: 1-8.

# Statistical Analysis of Climatic Seasonality on the Paving Conditions of a Brazilian Highway

**Saulo Rodrigues Castro[1], Samuel Rodrigues Castro*[2], Paulo José Silva[1] and Liséte Celina Lange[2]**

[1]*University Center of Formiga, UNIFOR/MG Dr. Arnaldo de Senna Av, 328, Água Vermelha 35.570-000, Formiga, MG, Brazil*
[2]*Federal University of Minas Gerais, Engineering School, Pampulha Campus Antônio Carlos Av, 6627, Bl; 2, 4628 31.270-901, Belo Horizonte, MG, Brazil*

**Abstract**

Road transport in Brazil is the main mean of people and cargo transportation and it becomes increasingly intense. Traffic loads and environmental factors are the primary causes of failures in asphalt paving, once the asphalt overlay is completely exposed to environmental hazards. The aims of the present work were to monitor and record occurrences involving paving failures in the highway granted to the Brazilian Public Private Partnership (PPP) that has the MG-050 Highway as the main link connection between the midwest of Minas Gerais and São Paulo state, with 402 km long, serving as an integration axis of the economy developed in these regions. Data involving general events of paving failures and, more specifically, pits and potholes parameters were collected during the years 2010 and 2011. A statistical analysis was performed using non-parametric tests, evaluating seasonal climate differences - dry and rainy seasons - in the quantitative data of such events. With 95% confidence, it was possible to verify that the number of occurrences of pits and potholes parameter is significantly different to the investigated seasons, more intense on rainy periods, when the paving becomes more susceptible to the occurrence of this event. This work evaluates, statistically, the action of the environment in the paving structure as a function of recorded failures, a preliminary analysis that deserves further studies.

**Keywords:** Climatic seasonality; Statistical analysis; Asphalt paving; Brazilian highways; MG 050

## Introduction

Second Bernucci et al. [1], the main objective of paving is to ensure the trafficability at any time of year and weather conditions, and provide for users comfort and safety. Once the natural soil is not resistant enough to withstand the repetition of wheel loads without suffer significant deformations, it becomes necessary to construct a structure called paving. It is built on the sub grade to withstand the vehicles loads and to distribute the requests to its several layers, limiting the stresses and strains in order to ensure a proper performance of the paving for a long time.

Traffic loads and environmental factors are the primary causes of asphalt paving deformation and damage, the asphalt overlay is completely exposed to environmental hazards.

The effect of climate changes, such as solar radiation and wind speed, on the paving material is inevitable. In particular, when the temperature of the asphalt overlay reaches 50°C, the deformation resistance of the paving structure considerably weakens [2].

In other hand, with water penetrating into the interface of the asphalt and aggregate, excess pore water pressure will be produced under overloading, leading to the separation of asphalt membrane from the aggregate surface. This occurrence leaves the pavement with pits and grooves, which could form loose structures or cracks and rapidly decrease the resistance against shearing or sliding failure because the bonding force between the aggregates has been greatly weakened [3].

Analysis on the design of asphalt paving have shown that, for this special engineering structure, the consideration of environmental influence is not specific and appropriate and the corresponding countermeasures often cannot reach people's anticipation. Yang and Ning [4] verified that high temperature leads rutting disease in asphalt paving and low temperature leads paving cracking; water penetrated into the interface of asphalt and aggregate, and the asphalt membrane began to peel from aggregates, then aggregates loose and turn into pits.

For most road applications, conventional asphalts shown good behavior, satisfying fully the necessary requirements for the proper performance of asphalt mixtures under traffic loading and weather conditions. However, conditions for volume of commercial vehicle and axle weight increasing, year over year, on designated roads or airports, in heavy traffic corridors and to adverse weather conditions, with large temperature differences between winter and summer, these are situations that has been increasingly necessary the use of modified asphalts [1].

Second Botaro et al. [5] the use of polymer-modified asphalts to improve the performance paving has been observed. When there is a match between the polymer and the asphalt matrix, the properties of the formed blends can contribute effectively to the reduction of the coating breakdown, thermal cracking and formation of wheel tracks. The use of polymer-modified asphalt matrix also increases the fatigue life of the paving.

The present work aims to analyze statistically the quantitative data of occurrences on paving failures and evaluate significant differences due to climatic seasonality. It is a result of a monitoring occurred in the years 2010 and 2011 on the highway road that was ceded a Public-Private Partnership (PPP) involving the MG-050 Highway, located in Minas Gerais-Brazil, and smaller stretches of BR-491 and BR-265 Highways.

## Materials and Methods

This work was performed, in a first step, through a description and characterization stage of the main study object, the MG-050 Highway.

***Corresponding author:** Samuel Rodrigues Castro, Federal University of Minas Gerais, Engineering School, Pampulha Campus Antônio CarlosAv, 6627, Bl; 2, 4628 31.270-901, Belo Horizonte, MG, Brazil, E-mail: samuc2000@yahoo.com.br

This corridor was chosen because it is an important route of inter-regional integration, serving as an axis of the economic development. In a second step, analytical, a statistical analysis was performed focusing on quantitative data of general events of paving failures and pits and potholes recorded to the years 2010 and 2011, in order to identify significant differences related to the climatic seasonality, in a proper results interpretation.

The evaluated highway was divided into 20 segments with different length, but homogeneous in traffic load, to facilitate the data compilation. A work that was carried out through an engineering firm responsible for the supervision of this highway road, which implemented in 2010 a systematic and automated monitoring system for this purpose. Two groups covered whole stretch of the PPP highway during every week and the new failures occurrences were recorded.

The general events of paving failures were characterized as events related to the operational conditions and paving quality and understand the quantitative data of pits and potholes, depressions, deformations, exudations, cracks, ruts, breakdowns, ripples, upheavals and wheel tracks.

The classification of climatic seasonality adopted in the data analyses was made according to Table 1.

According to Siegel and Castellan [6], a statistical test is used to respond with certain confidence about the hypothesis that arises about particular theory. The data were processed using the statistical package of STATISTICA 6.1 Software, which allowed the application of tests and generation of graphical results.

As a preliminary analysis, time series of general events of paving failures and pits and potholes data, collected monthly from February 2010 to December 2011, are plotted in bar charts in order to evaluate the evolution of them over the monitored period. In addition, in this step of data recognition, an evaluation of descriptive statistics of the related parameters was performed.

Adherence tests to general events of paving failures and pits and potholes data was checked using the Shapiro-Wilk and Chi-Square tests at a significance level ($\alpha$) equal to 0.05, in order to evaluate the distribution.

According Callegari-Jacques [7], in parametric tests the evaluated data set should fit a normal distribution or normal approximation. In the other hand, non-parametric tests, also called free distribution tests, should be used when there is no requirement as to the knowledge of the distribution of the variable in the population.

One should avoid the use of non parametric tests in situations that a condition of use of parametric tests prevails. The parametric tests are applied under higher stringency conditions, thus they are also more potent than the non parametric [6].

Non parametric statistical test was used to check for significant differences between the medians of general events of paving failures and pits and potholes data sets on dry and rainy seasons, at significance level ($\alpha$) equal to 0.05. The median represents a better value of central tendency of the evaluated data, a distribution different of normal.

| Seasonality | |
|---|---|
| **Dry** | **Rainy** |
| April | October |
| May | November |
| June | December |
| July | January |
| August | February |
| September | March |

**Table 1:** Climatic seasonality by month: Dry and Rainy season

The Mann Whitney U test, non parametric, was used to test the following hypotheses:

- $H_0$: dry season median = rainy season median;
- $H_1$: dry season median ≠ rainy season median.

Finally, a detailed analysis from the obtained results allowed a conclusion with a significant level of statistical credibility about the tested hypothesis.

## Results and Discussion

### A Brazilian highway: MG-050

The road system of Minas Gerais is the largest in the country, totaling 273,162 km in extension. However, only 25,981 km correspond to federal and state paved roads. It covers a region with average temperatures ranging from 16 – 18°C in the dry season, and reaches 22-23°C in the rainy season. Moreover, it has a minimum rainfall value about 15 - 50 mm in the dry season, which reaches 240 - 280 mm in the rainy season, approximately [8].

This road network is an important element of integration of this federation unit with other urban centers in Brazil, with MERCOSUL countries and the major ports of the country. Analyzing the survey conducted by the National Department of Transportation-DENATRAN, in June 2012, it is possible to infer that Minas Gerais has a fleet of 7,957,586 vehicles, of which 13.98% are cargo vehicles [9].

The MG-050 highway is an important route of inter-regional integration, which connects Belo Horizonte to the midwest of Minas Gerais and northwest of São Paulo state, with 402 km long, serving as an integration axis of the economy developed in these regions. Its area of influence includes 50 municipalities with a population about 1.5 million inhabitants, which represents approximately 7.6% of the population's state.

By this highway travels much of the agricultural sector production of Minas Gerais, in particular regarding to the poultry industry, swine, dairy and derivatives; moreover, sugar, coffee, fruits and horticulture. It stands out even by industrial activity metallurgy, non-metallic materials, food, mining, textiles, clothing, footwear, rubber, hides and skins. Tourism on the shores of Furnas Lake is another activity much explored in the midwest region of the state, taking this highway as the main access road.

In 1995, Brazil started the Grant Program of Federal Highways to the private initiative, a modality that has been tested and approved. In May 2007, Minas Gerais signed with a private group the concession of the first road PPP in the country, benefiting 344.40 km of MG-050, the main studied highway, 4.65 km of BR 491 and 22.00 km of BR-265 highways [10].

Therefore, MG-050 highway, linking Minas Gerais to São Paulo state, occupies a preponderant place in the country's road network, leveraging the development and promoting integration through the transport of goods and people.

### Statistical analysis

This topic presents the statistical results concerning analyzes realized with the quantitative data of general events of paving failures and, specifically, pits and potholes.

For general events of paving failures data 456 occurrences were recorded in 2010 and 1391 for the year 2011. In other hand, for the parameter "pits and potholes", 177 events were recorded in 2010 and 794 records for the year 2011. The Figure 1 shows the time series with the evolution of the analyzed data for these parameters.

In general, the quantity of pits and potholes corresponds to 52% of general events of paving failures data. In both cases, the number of recorded data in 2011 is higher than the number of data for the year 2010, a considerable difference which may be due to this last one year be considered a implementation period of the computerized system of monitoring.

The distribution of the data must be checked before performing the statistical analysis, because it will determine the choice of statistical techniques to be used. The parametric analysis should be applied in the case of the normal distribution of data, and the non-parametric analysis when the distribution is asymmetrical, different from the normal distribution. Thus, as a preliminary analysis, a descriptive statistics of the evaluated parameters was performed with the recorded data for segments and its results are presented in Table 2.

The analysis of some parameters shown in Table 2 gives an indication that the data are asymmetric and, thus, do not fit in a normal distribution. The asymmetry coefficient has non-zero values, with a value equal to 2.4 for general events of paving failures and 4.1 for pits and potholes. Other points to consider, the standard deviation parameter, for the two sets, has numerical values higher than the arithmetic mean, moreover, the values for the mean, median and mode parameters are not coincident, a common fact to asymmetrical data.

Normality tests of Shapiro-Wilk and Chi-Square were performed to verify if the data fit significantly to a normal distribution with a significance level equal to 0.05. Results are shown in Figure 2.

Normality tests of Shapiro-Wilk and Chi-Square present p-values less than 0.05 for analysis of general events of paving failures and pits and potholes data, considering the years 2010 and 2011. In both cases it indicates that the data did not fit a normal distribution, which can be justified because there are no homogeneous characteristics with respect to the paving quality along the segments of the granted highway to PPP. Therefore, it is stated with 95% confidence level that the evaluated data, for such years, are asymmetric.

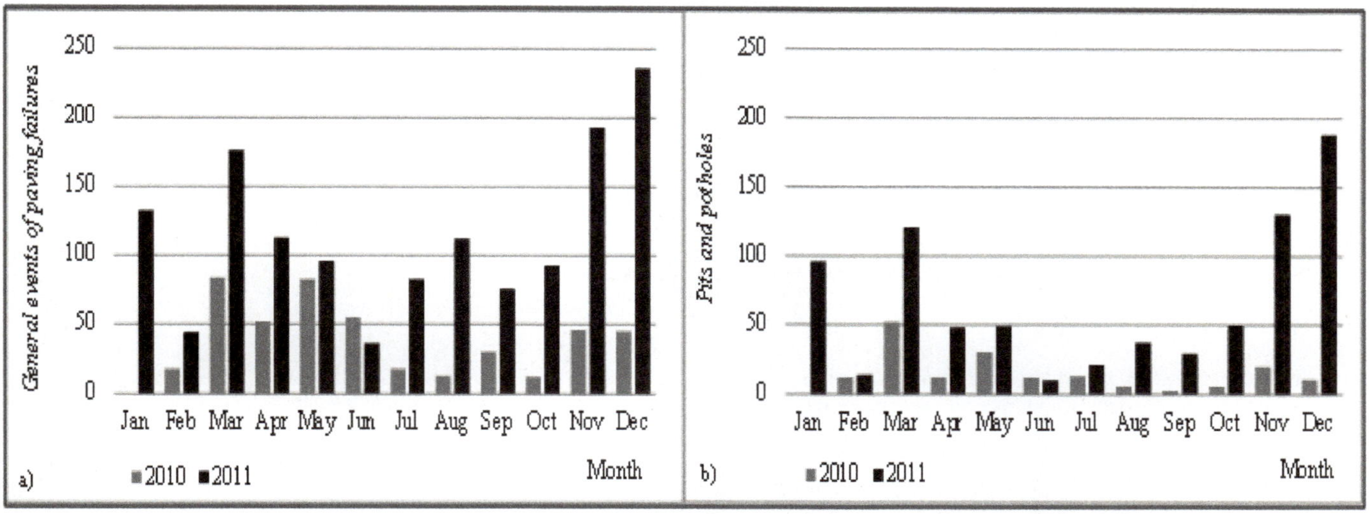


**Figure 1:** Time Series for analyzed data – years 2010 and 2011: a) General events of paving failures; and b) Pits and potholes.

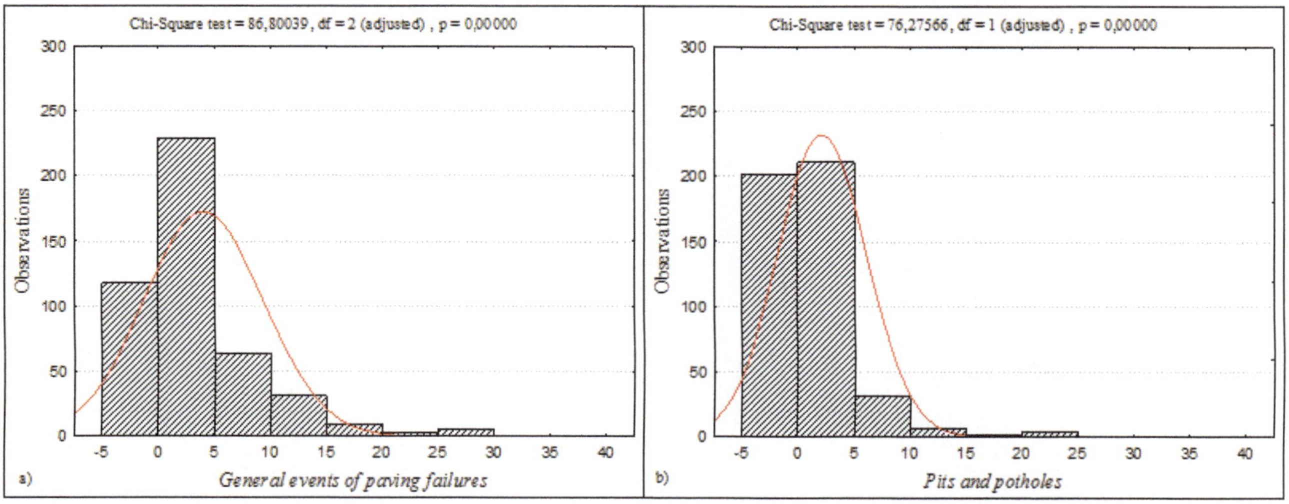


**Figure 2:** Normal analysis of standardized data distribution: a) General events of paving failures; and b) Pits and potholes.

| Parameters | General Events | Pits and potholes |
|---|---|---|
| Arithmetic Mean | 4.0 | 2.1 |
| Median | 2.0 | 1.0 |
| Mode | 0.0 | 0.0 |
| Standard Deviation | 5.3 | 4.0 |
| Variance | 28.4 | 15.7 |
| Asymmetry | 2.4 | 4.1 |
| Count | 460 | 460 |

**Table 2:** Descriptive statistics of the evaluated parameters: Dry and Rainy seasons

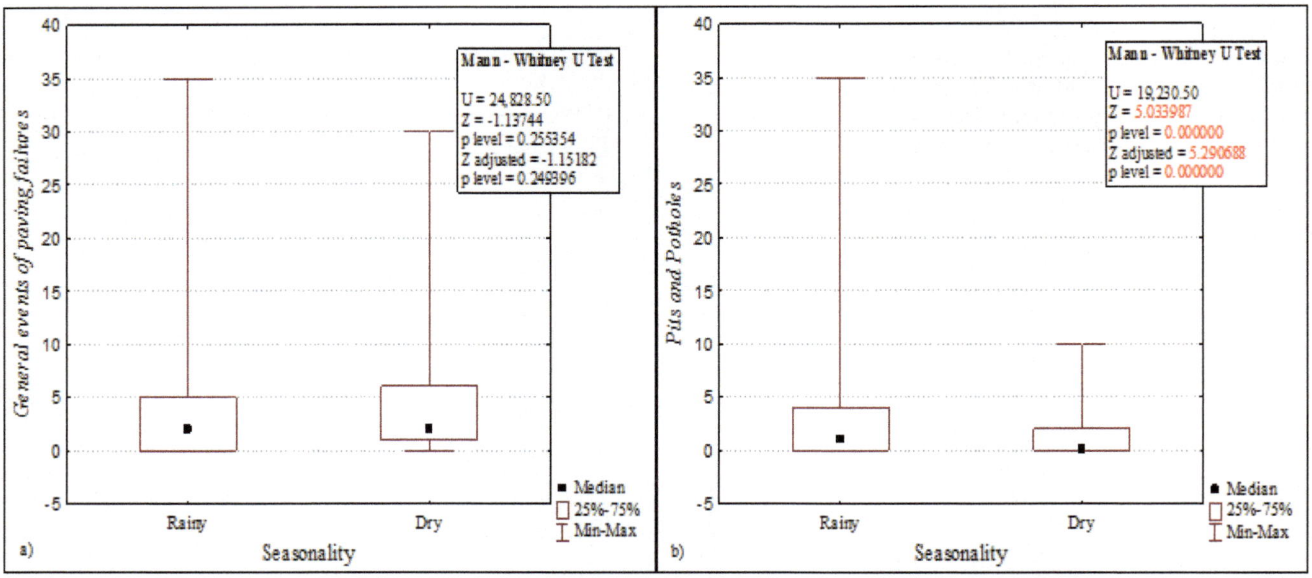


**Figure 3:** Mann-Whitney U Test – Dry vs. rainy seasons – a) General events of paving failures; and b) Pits and potholes.

Asymmetric data sets are commonly found in monitoring of environmental parameters or under conditions that are difficult to control, as the susceptible data to climate and other natural factors.

This is justified because the highway not shows homogeneous characteristics with respect to the paving quality and traffic load along the section granted to PPP.

As the data sets analyzed showed a distribution different to normal, with a considerable asymmetry observed graphically in Figure 2, the non parametric Mann-Whitney U test was chosen to assess significant differences between the quantitative data obtained in dry and rainy seasons for general events of paving failures and pits and potholes.

The Figure 3 shows the results for the Mann-Whitney U tests, besides the Box-Whisker plots, which allow a visual comparison between the data from dry and rainy seasons of the monitored parameters in 2010 and 2011.

The Mann-Whitney U test was chosen to allow a comparison between the medians of two independent samples, in this case, general events of paving failures and pits and potholes data sets for dry and rainy seasons.

By analyzing this statistical test results, the p-values for the parameter general events of paving failures were higher than 0.05, which indicates that there is no significant difference between the data obtained for the dry and rainy season, i.e., the hypothesis that the tested medians are equal cannot be rejected.

This result can be explained because this parameter includes data of factors that do not relate directly to the tested seasonal difference, for example, problems on paving construction. Another point is that general events data include factors that generate paving failures intensified in the rainy season, or even factors potentiated in the dry season. Thus, in an analysis that considers the median data of general events of paving failures for the two periods, a significant difference is not observed.

In the other hand, a different result was observed for the quantitative data of pits and potholes by comparing the dry and rainy seasons. P-values equal to zero is a result that allows a statistical conclusion that, with 95% confidence, the obtained data show significant differences for the tested periods.

This result is consistent considering the "pits and potholes" parameter, it is susceptible to an increase in rainy seasons and, according Yang and Ning [4], with water penetrating into the interface of the asphalt and aggregate, excess pore water pressure produces an over loading, leading to the separation of asphalt membrane from the aggregate surface.

Beyond the significant statistical difference in the median of the data, it is observed graphically that the upper limit for the rainy season in a random segment reaches 35 cases of occurrences and only 10 cases in the dry season.

In rainy periods related to the months from December to February there is also a possible increase in vehicle traffic load in Brazilian highways, which can be related to the festivities of end of year and holidays. It can explain this result and the quantitative data from this paving failure in rainy seasons.

## Conclusion

As a result of the monitoring work on the highway road involving the MG-050 Highway and smaller stretches of BR-491 and BR-265 Highways, it was concluded that, by a statistical analysis of the recorded occurrences in 2010 and 2011, the data distribution to general events of paving failures and pits and potholes parameters is asymmetric, different to a normal distribution. Therefore, non-parametric tests were used in the statistical analysis.

In evaluating the climatic seasonality differences, dry and rainy seasons, of quantitative data for general events of paving failures parameter, it was concluded that, there is no significant difference related to the recorded data during the analyzed period. Now, for pits and potholes parameter, a significant difference was observed between the quantitative data for dry and rainy seasons, in the last one season there is more recorded occurrences. So it is possible to infer in this case study:

- There are paving failures that are more pronounced in dry season and others, characteristic of rainy season;
- There are paving failures that can worsen during the rainy season, with the infiltration of water and traffic load, as pit and potholes, which increase the occurrences and quantitative data in this period.

This is an analysis that involves many factors and deserves further studies.

### Acknowledgements

The authors are grateful to FAPEMIG, CAPES and CNPq (Brazilian Government Agencies) for the financial support and also to "CONSOL - Engenheiros Consultores" for the cooperation during this work.

### References

1. Bernucci LB, da Motta LMG, Ceratti JAP, Soares JB (2006) Pavimentação asfáltica: Formação Básica para Engenheiros. Rio de Janeiro: Petrobras: Abeda.
2. Xue Q, Liu L, Zhao Y, Chen YJ, Li JS (2013) Dynamic behavior of asphalt pavement structure under temperature-stress coupled loading. Applied Thermal Engineering 53: 1-7.
3. Xue Q, LIU L (2013) Hydraulic-stress coupling effects on dynamic behavior of asphalt pavement structure material. Construction and Building Materials 43: 31-36.
4. Yang Q, Ning J (2011) The environmental influence of asphalt pavement and countermeasures. Energy Procedia 5: 2432-2436.
5. Botaro VR, Castro SR, Rodrigues JF, Cerantola AE (2006) Obtenção e caracterização de blendas de asfalto CAP 20, modificado com poliestireno reciclado, resíduos de pneu e lignina organossolve. Revista Escola de Minas 59: 117-122.
6. Siegel S, Castellan JN (2006) Estatística Não-paramétrica para Ciências do Comportamento - 2.ª Edição. Porto Alegre: Artmed.
7. Jacques C, Sidia M (2003) Bioestatística: Princípios e Aplicações. Porto Alegre: Artmed.
8. Instituto Nacional de Meteorologia – INMET (2013) Ministério da Agricultura, Pecuária e Abastecimento – Brasilia.
9. Confederação Nacional do Transporte – CNT (2012) Pesquisa CNT de rodovias 2012: relatório gerencial. Brasília: CNT: SEST: SENAT.
10. Campos Neto CAS, Soares RP, Ferreira IM, Pompermayer FM, Romminger AE (2011) 1592 Texto para discussão - Gargalos e demandas da infraestrutura rodoviária e os investimentos do PAC: Mapeamento IPEA de obras rodoviárias. Brasília: Instituto de Pesquisa Econômica Aplicada.

13

# Analysis of Masonry Infill in a Multi-Storied Building

**Niruba S***

*Civil Engineering Department, Erode Sengunthar Engineering College, Thudupathi, Erode, India*

**Abstract**

In many literatures, studies show that infill's modify the behavior of framed structures under lateral loads; the contribution of panels is generally neglected in common structural analyses. The structural effect of brick infill is generally not considered in the design of columns as well as other structural components of RC frame structures. The brick walls have significant in-plane stiffness contributing to the stiffness of the frame against lateral load. The lateral deflection is reduced significantly in the infilled frame compared to the deflection of the frame without infill. It is observed that frames with infill produce much smaller deflections as compared to frames without infill. The results reflect the significance of infill in increasing the strength, stiffness and frequency, of the entire system depending on the position and amount of infilling. Lower infilling is noted to provide more stiffness for the system as compared with upper locations. A number of non-linear static (pushover) analyses were performed on proper structural models of the building, considering both bare framed structure and the infilled one, in order to appraise the influence of infill walls on the failure mechanisms. In particular, a sensitivity analysis was performed by assigning different Partial Safety Factors (PSF) to the mechanical parameters of infill walls, in order to investigate their effect on the overall structural response of the building.

**Keywords:** Lateral loads; Inter-storey drift ratios; Storey displacements; Stiffness, Masonry infill

## Introduction

Masonry is the building of structures from which individual units laid in and bound together by mortar; the term masonry can also refer to the units themselves. The common materials of masonry construction are brick, stone, marble, granite, travertine, limestone, cast stone, concrete block, glass block, stucco, and tile. Masonry is generally a highly durable form of construction. Various advantages have been acquired on analyzing the structures with infill, they are,

- The use of materials such as bricks and stones can increase the thermal mass of a building.
- Most types of masonry typically will not require painting and so can provide a structure with reduced life-cycle costs.
- Masonry is very heat resistant and thus provides good fire protection.
- Masonry walls are more resistant to projectiles, such as debris from hurricanes or tornados.

Here also arise some of the disadvantages like,

- Extreme weather causes degradation of masonry wall surfaces due to frost damage. This type of damage is common with certain types of brick, though rare with concrete blocks.
- Masonry tends to be heavy and must be built upon a strong foundation, such as reinforced concrete, to avoid setting and cracking.
- Save for concrete, masonry construction does not lend itself well to mechanization, and requires more skilled labour than stick-framing.

In Euro code 6, six types of masonry units are defined: clay units, calcium silicate units, aggregate concrete units, autoclaved aerated concrete units, manufactured stone units and dimensioned natural stone units complied with the relevant European standards EN 771-1 to 6. However, brickwork can be combined from many of the following components: adobe, ashlars, blocks, bricks, bitumen, chalk, cement, lime and mortar. Depending on which materials are used, and how they are located, Reinforced Masonry (RM) walls can be divided into the following classes: confined masonry, reinforced cavity masonry, reinforced solid masonry, reinforced hollow unit masonry, reinforced grouted masonry and reinforced pocket type walls.

## Current Methods of Modeling Masonry

Two main types of masonry models can be distinguished into: the micro- and the macro-modeling.

In case of macro-models, the brick, the mortar and the joint is one homogeneous material that represents the behavior of the masonry.

Micro-models should be used when local actions need to be analyzed deeply. In case of macro-modeling, the global behavior of the structure is analyzed by a homogeneous material, so the behavior can be described by the global properties; thus, it is not attempted to make the calculations with local failures.

### Effect of infill

- The stresses, in the infill wall, however, were found to increase with the increase in Young's Modulus of elasticity [1] due to the increase in stiffness of the system, attracting more forces to the infill.
- The infill wall enhances the lateral stiffness of the framed structures; however, the presence of openings within the infill wall would reduce the lateral stiffness [1-3].
- The fundamental period only slightly increases as the infill wall thickness increases [2], since the increase in thickness only increases the mass of the structure rather than its stiffness.
- The infill was assumed to crack once the stress in the infill exceeded the ultimate compressive stress of the infill material [2].

***Corresponding author:** Niruba S, Civil Engineering Department, Erode Sengunthar Engineering College, Thudupathi, Erode
E-mail: nirubasn@gmail.com

- The strength if infill in terms of its Young's modulus (Ei) has a significant influence on the global performance of the structure [4]. The structural responses such as roof displacements, inter-storey drift ratios and the stresses in the infill wall decrease with increase in (Ei) values due to increase in stiffness of the model.
- The opening size of the infill [1,2,5] has a significant influence on the fundamental period, inter-storey drift ratios, infill stresses and the structural member forces. Generally, they increase as the opening size increases, indicating that the decrease in stiffness [5] is more significant than the decrease in mass.
- The specific weight of reinforced masonry is smaller, the thermal and somniferous conductivity of the masonry is worse and it also has better resistance against fire and chemicals.

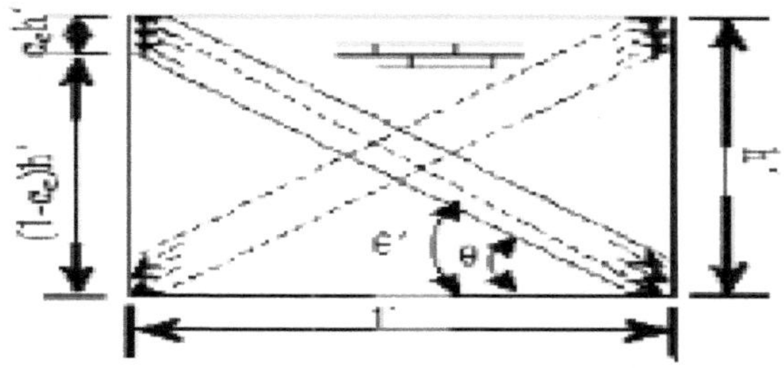

**Figure 1:** 'Equivalent Diagonal Strut' model for infill panel.

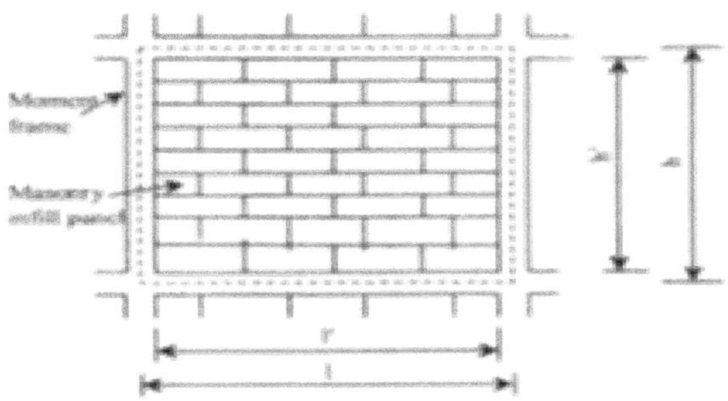

**Figure 2:** Masonry infill panels in framed structures.

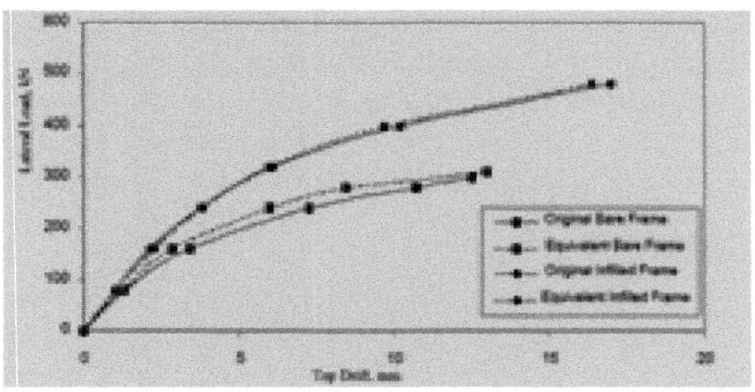

**Figure 3:** Load-Deflection characteristics for the original and equivalent systems.

- The wall can be loaded right after finishing the construction (Figures 1 and 2).

## Role of infill

Existence of infilling is noted to increase the ultimate lateral resistance of the system while resulting in less ultimate lateral deflection for lower infilling. The effect on both parameters is more pronounced for higher percentages of infilling. Two phenomena arise through the stage of loading and result in the response nonlinearity. First is to find the stiffness degradation of the reinforced concrete with load-induced orthotrophy depending on both the applied dynamic load [2,5] and the inherent deformational characteristics of the frame. Second is to find the progressive strength reduction of either of the diagonal struts, which is supposed to be sequential according to level of loading.

Conventional half-brick wall infilling is noted to affect nearly all of the dynamic parameters [4] of reinforced concrete frames. Infill influence on the kinetic and kinematic coefficients related to lateral excitation is found to depend on frame features such as number of stories and number of bays as well as infill amount and position. Lower location yields the higher strength, stiffness, and frequency of the system. Nonlinearity of the behavior is basically due to stiffness degradation, which consequently results in frequency attenuation [4] during the loading regime (Figure 3).

The sway characteristics of building are different for frames considering infill rather than frames without infill (Table 1).

## Effect of opening size

The infill wall enhances the lateral stiffness of the framed structures; however, the presence of openings within the infill wall would reduce the lateral stiffness [5]. The fundamental period increases as the opening size increases, as expected, due to reduction in stiffness of the model. Such variation of periods cannot be considered using the Code values. The fundamental period [2,5] of the fully infilled model was 54.87% higher than that of the bare frame model. There is no clear relationship between the opening size and the fundamental period, but the opening size does have an influence on the fundamental period of the structure [2].

The seismic response of in-filled frame structures is,

a. The strength of infill in terms of its Young's Modulus (Ei) has a significant influence on the global performance of the structure [1]. The structural responses such as roof displacements, inter-storey drift ratios and the stresses in the infill wall decrease with increase in (Ei) values due to increase in stiffness of the model [1,5]. It is therefore important to choose the right material for infill, know its properties and consider these in the analysis and design.

b. The minimum compressive strength of infill material required to maintain the structure in an un-cracked condition under a credible earthquake (with 0.2g PGA) varies with the height of the building [5]. It has been shown that under exposure to similar seismic hazards, medium rise buildings require higher strength infill material (compared to low rise building).

| Earthquake Load | Sway for Corner Column (mm) | Sway for Exterior Column (mm) | Sway for Interior Column (mm) | Maximum Value of Sway (mm) |
|---|---|---|---|---|
| From Left to Right without Infill | 54.3 | 54.2 | 54.8 | 56.6 |
| From Left to Right with Infill | 32.3 | 32.2 | 33.0 | 35.7 |
| From Front to Back without Infill | 73.2 | 73.2 | 73.1 | 73.2 |
| From Front to Back with Infill | 37.5 | 37.6 | 37.5 | 37.6 |

**Table 1:** Sway characteristics of different columns.

c. The opening size of the infill has a significant influence on the fundamental period, inter-storey drift ratios, infill stresses and the structural member forces. Generally they increase as the opening size increases, indicating that the decrease in stiffness [5] is more significant than the decrease in mass.

d. Under a particular level of PGA (0.2g), the increase in infill stress [4] is not very significant beyond infill strength of *Ei*=7500 MPa. This value could be considered to be the maximum limit of the Young's Modulus of the infill material if the infill walls are used for retrofitting old buildings.

e. The performances of buildings constructed with and without seismic provisions [2] are almost similar if the infill has a minimum value of 5000 MPa for its Young's Modulus (*Ei*). This is because the structural capacity is greatly influenced by the type of infill walls and the values of their Young's Modulus.

## Literature Review

Muyeed-Ul-Azamand Amanat [1] carried out a work on 'Effect of Infill as a Structural Component on the column Design of Multi-storied Building' in which various observations had been made on the infill characteristics. The lateral deflection is reduced significantly in the infilled frame compared to the deflection of the frame without infill. This leads to different steel requirements for frame structures considering infill. In order to understand the behavior of frames and steel requirements of column having brick masonry infill and without infill a finite element investigation is performed by modeling a 10-storied three dimensional building frame [5]. The investigation is done by using the equivalent strut model method. A detailed investigation is performed using various loads and load combinations of the building considering infill and without infill to find out steel requirements and to see the effect of infill in the sway characteristics of the building. It is observed that frames with infill produce much smaller deflections as compared to frames without infill. It is also observed that there is no significant difference in steel requirements of interior column but there is moderate difference in steel requirements in exterior column and significant difference in steel requirements in corner column.

Dorji and Thambiratnam [2] carried out the work on 'Modeling and Analysis of In-filled Frame Structures under seismic loads' in which the seismic response of in-filled frame structures had been studied. In-filled frame structures are commonly used in buildings, even in those located in seismically active regions. Present codes unfortunately, do not have adequate guidance for treating the modeling, analysis and design of in-filled frame structures. Finite Element time history analyses under different seismic records have been carried out and the influence of infill strength, openings and soft storey phenomenon are investigated. Results in terms of tip deflection, fundamental period, inter-storey drift ratio and stresses are presented and they will be useful in the seismic design of in-filled frame structures.

Mahumad et al. [5] carried out the work on 'Study of the Reinforced Concrete Frame with Brick Masonry Infill due to Lateral Loads' in which the behavior of Reinforced Concrete (R.C.) frames with brick masonry infill for various parametric changes have been studied to observe their influences in deformation patterns of the frame. Masonry infill panels have been widely used as interior and exterior partition walls for aesthetic reasons and functional needs. When infill walls are omitted in a particular storey, a soft storey is formed compared to much stiffer other stories. The present study is also aimed at finding out the effect of soft storey on frame structures due to horizontal loading. In both cases of wind and earthquake loads, if number of bay increases, then the deflection eventually decreases. As the storey level of building frame increases, deflection due to lateral loads naturally increases due to additional lateral loads. Deflection increases linearly if the span of bay increases linearly because of linearly increased loads [6].

Kodur et al. [4] carried out the work on 'Seismic design and analysis of masonry-infilled frames' in which a simple analytical procedure, which can be used by practicing engineers, for the seismic design of masonry-infilled frames is presented. The analytical procedure, based on the experimental and analytical studies reported in the literature, accounts for the effect of infill's in all three stages, namely, in computing seismic loading, in predicting response of the infilled frame, and in determining the strength of the infilled frame. Seismic loading is computed using the dynamic properties of the structure. Recommendations regarding the choice of infilled frame idealization, structural damping ratio, earthquake design spectrum, structural irregularity, and computational aids are made. Application of the proposed analytical procedure in a design situation is demonstrated through a numerical example, and it is shown that infill's can be accounted in the seismic design of frames during the normal course of design[7].

## Methodology

In most of the analytical works the infill in frame is not considered and it is treated as bare frame. By various literature studies, it is found that the infill has various characteristics like increasing the strength, reducing the deflection, and increasing the stiffness, etc. Hence the work to be done is,

a. By using ETABS software, the framed structures are analyzed including the infill material [8].

b. Computing the structures characteristics for both structures with and without infill

## Analysis Method

### Equivalent static analysis

Equivalent static analysis is the indirect method of considering the effect of ground motion and there is incorporation of dynamic properties of the structure in terms of fundamental period, response reduction factor, soil type, seismic zone and importance factor. Equivalent static analysis were carried out for all the models under the action of Dead Load (DL), Live Load (LL) and Earthquake Load (EQ) for different load combination as per IS 1893-2002 [9].This method is limited to regular type of structure whose response is governed by first mode of vibration. As per IS 1893-2002 regular structure up to 40m in height in zone IV and V and up to 90m in zone II and III can be analyzed by Time History and Response Spectrum analysis.

### Response spectrum method

The objective of response spectrum analysis is to obtain the likely maximum response of the systems. The response spectrum is a plot of the maximum response (maximum displacement, velocity, acceleration or any other quantity of interest) to a specified load function for all possible single degree-of-freedom systems. The abscissa of the spectrum is the natural period (or frequency) of the system and the ordinate is the maximum response. It is also a function of damping. The design response a spectrum given in IS 1893:2002 for a 5% damped system.

### Time history analysis

In order to examine the exact non-linear behavior of building structures, nonlinear time history analysis has to be carried out. In this method, the structure is subjected to real ground motion records. This makes this analysis method quite different from all of the other approximate analysis methods as the inertial forces are directly determined from these ground motions and the responses of the building either in deformations or in forces are calculated as a function of time, considering the dynamic properties of the building structure.

### Pushover analysis

Pushover analysis is a static, nonlinear procedure in which the magnitude of the lateral forces is incrementally increased, maintaining the predefined distribution pattern along the height of the building. With the increase in magnitude of the loads, weak links and failure modes are found. Pushover analysis can determine the behavior of a building, including the ultimate load and maximum inelastic deflection. Local nonlinear effects are modeled and the structure is pushed until a collapsed mechanism is developed. At each step, the base shear and roof displacement can be plotted to generate the pushover curve.

## Conclusion

- In all the analyses, the infill in the framed members is not considered, but the infill helps in the increase of ductility, stiffness and the flexural strength of the members.
- The behavior of an infilled frame is dependent on the properties of frame and infill; hence, the response of such frames should be based on overall frame to infill composite action rather than on isolated bare frame behavior.
- Effect of dynamic loading on the behavior of masonry infilled R.C. frame may be investigated to determine the characteristics with ease [10].
- The contribution of partial infill walls must be well identified so that while analyzing models for real structures, the composite action of the frame and infill would be realized.
- By considering the infill wall the roof displacement of the structure reduces and the stiffness of the structure increases. The masonry infill wall is more significant in small structures as the height of the structure increases the effect of masonry infill wall reduces.
- The inter-storey drift of all the structural members are within the code prescribed limit of 0.004. Since the structure becomes stiffer with the decrease in height of structure. By increasing the opening percentage the inter-storey drift also increases.
- In the case of infilled frame with a soft ground storey, the shear forces acting on columns are considerably higher than those obtained from analysis of the bare frame.
- It has been found that calculation of earthquake forces by treating RC frames as ordinary frames without regards to infill leads to underestimation of base shear.

### References

1. Muyeed-Ul-Azam HM, Amanat KM (2005) Effect of Infill as a Structural Component on the Column Design of Multi-storied Building. UAP Journal of Civil and Environmental Engineering.
2. Dorji J, Thambiratnam DP (2009) Modeling and Analysis of Infilled Frame Structures under Seismic Loads. The Open Construction and Building Technology Journal 3: 119-126.
3. Asteris PG (2003) Lateral stiffness of brick masonry infilled plane frames. ASCE J. Str. Engg., 129: 1071-1079.
4. Kodur VR, Erki MA, Quenneville JHP (1995) Seismic design and analysis of masonry-infilled frames. Canadian Journal of Civil Engineering 22: 576-587.
5. Mahmud K, Islam R, Al-Amin (2010) Study of the Reinforced Concrete Frame with Brick Masonry Infill due to Lateral Loads. IJCEE-IJENS.
6. Asteris PG, ASCE M (2003) Lateral Stiffness of Brick Masonry Infilled Plane Frames. Journal of Structural Engineering 129:1071.
7. Pradhan PM, Pradhan PL, Maskey RK (2012) A Review on Partial Infilled Frames under Lateral Loads. Kathmandu University Journal of Science, Engineering and Technology 8: 142-152.
8. ETABS nonlinear Version 9.5.0 Extended 3D analysis of the Building Systems Computers and Structures Inc. 1995 Berkeley, California.
9. Agarwal P, Shrikhande M, Earthquake Resistant Design of structures, Prentice Hall of India Pvt. Ltd., India (2006).
10. Mulgund Gv, Kulkarni AB (2011) Seismic Assessment of RC Frame Buildings with brick masonry infill. IJAEST 2: 140-147.

# 14

# Vertical Distribution of $NO_2$ in an Urban Area: Exposure Risk Assessment in Children

**Cheong KWD[1], Tham KW[1], Balasubramanian R[2] and Kalaiarasan M[3]***

*[1]Department of Building, National University of Singapore, Singapore*
*[2]Division of Environmental Science and Engineering, National University of Singapore, Singapore*
*[3]School of Building and Development, BCA Academy (Building and Construction Authority of Singapore), Singapore*

**Abstract**

The aim of the study is to perform a potential health risk assessment on children in contracting respiratory symptoms due to inhaling traffic-generated nitrogen dioxide ($NO_2$) in two typical high-rise naturally-ventilated residential building designs (slab and point block) located close to busy major expressways in a tropical climate. A total of six buildings were selected for the study. Ogawa passive samplers (PS-100) were used for $NO_2$ measurements in each building over a period of 5 weeks during the predominant monsoon seasons. Health risk assessment showed children residing at the mid floors of the buildings had the highest health risk regardless of their age .i.e. infants, children (1 year and under), children (8-10 years)compared to those residents residing at the high and low floors. This was expected since the highest concentration of traffic-generated $NO_2$ concentration occurred at the mid floors of the buildings. In a typical floor, children (1 year and under) had the highest followed by children (8-10 years) whilst new born infants had the least potential health risk in contracting respiratory symptoms. The reason might could be new born infants obtain passive immunity from their mothers and in children (1 year and under), the passive immunity fall during this age period as they are developing their very own immunity against respiratory symptoms. Children (8-10 years) had the their potential health risk to respiratory symptoms in between the other two age groups as these children could have developed more immunity against respiratory symptoms compared to the children (1 year and under) but less immunity compared to infants. Based on the mean overall HR values, children living in a slab block has about 1.27 times more risk in contracting a respiratory symptoms due to $NO_2$ inhalation compared to those living in a point block.

**Keywords:** Air pollution; Outdoor $NO_2$; Indoor $NO_2$; Exposure risk assessment; Slab block; Point block

## Introduction

In urban air, $NO_2$, a secondary air pollutant, is mostly derived from local vehicular traffic through the oxidation of NO by $O_3$ in the atmosphere [1-3]. $NO_2$ is a highly reactive oxidant and corrosive in nature, but it is also a good surrogate indicator of traffic-generated pollutants [4]. $NO_2$ acts mainly as an irritant affecting the mucosa of the eyes, nose, throat, and respiratory tract. Prolonged exposure to high $NO_2$ concentration levels may contribute to the development of acute or chronic bronchitis whilst low concentration levels of $NO_2$ exposure may cause increased bronchial reactivity in some asthmatics, decreased lung function in patients with chronic obstructive pulmonary disease and increased risk of respiratory infections, especially in young children [5,6]. In view of the adverse health effects of $NO_2$, the World Health Organization set guideline values for $NO_2$ (a 1-hour level of 200 µg/m³ and an annual average of 40 µg/m³) [7].

There is limited study on the potential health risk assessment on children in contracting respiratory symptoms due to inhaling traffic-generated nitrogen dioxide ($NO_2$) in high-rise naturally-ventilated residential building located close to busy major expressways in a tropical climate. The US EPA has taken some initiative in developing a total risk integrated methodology [8] which is a set of models for assessing exposure and risk for criteria air pollutants and hazardous air pollutants. Most human exposure assessment models are often based on the outdoor pollutant concentration levels which are used as the input parameter for predicting total human exposure [9]. The U.S. EPA has developed several computer-based exposure models applicable to pollutants which are directly or indirectly linked to mobile sources [10-12]. A comparison of some of the models for human exposure to air pollution is discussed by Duan [13]. Some indirect exposure methods for estimating population exposures to indoor pollutants have been developed [14].

## Motivation of the study

In Singapore, about 1 out of 5 children are asthmatic and Singapore ranks number one in the Asia Pacific region in terms of the number of asthmatic kids between the ages 13-14 Y [15]. Singapore is a small nation with a hot and humid climate all year round, and has a land mass area of 712.4 km². The island has a current population of 5.18 million and a very high population density of 7126 persons/ km² as at Jun 2011. Singapore is ranked third in the world population density. About 83% of the residents live in high-rise public housing which is naturally-ventilated [16]. The land constraints and the very high population density are the main reasons for these buildings to be usually high-rise and in close proximity to each other. The point and slab block designs are typical residential housing designs. The point block is generally 'H' shape on plan which has 4 homes in each horizontal storey with 2 homes facing the expressway while the other two are facing away from the expressway. The slab block is rectangular on plan and normally has 18-20 homes, in each horizontal storey with their living rooms facing the expressway. Since Singapore does not have natural resources, the citizens including children have become a vital asset to the government. Motor vehicle population has inched upwards to 926,000 of which about 62% are due to car ownership [16]. The motor vehicle population continues to grow every year despite congestion pricing measures which could be due to the increasing number of new

***Corresponding author:** Mano Kalaiarasan, School of Building and Development, BCA Academy, 200 Braddell Road, Singapore-79700
E-mail: Mano_SELLAPPA@bca.gov.sg

residents. The increasing motor vehicle and resident population have caused concerns on the indoor and outdoor air quality in the urban areas, especially in high-rise residential buildings located near major roads and expressways as on-road motor vehicles are main sources of traffic-generated $NO_2$ in urban areas. To address the health impact issues among children who are living in naturally-ventilated high-rise residential buildings in contracting respiratory symptoms due to inhalation of traffic-generated $NO_2$, a potential health risk assessment study was conducted in the two typical building designs, the slab block (3 buildings) and point block (3 buildings) design using an established health risk model [17]. This paper will form a scientific basis for town planning of new urban developments in terms of mitigating health impacts in a tropical climate.

## Sampling strategy

The locations of the measurement sites are shown in Figure 1. Three buildings were located in close proximity to the Central Expressway (CTE). CTE is one of the most highly-utilized expressways which link many residential towns to the Central Business District of Singapore. Two buildings were located along the Ayer Rajah Expressway (AYE). The AYE is located at the southern-western part of Singapore. It enables traffic movement from the city to the residential area along the west, and ends at Tuas checkpoint before going to Malaysia. One of the buildings was located along the Pan Island Expressway (PIE). The PIE is the oldest and longest expressway in Singapore. It extends along the length of the island, connecting Tuas in the west to Singapore Changi Airport in the east. All expressways in Singapore are dual carriageways with each carriageway having 3 lanes [18].

Once the building has been selected, sampling locations in each building were identified for the measurement of $NO_2$ and environmental parameters such as temperature, relative humidity (RH), wind speed and direction at the different heights of the building. The measurements were made at the windward face of the building. All the buildings selected for the study were naturally-ventilated high-rise residential buildings and whose proximity to expressway are less than or equal to 30 meters downstream of the expressway. The ground on which the buildings stood was reasonably flat and almost the same level as the expressways. The predominant winds blew almost perpendicularly towards the building facades where the living rooms were located. As a result, most of the traffic-generated $NO_2$ was infiltrate into the buildings. All the measurement sites were selected such that the adjacent buildings nearby the site did not interfere with the predominant wind flow direction i.e. these measurements can be considered as being conducted in an open street geometry and street-canyon effect is negligible. The main obstacles for airflow towards the buildings were the trees and hedges planted along the expressways. In each expressway, the trees were planted in a single row and had drooping branches with dense and complex canopy structures as shown in Figure 2. This type of greenery has become a unique feature in Singapore for naturally-ventilated high-rise residential buildings that are located in close proximity to expressways. The average tree canopy spans up to about 4-5 storey high. Facing the tree canopy is 1 m tall hedges which are about 0.7 m wide.

## Instrumentation and measurement

For each building, indoor and outdoor $NO_2$ exposure levels were made using Ogawa passive samplers (PS-100) at every third floor of the buildings over a period of 5 weeks during the predominant Northeast (December-early March)/Southeast (June-September) monsoon seasons. Households with no/minimal smoking activity were selected for the measurement. Cooking hours per day ranged from 0.5-3hr mostly in the evenings as most of the adults were working during the day and children were at schools. All the measurements were done from the year 2005 to the year 2008. The location of instruments in a typical point and slab block is shown in Figures 3 and 4, respectively. Ogawa PS-100 passive samplers were used to measure indoor and outdoor $NO_2$ concentration levels of building. Sampling and analysis

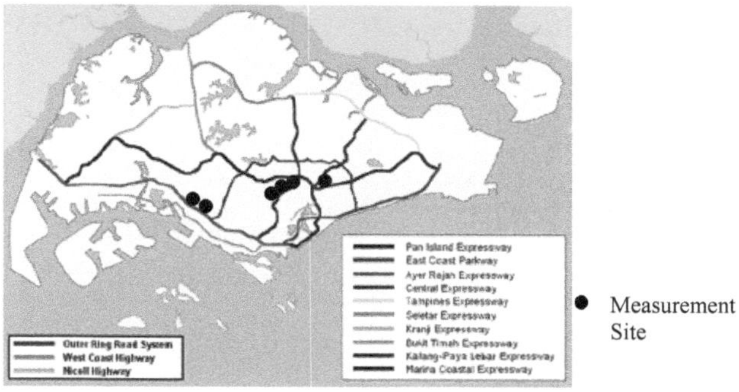


Source: Wikipedia [18].

**Figure 1:** Location of Measurement Sites along the Different Expressways.

**Figure 2:** Dense Complex Canopy Structure in Front of the Slab Block along the CTE.

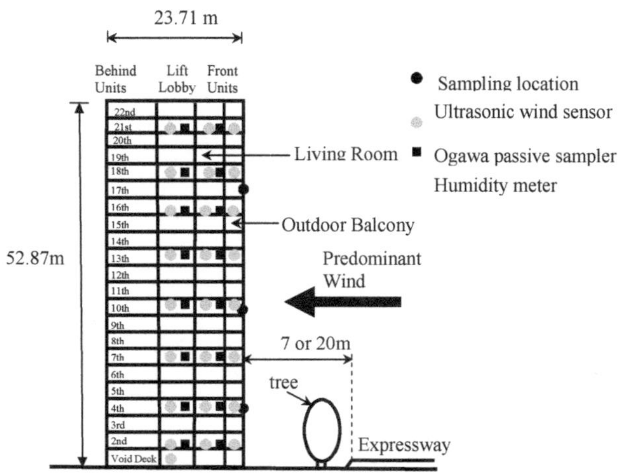


**Figure 3:** Experimental Setup in a Typical 22-Storey Point Block.

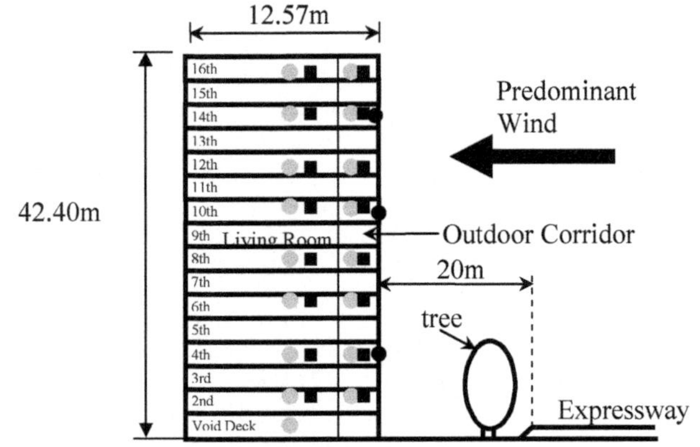


**Figure 4:** Experimental Setup in a Typical 16-Storey Slab Block.

was performed according to manufacturer protocols [19]. The passive samplers were exposed in duplicate together with a field blank. Each sampler was transported to and from the site in a sealed plastic bag in an airtight container. The methodology used for traffic and the indoor and outdoor $NO_2$ measurements has been published elsewhere [20]. $NO_2$ exposure models were derived using the data from field measurement and the results from these exposure models were fed into the established health risk model as input parameters. For the indoor measurements, the passive samplers were placed in the living rooms about 1.5-1.8 m high [21] from the floor (breathing zone) and at least 1m away from the any wall obstructions. For outdoor measurements, the passive samplers were located at the outdoor balcony and the lift lobby. The upstream background $NO_2$ levels were also measured using the passive samplers. All the passive samplers were exposed to exposure protection air for 1 week and kept refrigerated after each test. The samplers were sent back to the laboratory for chemical analysis within three days after the samples were collected. Field blanks were collected at the measured locations to determine any potential background contamination during sampling, transport and storage. 24hr temperature and relative humidity data were obtained from both indoors and outdoors using HOBO® Data Loggers placed at the living room, lift lobby areas and outdoor balconies of the selected households. The outdoor wind speed and direction were obtained using Ultrasonic Vaisala Wind Sensors (WS425) at three representative floors (low, mid and high floors) from 1000-2000 hr.

## Health risk model

The potential health risk of children in contracting respiratory symptoms due to inhalation of traffic-generated $NO_2$ was based on the health risk model described by Pandey et al. [17]. The model account for age specific dose rates, age-specific breathing rates, age-specific body weights, diurnal concentration of the pollutant, occupancy factor (percentage of population likely to be in the building at a given interval of time) and the lowest observed adverse effect levels (loael). Loael is defined as the lowest tested dose of a substance that has been reported to cause harmful (adverse) health effects on people or animals. Exposure occurs when a person comes in contact with a pollutant. A person's exposure to a pollutant is defined as the contact of the pollutant at one or more boundaries of the person i.e. mouth or skin at specific concentrations over a period of time. Dose occurs only when the pollutant crosses the physical boundary of a person. The analysis divides the children into three age-specific categories namely, infants, children (1 year and below) and children (8-10 years).

The dose rate of a pollutant in a building has been estimated through the following expression over a day as shown in Equation (1).

$$Dose\,rate(D) = [BR/BW]\int_0^{24} C(t)OF(t)\,dt \qquad (1)$$

where, D is the age-specific dose rate (μg kg$^{-1}$), BR is age-specific breathing rate (L min$^{-1}$), BW is age-specific body weight (kg), C(t) is diurnal concentration of the pollutant (μg m$^{-3}$), and OF($t$) is occupancy factor of zone (percentage of population likely to be in the zone at a given interval of time).

For the estimation of the potential health risks of a child due to

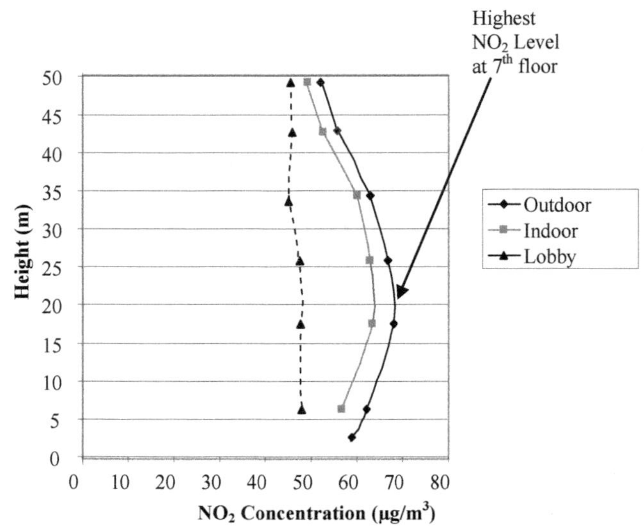


**Figure 5:** Vertical Distribution Profile of Weekly Mean $NO_2$ Concentration at AYE Point Block 39 along AYE.

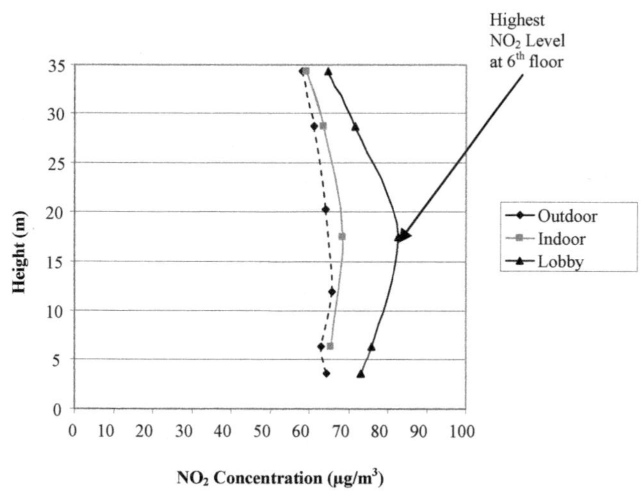


**Figure 6:** Vertical Distribution Profile of Weekly Mean $NO_2$ Concentration at Slab Block 401 along AYE.

| Age Group | Inhalation Volume (m³/day) | Body Weight (kg) |
|---|---|---|
| Children 8-10 year old | 10 | 30 |
| Children under 1-year old | 3.8 | 10 |
| New born | 0.8 | 3 |

Source: Cerna et al. [22]

**Table 1:** Breathing Rates and Body Weights of Children.

inhalation of $NO_2$, the following expression as shown in Equation (2) was used.

$$Health\ risk\ (HR) = [Dose\,rate\ /\ pollutant\ specific\ loael] \quad (2)$$

Information on the inhalation volume and body mass is obtained from Cerna et al. [22] as shown in Table 1. A loael value of 1.5 µg/kg per day for $NO_2$ as used by Pandey et al. [17] was adopted for this study. All these information is not available in Singapore. For the $NO_2$ loael value, it is the dose value at which expiratory flow rate becomes lower than 100%. This value was estimated, based on the dose-response model constructed on the basis of data available in Neuberger et al. [23]. The HR value is dimensionless and is useful for making relative comparisons.

## Results and Discussion

### Vertical distribution profile of $NO_2$

The indoor / outdoor (I/O) ratio values of $NO_2$ concentration in the apartments of all the point and slab blocks turned out to be less than unity, which suggests that the transport of traffic-generated $NO_2$ was mostly from outdoors to indoors rather than from within internal sources. The major outdoor source was the nearby traffic. For the point block, the I/O ratio of $NO_2$ concentration ranged from 0.84-0.98 (0.93 ± 0.04). An I/O ratio close to unity indicates that the outdoor air pollution made a significant contribution to the indoor air quality of the building.

The mean weekly mass concentration at the various floors of a building exceeded the WHO [7] maximum allowable annual mean value of 40 µg/m³. This is of major concern since the health of residents including that of "sensitive" populations such as asthmatics, children, and the elderly is likely to be adversely affected. The vertical distribution profile of traffic-generated $NO_2$ of a typical point block has been published elsewhere [20]. The vertical distribution profile of traffic-generated $NO_2$ of a typical slab block is similar to that of a typical point block. Figures 5 and 6 show the typical vertical distribution profiles of traffic-generated $NO_2$ in a point and slab block, respectively. The only difference between the point block and slab block configurations is that at corresponding floors, the $NO_2$ mass concentration for slab block was much higher than that of point block under similar traffic and meteorological conditions. This difference could be attributed to the slab block configuration which tends to slow down the wind speed thus allowing the accumulation of $NO_2$ mass concentration in front of the building.

### Health risk assessment study

**Exposure model:** The $NO_2$ exposure models for $NO_2$ mass concentration developed for a typical a 22-storey point block and a 16-storey slab block is shown in Equations (1)-(2) based on the observational data collected over different measurement periods. Find Graph version 2.01 was used for the regression analysis. It was found that the Gaussian plus a line model function was the best function to fit the measured daily mean vertical distribution profile of $NO_2$ ($r^2$ = 0.91-0.94; $p < 0.04$) for the point and slab blocks. For the mean estimated dose rate and the dimensionless health risk values, conventionally employed somatic and respiration parameters for the various age groups were used as shown in Table 1.

For the 22-storey building, the generalized exposure model is shown in Equation (1):

$$Weekly\ mean\ indoor\ NO_2\ concentration$$

$$= 55.074 + 12.212e^{-0.5\left[\frac{H-22.274}{14.456}\right]^2} - 0.071H$$

$$(\mu g/m^3) \qquad (r^2 = 0.94) \qquad (1)$$

For the 16-storey building, the generalized exposure model is shown in Equation (2):

| Floor | Dose Rate(µg kg⁻¹) | | | HR(dimensionless) | | |
|---|---|---|---|---|---|---|
| | New born | Children (1 yr and under) | Children (8-10 yr) | New born | Children (1 yr and under) | Children (8-10 yr) |
| 22 | 8.32 ± 1.30 | 11.86 ± 1.08 | 10.40 ± 1.05 | 0.79 ± 0.06 | 1.13 ± 0.07 | 0.99 ± 0.02 |
| 21 | 8.41 ± 0.98 | 11.98 ± 1.23 | 10.52 ± 1.27 | 0.80 ± 0.05 | 1.14 ± 0.05 | 1.00 ± 0.06 |
| 20 | 8.53 ± 1.16 | 12.15 ± 1.36 | 10.66 ± 1.42 | 0.81 ± 0.03 | 1.16 ± 0.03 | 1.01 ± 0.07 |
| 19 | 8.68 ± 0.75 | 12.35 ± 1.01 | 10.83 ± 1.13 | 0.83 ± 0.01 | 1.18 ± 0.08 | 1.03 ± 0.05 |
| 18 | 8.83 ± 1.04 | 12.58 ± 2.02 | 11.04 ± 1.16 | 0.84 ± 0.06 | 1.20 ± 0.05 | 1.05 ± 0.07 |
| 17 | 9.02 ± 1.31 | 12.86 ± 1.79 | 11.28 ± 0.73 | 0.86 ± 0.09 | 1.22 ± 0.06 | 1.07 ± 0.04 |
| 16 | 9.23 ± 1.70 | 13.16 ± 0.96 | 11.54 ± 0.84 | 0.88 ± 0.03 | 1.25 ± 0.09 | 1.10 ± 0.05 |
| 15 | 9.46 ± 1.08 | 13.48 ± 1.45 | 11.82 ± 0.89 | 0.90 ± 0.06 | 1.28 ± 0.07 | 1.13 ± 0.08 |
| 14 | 9.69 ± 0.99 | 13.81 ± 2.22 | 12.11 ± 1.45 | 0.92 ± 0.03 | 1.31 ± 0.04 | 1.15 ± 0.02 |
| 13 | 9.92 ± 1.98 | 14.14 ± 2.35 | 12.40 ± 1.96 | 0.94 ± 0.02 | 1.35 ± 0.03 | 1.18 ± 0.06 |
| 12 | 10.13 ± 2.15 | 14.43 ± 1.92 | 12.66 ± 0.91 | 0.96 ± 0.07 | 1.37 ± 0.09 | 1.21 ± 0.04 |
| 11 | 10.3 ± 1.88 | 14.68 ± 1.73 | 12.88 ± 1.14 | 0.98±0.08 | 1.40 ± 0.04 | 1.22 ± 0.03 |
| 10 | 10.43 ± 1.27 | 14.86 ± 2.02 | 13.04 ± 1.49 | 0.99 ± 0.05 | 1.42 ± 0.07 | 1.24 ± 0.05 |
| 9 | 10.50 ± 2.01 | 14.97 ± 1.18 | 13.12 ± 1.24 | 1.00 ± 0.09 | 1.43 ± 0.03 | 1.25 ± 0.01 |
| 8 | 10.52 ± 1.67 | 14.99 ± 1.64 | 13.14 ± 1.03 | 0.97 ± 0.04 | 1.43 ± 0.05 | 1.25 ± 0.04 |
| 7 | 10.48 ± 0.75 | 14.93 ± 1.08 | 13.10 ± 0.68 | 0.94 ± 0.03 | 1.39 ± 0.07 | 1.20 ± 0.06 |
| 6 | 10.3 9± 1.39 | 14.80 ± 1.15 | 12.98 ± 0.59 | 0.92 ± 0.06 | 1.36 ± 0.06 | 1.15 ± 0.03 |
| 5 | 10.25 ± 1.44 | 14.61 ± 2.38 | 12.81 ± 1.17 | 0.91 ± 0.05 | 1.33 ± 0.04 | 1.12 ± 0.02 |
| 4 | 10.09 ± 1.91 | 14.39 ± 1.67 | 12.62 ± 1.39 | 0.89 ± 0.08 | 1.29 ± 0.06 | 1.09 ± 0.04 |
| 3 | 9.92 ± 0.96 | 14.13 ± 1.35 | 12.40 ± 0.83 | 0.87 ± 0.05 | 1.25 ± 0.04 | 1.08 ± 0.05 |
| 2 | 9.74 ± 0.78 | 13.89 ± 2.06 | 12.18 ± 0.74 | 0.86 ± 0.02 | 1.24 ± 0.03 | 1.07 ± 0.06 |
| 1 | 9.58 ± 1.20 | 13.65 ± 1.51 | 11.98 ± 1.07 | 0.85 ± 0.09 | 1.22 ± 0.02 | 1.04 ± 0.05 |

**Table 2:** Predicted Dose Rates and HR Values Due to $NO_2$ Inhalation at the Various Floors of a Typical 22-Storey Point Block.

| Floor | Dose Rate(μg kg$^{-1}$) | | | HR(dimensionless) | | |
|---|---|---|---|---|---|---|
| | New born | Children (1 yr and under) | Children (8-10 yr) | New born | Children (1 yr and under) | Children (8-10 yr) |
| 16 | 10.61 ± 1.13 | 15.13 ± 1.55 | 13.27 ± 1.24 | 1.01 ± 0.04 | 1.44 ± 0.03 | 1.26 ± 0.05 |
| 15 | 10.74 ± 1.28 | 15.30 ± 1.36 | 13.42 ± 1.48 | 1.02 ± 0.02 | 1.46 ± 0.01 | 1.28 ± 0.02 |
| 14 | 10.90 ± 0.96 | 15.53 ± 1.19 | 13.62 ± 1.36 | 1.04 ± 0.06 | 1.48 ± 0.04 | 1.30 ± 0.04 |
| 13 | 11.12 ± 1.35 | 15.84 ± 1.28 | 13.90 ± 1.55 | 1.06 ± 0.05 | 1.51 ± 0.08 | 1.32 ± 0.03 |
| 12 | 11.43 ± 1.49 | 16.27 ± 2.62 | 14.27 ± 0.76 | 1.09 ± 0.04 | 1.54 ± 0.04 | 1.36 ± 0.03 |
| 11 | 11.80 ± 1.16 | 16.82 ± 1.71 | 14.75 ± 1.86 | 1.12 ± 0.03 | 1.60 ± 0.06 | 1.41 ± 0.06 |
| 10 | 12.25 ± 1.81 | 17.45 ± 2.55 | 15.31 ± 1.69 | 1.17 ± 0.05 | 1.66 ± 0.05 | 1.46 ± 0.02 |
| 9 | 12.71 ± 1.12 | 18.11 ± 1.94 | 15.88 ± 1.51 | 1.21 ± 0.02 | 1.72 ± 0.08 | 1.51 ± 0.03 |
| 8 | 13.10 ± 1.61 | 18.67 ± 2.33 | 16.37 ± 1.90 | 1.25 ± 0.07 | 1.78 ± 0.09 | 1.56 ± 0.07 |
| 7 | 13.35 ± 0.98 | 19.03 ± 1.88 | 16.69 ± 1.05 | 1.27 ± 0.04 | 1.81 ± 0.05 | 1.59 ± 0.04 |
| 6 | 13.42 ± 1.81 | 19.13 ± 2.01 | 16.78 ± 1.30 | 1.28 ± 0.06 | 1.82 ± 0.06 | 1.53 ± 0.07 |
| 5 | 13.31 ± 1.86 | 18.97 ± 1.95 | 16.64 ± 1.64 | 1.22 ± 0.02 | 1.75 ± 0.02 | 1.49 ± 0.03 |
| 4 | 13.07 ± 1.37 | 18.63 ± 1.44 | 16.34 ± 1.77 | 1.18 ± 0.03 | 1.65 ± 0.05 | 1.56±0.02 |
| 3 | 12.77 ± 1.52 | 18.20 ± 2.27 | 15.96 ± 1.81 | 1.16 ± 0.04 | 1.62 ± 0.05 | 1.46 ± 0.04 |
| 2 | 12.48 ± 1.65 | 17.79 ± 1.96 | 15.60 ± 1.47 | 1.13 ± 0.02 | 1.59 ± 0.08 | 1.43 ± 0.05 |
| 1 | 12.26 ± 1.70 | 17.47 ± 1.42 | 15.32 ± 1.63 | 1.11 ± 0.05 | 1.53 ± 0.06 | 1.41 ± 0.06 |

**Table 3:** Predicted Dose Rates and HR Values Due to $NO_2$ Inhalation at the Various Floors in a Typical 16-Storey Slab Block.

| | HR for $NO_2$ (Dimensionless) | | | | | |
|---|---|---|---|---|---|---|
| | Point Block | | | Slab Block | | |
| Floor | New born | Children (1 yr and under) | Children (8-10 yr) | New born | Children (1 yr and under) | Children (8-10 yr) |
| Low floor | 0.90 ± 0.02 | 1.29 ± 0.04 | 1.10 ± 0.01 | 1.14 ± 0.04 | 1.62 ± 0.03 | 1.44 ± 0.05 |
| Mid floor | 0.98 ± 0.02 | 1.37 ± 0.03 | 1.20 ± 0.02 | 1.21 ± 0.04 | 1.73 ± 0.02 | 1.52± 0.04 |
| High floor | 0.83 ± 0.03 | 1.18 ± 0.02 | 1.04 ± 0.03 | 1.04 ± 0.03 | 1.49 ± 0.04 | 1.30 ± 0.03 |
| $HR_{mid\ floor}/HR_{high\ floor}$ | 1.16 | 1.15 | 1.24 | 1.15 | 1.16 | 1.17 |
| $HR_{low\ floor}/HR_{high\ floor}$ | 1.09 | 1.06 | 1.13 | 1.10 | 1.08 | 1.11 |

**Table 4:** Predicted HR Values at the Different Floors of the 22-Storey Point Block and 16-Storey Slab Block.

*Weekly mean indoor $NO_2$ concentration* =

$$74.057 + 13.042e^{-0.5\left[\frac{H-16.836}{8.207}\right]^2} - 0.195H \quad (\mu g/m^3) \quad (r^2 = 0.91) \quad (2)$$

Where H represents the storey in meters as measured from ground floor.

**Comparison of predicted health risk (hr) results in buildings:** The predicted potential health risk (HR) values of children living in a typical 22-storey point block and a 16-storey slab block are shown in Table 2-4. For both the blocks, the health risk analysis indicated that children living at the mid floors of the buildings had the highest health risk in contracting respiratory symptoms for all age categories: infants, children (1 year and below) and children (8-10 years) compared to the high (lowest) and low floors (second highest) due to $NO_2$ inhalation. This was expected since the highest concentration of to $NO_2$ concentration occurred at the mid floors of the buildings. In a floor, new born infants had the least potential health risk to respiratory symptoms which could be due the passive immunity obtained from their mothers whilst children (1 year and under) had the highest potential health risk to respiratory symptoms. The reason might be these children's passive immunity fall during this age period and are developing their very own immunity against respiratory symptoms. Children (8-10 years) had the their potential health risk to respiratory symptoms in between the other two age groups as these children could have developed more immunity against respiratory symptoms compared to the children (1 year and under) but less immunity compared to infants. For the point block, the mean overall mean HR value for $NO_2$ is 1.06 ± 0.18 whereas for the slab block, the meal overall HR value for $NO_2$ is 1.35 ± 0.21. Based on the mean overall HR values, children living in a slab block has about 1.27 times more risk in contracting a respiratory symptom due to $NO_2$ inhalation compared to those living in a point block.

## Conclusion

The study shows children residing at the mid floors of the buildings had the highest health risk regardless of their age. In a typical floor, new born infants had the least potential health risk to respiratory symptoms which could be due the passive immunity obtained from their mothers whilst children (1 year and under) had the highest potential health risk to respiratory symptoms. The reason might be these children's passive immunity fall during this age period and are developing their very own immunity against respiratory symptoms. Children (8-10 years) had the their potential health risk to respiratory symptoms in between the other two age groups as these children could have developed more immunity against respiratory symptoms compared to the children (1 year and under) but less immunity compared to infants. Based on the mean overall HR values, children living in a slab block has about 1.27 times more risk in contracting a respiratory symptoms due to $NO_2$ inhalation compared to those living in a point block. Hence in terms of health concern, the point block is of a better design compared to the slab block. It is recommended the point block design is adopted for future housing projects located very close to major roads or expressways since the point block design pose lesser potential health risk in terms of contracting respiratory symptoms compared to the slab block design. To house more residents, taller point block designs could be considered. Other measures to mitigate $NO_2$ exposure to children is to introduce more free air passages like intermediate roof garden close to the mid and lower floors so as to suppress any $NO_2$ peaks at the mid floors.

**Acknowledgement**

I wish to express my deepest gratitude and acknowledgement to World Future

Foundation (WFF) especially to Dr Feng Lun and the dedicated staff whose kind and generous support has made this paper possible. WFF is the first philanthropic foundation in Singapore funded by entrepreneurs from mainland China, and is professionally managed by an international team. WFF is dedicated to promote research on environmental protection and sustainability, especially focusing on the development models of future cities.

## References

1. Perez P, Trier A (2001) Prediction of NO and $NO_2$ concentrations near a street with heavy traffic in Santiago, Chile. Atmos Environ 35: 1783-1789.

2. Abam FI, Unachukwu GO (2009) Vehicular Emissions and Air Quality Standards in Nigeria. European J Sci Res 34: 550-560.

3. Ali M, Athar M (2007) Air pollution due to traffic, air quality monitoring along three sections of National Highway N-5, Pakistan. Environ Monit Assess 136: 219-226.

4. Jerrett M, Arain A, Kanaroglou P, Beckerman B, Potoglou D, et al. (2005) A review and evaluation of intraurban air pollution exposure models. J Expo Anal Environ Epidemiol 15: 185-204.

5. US EPA (2012) Air Quality Criteria Document, US Environmental Protection Agency.

6. Gillespie-Bennett J, Pierse N, Wickens K, Crane J, Howden-Chapman P, et al. (2010). The respiratory health effects of nitrogen dioxide in children with asthma. Eur Respir J 38: 303-309.

7. WHO (2005) Air Quality Guidline.

8. US EPA (2005) Total risk integrated methodology (TRIM) air pollutants exposure model documentation (TRIM.Expo /APEX Version 4) Volume I: User's guide. Research Triangle Park, NC, Office of Air Quality Planning and Standards.

9. Colls JJ, Micallef A (1997) Towards better human exposure estimates for setting of air quality standards. Atm Env 31: 4253-4254.

10. Graham SE, Burke JM (2004) Human exposure modeling for air pollutants using SHEDS. Washington State University.

11. Ozkaynak AH, Zufall MJ, Burke JM, Xue J, Zidek J (1999) A probabilistic population exposure model for $PM_{10}$ and $PM_{2.5}$. Epidemiology 10: S79-S79.

12. McCurdy T (1995) Estimating human exposure to selected motor vehicle pollutants using the NEM series of models: Lessons to be learned. J Expo Anal Environ Epidemiol 5: 533-550.

13. Duan N (1982) Models for human exposure to air pollution. Environ Int 8: 305-309.

14. Klepeis NE (1999) An introduction to the indirect exposure assessment approach: modeling human exposure using microenvironmental measurements and the recent National Human Activity Pattern Survey. Environ Health Perspect 107: 365-374.

15. (1998) International Study of Asthma and Allergies in Childhood (ISSAC).

16. (2012) Department of Statistic, Singapore.

17. Pandey JS, Kumar R, Devotta S (2005) Health risks of $NO_2$, SPM and $SO_2$ in Delhi (India). Atmos Environ 39: 6868-6874.

18. (2012) Wikipedia. Expressways of Singapore.

19. Ogawa (1997) NO, $NO_2$, $NO_X$ and $SO_2$ sampling protocol using the ogawa sampler.

20. Kalaiarasan M, Balasubramanian R, Cheong KWD, Tham KW (2008) Field-based investigation of traffic pollutants in multi-storey buildings in Singapore. Proceedings of IA2008, Copenhagen, Denmark.

21. Kalaiarasan M, Balasubramanian R, Cheong KWD, Tham KW (2009) Traffic-generated airborne particles in naturally ventilated multi-storey residential buildings of Singapore: Vertical distribution and potential health risks. Special Issue of Building and Environment 44: 1493-1500.

22. Cerna M, Jelinek R, Janoutova J, Kotesovec F, Benes I, et al. (1998) Risk assessment of the common air pollutants in Teplice, Czech Republic. Toxicol Lett 96: 203-208.

23. Neuberger M, Moshammer H, Kundi M (2002) Declining ambient air pollution and lung function improvement in Austrian children. Atmos Environ 36: 1733-1736.

# Seismic Analysis of Indoor Auditorium

**Dilipan Bose S[1*] and Aravindan S[2]**

[1]PG Student, M.tech Structural Engineering, Bharath University, India
[2]Assistant Professor, Department of Civil Engineering, Bharath University, India

**Abstract**

The project titled "Seismic analysis of Indoor Auditorium" has been taken up with an objective to determine the seismic response and behaviour of an Auditorium constructed in Chennai area. Even though Chennai is considered as least prone to major earthquake, it is expected that a structure would survive major earthquakes without collapse that might occur unexpectedly during the life of the building. It should also be noted that after the Bhuj earthquake, Indian Standard IS: 1893 was revised and Chennai city was upgraded from zone II to zone III which leads to a substantial increase of the design ground motion parameters. Hence, this project presents an exploratory analysis of the seismic performance of multi-storey buildings system built in the specified area with a comparative study of the structures under past major earthquakes. Computer modeling is undertaken using Ansys is a multi-purpose software which enables to run and simulate tests or working conditions. It also helps in determining and improving weak points, computing by 3D simulations in virtual environment.

**Keywords:** Auditorium; Analysis; Ansys; Earthquake; Seismic; Finite element analysis; Static analysis

## Introduction

An auditorium is a room built to enable the audience to hear and watch performances at venues such as theatres. Generally, an auditorium has an increasing sloped-styled seating, So as to allow the audience at the back of the auditorium to see the stage without any disturbances. The structural design of an auditorium should be a sloped foundation so as to follow the seating layout. Seismic analysis of an auditorium is done using Ansys and various seismic modes and data are collected. The member with the highest deflection is pointed out and strenghtening methods can be carried out in the next project [1,2].

## Objective of the Project

- To create a 3-D finite element model
- Analyse the element using Static structural analysis, Modal analysis and Response spectrum analysis
- Find out the member with maximum displacement due to seismic activity.

## Methodology

### Finite element program

The ANSYS program is a computer program for Finite element analysis and design. Also used to find out how a given design works under operating conditions. It has its own integrate pre and post processor. Ansys program can be used in all disciplines of engineering-structural, mechanical, electrical, electromagnetic, electronic, thermal, and fluid.

### Basic concepts in FEA

Stress analysis one of the engineering discipline helps in determining stress in materials and structures that are subjected to static or dynamic forces.

The aim of this analysis is to determine the collection of elements, which is also referred as a structure, can safely withstand the specified forces. This can be achieved when the determined stress from the applied forces are less than the ultimate tensile strength and the ultimate compressive strength the material is known to be able to withstand, inspite of that a factor of safety is applied in design.

The factor of safety is one of the important design requirements for any structure based on uncertainty in loads, material strength (yield and ultimate), and consequences of failure. The factor of safety is to prevent detrimental deformations and the factor of safety on ultimate strength is to prevent collapse. The factor of safety is used to calculate the maximum allowable stress [3].

Factor of Safety = Ultimate Tensile Strength/Maximum allowable stress

When performing stress analysis, a factor of safety is calculated to compare with the required factor of safety. The factor of safety is a design requirement given to the stress analyst. The Analyst calculates the design factor. Margin of safety is another way to express the design factor.

Design Factor = Ultimate Tensile Strength/Maximum Calculated Tensile Stress. A key part of analysis involves determining the type of loads acting on a structure, including tension, compression, shear, torsion, bending, or combinations of such loads.

### Problem definition

The finite element method is a numerical method for solving problems of engineering and mathematical physics. It is useful for problems with complicated geometries, loadings, and material properties where analytical solutions cannot be obtained [4].

FEA is a technique that discretizes a given physical or mathematical problem into smaller fundamental parts called elements. An analysis of each element is conducted. A solution to the problem as a whole is obtained by assembling the individual solutions of the elements. Complex problems can be tackled by dividing the problem into smaller and simpler problems that can be solved using existing mathematical tools.

***Corresponding author:** Dilipan Bose S, PG Student, M.tech Structural Engineering, Bharath University, India, E-mail: dlbose143@yahoo.com

### Steps in finite element modeling

- Discretize and Select Element Types
- Select a Displacement Function
- Define the Stress/Strain Relationships
- Derive the Element Stiffness Matrix and Equations
- Assemble the Element Equations to obtain the Global Equations
- Solve for the Unknown Displacements
- Solve for the Element Strains and Stresses
- Interpret the Results.

### Finite element meshing of auditorium

Grid generation is the first step of analysis. Grid generation usually requires simplification and idealization of the design model. This requirement is the most cumbersome aspect of grid generation process. Therefore, the analysis model is often rebuilt from scratch, based upon the judgment of skilled analysts in removing details from the design, and duplicating much of the work in creating the geometry. Often, integrated tools are interactive and require the design engineer to provide complex input.

The geometrical question in a finite-element analysis is represented by collection of finite elements used and is known as mesh. Creating the mesh is often the most difficult part of finite element modelling. As a result, the grid generation process is not yet a 'push button" process; it is the most labor- intensive and time-consuming aspect of the computational structural dynamics. It takes too many man-hours and calendar days, and it requires a grid specialist. Mesh generation could be performed graphically, drawing lines on a computer to form elements. Techniques such as extruding a shell mesh to create a solid mesh were developed for creating complex geometric models. With use of solid modelling CAD programs, the geometry to be used in finite element analysis already exists in some format. This geometry can be used for simulation, only if the programs being used have the interface capabilities. Once we have created the geometry in the system, the finite element programs create some type of automatic meshing. In beam elements it is the process of breaking up bigger line entities into smaller line entities. In 2D elements, flat surfaces must be broken into the elements having three or four sides. In shell elements, 3D surfaces must be equally divided into the shell elements. In solid elements, a particular volume must be sectioned into elements. Automatic meshing capabilities directs the user place the elements. Most automatic mashers create tetrahedral elements in solid volumes. Tetrahedral elements can be less in accuracy compared to brick elements. But structural dynamics analysis tetrahedral mesh is sufficient. Structural analysis using Finite Element Method gives a preliminary assessment of dynamic behavior of mechanical structures in order to establish the guidelines for designing the experiment. The main objective of this study was to develop a methodology of seismic analysis of electronic cabinets using ANSYS. Generally has two approaches i.e. transient analysis and response spectrum analysis are adopted for the seismic analysis in ANSYS.

## Meshing

Meshing is used in a Workbench is to provide robust, ease in the use of meshing. These tools are highly automated and it has a moderate to high degree of user control. To carry out a finite element analysis, the model must be divided into a number of small pieces and those pieces are known as finite elements. After the model has been divided into a number of discrete parts, Finite Element Analysis can be described as a discretization technique.

If the system we investigate is one-dimensional in nature, we may have to use line elements to represent our geometry and carry out our analysis. If the system is two-dimensional, then a 2D mesh is required If the system is a complex and 3D representation of the continuum, then we use a 3D mesh.

### Mesh requirements

The Finite Element Method (FEM) has certain requirements on a mesh:

- The mesh must be valid, (no holes, self-intersections, or faces joined at two or more edges).
- The mesh must conform to the boundary of the domain. This is an obvious requirement, but some schemes such as a Delaunay triangulation may not satisfy this condition.
- The density of the mesh must be controllable, to allow trade-off between accuracy and solution time.
- The grid density will vary depending on local accuracy requirements, but any variations must be smooth to reduce or eliminate numerical diffusion/refraction effects.
- There are some requirements on the shape of elements. In general, the elements should as equiangular as possible in equilateral triangles & regular tetrahedral. Highly distorted elements (long, thin triangles, squashed tetrahedral) can lead to numerical stability problems caused by round-off errors. This requirement is modified for boundary layers, where highly stretched elements are desired and facilitated in the FEM formulation. The min-max-angle property is still required in this case.

Figures 1 and 2 shows the FEA Model with Meshing which includes 320 columns and 494 beams.

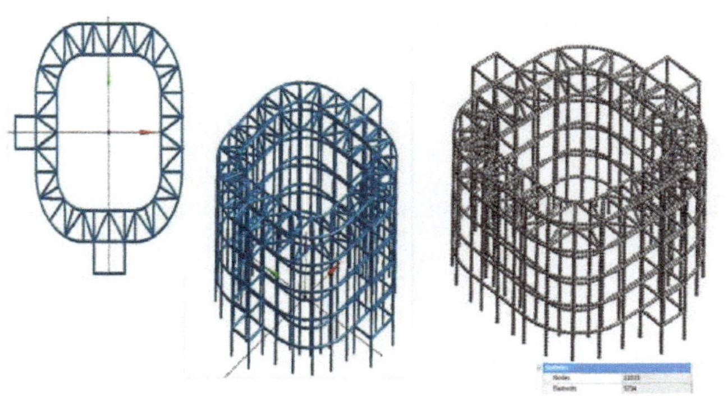

**Figure 1:** FEA Model with Meshing-320 columns and 494 beams.

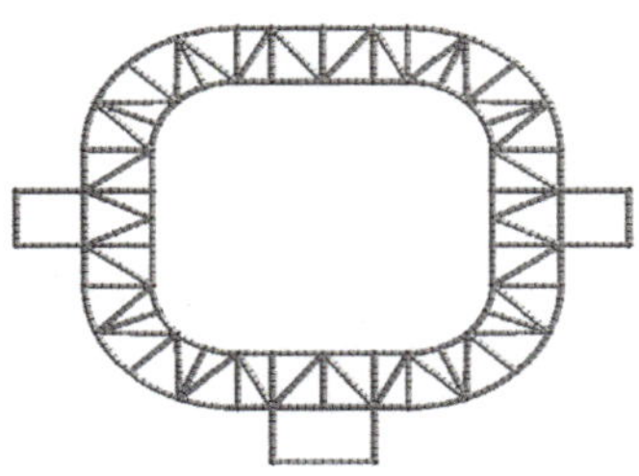

**Figure 2:** Meshing Top View.

## Static Structural Analysis

Static analysis calculates the effects of *steady* loading conditions on a structure. A static analysis includes *steady* inertia loads and time-varying loads that can be accounted as static equivalent loads [5]. Static analysis is used to determine the displacements, stresses, strains, and forces in structures or components. These are the kinds of loading that can be applied in a static analysis:

- Externally applied forces and pressures
- Steady-state inertial forces (gravity or rotational velocity)
- Imposed (non-zero) displacements
- Temperatures (for thermal strain)
- Fluences (for nuclear swelling)

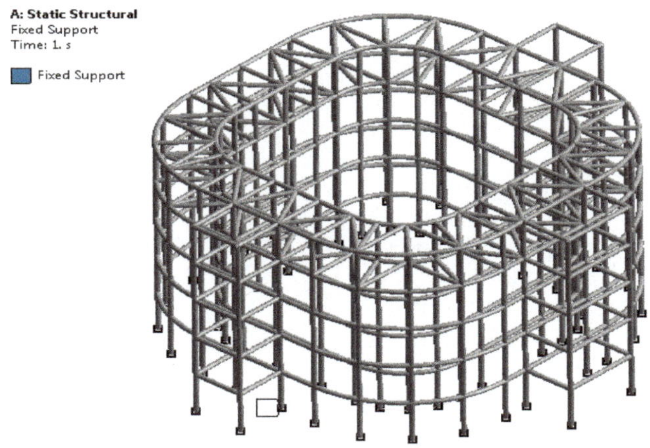


**Figure 3:** Fixed Support BC's.

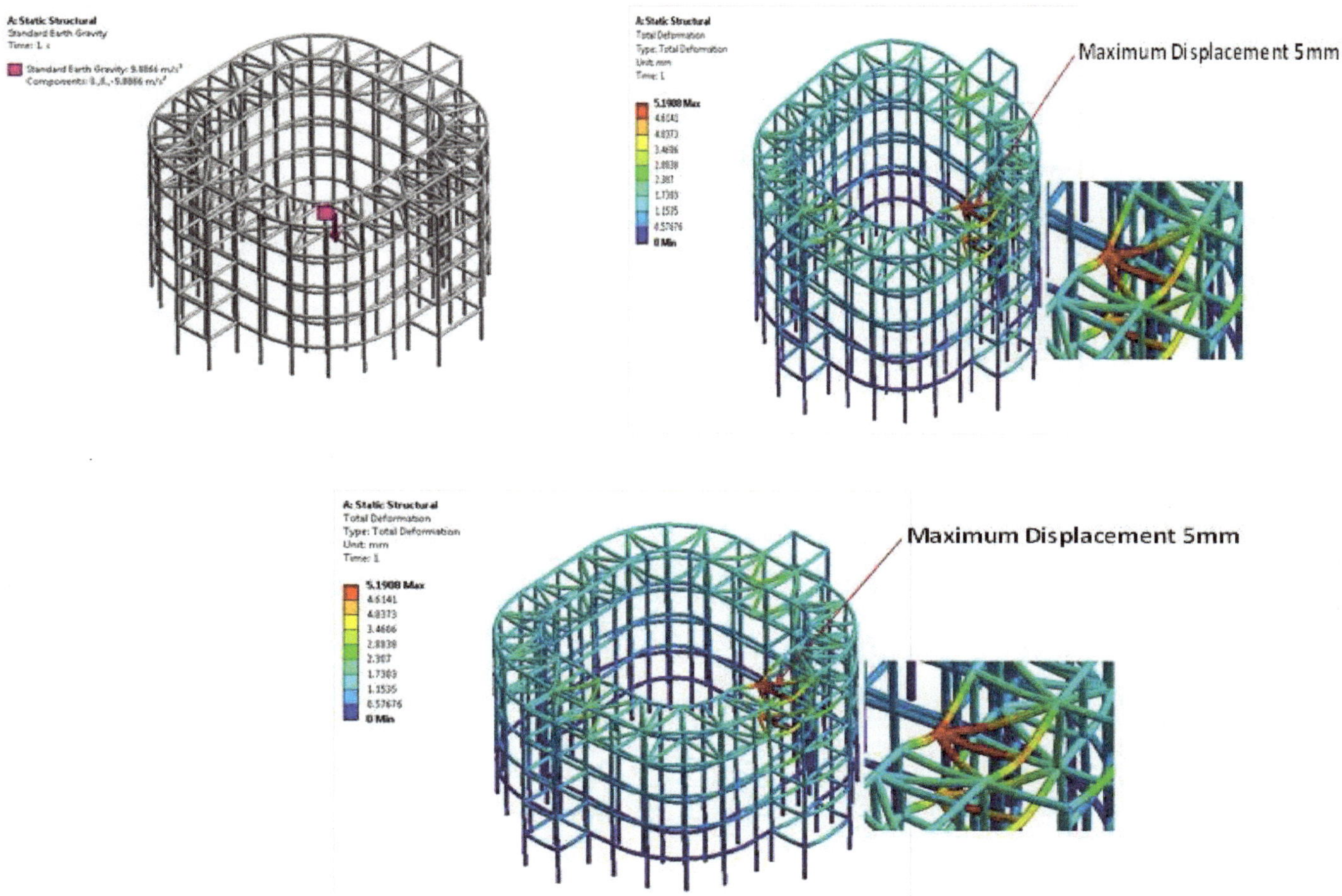


**Figure 4:** Displacement Contours.

## Overview of steps in a static analysis

The procedure for a static analysis consists of three main steps:

- Building the model.
- Applying loads and obtaining the solution.
- Reviewing the results. (Figures 3-8)

The first mode of the auditorium is rotating about an axis parallel to the x-axis in the E-W direction and the building swings in the N-S direction (Figure 9). The second and third modes are both twisting, in clockwise and counter-clockwise, respectively (Figure 9a and 9b). The

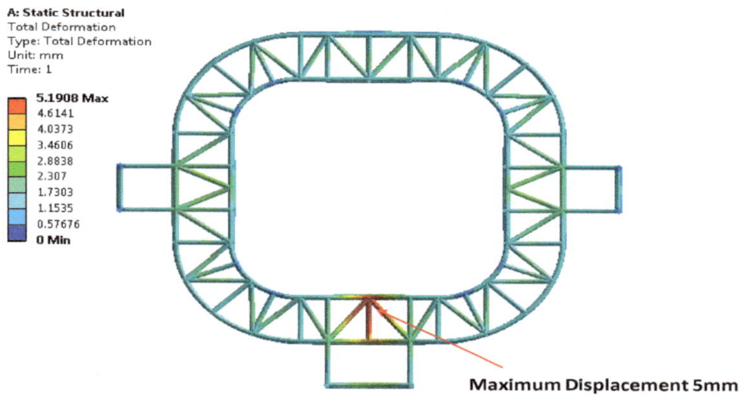


**Figure 5:** Displacement Contours-Top View.

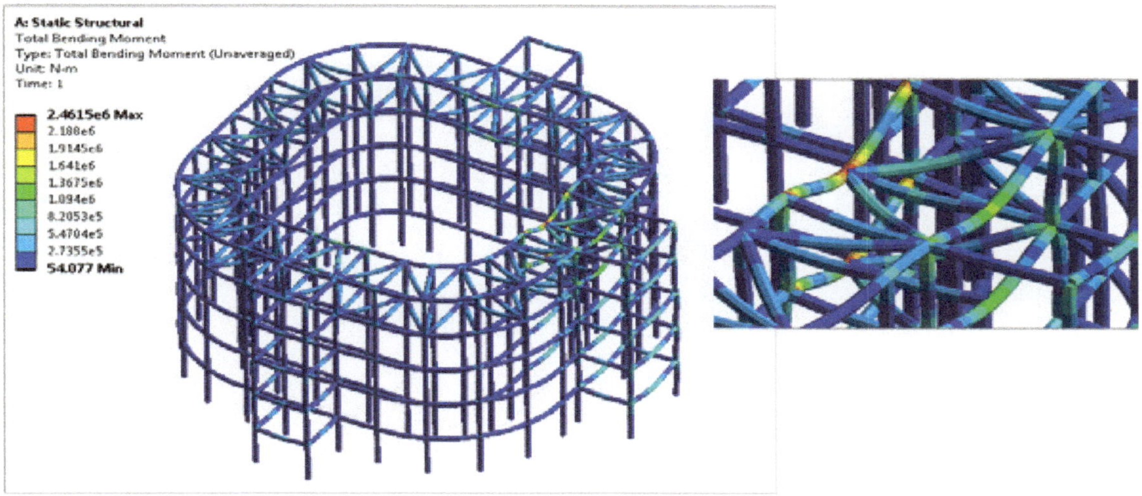


**Figure 6:** Total Bending Moment.

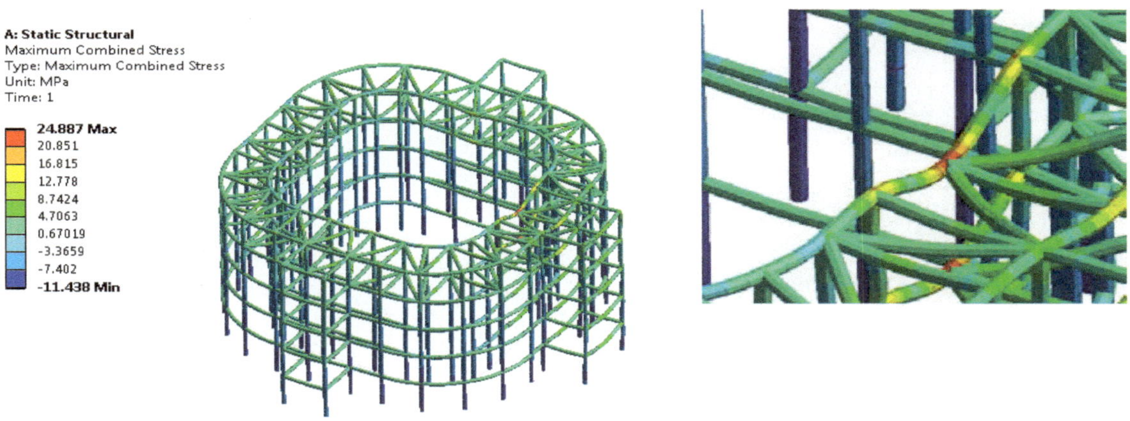


**Figure 7:** Maximum Combined Stress.

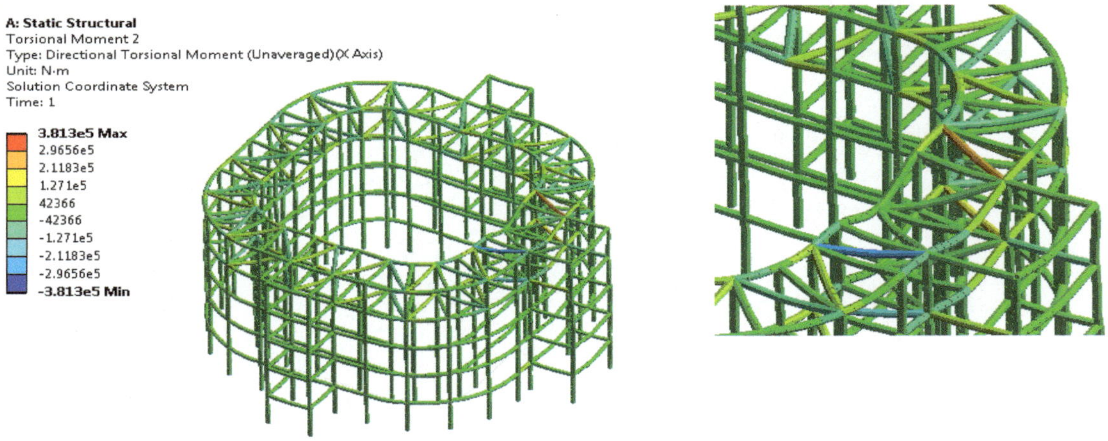


**Figure 8:** Torsional Moment.

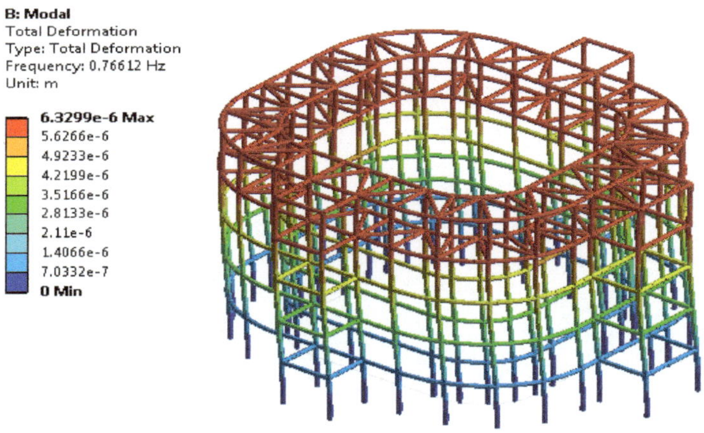


**Figure 9:** Natural Frequencies of Indoor Stadium: Mode-1.

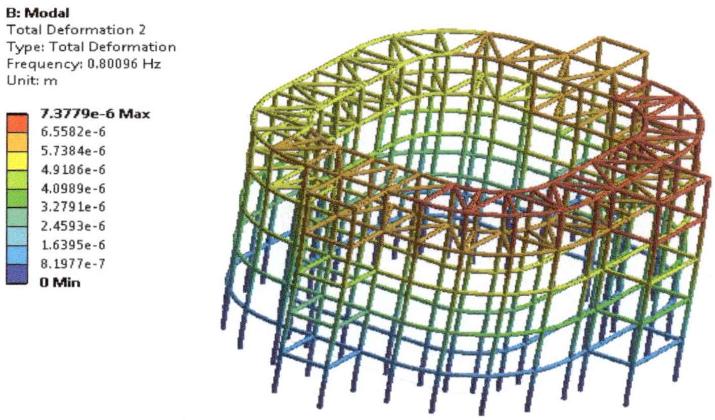


**Figure 9a:** Natural Frequencies of Indoor Stadium:Mode-2.

fourth mode is bending in the N-S direction (Figure 9c). The last mode shows severe vertical motion at the northwest part of the auditorium and some bending (Figure 9d).

Spectrum analysis yielded a lot of information. For the excitation, there are three displacement response values at each node: translational displacements in the x, y and z direction, rotations about the x, y and z axis, and total displacement and rotation [6]. The same can be said about the acceleration. Therefore, a total of 3 displacement graphs can be generated. In addition, vector graphs can also be created (Figures 10-14).

## Conclusion

A 3-D finite element model was built and analyzed in the study.

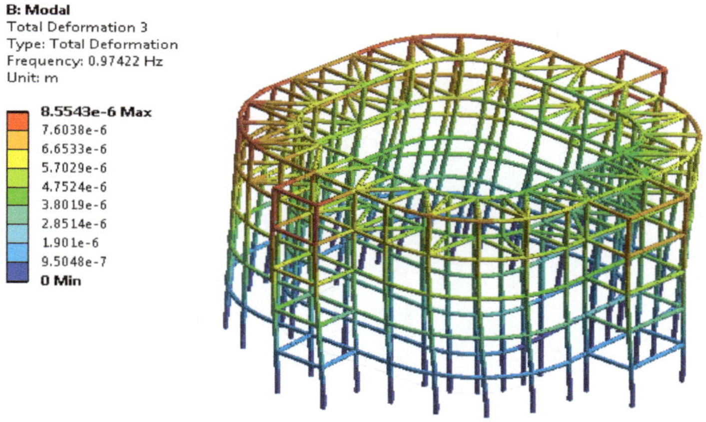


**Figure 9b:** Natural Frequencies of Indoor Stadium: Mode-3.

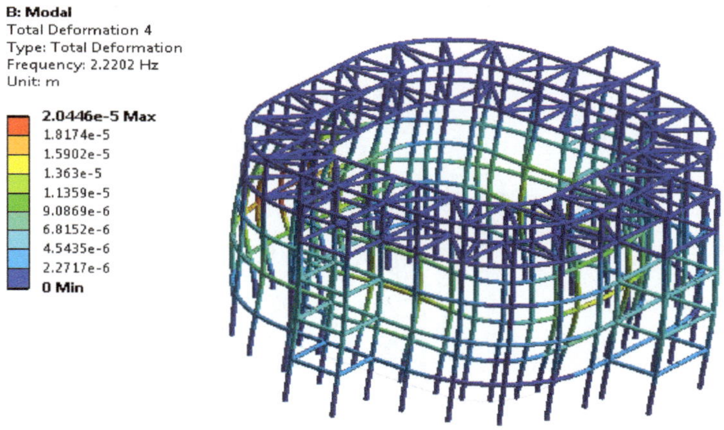


**Figure 9c:** Natural Frequencies of Indoor Stadium: Mode-4.

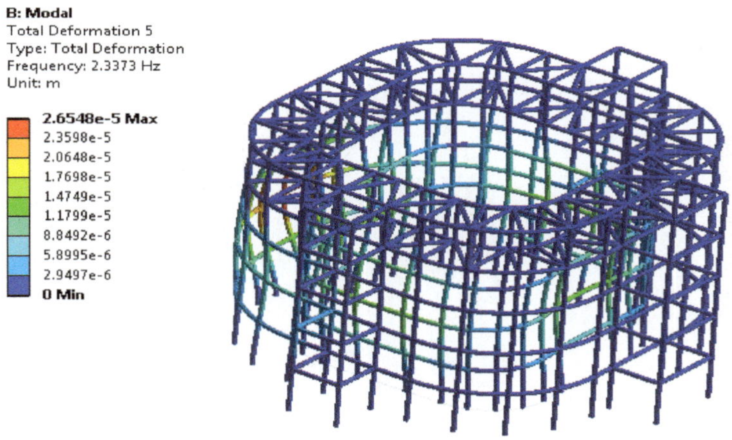


**Figure 9d:** Natural Frequencies of Indoor Stadium: Mode-5.

The model is based on the actual dimensions from the blueprint and includes all structural components. Static and Modal and spectrum analyses were performed with the FE package, ANSYS. The fundamental frequency obtained from the modal analysis compares well with the empirical value. Displacement, acceleration and stress distributions are generated from spectrum analysis with the response spectrum. It is suggested from the analyses that the most sensitive areas for the accelerate meters to pick up structural motions are the top level and the northwest part of the building. .

## Acknowledgment

My foremost gratitude goes to all that have contributed in the achievement of this study. I am Grateful to the Honourable Chancellor, **Dr. J. Sundeep Anand**, Bharath University, Chennai.

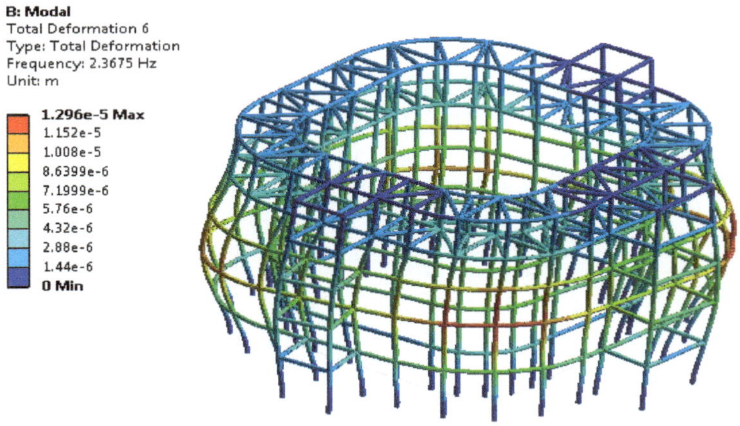


**Figure 9e:** Natural Frequencies of Indoor Stadium: Mode-6.

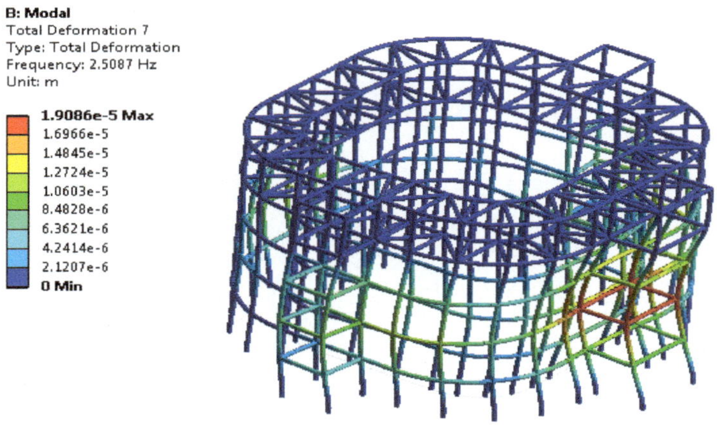


**Figure 9f:** Natural Frequencies of Indoor Stadium: Mode-7.

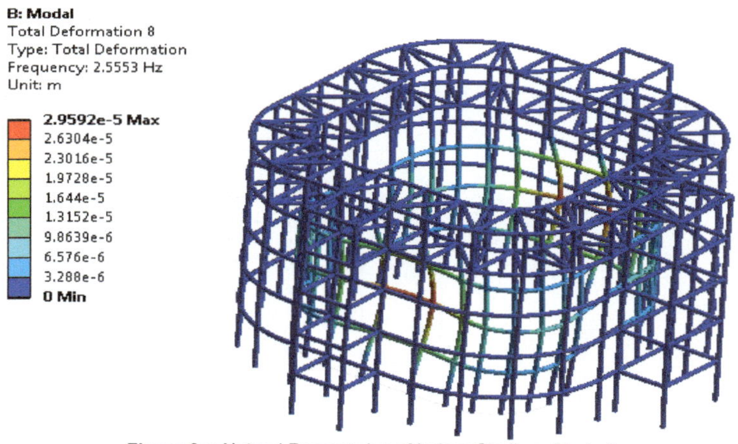


**Figure 9g:** Natural Frequencies of Indoor Stadium: Mode-8.

I am Grateful to **Dr. v. Tamizharasan**, Head Of the Department, Civil Engineering, Bharath University, Chennai. I am deeply indebted to my supervisor, Professor **S. Aravindan**, Assistant professor, Civil Engineering, Bharath University, Chennai.

I am sincerely grateful to Library of Structural Engineering Research Centre, Taramani, Chennai, for their support in providing me with Journals.

The Experimental work was carried out in the laboratory of Civil Engineering Department, Bharath University, Chennai. I am grateful to the support and assistance provided by the team of talented and dedicated technical staff.

## References

1. Iqbal AS, Pampanin S, Buchanan a, Palermo A (2007) Improved Seismic Performance of LVL Post-tensioned Walls Coupled with UFP devices. ACI Structural Journal 62: 348-352.

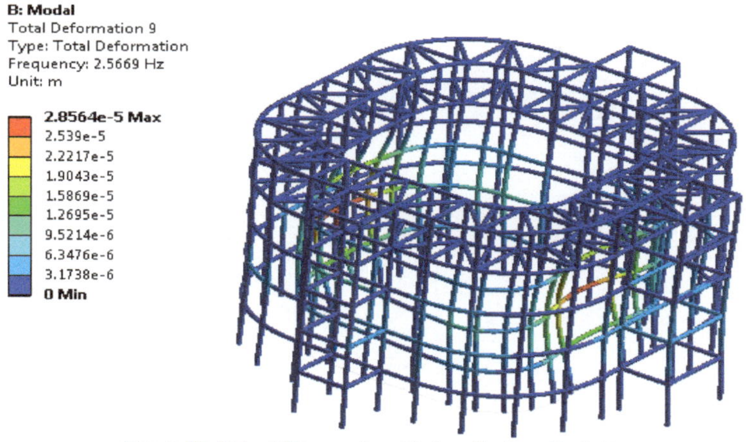


**Figure 9h:** Natural Frequencies of Indoor Stadium: Mode-9.

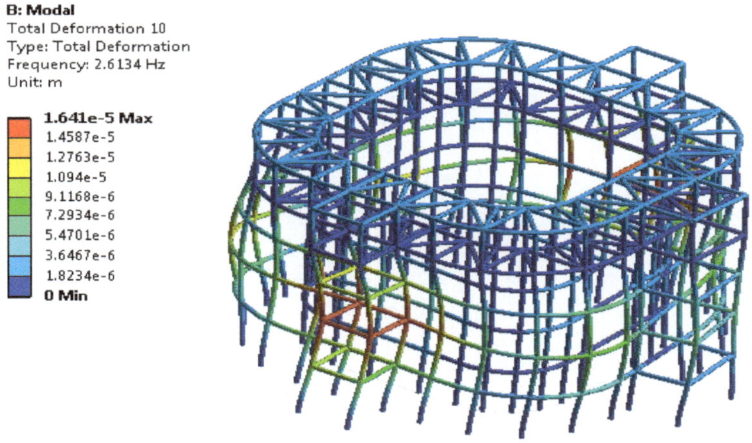


**Figure 9i:** Natural Frequencies of Indoor Stadium: Mode-10.

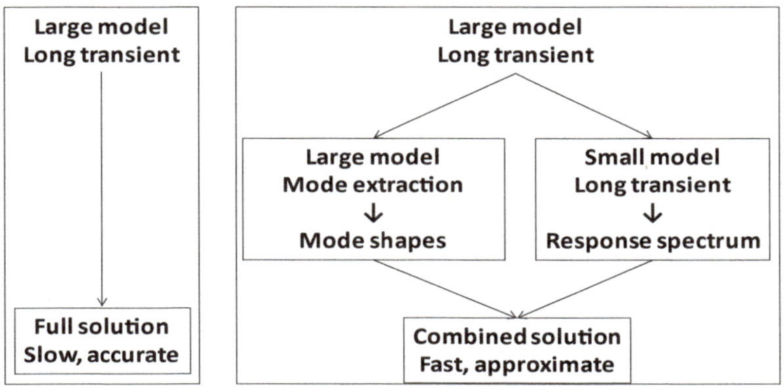


**Figure 10:** Response Spectrum Analysis.


2. Newcombe MP, Pampanin S, Buchanan A, Palermo A (2008) Section Analysis and Cyclic Behavior of Post-Tensioned Jointed Ductile Connections for Multi-Story Timber Buildings. Journal of Earthquake Engineering 12: 83-110.

3. Pampanin S, Christopoulos C, Nigel Priestley MJ (2003) Performance-Based Seismic Response of Frame Structures Including Residual Deformations. Part II: Multi-Degree of Freedom Systems. Journal of Earthquake Engineering 7: 119-14.

4. Durrani AJ, Wight JK (1985) Behavior of Interior Beam-to-Column Connections under Earthquake-Type Loading. ACI Structural Journal 82: 343-349.

5. Carr AJ (1994) Dynamic Analysis of Structures. Bulletin, New Zealand National Society for Earthquake Engineering 27: 129-146.

6. Priestley MJN, Kowalaky MJ (2000) Direct Displacement-Based Seismic Design of Concrete Buildings. Bulletin, New Zealand National Society for Earthquake Engineering 33: 421-444.

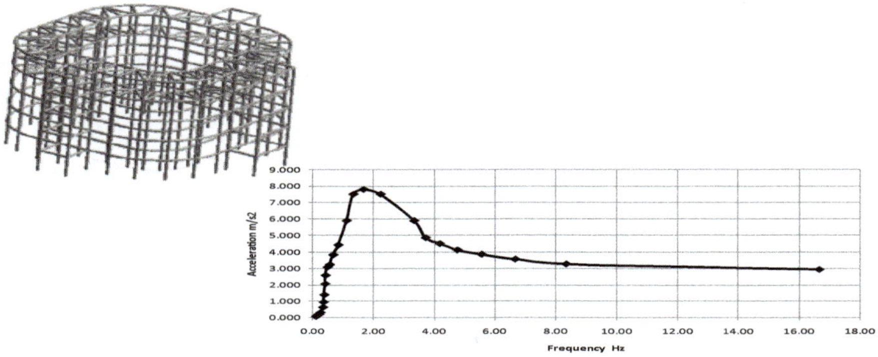


**Figure 11:** Seismic Load.

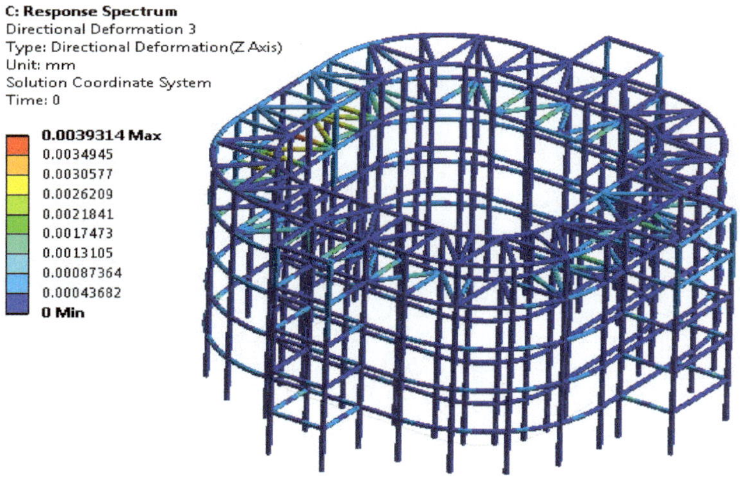


**Figure 12:** Deformation –X.

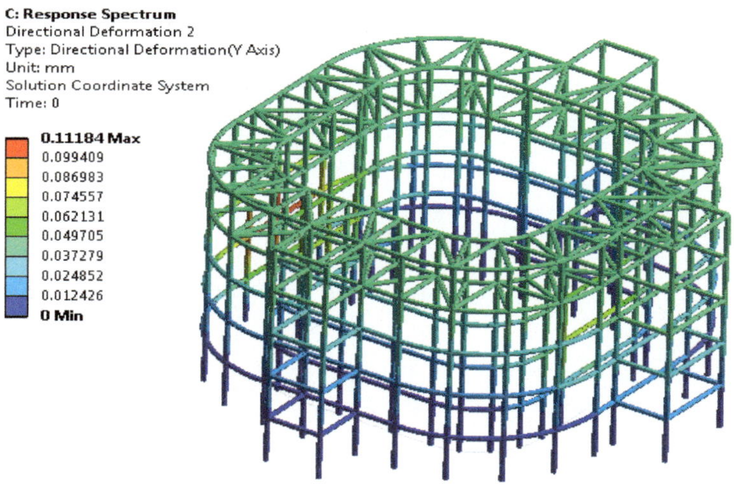


**Figure 13:** Deformation –Y.

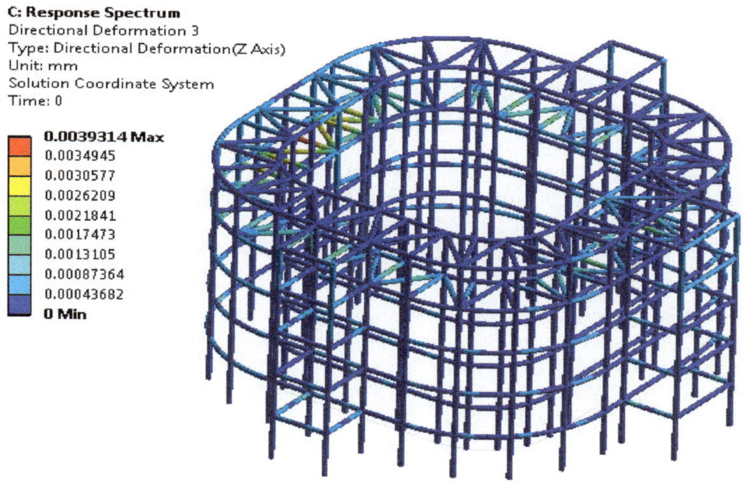


**Figure 14:** Deformation –Z.

16

# Emission Inventory for Criteria Pollutants of a Thermal Power Plant

**Anand Kumar Varma S* and Srimurali M**

*Department of Civil Engineering, S.V. University, Tirupati, AP, India*

**Abstract**

The emissions inventory is the basic requirement for modeling of air quality, environmental monitoring and impact assessment. For any air pollution control program emission inventory is a key component. Emission inventory should be developed for acquiring emission data of significant amount. Fulfillment of air emissions inventory for a power plant identifies the source of air pollution. The air pollutants released from the entire source are measured to estimate the emission rates and emission concentrations.

In the present study, we made an emission inventory for criteria pollutants of Suspended Particulate Materials (SPM), sulfur oxides ($SO_x$), nitrogen oxides ($NO_x$) and carbon monoxide (CO) from stationary sources of a thermal power plant. Fuel consumption and emission database was developed for Rayalaseema thermal power project (1050 MW). Emission factors are calculated for the pollutants. The project lies on latitude of 14°42'52"N and longitude of 78°27'29"E located at VV Reddy Nagar, Kadapa, Andhra Pradesh, India. The methods used for inventory presentation are Manufacturers' design specifications and direct measurement. Results are compared from two methods.

**Keywords:** Thermal power plant; Emission inventory; Criteria pollutants; Modeling; Impact assessment

## Introduction

### Emissions inventory

The emissions inventory [1] is the primary task to identify the various sources of pollution from the particular project. It will give the physical and geographical conditions of the source. More reliable data can be produced regarding emission rates and emission concentrations. Systematic procedure is provided for emission inventory as follows:

### Source identification

For the development of emission inventory [2] clear understanding of the project is essential. Physical site visit of the project and complete study on processing is essential to locate the required air pollution sources. Detailed review of maps or engineering drawings of the project site is essential; this can be obtained from review of literature on the industry. The sources of air pollution at the project identified the following:

- Release Type
- Location
- Emission of criteria air pollutants

#### Release type

Point sources, five stacks are identified with 220 m. tall and 4 m. diameter at the exit. These stacks emit flue gases into the atmosphere forcefully above ground level [3]. These are considered as tall point sources [2].

#### Wake: Affected Point Sources

There are no nearby buildings or other structures to interfere with the plume rise and dispersion. The point source is absolutely wake free source.

#### Location

The site of Rayalaseema Thermal Power Project [4] at Kalamala is located at a distance of 15 Km from Yerraguntla, 20 Km from Proddutur, 45 Km from Jammalamadugu and 8 Km from Muddanur railway station which is situated on Chennai-Guntakal Broad gauge line. The project site is at 180 m above MSL and is far away from any forest boundary. It lies on latitude of 14°42'52"N and longitude of 78°27'29"E. The project is located in an area of 1650 Hectares. The following Figure 1 depicts the location of the power plant where the study is carried out.

#### RTPP: Background

Andhra Pradesh state electricity board set up the Thermal power station at Kalamala with an installed capacity of 1050 Mw. It is established with five units under three stages. Each unit generates 210 mw of power. Table 1 gives the basic details regarding the capacity of the plant.

### Criteria air pollutants emitted

Criteria air pollutants [5] from the source are: Suspended

**Figure 1:** Rayalaseema TPP - Location map.

| Stage / Stations | Units | Capacity (MW) | Date of Commissioning |
|---|---|---|---|
| Stage I | Unit 1 | 210 | 31-03-1994 |
| Stage I | Unit 2 | 210 | 25-02-1995 |
| Stage-II | Unit3 | 210 | 25-01-2007 |
| Stage-II | Unit 4 | 210 | 20-11-2007 |
| Stage III | Unit 5 | 210 | 29-11-2010 |
| **Total** | | **1050** | |

**Table 1:** Installed capacity of power plant.

***Corresponding author:** Anand Kumar VS, Research scholar, Department of Civil Engineering, SV University, Tirupati, AP, India
E mail: manju_sakv@yahoo.co.in

Particulate Materials (SPM), sulfur oxides ($SO_x$), nitrogen oxides ($NO_x$) and carbon monoxide (CO) from stationary source of a thermal power plant. Ozone also produced as secondary pollutant from its precursors. All the pollutants produced are considered as photo toxicants [6]. They will show severe environmental impact on man, material, livestock and vegetation.

## Methods for the Estimation of Emission Rates

There are number of methods [7] for the estimation of emission rate and emission concentration from each source. In the present investigation direct measurement, Manufacturer's design specifications and Emission factors [5] are used for emission inventory.

### Manufacturer's design specifications

Manufacturer's design specifications are developed from composition of Singareni coal to estimate the emission rates of air pollutants from identified sources. Here quality of coal is taken into consideration to estimate emission rates and emission concentrations.

### Fuel (Coal)

RTPP is linked with Singareni Collieries Co., Ltd., for supply of coal. This company has a proven coal resource of around 4000 million tones with 70% extractability. This company is sufficient to produce 30000 MW of electricity for 25 years. Annual coal requirements are calculated on the basis of the following and presented in Table 2.

Coal consumption 137 Tph. per unit

Coal consumption for 5 units = 137 × 5 = 685 Tph.

Coal consumption for 5 units = 685 × 24 × 365 = 6000600 Tones

| Parameter | Mean Level |
|---|---|
| Hydrogen | 2.74 % |
| Carbon | 37.73 % |
| Nitrogen | 0.74 % |
| Sulfur | 0.44 % |
| Ash | 44.84 % |
| Moisture | 6.17 % |
| Oxygen | 7.34 % |
| Gross calorific value | 3686 K Cal/Kg |

**Table 2:** Typical Ultimate Coal Analysis.

| Combustion Product | Total Emission Rate |
|---|---|
| $SO_2$ | 3973 kg/h |
| $NO_X$ | 254.1 kg/h |
| $H_2O$ | 9.232 % |
| $O_2$ | 3.837 % |
| GCV | 3698 |

**Table 3:** Calculated combustion products.

### Coal combustion calculation results

The following Table 3 shows the results obtained from coal combustion calculated for the 685 Tons of coal per hour.

### Direct measurement

In Rayalaseema thermal power project Emission rates and source release parameters are established from the results of source emission online flue gas monitoring system direct pc based automation system for monitoring & recording facility capable of monitoring $O_2$, CO, $SO_2$, $NO_x$, $CO_2$ gas, stack temperature, excess air, combustion efficiency system provided with all the required accessories as required install after economizer and near id fan, to make the fully functional.

Automatic flue gas sampling & filtration system, mounted in weather proof ip55 protected enclosure

SPM is estimated by gravimetric method.

Pollutant emission rates are directly measured from the stack by using continuous monitoring system [8]. All the parameters required for the presentation of emission inventory are measured directly at the source. This method is more appropriate and reliable in emission inventory [9]. Direct measurement includes all stack release parameters specified in Table 4.

### Calculation of flue gas concentration for point sources

The equation used for the calculation of pollutant concentration released from a source is:

$$Cp = \frac{ERp}{FR} \qquad .........1$$

Where:

$C_p$ = the concentration of pollutant p emitted in mg/m$^3$

ERp = the rate of emission of pollutant p in mg/s

FR = the flow rate of flue gas in m$^3$/s

The emission inventory can be expressed in two types of emission concentrations:

1. Actual concentration [10] of a pollutant released from a source in mg/Am$^3$. It is calculated from the measured gaseous volumetric flow rate (Am$^3$/s) and measured emission rate (Table 5).
2. Actual pollutant concentration released from a source is corrected to the normal conditions in mg/Nm$^3$ [10]. This is calculated using the gaseous volumetric flow rate corrected to normal conditions (dry, 273K, 101.3 KPa) (Table 5).

Total emission rates and concentrations from all stacks in Rayalaseema thermal power plant is presented in Table 6.

Emission Factors

An emission factor [5] is a value derived to calculate the emission of a pollutant throughout the process. These factors will give average

| Source | Source type | Height of the Stack (m) | Exit Temperature of Flue gas (°C) | Diameter of the stack (m) | Exit Velocity (m/S) | $O_2$(%) | $H_2O$ (%) | Flow rate (Am³/S) | Flow rate (Nm³/S) |
|---|---|---|---|---|---|---|---|---|---|
| Stack - 1 | Wake-free point source | 220 | 140 | 4 | 28.1 | 4.5 | 9 | 3.37 | **330** |
| Stack - 2 | Wake-free point source | 220 | 140 | 4 | 28.52 | 4.6 | 10 | 3.41 | **328** |
| Stack - 3 | Wake-free point source | 220 | 140 | 4 | 29.9 | 4.2 | 9.3 | 3.46 | **326** |
| Stack - 4 | Wake-free point source | 220 | 140 | 4 | 28.3 | 4.4 | 9.8 | 3.35 | **324** |
| Stack - 5 | Wake-free point source | 220 | 140 | 4 | 28.7 | 4.4 | 9.6 | 3.37 | **329** |

**Table 4:** Stack source release parameters.

| Source | Pollutant | Emission Rate (gm./s) | Corrected Concentration (mg/Nm³) | Regulated Concentration (mg/Nm³ at stack reference conditions) |
|---|---|---|---|---|
| Stack -1 | SPM | 32.67 | 99 | 150 |
| | Sulfur dioxide | 224 | 679 | NA |
| | Nitrogen oxides | 13.53 | 41 | NA |
| | Carbon monoxide | 0.73 | 2.22 | NA |
| Stack -2 | SPM | 32.8 | 100 | 150 |
| | Sulfur dioxide | 221.7 | 676 | NA |
| | Nitrogen oxides | 14.4 | 44 | NA |
| | Carbon monoxide | 0.73 | 2.24 | NA |
| Stack -3 | SPM | 32.6 | 100 | 150 |
| | Sulfur dioxide | 221.6 | 680 | NA |
| | Nitrogen oxides | 14.67 | 45 | NA |
| | Carbon monoxide | 0.74 | 2.29 | NA |
| Stack -4 | SPM | 33 | 102 | 150 |
| | Sulfur dioxide | 220.9 | 682 | NA |
| | Nitrogen oxides | 13.6 | 42 | NA |
| | Carbon monoxide | 0.71 | 2.2 | NA |
| Stack -5 | SPM | 33.2 | 101 | 150 |
| | Sulfur dioxide | 206.2 | 679 | NA |
| | Nitrogen oxides | 13.1 | 40 | NA |
| | Carbon monoxide | 0.75 | 2.3 | NA |

**Table 5:** Stack Emission Concentrations.

| Source | Pollutant | Total Emission Rate (kg/h) | Emission Concentration (kg/Nm³ at stack reference conditions) |
|---|---|---|---|
| Stacks - 1,2,3,4 and 5 | SPM | 591.31 | 0.502 |
| | $SO_2$ | 3939.84 | 3.396 |
| | $NO_X$ | 249.48 | 0.212 |
| | CO | 13.176 | 0.011 |

**Table 6:** Total emission rates and concentrations.

| Emission product | Emission Rate (Manufacturers' design specifications) | Emission Rate (Direct measurement) |
|---|---|---|
| SPM | ---- | 591.31 kg/h |
| $SO_2$ | 3973 kg/h | 3939.84 kg/h |
| $NO_X$ | 254.1 kg/h | 249.48 kg/h |
| CO | ---- | 13.176 kg/h |
| $H_2O$ | 9.232% | 9.54% |
| $O_2$ | 3.837 % | 4.42% |
| GCV | 3698 | 3686 |

**Table 7:** Comparison of emission rates.

value of available data of acceptable quality, and these are generally representing long-term averages of the source type. Emission factors are used when other information is not available. Emission factors for criteria pollutants of Rayalaseema Thermal Power Project are calculated [11] as follows:

Emission Factor (EF) for SPM = ER/Activity .....2 (Emission control 99.5%)

591.31/685 = 0.863

Annual emission rate (ER annual) = EF × Activity annual ......3

Annual emission rate (ER annual) = 0.863 × 6000600 = 5178 Tpy

Emission Factor (EF) for $SO_2$ = ER/Activity .....4

3939.84/685 = 5.75

Annual emission rate (ER annual) = 5.75 × 6000600 = 34503 Tpy

Emission Factor (EF) for $NO_X$ = ER/Activity .....5

249.48/685 = 0.364

Annual emission rate (ER annual) = 0.364 × 6000600 = 2184 Tpy

Emission Factor (EF) for CO = ER/Activity .....6

13.176/685 = 0.019

Annual emission rate (ER annual) = 0.019 × 6000600 = 114 Tpy

The emission factors calculated can be used to quantify the emissions of various power plants using coal from Singareni Collieries Co., Ltd., depending on the activity.

## Presentation of Emissions Inventory

Comparison of emission inventory by Manufacturers' design specifications and direct measurements calculation results of total emission rates [7] of various combustion products are represented in Table 7.

## Conclusion

This study provided all basic information and data required for air pollution modeling, and clearly described the details of site specific conditions. Emission rates of criteria pollutants and other combustion products are presented from two methods, Manufacturers' design specifications and direct measurement. Direct measurement is more accurate methodology for the development of emission inventory for coal combustion thermal power plants.

In this emission inventory presentation the results observed from Manufacturers' design specifications and direct measurement are comparatively equal. Emission rates provided can be applied for environmental modeling and for environmental impact assessment criteria. Emission factors also calculated for criteria pollutants for future estimations, if emission data is not available for that particular power project. Annual emission rates are provided from emission factors and annual activity.

### References


1. EMEP/EEA air pollutant emission inventory guidebook.
2. Barclay J, Scire J, (2005) Approved Methods for the Modeling and Assessment of Air Pollutants in New South Wales - NSW Government Gazette, Sydney.
3. Bonnie N (2004) Alberta Environment Summary Report on 2004 NPRI Air Emissions.
4. Sastry CA (1993) Environmental impact assessment report of RTPP.
5. Compilation of Air Pollutant Emission Factors (1995) U.S. Environmental Protection Agency, Office of Air Quality Planning and Standards, Office of Air and Radiation: Research Triangle Park, NC.
6. Mittal LM, Sharma C, Singh R Estimates of Emissions from Coal Fired Thermal Power Plants in India.
7. Emission estimation technique manual for fossil fuel electric power generation (1999) NPI, Environment Australia.
8. Continuous Emissions Monitoring Systems Data. U.S. Environmental Protection Agency. Clear Air Markets Division.
9. Technology Transfer Network: Clearing house for Inventories & Emissions Factors, Emissions Inventories, U.S. Environmental Protection Agency.
10. Barclay J, Scire j (2005) Approved Methods for the Sampling and Analysis of Air Pollutants in New South Wales, Department of Environment and Conservation NSW, Sydney.
11. Granholm JM, Steven EG, Chester, Calculating air emissions for the Michigan air emissions reporting system (MAERS).

# Life Cycle Cost Analysis (LCCA) Delivery Model for an Urban Flexible Pavement

**Audu HAP[1], Aniekwu AN[2] and Oghorodje EE[3]***

*[1,2]Department of Civil Engineering, Faculty of Engineering, University of Benin, Benin City, Edo State, Nigeria*
*[3]Department of Highway Engineering, Ministry of Works, Asaba, Delta State, Nigeria*

**Abstract**

Pavement alternative evaluation is a key aspect in decision making process in transportation system planning, design and delivery. This study examined the Life-Cycle Cost Analysis (LCCA) model for flexible road pavement infrastructure delivery in a developing economy using 1.9 km urban road in Effurun Metropolis, Uvwie Local Government Area of Delta State of Nigeria as a case study. The overview of existing knowledge, the importance and evaluation of LCCA of pavement delivery is done. The Pavement delivery and design data used for this study, which were obtained from the Office of the Ministry of Works, Effurun, Delta State of Nigeria, included soil sub-grade CBR results, water content condition, soil type and grain size distribution, the reports of route investigation, vehicular traffic counts, and previous maintenance, material and market survey reports. The pavement design data previously used for the design and procurement of urban roads in the area were used for the design of three competing pavement alternatives models, viz: Hot Rolled Asphalt (HRA), Interlocking Concrete Block Pavement (ICBP) and a third; Do-nothing scenario pavement alternative. The Bill of Engineering Measurements and Evaluation (BEME) was developed for the designed alternative to compute their Life-Cycle Costs (LCCs) using present worth cost (PWC) method with varying interest rates in MS Excel software spread sheets for a design life of 20 years. Regression modeling for the three scenarios was done with computer software's (SPSS and MS Excel) packages. The model's data robustness was determined and found to be between 0.95-0.99. The model's validity and sensitivity were also carried out. The study indicates a 43.8% cost savings with the ICBP alternative over the HRA option. The Do-nothing alternative is non–aesthetical and costliest due to the maintenance cost of re-instating the road structure. The models have significance value of less than 0.001(F) respectively. The developed LCCA template and model are applicable for investment studies and decision making as well as cost effectiveness assessment of pavement and bridge delivery.

**Keywords:** Flexible road pavement; Transport infrastructure delivery; Life cycle cost analysis; Present worth cost

## Introduction

Transportation system is the planned networks of real and abstract components that interacts and play different roles in the process of physical movement of goods and persons from an origin or source to a destination point through a motorized or non-motorized or combined means. The planning and design of flexible road pavement continue to be the core of the transportation engineering especially in areas such as operations, logistics, network analysis, road project financing and policy analysis [1,2]. Life Cycle Cost Analysis (LCCA) is the project design, delivery and asset management strategy that evaluates the total cradle to grave cost involvements or attributes to arrive at the best-value choice considering the entire life span, discount rate and specific details. In developing countries like Nigeria, this tool is inadequate. A simplified Life-Cycle Cost economic evaluation tool for determining most cost effective flexible pavement procurement [3] is lacking. The high cost of hot rolled asphalt pavement, early failures, poor pavement performance, and poor sub-structural drainage amongst others are common. These inadequacies are generally overbearing in the drive towards sustainable flexible road pavement procurement in Nigeria and especially in the Niger-Delta region with swampy terrain and poor drainages [4,5].

Most transportation infrastructure problems existing presently in developing countries are all indicative of a poor-performing transportation infrastructural system delivery [6]. Capital projects such as expressways, urban and rural roads, railways, public transportation facilities and allied systems will be most suitably delivered using an economic evaluation model [7-10]. Besides, it will guarantee an optimum project performance delivery that meets design Level of Service (LOS) and confidence of investors and all stakeholders. This study evaluates Life Cycle Cost of flexible road pavement design alternatives in a typical Niger Delta urban town to determine the most economic pavement choices and models using a case study [11]. This study is aimed at achieving a model that will aid optimum total asset value in pavement infrastructure investment, planning, design and sustained service delivery in the now-competitive road infrastructural planning, budgetary and policy formulation in developing countries. The public and private sectors will have improved confidence in decision making.

## Literature Review

### Road pavement design and challenges

Based on the type of construction, material used, structural design criteria or load distribution modes, pavement can be classified as Rigid or Flexible pavement or even composite. Most pavement surface alternatives fall into these three categories of pavement. Rigid pavements are specifically known to have high flexural strength and are able to distribute concentrated loads over a larger area. Cement reinforced or mass concrete pavements are classified as rigid while others are classified as flexible pavement. Flexible pavements have some intrinsic ability to adjust to effects due to traffic loading with serious deformation. Essentially, flexible pavement consists of the sub-

***Corresponding author:** Oghorodje EE, Department of Highway Engineering, Ministry of Works, Asaba, Delta State, Nigeria, E-mail: eoghorodje@gmail.com

base and base layers made of granular materials which distribute the load to the sub-grade mainly by inter-granular pressures. These are known to have low flexural strengths.

According to American Association of State Highway and Transportation Officials (AASHTO), Guide for Design of Pavement Structure (1993); Pavement type selection is a three-part process which includes a pavement design analysis, life cycle cost analysis and evaluation of project specific details. The pavement design should be performed first, since the results may preclude the need to continue with the remainder of the pavement type selection process which includes life cycle cost analysis and project specific details [12,13]. The alternative resulting in the lowest present worth over a given analysis period is considered the most cost effective. The pavement design analysis includes the review and analysis of sub-grade competency, geometrics, traffic analysis, materials, climate, drainage, environment, construction considerations, and any other pavement design factors. Basically, the design and selection of a sustainable road pavement or surface type is determined by these design parameters. The design criteria for flexible pavement are to have a limited sub-grade deformation under load and limited sub-grade deformation at the underside surface of the surface wearing course. Pavement design HRA overlay design can be accomplished either by use of the mechanistic-empirical based scheme used in the Everpave©1 computer program or the AASHTO Guide for Design of Pavement Structures. The Everpave© program is for use with flexible pavements. The AASHTO procedure can be applied to either flexible or rigid pavement structures.

In this study, the Hot Rolled Asphalt (HRA) pavement, flexible precast Interlocking Concrete Block Pavement (ICBP) and a do-nothing (no wearing course) depending on the construction method and strategy are considered. The flexible precast Interlocking Concrete Block Pavement (ICBP) is considered as a flexible pavement due to its advantages over HRA in terms of its behavior under flexural load due to vehicular traffic [14,15]. The Mechanistic-Empirical Pavement Design Guide (MEPDG) by the Transportation Research Board [16]; AASHTO, Guide for Design of Pavement Structure (1993) design procedures amongst others are applicable for flexible road pavement design. The mechanistic based design procedures incorporates the treatment of life-cycle costs and design reliability while the empirical design approach rely more on empirical correlations with past performance, index-value-based characterizations of material properties layer coefficient, R-value, California Bearing Ratio (CBR), etc., and adopted engineering design strategy [17,18].

The highway and transportation engineer must therefore begin to examine the various ways of improving the sector through innovations in the selection of geometrical design parameters, material research, traffic flow assessment, work method, pavement design approach, maintenance strategy and economics. Other areas include geotechnical engineering of in-situ materials, application of geo-textiles challenged terrains and local content utilization, as these design variables affect the life cycle cost of the infrastructure delivered [19-21].

## Pavement life cycle cost analysis models

A considerable number of past studies on the application of Life Cycle Cost Analysis (LCCA) and Life Cycle Analysis (LCA) and Sustainability Assessments provide useful information and results to assist the pavement engineer in the pavement selection. The LCCA and LCA applications in infrastructure planning and delivery helps the stakeholders and decision makers deal with costs implication and sustainability considerations in a project life span basis using project cost evaluation methodologies and BS/ISO 15686-5 Standards [22].

Key amongst these works of these researchers is their aim at optimizing the benefits of the Life Cycle Cost Analysis (LCCA) and Life Cycle Analysis (LCA) and Sustainability objectives for buildings using various economic evaluation techniques with sets of best design alternatives, variables and assumptions [23,24]. The existing models obtainable in the United States, Europe, and Canada are complex and unsuitable for application in developing countries because of the climatic, information, technological, local systems development pattern and procurement as well as the cultural differences from that of the developed countries. Road authorities in the developed economies have adopted models for Life Cycle Cost Analysis (LCCA) with the intention to reduce the total costs for road infrastructure and maximize the socio-economic benefits with a lowered social economic cost and environmental impact. These models have been mainly used for the selection of road construction types or pavement types and other road infrastructure, such as bridges [25-28].

These models cannot be used as standard models in developing nations since they are developed according to requirements for particular road projects in some particular environment in developed countries. Moreover, the limitations of these existing models include the use of unrealistic and roughly calculated maintenance costs and user's costs, lack of regional attributes and tropical road design. Figure 1 illustrates the analysis period-pavement condition index of a

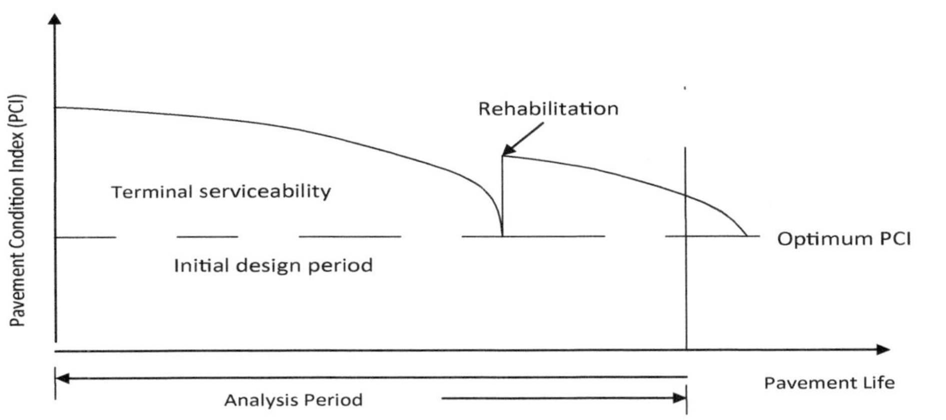


**Figure 1:** Analysis period for a pavement designed alternative (Adapted: FHWA-SA-98-079, Life cycle cost analysis in pavement Division interim Technical bulletin, 1998).

pavement design alternative.

### Economic evaluation of transportation projects

One of the criteria in measuring any road infrastructure projects performance is the cost efficiency or life cycle cost [29]. The life cycle cost is evaluated using any of the following methods viz; Present Worth Cost (PWC); Equivalent Uniform Annual Cost (EUAC); Benefit-Cost Ratio (BCR) and Rate of Return (ROR). Using the same baseline design parameters, design life and service life, all the options mentioned above give the same results that enhance the highway engineer's decision making. In this study, the Present Worth Cost (PWC) was employed.

The present Worth Cost (PWC) method is one of the economic evaluation methods for transportation infrastructures and systems, it involves the conversion of all of the present and future estimated expenses to a base of today's costs using an appropriate discount rate. The totals of the present value costs are then compared with one another. The present worth of costs method is directly comparable to the equivalent uniform annual cost method for comparable benefits [30]. In this presentation, a limitation to the present worth cost method is made for clarity using tropical information from field practice, experience and agency costs. The general expression for the Present Worth Cost of a transportation infrastructural project, given by Kumares et al. [31]:

$$PWC = \sum_{n=0}^{N} \frac{P^n}{(1+i)^n} \quad (1)$$

Where,

P= Cost of the agency and user cost in year n

i=interest rate

n= Number of years to when the sum will be expanded or saved.

N=Service life of the facility (in years).

The factor for discounting, either costs or benefits, is:

$$Pwcf = \frac{1}{(1+i)^n} \quad (2)$$

Where,

Pwcf=Present worth cost factor for a particular i and n

Based on the preceding considerations, a possible model of present worth cost of an infrastructure (pavement) is given by:

$$P = Pi + Pwcf\,(AE) \quad (3)$$

where;

Pi=Agency's Initial Cost

Pwcf=Present worth cost factor for particular i and n.

AE=Annual Expenditure (Post-Construction maintenance costs)

This is a deterministic approach whereby all cost inputs are known or estimated; where not well documented from past experience and available data are given a single fixed value within the available pavement design output, estimates, relevant tender documents and relevant design and field data on the project is used using basically equation (3) above. Interestingly, a deterministic relationship exists between the cost elements and the infrastructure age during the life-cycle exists as contained in equation (3) above, which can be examined through the use of linear regression model in the form;

$$Y = b + mx + e \quad (4)$$

where;

Y=Dependent variable, deterministic Present worth Cost of alternative Pavement or infrastructure or scheme in Million Naira/km.

b= Independent variable, regression constant or vertical intercept on the cost axis Agency's (Pre-construction+Initial Construction) in Million Naira/km.

m=Regression coefficient or slope of trend line relating the cost and the entire project life (a function of the annual expenditure, salvage value, analysis period, discount rate)

x=Independent variable, particular period or year of interest of projection of the present worth cost in year.

e=Independent variable, other costs (such as socio-economic and political cost element and error term due to uncertainty in data analyses.

## Methodology

The work specifically investigates a road structure with sub-grade and base formation common to the three competing flexible pavement alternatives in a 1.9 km urban stretch single carriageway in Effurun Metropolis in Uvwie Local Government Area of Delta State, Nigeria using the case study research approach [11]. The study area is particularly defined by Geographical coordinates 5°43' 00"N to 5° 48' 00"N and 5° 31'30"E to 5° 34'00'E. The pavement delivery and design data were obtained from the Office of the Ministry of Works, Effurun, Delta State, Nigeria.

The obtained data include soil sub-grade CBR results, water content condition, soil type and grain size distribution, the reports of route investigation, traffic vehicular traffic counts, previous maintenance, material and market survey reports. The pavement design data were processed and used for the design of three competing pavement alternatives, viz: Hot Rolled Asphalt (HRA), Interlocking Concrete Block Pavement (ICBP) and a third; Do-nothing scenario pavement surface courses. The urban road pavement thickness is on a wet sub grade. In this study, the low traffic urban road was designed to meet an average of 400 vehicles per day on both directions. Using the relevant Design Codes (such as AASHTO1993), Guides and Manuals, the Bill of Engineering Measurements (BEME) was developed for the designed alternative sectional full depth (Table 1). It was later used to compute the Life-Cycle Costs (LCCs) using Present Worth Cost (PWC) method and the varying interest rates using Microsoft (MS) Excel software, for a design life of 20 years.

The regression modeling for the three scenarios was done with computer software's packages [Statistical Package for Social Science (SPSS) and MS Excel]. The computed cost data from the various design alternatives inputs were analyzed using regression analysis with SPSS and MS Excel software packages.

The goodness of fit of the economic models or robustness were determined from the R-square values which range from 0.95 to 0.99 (See regression models statistical values in Figure 2). The model's validity was also carried out.

## Results and Discussion

### Results

The summary of the full depth of the flexible pavement is presented in Table 1; Figure 2 shows the combined Life Cycle Cost models for the

| HRA (Alternative 1) | ICBP (Alternative 2) | Do-Nothing (Alternative 3) |
|---|---|---|
| 2 layer HRA (90 mm) inclusive of tack and priming coats | 80 mm ICPB<br>50 mm bedding sand | Nil Surface layer |
| common 150 mm thick of base course (crushed stone) on minimum 750 mm depth sub-base (sharp sand) to the 3 competing flexible pavement alternatives | | |

**Table 1:** Summary of the full depth of pavement for the three flexible pavement alternatives.

| Interest Rates | Model Statistics | Hra (Alternative 1) | Interlocking Conc. Block Pavement (Alternative 2) | A Do-Nothing Scenerio (Alternative 3) | General Remarks |
|---|---|---|---|---|---|
| 4.6 | Model | y=6x+50 | y=4x+30 | y=5x-10 | Initial procurement cost of Alternative 2 (ICBP) is 65.73% cheaper than the Alternative 1 (HRA), while alternative 3 is the Do-Nothing without any pavement course.<br>N38.23 million, 43.4% cost savings made with Alternative 2 (ICBP) at 4.6% interest rate |
| | R square value | 0.96 | 0.94 | 0.98 | |
| | LCC (Million Naira/Km) | 88.08 | 49.85 | 53.26 | |
| 4.8 | Model | y=6x+50 | y=4x+40 | y=5x-10 | |
| | R square value | 0.96 | 0.95 | 0.98 | |
| | LCC (Million Naira/Km) | 86.79 | 49.03 | 51.92 | |
| 5.0 | Model | y=6x+50 | y=4x+40 | y=5x-10 | |
| | R square value | 0.96 | 0.94 | 0.98 | |
| | LCC (Million Naira/Km) | 85.54 | 48.24 | 50.88 | |
| 5.20 | Model | y=6x+50 | y=4x+40 | y=5x-10 | |
| | R square value | 0.96 | 0.94 | 0.98 | |
| | LCC (Million Naira/Km) | 84.31 | 47.47 | 49.74 | |
| 5.4 | Model | y=6x+50 | y=3x+40 | y=5x-10 | N36.4 million, 43.8% cost savings made with Alternative 2 (ICBP) at 5.4% interest rate. The Do-Nothing is even costlier due to the cost maintenance of road structure without a pavement surface. |
| | R square value | 0.96 | 0.94 | 0.98 | |
| | LCC (Million Naira/Km) | 83.12 | 46.72 | 48.64 | |
| 5.6 | Model | y=6x+50 | y=3x+40 | y=4x-10 | |
| | R square value | 0.96 | 0.95 | 0.99 | |
| | LCC (Million Naira/Km) | 81.96 | 45.99 | 47.56 | |
| 5.8 | Model | y=6x+50 | y=3x+40 | y=4x-10 | |
| | R square value | 0.96 | 0.95 | 0.99 | |
| | LCC (Million Naira/Km) | 80.96 | 45.28 | 46.52 | |
| 6 | Model | y=6x+50 | y=3x+40 | y=4x-10 | |
| | R square value | 0.96 | 0.95 | 0.99 | |
| | LCC (Million Naira/Km) | 79.74 | 44.59 | 45.51 | |
| 6.2 | Model | y=6x+60 | y=3x+40 | y=4x-10 | N34.75million, 44.2% cost savings made with Alternative 2 (ICBP) at 6.2% interest rate |
| | R square value | 0.96 | 0.95 | 0.99 | |
| | LCC (Million Naira/Km) | 78.67 | 43.93 | 44.52 | |

**Table 2:** Life cycle cost (LCC) for the 3 pavement alternatives for the 6m-lane urban road in NAIRA/km at various interest rates.

three competing pavement design alternatives at 6.2% interest rate with their respective regression equations. The life cycle cost for the three pavement alternatives for the 6 m -lane urban road in Million-Naira/km at various interest rates is shown in Table 2.

## Discussion

From the first flexible pavement alternative (2 layers 90 mm HRA surface course) shown in Table 1 and Figure 2, the Life-Cycle Costs for the HRA (alternative 1), Flexible urban road stretch using 5.4% and 6.2% interest rates (Table 2) are N83.12 million and N78.67 million per km respectively. The Life-Cycle Costs for the second pavement alternative (80 mm ICBP course) studied with a 20-year analysis period using 5.4% interest (shown in Table 2) is N46.72 million per km. This gives 43.8% savings over HRA (alternative 1), although both of the alternatives meet the design Level Of Service (LOS). The Do-Nothing alternative (alternative 3) has nil initial pavement surface construction cost but the maintenance cost of re-instating and replacing lost, eroded or weathered earthworks frequently to keep road passable is enormous. It does not meet the design Level Of Service (LOS), and it is uneconomical and non–aesthetical.

From Table 2, it was observed that the life cycle cost of the three alternatives decreases with increasing interest rates. The LCC of the alternatives at the 6.2% interest rate was found to be the most cost effective when compare with the 5.4% interest rate which was used as the baseline interest rate. The Do-nothing alternative was the costliest due to the cost of maintenance of the road structure without a pavement surface. The result also indicate that the Interlocking Concrete Pavement alternative gives better value for money apart

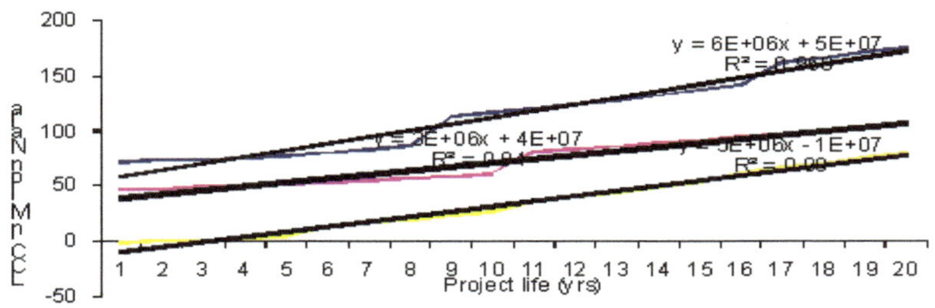


**Figure 2:** Combined LCC model for competing pavement design alternatives at 6.2% interest rate.

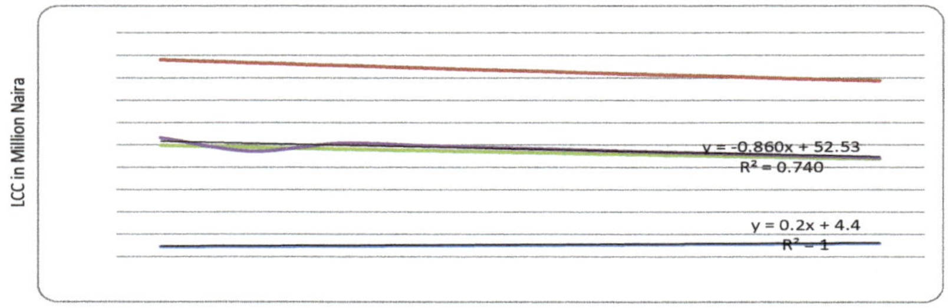


**Figure 3:** Interest rates-LCC plot fit plot for HRA (ALT.1), ICBP (Alt.2) and Do-nothing (Alt.3) alternatives.

from being able to offer better drainage properties [31], while the Do-Nothing has a higher LCC that does not satisfy design level of service. The Life-Cycle Cost Analysis (LCCA) template and model which are developed in this study are applicable for investment studies and decision making in bridge development and other aspects of the transport sector, telecommunication, power, and water infrastructural planning, procurement, delivery and maintenance.

**Goodness of fit of the models:** The goodness of fit of the developed economic models is robust as observed from the R-square values of 0.96 and 0.95 for the HRA and ICBP respectively (See regression models statistical values in Figure 2), while the SPSS analysis has a Durbin-Watson statistic coefficient of 0.95. This shows that the LCC model is highly robust in replicating the project data used in the analysis.

**Model validity:** Statistical parameters in Table 1, Table 2 and Figure 2 indicate that the cost models mean square value ($R^2$) for HRA alternative is 0.96 while that of ICBP alternative is 0.95. The models have significance value (F) of $1.5 \times 10^{-15}$ and $1.3 \times 10^{-13}$ respectively for the HRA and ICBP alternatives, which are less than 0.001 in both cases. Statistically, the parameters obtained in this study are in conformity with the result of study on "Cost Model for Pre-and Post-Haulage Road Freight Transport to and from the Intermodal Terminal" in Sweden [32], which had a mean square value ($R^2$) of 0.96 with a significance value of (F) of 0.001.

**Sensitivity analysis:** Sensitivity Analysis (SA) of on the Life-Cycle Costs models for the three alternatives respectively indicates a decrease with increasing interest rates. The Sensitivity Analysis (SA) carried out in this study has shown that a higher interest rate of 6.2% is most cost effective compared with the 5.4% baseline interest rate which was initially used in the study.

## Conclusion

In this study, a combined Life Cycle Cost model for the delivery of three urban flexible pavement design alternatives using different interest rates was developed. This model will be of immense benefits to developing economy like Nigeria where great investment utilization, optimization and prioritization challenges are involved. The study shows that the model's independent variable (pavement age) significantly predicts the dependent variable, Life-Cycle Cost (LCC) in Million Naira per km of urban 2-lane road. The mean square value ($R^2$) and significance value (F) of the model for HRA and ICBP alternatives, in this study, are in conformity with the result of a similar work done in Sweden. The results of the study shows that the Interlocking Concrete Pavement alternative gives better value for money on the long term, while the Do-Nothing alternative has a higher Life-Cycle Cost (LCC) that does not satisfy the required design level of service. The results got in this study will aid transportation economists, planners and other stakeholders in budgetary, project design, finance allocation, policy making and management strategy for pavement planning delivery in developing countries.

## References

1. Garrison W, Ward J (2000) Tomorrow's Transportation: Changing Cities, Economics and Lives. Artech House, Boston 335.
2. Kumares C, Darcy B, Chris T, Herbert S, Richard W, et al. (2002) Development of Transportation Engineering Research, Education, and Practice in a Changing Civil Engineering World. J Transport Eng, ASCE, USA.
3. Abam T (2001) Regional Hydrological Research perspectives in the Niger Delta. Niger Delta Environmental Survey, 20 Marina Street, NAL Towers, Lagos.
4. Nigerian Vanguard (2009) Nigerian Vanguard, 29th May.
5. Nigerian Vanguard (2011) Nigerian Vanguard, 1st October.
6. Desai V, Raval P (2008) Sustainable Road Asset Management Plan. Proc of Int Conf, Best practice of relief congestion on mixed traffic urban streets in developing Countries, IIT Madras, Chennai, India.
7. Mbelede C (2010) Cost Engineering Practice in 21st Century Nigeria –The Appraise and Cost Engineer. J Inst Appraises Cost Eng (IACE).
8. John M (2008) Life Cycle Delivery of Public Infrastructure. Precedents and Opportunities for the Commonwealth. Barchan Foundation, Inc., University of Southern California, USA.
9. World Bank (1999) World Bank Annual Report Infrastructure for Development.

10. Ardit D, Messiha M (1999) LCCA in Municipal Organizations. J Infrastruc Sys 5: 1-10.

11. Yin R (2003) Case Study Research: Design and Method. SAGE Publications, California, USA.

12. AASHTO (1993) AASHTO Guide for Design of Pavement Structures. American Association of State Highway and Transportation Officials, Washington, D.C. 1: 44.

13. Clerk A (1987) Block paving research and development. Interlocking Concrete Pavement Institute, British Cement Association.

14. Portland Cement Association (2009) PCA Monitor Flash report/Update, Paving-the New realities.

15. Peterson E (1985) Life cycle Cost Analysis of Pavements. NCHRP Synthesis of Highway Practice No. 122 Highway Research Board. National Research Council, Washington, D.C.

16. TRB (2004) Design procedures Transportation management and public policy, Transportation Research Board. National Research Council, USA.

17. Timm D, Birgisson B, Newcomb D (1998) Mechanistic-Empirical Flexible Pavement Design. Research Edition (unpublished). Minnesota Department of Transportation, Minneapolis, USA.

18. Thompson M, Barenberg E, Carpenter S, Darter M, Dempsey B, et al. (1990) Calibrated Mechanistic Structural Analysis Procedures for Pavements. Final Report (unpublished). Flexible Pavement Design 7.NCHRP Project 1: 1-26. TRB, National Research Council, Washington, D.C.

19. Suzy E (2003) Construction Products and Life-Cycle thinking" Principal Consultant, Centre for Sustainable Construction, BRE, Garston, Herts WD25 9XX, UK.

20. Walls J, Smith R (1998) Life Cycle Cost Analysis in Pavement Design. Interim Technical, Report No. DHWA-SA-98-079 FHWA. U. S. Department of Transportation, Washington, D.C.

21. Vidya N (2012) Economic Life Cycle of the National Games, India. Civil Engineering Department, AISSM. Institute of Information Technology, Kennedy, USA.

22. Lockie S (2015) Life Cycle Costing and sustainability Assessments: An Industry perspective with Case Studies. Design Economics for the built Environment Impact of Sustainability on Project Evaluation 262.

23. Ren J, Manzardo A, Zuliani F, Scipioni A (2015) Prioritization of bioethanol production pathways in China based on Life Cycle Sustainability assessment and multicriteria decision-making. Int J Life Cycle Assess 1-12.

24. Islam H, Jollands M, Setunge S (2015) Life Cycle Assessment and life Cycle Cost Implications of wall assemblages designs. A review, Renewable and Sustainable Energies Review 42: 129-140.

25. Labi S, Sinha C (2003) Measures of Short-Term Effectiveness of Highway Pavement Maintenance. ASCE J Transport Eng 129: 673-683. TRB. 7$^{th}$ International Conference on Managing Pavement Asset.

26. FHWA (1998) Life-Cycle Cost Analysis in Pavement Design-Search for Better Investment Decisions. Interim Technical Bulletin on LCCA. Washington, DC: United States Department of Transportation-Federal Highway Administration.

27. FHWA (2010) Improving Transportation Investment Decisions through Life-Cycle Cost Analysis. LCCA-EEI-Asset management-FHWA, United States Department of Transportation-Federal Highway Administration.

28. Smith R, Freeman T, Pendleton O (1993) Pavement Maintenance Effectiveness" Strategic Highway Research Program, National Research Council, Washington D.C., USA.

29. Dell'Isola A, Kirk S (1981) Life Cycle Costing for Design Professionals. McGraw-Hill, New York.

30. Nicholas J, Lester A (1999) Traffic and Highway Engineering for Undergraduate Studies in Transportation, Highway and Traffic Engineering (2ndedn.) USA.

31. ICPI (2008) A Comparison Guide to Porous Asphalt and Pervious Concrete Technical Specification No.4, Interlocking Concrete Pavement Interlocking Concrete Pavement Institute (ICPI), USA.

32. Peter K, Mingzhou J (2014) Development of performance Measurement for Freight Transportation, Louisiana Department of transportation and development, Louisiana Transportation research Centre, National Centre for Intermodal Transportation for Economic Competiveness, (NCITEC).

# 18

# Study of Equivalent Strength and Effects of the Type of Multi-story Steel Braced Frames

**Takayuki Kinoshita* and Takumi Ito**
*Department of Architecture, Tokyo University of Science, Japan*

**Abstract**

In recent years, the terrible earthquake events that are not expected in the design process have occurred in the world. So, it is necessary to estimate the seismic performance accurately. This paper focuses on the energy absorbing efficiency of steel braced frames. Steel brace members in frames show the complicated and unstable behavior because many kinds of buckling and fractures are combined. And evaluation method of restoring force characteristics and capacity have been studied in enormous past researches, however, there are some problems to adopt for seismic design procedure. Also the response characteristics of steel braced frames are affected by the interaction between frames and braces. This paper suggests the analytical method to evaluate the energy absorbing efficiency of steel braced frames as equivalent strength. The equivalent strength is formulated with yield shear coefficient of the frame and slenderness ratio of the brace. Also, to verify the applicability of multi-story steel braced frames response analysis of two types of 5-story steel braced frames (X bracing, inverted-V bracing) is conducted. As compared with energy absorbing efficiency of the response analysis result, it is confirmed that the equivalent strength can be accuracy of plus or minus 20 percent or so.

**Keywords:** Steel braced frames; Restoring force characteristics; Buckling; Seismic performance; Equivalent strength

## Introduction

Japanese seismic design code [1] prescribes that building structures must keep sufficient seismic performance such as ultimate strength, ductility, energy absorbing capacity under terrible earthquake input. A lot of past researches have investigated buckling resistant capacity, and energy absorbing capacity. Furthermore, evaluation methods on inelastic behavior and buckling strength, post-buckling stable strength and energy absorbing capacity until the braces reaches the ultimate state of braces have been established. According to design guideline of building structures based on energy absorbing capacity [2], which the energy absorbing capacity is important index to evaluate seismic performance of steel braced frames. Steel braces are one of important components of seismic resistant members of steel structures, and they increase rigidity and strength of stories. However, it is commonly recognized that the braces during ultimate state show unstable behavior caused by buckling and crack. Therefore, in past researches, several restoring force and hysteresis characteristic models of steel compression members [3,4] have been proposed, however, simple methods to evaluate seismic performance are required in general seismic design. In consequence, it is necessary to quantitatively evaluate seismic performance. In recent researches in the world, the behavior of braces and steel constructions have been treated, e.g. model for cyclic inelastic buckling of steel braces is presented [5] a useful method to predict the behavior of steel braces is proposed [6], and the robustness assessment methods of steel framed buildings under catastrophic events are treated [7]. This paper suggests the method to evaluate energy absorbing efficiency of steel braces as equivalent strength [8] based on energy absorption of steel braces after seismic response. Moreover, to clarify the applicability of equivalent strength, response analysis of two types of five-story three-bay steel braced frames have been conducted and compares evaluation values of suggested equivalent strength with analytical results.

## Energy Absorbing Efficiency of Steel Braced Frames

The inelastic behavior of steel structures under earthquake is influenced by vibration characteristics of structures and phase characteristic of the input earthquake motion. Regarding braced frames, this behavior is generated by buckling of braces. In this study, it is assumed that the displacement history of the braces is almost same with moment resisting frame. The accuracy of random vibration is lost; however, this supposition takes account of the inelastic response characteristics of braces generally. The analytical procedure to evaluate the energy absorbing efficiency of braces is as follows:

1) Response displacement history is estimated from the result of response analysis of moment resisting frames (which braces are not installed).
2) Hysteresis loop of a pair of braces is estimated from the above response displacement history.
3) This hysteresis loop of a pair of braces is translated into equivalent perfect elasto-plasticity. Energy absorbing efficiency of a pair of braces is expressed as the strength of elastic perfectly plastic (herein after referred, this is called "equivalent strength").

## Response displacement history of moment resisting frames

Regarding vibration systems which has spindle- shaped restoring force characteristics such as moment resisting frames, seismic response displacement corresponded to the coefficient of structural characteristic (Ds, which is defined as standard strength of frames on Japanese seismic design code) [1] is estimated. Figure 1 shows an analytical model. The storing force model of moment resisting frames is Skeleton Shift Model [9]. Figure 2 shows skeleton curve and hysteresis rule of Skeleton Shift Model. Table 1 shows analytical variables. Herein, total seismic input energy of RDs030 is a standard model, and seismic input energy of the other models is adjusted to

***Corresponding author:** Takayuki Kinoshita, Department of Architecture, Tokyo University of Science, Japan, E-mail: 4116619@ed.tus.ac.jp

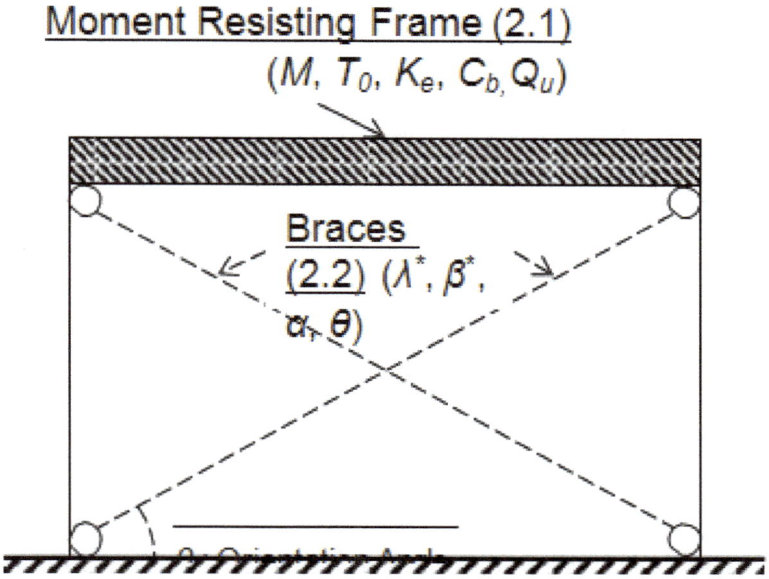


Figure 1: Analytical model.

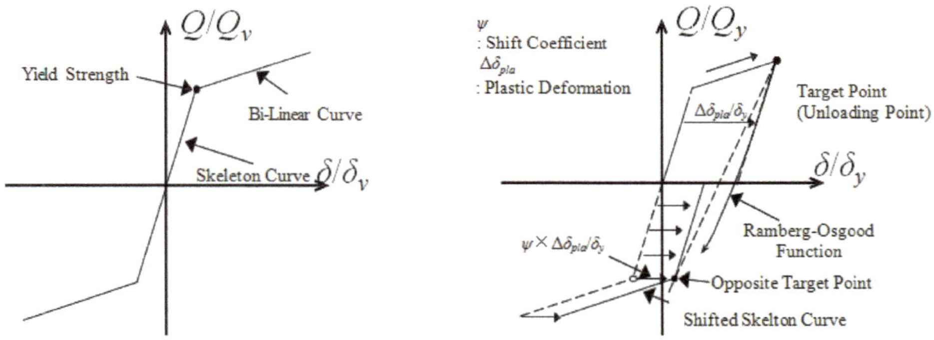


Figure 2: Skeleton curve and hysteresis rule of the skeleton shift model [9].

| Analytical model | Mass $M$ (ton) | Natural period $T_0$ (sec) | Elastic rigidity $K_e$ (kN/mm) | Yield base shear coefficient $C_b$ (-) | Ultimate lateral strength $Q_u$ (kN) |
|---|---|---|---|---|---|
| RDs025 | 2000 | 1.00 | 79 | 0.25 | 4900 |
| RDs030 | | | | 0.30 | 5880 |
| RDs035 | | | | 0.35 | 6860 |
| RDs040 | | | | 0.40 | 7840 |
| RDs045 | | | | 0.45 | 8820 |
| RDs050 | | | | 0.50 | 9800 |
| RDs055 | | | | 0.55 | 10780 |

Table 1: Analytical variables related to moment resisting frames.

the same as RDs030. In addition, input motions are El Centro NS, Hachinohe EW, and Fukiai NS.

## Hysteresis loop of steel braces

Restoring force characteristics of steel braces is as shown in Figure 3 [10]. This model is modified based on the "Wakabayashi model" [3]. In addition, the wide range of analytical variables is allowable, which is formulated by reference of enormous past test data. Herein, the response displacement history of moment resisting frames in the preceding paragraph is translated into that of a pair of braces. Hysteresis loop of a pair of braces is estimated with simulated from the response displacement history. If a pair of braces is connected with the moment resisting frames ($C_b$ = 0.25), for example, the displacement history of RDs025 is translated into the displacement history of the pair of braces. Table 2 shows analytical variables related to braces. Analytical variables are as follows: non-dimensional slenderness ratio $\lambda^*$, non-dimensional width-thickness ratio $\beta^*$, yield deformation ratio $\alpha$ of the yield deformation of braces to the yield deformation of moment resisting frames, and orientation angle of braces $\theta$. In addition, local buckling occurs in the flange part first in regard to H section, therefore, the width thickness ratio of flange part is used (Figure 4). Where $\lambda^*$ and $\beta^*$ are given as following:

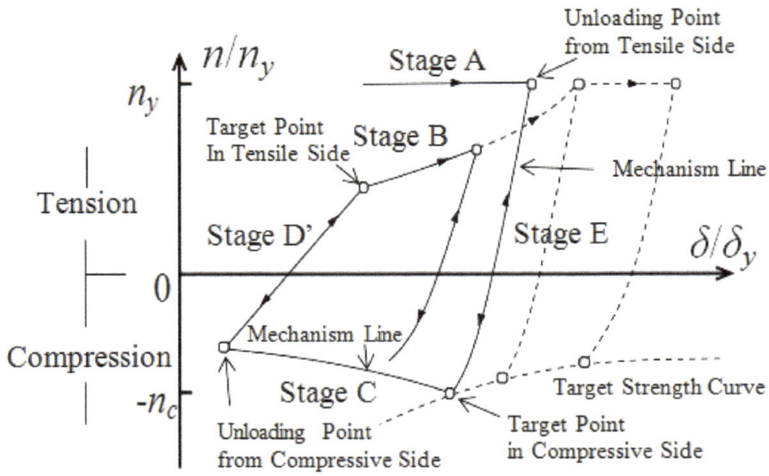


**Figure 3:** Restoring force characteristics of braces [10]. (Stage A: Full plastic state in tension. Stage B: The section near the center of the member is in the yield state by a tensile force and a moment. The member is a stage where it has been extended from the state where it was crooked. Stage C: The section near the center of the member is in the yield state by a compressive force and a moment. The member is a stage where it is crooked rapidly. Stage D: Elastic unloading state from the compressive side. Stage E: Elastic unloading state from the tensile side).

| Section | λ* | B* | α | θ |
|---|---|---|---|---|
| H section | 0.3~2.0 | 0.3 | 4 | 45° |
| Box | | 0.7, 1.1 | | |
| Circular | | 0.04, 0.06 | | |

**Table 2:** Analytical variables related to braces.

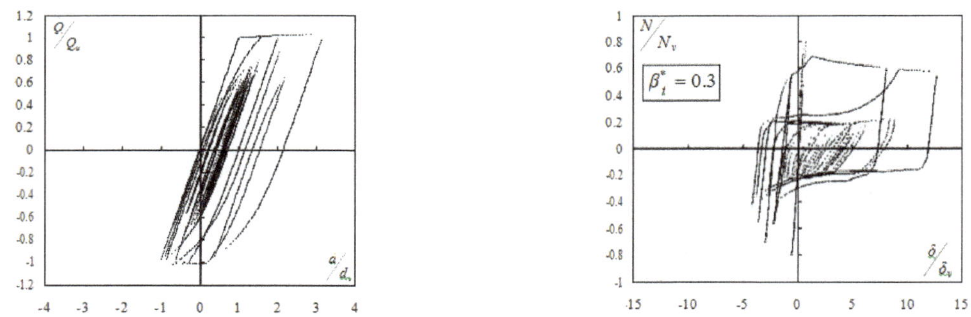

(a) hysteresis loop of moment resisting frames (b) hysteresis loop of a pair of braces

**Figure 4:** Examples of analytical results.

Non-dimensional slenderness ratio $\lambda^*$

$$\lambda^* = (l_b / i) \cdot \left(\sqrt{\varepsilon_y / \pi}\right)$$

Where, $l_b$: effective buckling strength, i: radius strength, $\varepsilon_y$: yield strain

Non-dimensional width-thickness ratio $\beta^*$

(H section) $\beta_f^* = b / t_f \cdot \sqrt{\varepsilon_{yf}}$

Where, b: half of flange width, $t_f$: flange thickness, $\varepsilon_{yf}$ : yield strain of flange

(Box) $\beta_f^* = B / t \cdot \sqrt{\varepsilon_y}$

Where, B: width, t: thickness, $\varepsilon_y$: yield strain

(Circular) $\beta^* = D / t \cdot \varepsilon_y$

Where, D: diameter, t: thickness, $\varepsilon_y$: yield strain

### Estimation of equivalent strength of a pair of braces

Equivalent strength Qeq B [8] is defined as follows:

$$Q_{eq\_B} = \frac{\text{(Plastic Strain Energy)}}{\text{(Cumulative Plastic Deformation)}}$$

Herein, Qeq_B is the strength of perfect elasto-plasticity, which is equivalent to the hysteresis loop of a pair of braces with regard to energy absorption. In addition, Qeq_B is converted into normalized axial force per 1 brace by dividing Qeq_B by 2Nycosθ (Hereinafter, this normalized axial force is written as Neq/Ny). Figures 5 and 6 illustrates the concept of equivalent strength and Figure 6 shows the definition of cumulative deformation in this study [8].

### Analytical result and formulization of equivalent strength

Analytical results are shown in Figures 7 and 8. It is confirmed that the equivalent strength becomes small in inversely proportion as the general slenderness ratio increases. In addition, it is confirmed that equivalent strength is mostly not affected by maximum plastic deformation of braces (Here-in-after, maximum plastic deformation is written as $\mu_{max}$). From the results, equivalent strength is formulized by regression analysis as follows:

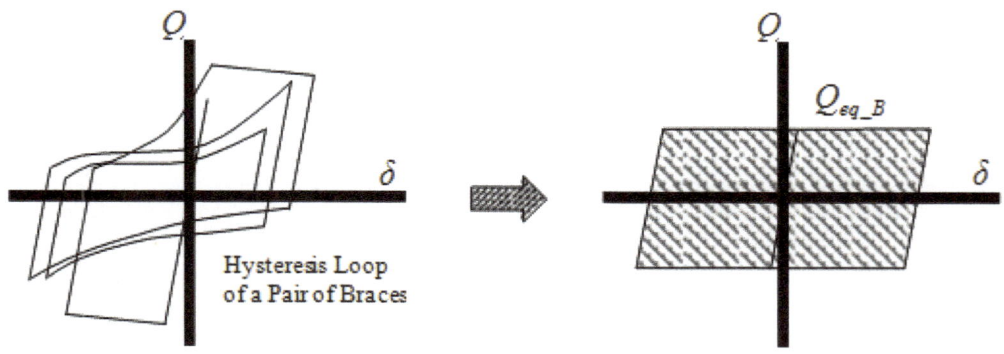


**Figure 5:** The concept of equivalent strength [5].

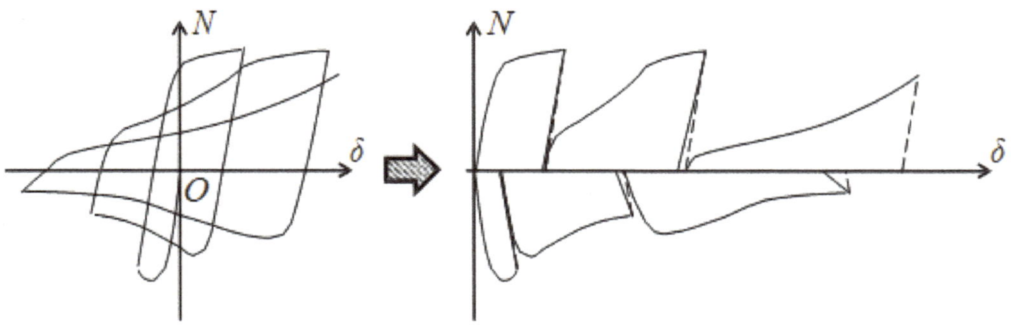


**Figure 6:** Definition of cumulative deformation [5].

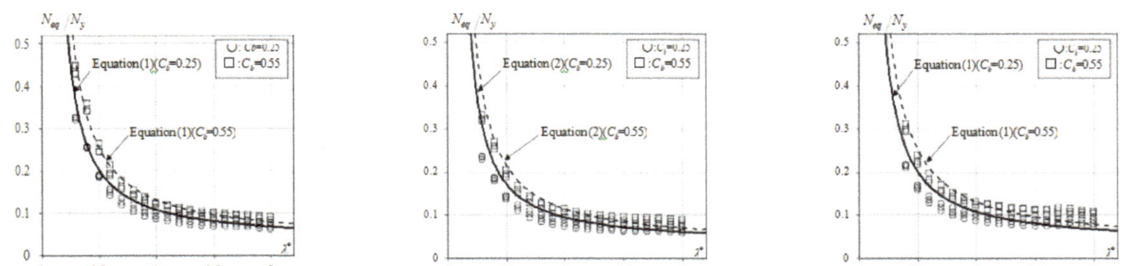

(a) H section ($\beta_f^*$=0.3) (b) Box-type ($\beta^*$=0.7) (c) Circular ($\beta^*$=0.04)

**Figure 7:** Relationship between *Neq/Ny* and $\lambda^*$

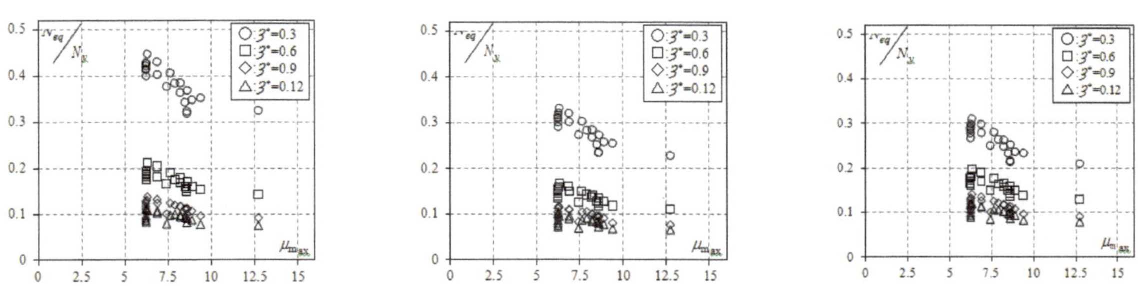

(a) H section ($\beta_f^*$=0.3) (b) Box-type ($\beta^*$=0.7) (c) Circular ($\beta^*$=0.04)

**Figure 8:** Relationship between *Neq/Ny* and ductility factor.

(H-section, Circular)

$$\frac{N_{eq}}{N_y} = a_{cb} \cdot \frac{1.8}{25\lambda^* - 2} + 0.03 \qquad (1)$$

(Box-type)

$$\frac{N_{eq}}{N_y} = a_{cb} \cdot \frac{1.5}{25\lambda^* - 2} + 0.03 \qquad (2)$$

$a_{cb}$: the constant determined by $C_b$

$$C_b = 0.25 \quad \Rightarrow a_{cb} = 1.0$$

$$C_b = 0.30 \sim 0.35 \quad \Rightarrow a_{cb} = 1.1$$

$$C_b = 0.40 \sim 0.45 \quad \Rightarrow a_{cb} = 1.2$$

$$C_b = 0.50 \sim 0.55 \quad \Rightarrow a_{cb} = 1.3$$

Equations (1) and (2) are illustrated in Figure 7. The evaluation has good agreements with analytical results.

## Equivalent Strength of Multi-story Steel Braced Frames

In X braced frames, braces are connected to the nodes of columns and beams. Therefore, even though braces of compressive sides buckle, braces of tensile sides keep the strength. In addition, stress of braces is transferred to columns and beams as axial force, therefore, mechanical characteristics of X braced frames are relatively clear. In contrast, in inverted-V braced frames (Figure 9), after the strength deterioration caused by buckling or fracture is occurred on compression member, an additional shear force is exerted in the middle of the beam. So then, unexpected failure mode will be generated. Moreover, the center of the beam is displaced vertically downward (Figure 9) and displacement history is shifted to the side of compression. From the result, residual buckling deformation of braces occurs, therefore, it is expected that energy absorbing capacity of a pair of braces decreases. In this chapter, seismic response analysis is performed on two types of multi-story braced frames (X-braced, inverted-V braced) and the energy absorbing efficiency as equivalent strength are evaluated. In addition, the results are compared with the result in Chapter 2.

### Outline of analysis

**Analytical frame model:** Herein, the low-rise and medium-rise buildings are analyzed, which is designed by reference [11]. And also, the two bracing types of five-story three-bay models (X- bracing, inverted-V bracing) are modeled. Figure 10 shows two types of model buildings. Here, to guarantee to form the whole story collapse mode, the section size of members is decided (Figure 11). Table 3 shows material properties, and Table 4 shows member lists. Effective length factor of braces is 0.75 or 1.0 to take effect of the members related to braces into account.

**Restoring force characteristics model:** Restoring force characteristics model of columns and beams is assumed as perfect elasto-plasticity model, and break point is full plastic moment without axial force. Restoring force model of braces under axial load is the model of Figure 3. Table 5 shows the list of input motions. Maximum velocity of input motions is normalized to 100 kines.

### Analytical result

Equivalent Strength of a pair of braces is estimated if maximum story drift angle is larger than 1/200 rad. Herein, in according with Japanese seismic design code, ultimate maximum story drift angle is provisioned 1/200 rad. Figure 11 shows comparison between Equivalent Strength and equation (1) in each input motion. From Figure 11, it is confirmed that effective length factor has little influence on equivalent strength of two types of braced frames. In addition, the mean value (λ) and the standard deviation (σ) show similarity in each type. In regard to X-braced, the upper limit of the standard deviation 1σ has good agreements with equation (1) (Figure 12). By contrast, in regard to inverted-v braced, equivalent strength is lower than equation (1). Figure 13 show hysteresis loop of a brace in inverted-V braced frames. It is confirmed that load history is shifted to the side of compression. Therefore, equivalent strength is lower than equation (1).

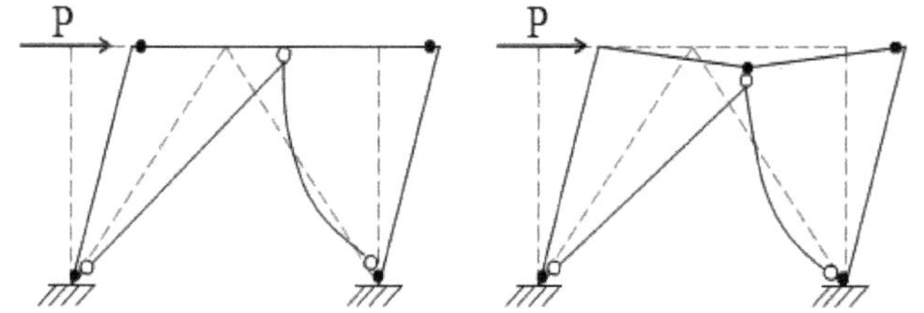


(a) normal mechanism (b) beam-hinging mechanism

Figure 9: Collapse mechanism of inverted-V Braced frames.

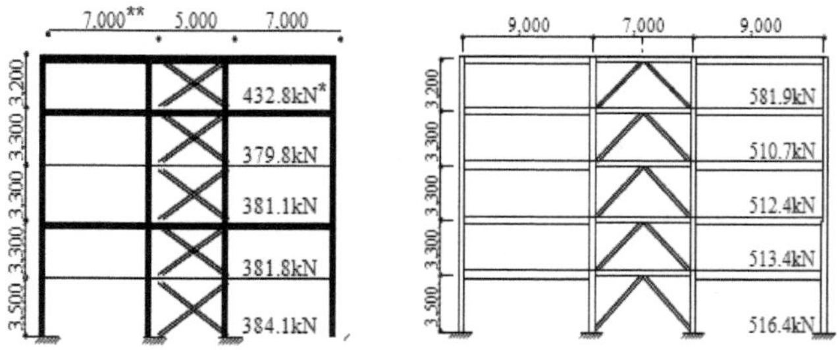


(a) X Braced Frames (b) Inverted-V Braced Frames

Figure 10: Model buildings. (*The figure in Figure.10 (eg. 432.8 kN) mean the dead load + live load of the stories. **The unit of measurement is "mm").

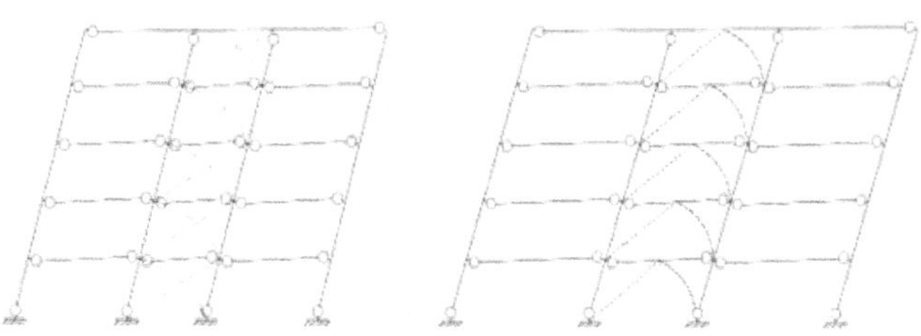

(a) X Braced Frames (b) Inverted-V Braced Frames

Figure 11: Collapse mode of examined frames. (*Broken lines show braces; the straight one shows the tensile yielding braces and the curve one shows the buckling braces).

| Members | Steel grades | Yield stress (σy) (N/mm²) | Young's modulus (E) (N/mm²) |
|---|---|---|---|
| Columns | BCR295 | 295 | 205000 |
| Beams | SN400 | 235 | |
| Braces | | | |

Table 3: Material property.

| (a) X braced frames | | | | | |
|---|---|---|---|---|---|
| **Column** | | | | | |
| **Story** | **Section (mm)** | **A (cm²)** | **$I_x$ (cm⁴)** | | **$Z_p$ (cm³)** |
| 5 | H-400 × 400 × 12 | 186.2 | 46774 | | 2711 |
| 4 | | | | | |
| 3 | H-400 × 400 × 12 | 245.8 | 60503 | | 3541 |
| 2 | | | | | |
| 1 | | | | | |
| **Beam** | | | | | |
| Story | Section (mm) | A (cm²) | $I_x$ (cm⁴) | | $Z_p$ (cm³) |
| 5 | H-350 × 175 × 7 × 11 | 61.5 | 13124 | | 841 |
| 4 | | | | | |
| 3 | H-500 × 200 × 10 × 16 | 110.8 | 46037 | | 2096 |
| 2 | | | | | |
| 1 | | | | | |
| **Brace** | | | | | |
| Story | Section (mm) | A (cm²) | $\lambda^*$ | | $\beta_f^*$ |
| | | | K=0.75 | K=1.0 | |
| 5 | H-125 × 125 × 6.5 × 9 | 29.5 | 1.52 | 2.03 | 0.235 |
| 4 | | | 1.54 | 2.05 | |
| 3 | H-150 × 150 × 7 × 10 | 29.1 | 1.28 | 1.7 | 0.254 |
| 2 | | | 1.28 | 1.7 | |
| 1 | | | 1.3 | 1.73 | |

| (b) Inverted-V braced frames | | | | | |
|---|---|---|---|---|---|
| **Column** | | | | | |
| **Story** | **Section (mm)** | **A (cm²)** | **$I_x$ (cm⁴)** | | **$Z_p$ (cm³)** |
| 5 | H -350 × 350 × 12 | 162.2 | 30932 | | 2057 |
| 4 | | | | | |
| 3 | H -400 × 400 × 12 | 186.2 | 46774 | | 2711 |
| 2 | | | | | |
| 1 | H -400 × 400 × 12 | 245.8 | 60503 | | 3541 |
| **Beam** | | | | | |
| Story | Section (mm) | A (cm²) | $I_x$ (cm⁴) | | $Z_p$ (cm³) |
| 5 | H-340 × 250 × 9 × 4 | 98.1 | 20888 | | 1360 |
| 4 | | | | | |
| 3 | H-482 × 300 × 11 × 15 | 139.7 | 57552 | | 2663 |
| 2 | | | | | |
| 1 | | | | | |
| **Brace** | | | | | |
| Story | Section (mm) | A (cm²) | $\lambda^*$ | | $\beta_f^*$ |
| | | | K=0.75 | K=1.0 | |
| 5 | H-150 × 150 × 7 × 10 | 39.1 | 1.01 | 1.35 | - |
| 4 | - | - | 1.03 | 1.37 | 0.254 |
| 3 | H-200 × 200 × 8 × 12 | 62.1 | 0.77 | 1.02 | - |
| 2 | - | - | 0.77 | 1.02 | 0.282 |
| 1 | - | - | 0.79 | 1.05 | - |
| A: Cross section<br>Ix: Geometrical moment of inertia in the strong axis direction<br>Zp: Plastic section modulus | | | | | |

**Table 4:** Members lists of two types of frames (a) X braced frames and (b) Inverted-V braced frames.

| Input motion | Observation date | Magnitude | Maximum acceleration (gal) | Normalized maximum velocity (kine) | Phase characteristics |
|---|---|---|---|---|---|
| El Centro NS | 1940/5/19 | 7 | 900.7 | 100 | Middle distance |
| Fukiai NS | 1995/1/17 | 6.9 | 994.9 | | Short distance |
| Hachinohe EW | 1968/5/16 | 7.9 | 468.4 | | Long distance |
| Taft EW | 1995/1/17 | 7.4 | 999.0 | | Middle distance |
| BCJ-L2* | - | | 457.9 | | Oceanic |
| *Herein, BCJ-L2 is simulated earthquake motion suggested by the building center of Japan. | | | | | |

**Table 5:** Input motions.

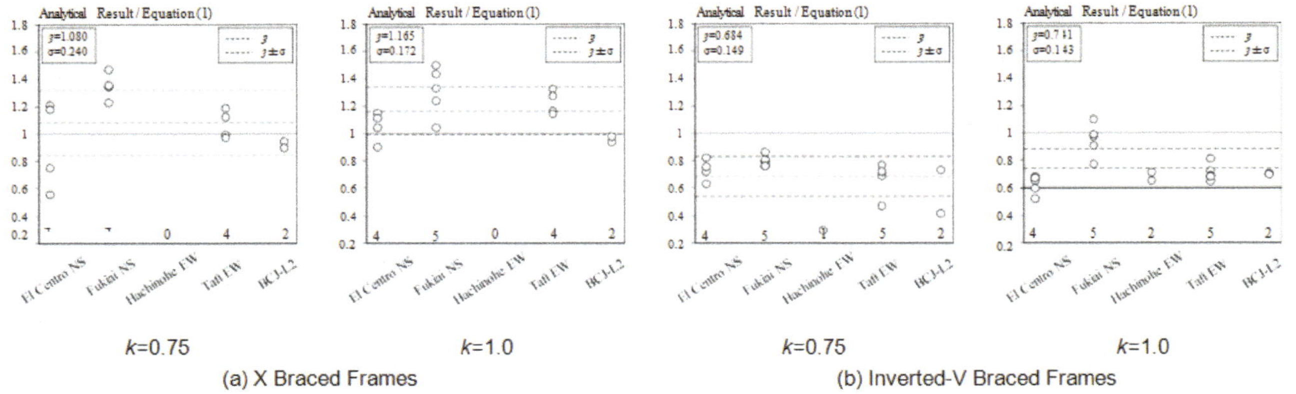


**Figure 12:** Comparison between equation (1) and equivalent strength in each input motion (*The figures mean the number of the stories where maximum story drift angle is larger than 1/200 rad).

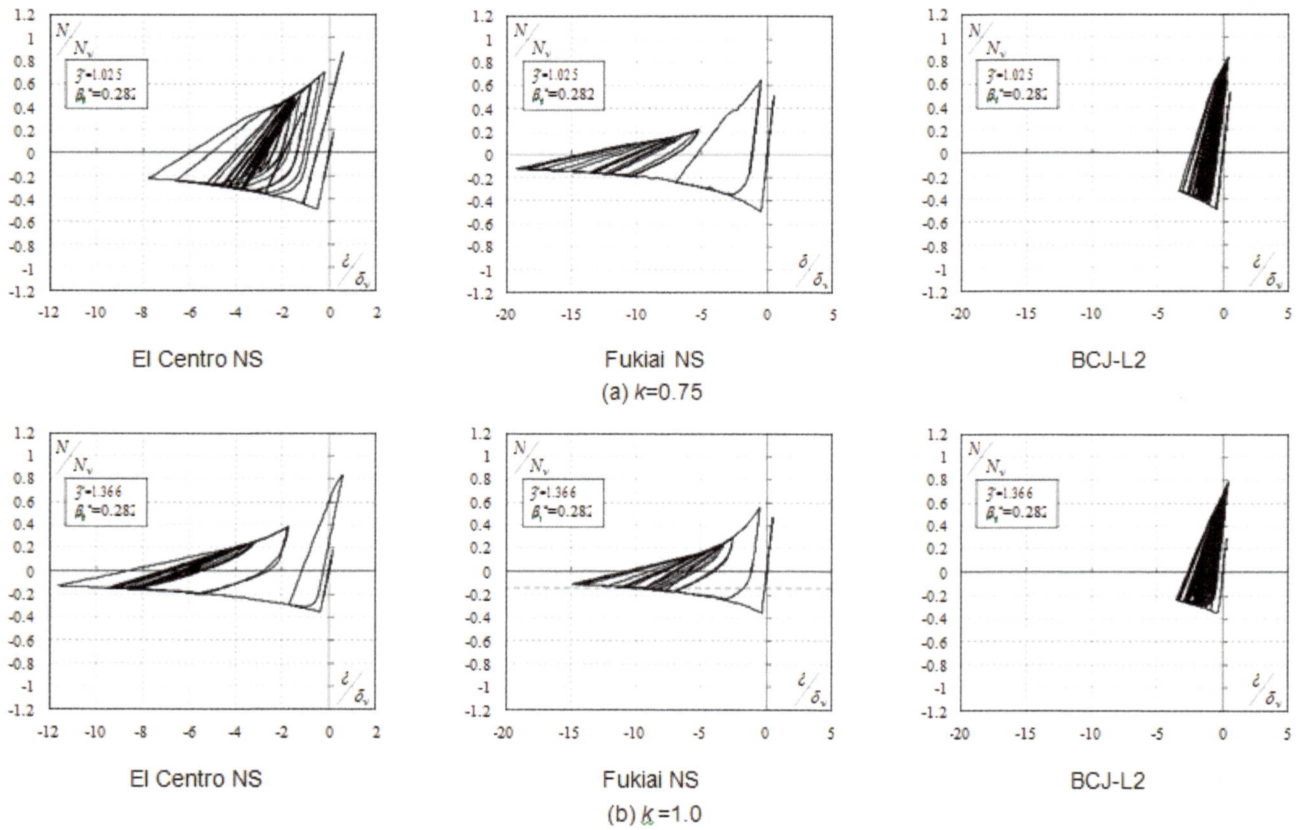


**Figure 13:** Hysteresis loop of braces in inverted-V braced frames.

Besides, it is confirmed that the accuracy of equation (1) is ±20% or so of the response.

## Conclusions

In this study, the energy absorbing efficiency of steel braced frames is formulized. In addition, seismic response analysis of five-stories three-bay steel braced frames was conducted to clarify the applicability to multi-story braced frames. Conclusions as following:

1) Energy absorbing capacity is translated as equivalent strength with normalized slenderness ratio $\lambda^*$ of braces and yield base shear coefficient $C_b$ of moment resisting frames.

2) From response analysis, equivalent strength has a tendency to vary inversely with $\lambda^*$ based on Figure 7. By contrast, equivalent strength is mostly not affected by maximum plastic deformation of braces based on Figure 8.

3) To clarify the applicability of equation of equivalent strength to multi-story braced frames, seismic response analysis is conducted in two types of braced frames.

4) The type of bracing has the effect on energy absorbing efficiency of brace. In particular, in case of inverted-V braced frames, energy absorbing efficiency of braces is lower than equation because of the vertical displacement caused by additional axial shear force.

## References


1. Government Publication in Japan (2015) A guideline of building construction technologies.
2. The Building Center of Japan (2006) A guideline of seismic evaluation method based on energy absorbing capacity (2ndedn). IITK-GSDMA Project on Building Codes.
4. Shibata M, Nakamura T, Wakabayashi M (1982) Mathematical expression of hysteric behavior of braces: Part 1, Derivation of Hysteresis Functions. Transactions of the Architectural Institute of Japan 316: 18-24.
3. Kato B, Akiyama H (1977) Restoring force characteristics of steel frames equipped with diagonal bracings. Transactions of the Architect Institute of Japan 260: 99-108.
4. Uritz P, Filippou FC, Mahin SA (2008) Model for cyclic inelastic buckling of steel braces. J Struct Eng.
5. Faggianno B, Fiorino L, Formisano A, Macillo V, Castaldo C, et al. (2014) Assessment of the design provisions for steel concentric X bracing frames with reference to Italian and European codes. The Open Construct and Building Technol J 208-215.
6. Formisano A, Landolfo R, Mazzolani FM (2015) Robustness assessment approaches for steel framed structures under catastrophic events. Computers and Struct 147: 216-228.
7. Matsumoto Y, Ito T, Goto M, Inoue S, Murayama N, et al. (2015) A study on post-buckling behavior and energy dissipation capacity of braced moment frames: Part 1–3. Annual Meeting of Architectural Institute of Japan, Kanto, Japan 1127-1132.
8. Meng L, Ohi K, Takanashi K (1992) A simplified model of steel structural members with strength deterioration used for earthquake response analysis. J Struct of the Architect Institute of Japan 437: 115-124
9. Ito T, Sugiyama S, Inoue S (2015) Restoring force characteristics model considering deterioration behaviour of steel compression members subjected to cyclic loading. J Structure of the Architectural Institute of Japan 710: 715-725.
10. Consortium for Building Research & Development (2012-2015) Report of study of box-column steel construction connected with steel braces.

# Partial Replacement of Wood Ash with Ordinary Portland Cement and Foundry Sand as Fine Aggregate

**Amir Shafi Batt* and Anshul Garg**

*Department of Civil Engineering, Lovely Professional University, Punjab, India*

**Abstract**

Wood ash is generated as residual/waste from combustion done in boilers at pulp and paper mills, steam power plants, and other thermal power generating facilities. Since wood is a renewable resource for energy and an environmentally friendly material, there is an increased requirement of using waste wood for the purpose of energy production thus leading to formation of more wood ash waste. The study focuses on incorporation of wood ash in combination with ordinary Portland cement while using it for various structural works. A critical review study in sieve analysis, consistency, and water absorption, setting time and slump tests of wood ash added to OPC will produce significant results to emphasize the detailed study process. Uncontrolled burning of saw dust to form wood ash is used as a partial replacement of cement, thereby changing its physical and chemical properties. These properties are found somewhat similar to fly ash. The concrete mixes are replaced with the amorphous wood ash as an admixture of cement having grain size less than 75 microns in proportions of 5%, 10%, 15%, 20%, 25% and 30% by weight of cement. In this Study, a research work is conducted for determining the change in workability or consistency of concrete mix, compressive strength, split tensile strength, flexural strength, etc., along with review of some durability properties. Foundry Sand is used as fine aggregate to obtain efficient results.

**Keywords:** Compressive strength; Wood ash; Carbonation; ASTM

## Introduction

Apart from historical evidences about demand of renewable energy in the present modernising world, in the era of urbanisation now the demand for renewable energy resources have further increased. A part of these resources is made by biomass resources including forestry and agricultural wastes. Forestry and agricultural biomasses are considered as efficient and favourable sources of fuel for energy production as their availability is in abundance and are cheap [1]. In the current period of energy production, power plants which run from biomass have low operational cost and have continuous supply of renewable fuel. The wastes generated from the biomass industries like sawdust, woodchips, wood bark, and hard chips) can be used as fuel offer a better way for their safe and efficient disposal [2]. The true residual solid wastes of these agricultural components are produced by thermal incineration which is environmentally safe and much economic. These wood ash wastes can be obtained in abundance from industries which require wood as their fuel for operating their boiler units [3]. The technologies available for production of wood ash in bulk are the fluidized bed and grate furnaces [4]. In grate furnaces biomass wastes are incinerated under high temperatures resulting in vitalization of organic complex compounds [5,6]

The wood ash incorporated as partial replacement of cement helps avoiding insignificant and bulk consumption of pure cement. Besides this the production of cement material all alone results in increased emission of certain greenhouse gases and much more pollutants. Hence replacing with wood ash leads to less production of cement, thus proving environmentally safe. Apart from this, this replacing technique reduces the cost also. The characteristic of ashes will be different for different types of agricultural wastes, timber, etc. These characteristics depend upon the (1) biomass characteristics (2) Incinerating technology (3) & location from where wood ash is obtained.

At present most of the biomass ash produced from power plants and timber factories is disposed using landfilling technique which accounts for 70% [5] ash generated and some part remains uncontrolled i.e., 20% [4]. Besides this the costs for disposal keep increasing day by day. So it is better to include that waste in certain manufacturing process. But before utilizing it needs to be treated well as it contains certain contaminants and compounds of C and N. Due to presence of higher contaminants these need not to be treated in an open environment.

A detailed study or research is being done to obtain different replacement materials for cement. As the cement industry at present is on a booming rise, the demand for cement is increasing but the cost should remain normal. So utilizing the wood ash by efficient means solves a twofold problem of their disposal as well as providing an efficient replacer for cement. A number of tests are carried by researchers which show positive effects of wood ash in cement. Hence using these wastes will be beneficial economically and environmentally.

The ultimate aim of this study is to analyze the effect of wood ash on different major properties of concrete.

## Objectives

The study focuses on the characteristics of wood ash/ saw dust and the properties incurred due to replacement of cement with wood ash. The objectives are:

1. To study the mechanical strength (compressive and tensile strength) of concrete along with the wood ash as partial replacement for cement.
2. To study the carbonation and drying shrinkage.
3. To study the effect on bulk density.

### Physical properties of wood ash

Wood ash particles are of different sizes and shapes. To obtain the

***Corresponding author:** Amir Shafi Batt, Student, Department of Civil Engineering, Lovely Professional University, Punjab-144411, India E-mail:amirbatt3@gmail.com, anshul.18374@lpu.co.in

fineness for replacement with cement sieve analysis is carried. Udeyou et al. [7] observed the physical properties of wood ash as of varying shapes. The average loss on ignition was found out to be equal to 10.46.

### Chemical composition

Some of the major components of wood ash are lime, (ca(OH)$_2$), $CaCO_3$ and calcium silicate [6,8]. Nike et al. [9] proposed some of the chemical characteristics wood ash with different types of wood. The % age loss on ignition was between 6.5% to 58.1% and moisture content of 0.5% to 3.3%. The chemical properties of five sources of wood are in Table 1.

## Materials and Methods

### Materials

**Cement:** Ordinary Portland cement was used having particle size of 3.9 μm, specific gravity of 3.01. The physical and chemical analysis properties are provided in Table 2.

**Wood ash:** wood ash was made available from wood furnishing factory Phagwara Punjab. The wood ash was obtained by incineration of carpentry waste and other agricultural wastes like rotten wood.

**Aggregates:** Foundry sand having grain size of 4.75 mm along with specific gravity as 2.6. The coarse aggregates used were crushed gravel of size about 10 mm and specific gravity of 2.6. The grain size or particle size distribution was according to ASTM C33/ C33M-08 (Figure 1).

| Particulates (%) | Type 1 | Type 2 | Type 3 | Type 4 | Type 5 |
|---|---|---|---|---|---|
| $SiO_2$ | 31.4 | 13.1 | 50.7 | 30.0 | 8.1 |
| $Al_2O_3$ | 17.3 | 7.8 | 8.2 | 12.3 | 7.5 |
| $Fe_2O_3$ | 9.7 | 2.6 | 2.1 | 14.2 | 3.0 |
| CaO | 3.6 | 13.7 | 19.6 | 2.2 | 25.3 |
| MgO | 0.7 | 2.6 | 6.5 | 0.7 | 4.5 |
| $K_2O$ | 1.1 | 0.4 | 2.8 | 2.0 | 2.7 |
| $Na_2O$ | 0.8 | 0.6 | 2.1 | 0.5 | 3.3 |
| Moisture | 2.3 | 0.5 | 0.2 | 0.4 | 3.3 |
| LOI | 30.6 | 58.1 | 6.7 | 35.3 | 32.8 |

**Table 1:** Chemical composition of wood ash [12].

| Constituents (% Age) | Values |
|---|---|
| Chemical properties | |
| $SiO_2$ | 21.25 |
| $Al_2O_3$ | 5.04 |
| $Fe_2O_3$ | 3.24 |
| CaO | 63.61 |
| MgO | 4.56 |
| Losson Ignition | 3.26 |
| Physical properties | |
| Specific Gravity | 3.1 |
| Mean Size | 23 μm |

**Table 2:** Constituents in OPC.

**Figure 1:** Typical foundry sand.

| % age of wood ash added | 7 days | 28 days | 56 days |
|---|---|---|---|
| 0 | 18.39 | 32.41 | 36.05 |
| 5% | 16.16 | 25.25 | 30.00 |
| 10% | 16.69 | 26.32 | 31.12 |
| 15% | 17.32 | 28.14 | 34.62 |
| 20% | 16.34 | 27.14 | 32.71 |

**Table 3:** Compressive strength (N/mm$^2$) test result (wood ash as partial replacement of cement).

### Methods

**Selection of mix proportion:** The target compressive strength for M20 mix was calculated according to IS: 10262 2009 [9] as 27.8 N/mm$^2$ for 28 days. The water-cement ratio, fine aggregate content and coarse aggregate content was accordingly then carried as per the mentioned IS code [ 9]. Thus mix proportion obtained was:

Cement content = 409.9 kg/m$^3$

Fine Aggregate Content = 545 kg/m$^3$

Coarse Aggregate Content = 1163 kg/m$^3$

**Preparation of mix:** Preparation of control mix (M20 Design mix Considered) was done first for which the trial mix having water-cement ratio of 0.47 was prepared. Mix was checked for workability (to obtain consistent mix) and Slump Cone Tests to validate the results. Control specimen was casted for 7 days, 28 days and 56 days having foundry sand were casted.

Secondly the preparation of mix for same 0.47 water cement ratio was done along with wood ash in different proportions of 5%, 10%, 15%, 20% and 25% by weight of OPC. The specimens were casted for 7days, 28 days and 56 days.

Cubes of 150 mm × 150 mm × 150 mm were casted to check compressive strength. Cylindrical Specimens were casted to check Split Tensile strength and Beams of dimensions 300 mm × 10 mm were casted to check flexural strength.

A minimum of 3 specimens were casted for a single test to analyse. Whole procedure was carried at room temperature. Compacting of concrete was carried with the help of vibrating machine.

**Curing of blocks:** Curing of all specimens was done after 2 days.

**Tests carried:** Tests were carried for compressive strength, split tensile strength, flexural strength, workability, water absorption, carbonation, drying shrinkage and for acid attacks like $H_2SO_4$ and HCL.

## Results and Discussion

### Compressive strength

The replacement percentage was 5%, 10%, 15% & 20% by weight of cement. Tests were conducted on 7 days, 28 days and 56 days using the digitalized CTM (compression Testing Machine), so the accumulation of errors can be said to be minimum in this research. The results are provided in Table 3.

The conclusions drawn based on the results are:

1. The compressive strength for the control mixes for 7 days, 28 days and 56 days obtained were 18 N/mm$^2$, 32 N/mm$^2$ and 36 N/mm$^2$ respectively.

2. With the use of wood ash there was an increase in the compressive strength but that increase was not up to the control specimens. The compressive strengths obtained for respective days are given in Table 3:

3. The Optimum results were obtained at 15% replacement.

4. There was increase in the ductile behavior of concrete for wood ash replacement when tested under CTM. The time required to break the wood ash specimen was long enough as compared to break the control specimen, as the development of cracks started increasing slowly under same rate of loading (Figures 2 and 3).

Rajamma et al. [1] observed the effects of wood ash on the compressive strength of concrete blocks. Wood ash was used to replace cement in percentages 10%, 20% and 30% by weight of cement, the optimum and significant results were shown on 10% of replacement exhibited higher 28 days strength. But with the replacement of 20% and 30% of by weight of cement, the 28days strength was insignificant and much reduced.

Udoeyo et al. [10] observed the compressive strength of concrete with replacements of 5%, 10%, 15% up to 30% and 30 by weight of cement).

Abdullia et al. [11] observed the compressive strength for 10%, 15%, 20% and 25%. The optimum results were observed at 15%. But the curing was done for 7 days, 14 days and 21 days. This showed the variation in results.

## Split tensile strength

The procedure was carried according to IS: 5816-199 [12]. The split tensile strength of the concrete with wood ash increased. The enhancement in split tensile strength was because of proper binding and insignificant surface area of binding particles (Table 4). But the increase was not higher than the control mix. We can say almost same results were obtained for 15% as the control specimen (Figures 4 and 5).

(a) (b)

**Figure 2:** Compressive strength testing.

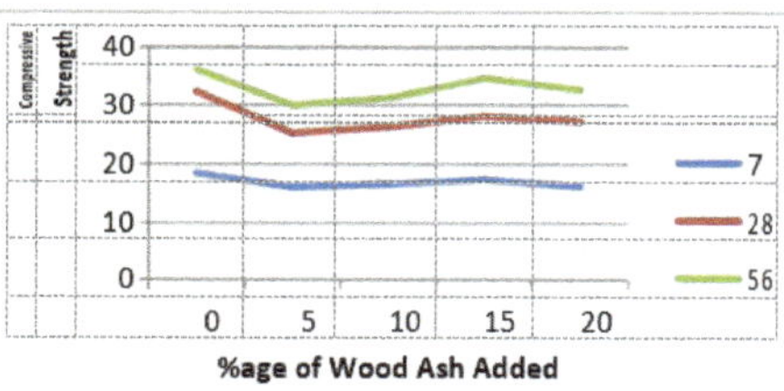


**Figure 3:** Comparison of compressive strength along wood ash percentage.

| % Age of wood ash | Split tensile strength at 7 days (N/mm²) | Split tensile strength at 28 days (N/mm²) | Split tensile strength at 56 days (N/mm²) |
|---|---|---|---|
| 0 | 2.32 | 3.64 | 4.16 |
| 5 | 1.37 | 2.47 | 3.14 |
| 10 | 1.82 | 2.85 | 3.79 |
| 15 | 2.19 | 3.29 | 4.02 |
| 20 | 1.92 | 2.77 | 3.96 |

**Table 4:** Split tensile strength test (wood ash as partial replacement of cement).

**Figure 4:** Failure of control specimen.

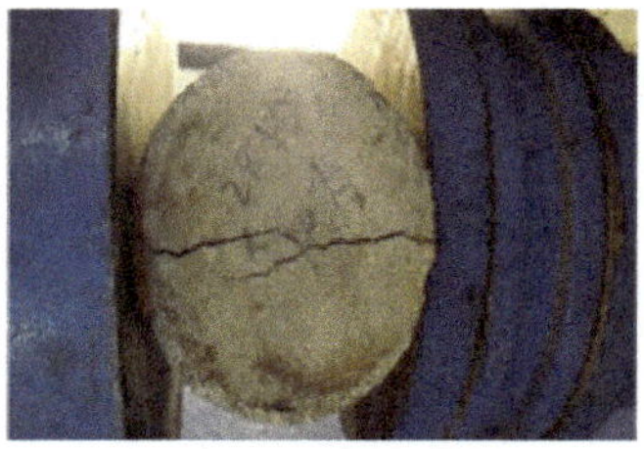

**Figure 5:** Failure of specimen along with wood ash.

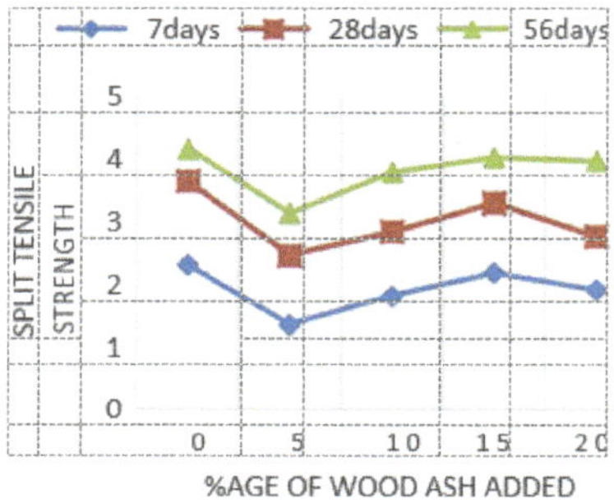


**Figure 6:** Comparison of split tensile strength along wood ash percentage.

(1) The average split tensile strength achieved by the control specimens for 7 days, 28 days and 56 days were 2.32 N/mm², 3.64 N/mm²and 4.16 N/mm² respectively.

(2) The strength of concrete along with the wood ash varied accordingly given in Table 4. The Optimum results were obtained on 15% replacement.

(3) Increase in the split tensile strength is because of the enhanced quality of cement paste due to addition of wood ash.

The failure for control specimen was brittle which resulted its splitting in two equal halves whereas when wood ash was incorporated the failure observed was not sudden but quite uniform upon load condition as given in Figure 6.

## Flexural strength

Beams were casted having dimensions of 300 mm × 10 mm × 10 mm to check flexural strength as per IS: 516 – 1959 [13]. Vibration of mix was done on vibrating machine according to IS recommendations. A system of 2 point loading was used with CTM to test the specimens. It was observed that there was an increase in the flexural strength

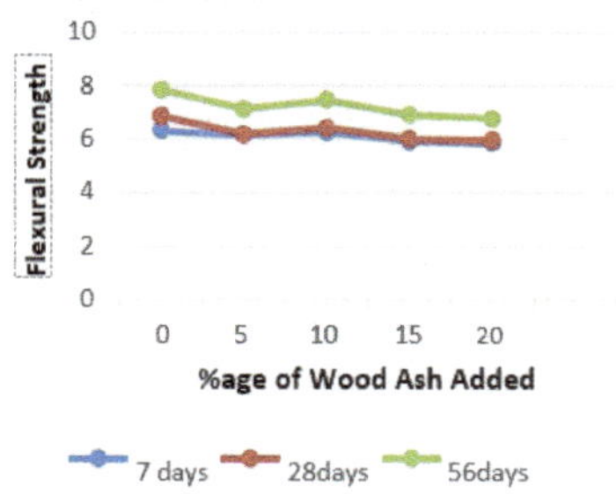


**Figure 7:** Comparison of flexural strength along wood ash percentage.

| % Age of wood ash | Flexural strength at 7 days (N/mm²) | Flexural strength at 28 days (N/mm²) | Flexural strength at 56 days (N/mm²) |
|---|---|---|---|
| 0 | 6.34 | 6.87 | 7.84 |
| 5 | 6.17 | 6.19 | 7.13 |
| 10 | 6.26 | 6.43 | 7.47 |
| 15 | 5.91 | 6.01 | 6.91 |
| 20 | 5.84 | 5.96 | 6.77 |

**Table 5:** Flexural strength test.

specimens with wood ash but not greater than control specimens. Optimum results were obtained at 10% replacement (Figure 7). The results are given in Table 5.

## Water absorption

An increase in water absorption was observed maximum for 15% at 28 days which indicates that rate of pozzolonic reaction is much higher at 15% replacement. Beyond 15% replacement the water absorption significantly decreased.

Udoey et al. [7,10] According to this study the t-test carried showed certain results using wood ash as replacement of cement in the level of 5%, 10% and 25%, there was increase in water absorption for 28 days which was observed as significant. But the water absorption values uptill 10% were observed significant which produced positive results both with a without wood ash.

## Soundness

The soundness of concrete mix was studied and observed to be increase while increasing the wood ash % age. Similar results were obtained from other studies when the replacements were done from 5% to 30%. At 30% concrete obtained produced much soundness. Thus in short soundness increases in direct proportions of increased wood ash incorporation [14,15].

## Carbonation

The carbonation results along with the wood ash as partial replacement when water cement ratio of 0.50 was observed to decrease, providing optimum results at 5% of replacement. From other researches similar results were observed. The tests for carbonation showed that with 5% incorporation of wood ash in cement resulted in reduction in depth of carbonation. With the 10%, 15% and 25% the reduction in depth was insignificant [16].

## Bulk density

According to a study [17-19] bulk density was observed to decrease with increasing % age of wood ash as replacement for cement. Much more significant results were obtained at 40% replacement.

Reduction in bulk density at 40% was 2281 kg per cub. Meter and at 0% was 2482 kg per cub meter. This shows that wood ash observed is of low specific gravity.

## Drying shrinkage

Naike et al. [8] observed the shrinkage in concrete with wood ash. Replacement was done for 5%, 8% & 10% by weight of cement as a binder. The following results were obtained:

1. Shrinkage of concrete cube was 0.0092% at 7 days and 0.052% at 232 days.

2. At 5% replacement of cement with wood ash shrinkage was 0.012% at 7 days and 0.027% at 232 days.

3. At 8% replacement of cement with wood ash shrinkage was 0.014% at 7 days and 0.014% at 232 days.

4. At 12% replacement of cement with wood ash shrinkage was 0.0051% at 7 days and 0.044% at 232 days.

## Conclusions

Following are some of the investigated conclusions from the above study:

1. Wood ash may vary in quantity and quality because of many factors like temperature, type of wood or biomass, combustion type, etc. So it is quite necessary to analyze the wood ash before using.

2. The strength parameters obtained were quite better than the attaining target of M20. The results for compressive strength were much significant. The optimum level of replacement with wood ash produced positive results.

3. The incorporation of wood ash resulted in increase in the water absorption

4. Bulk density of concrete was observed to decrease with the increasing % age of wood ash.

5. Incorporation of wood ash concrete resulted in increase in mass at initial stages when immersed in acids.

6. Incorporation of wood ash made concrete ductile enough. It means that concrete was able to bear loads for longer time as the failure was not sudden.

7. Incorporation of wood ash enhanced the quality of paste, thereby increasing both split tensile strength and flexural strength of concrete.

## References

1. Rajamma R, Ball RJ, Luis AC, Tarelho S, Allen GC, et al. (2009) Characteristics and use of biomass fly ash in cement based materials. J Hazard Mater 172: 1049–1060.
2. Chee CB, Ramli M (2012) Mechanical strength: Durability and drying shrinkage of structural mortar containing HCWA as partial replacement of cement. Constr Build Mater 30: 320–329.
3. Loo SV, Koppejan J (2003) Handbook of biomass combustion and Co firing. Twente University Press, The Netherlands.
4. Etiegni L, Campbell AG (1991) Physical and chemical characteristics of wood ash. Bioresour Techno 37: 173–178.
5. Werther J, Saenger M, Hartage EU, Ogda T, Siagi Z (2000) Combustion of agricultural residues. Prog Energy Combust Sci 26: 1–27.
6. American Standard Specification for Concrete Aggregates (2000) The American Society for Testing and Materials. West Conshohooken, USA.
7. Udoeyo FF, Inyang H, Young DT, Oparadu EE (2006) Potential of wood ash waste as an additive in concrete. J Mater Civ Eng 18: 605–611.

8. Naik TR, Kraus RN, Siddique R (2002) Demonstration of manufacturing technology for concrete and CLSM utilizing wood ash from Wisconsin, Wisconsin Department of Natural Resources (Madison, WI) for project #1-06 UWM report no. CBU-2002-30, Center for By-products Utilization, Department of Civil Engineering and Mechanics, University of Wisconsin-Milwaukee, Milwauke.
9. Indian Standard Concrete Mix Proportionng – Gltideiliiyes (First Revision) ICS 91.100.30O BIS
10. Udoeyo FF, Dashibil PU (2002) Sawdust ash as concrete material. J Mater Civ Eng 14: 173–176.
11. Abdullahi M (2006) Characteristics of wood ash/OPC concrete. Leonardo Electr J Pract Technol 8: 9–16.
12. IS 5816 (1999) Method of test splitting tensile strength of concrete [CED 2: Cement and Concrete].
13. IS: 516-1959 Method of flexural strength of Concrete [CED 2: Cement and Concrete].
14. Wang S, Baxter L, Fonseca F (2008) Biomass fly ash in concrete: SEM, EDX and ESEM analysis. Fuel 87: 372–379.
15. Etiegni L (1990) Wood ash recycling and land disposal, Ph.D. thesis, Department of Forest Products, University of Idaho at Moscow, Idaho, USA.
16. Khan MI, Lynsdale CJ (2002) Strength, permeability, and carbonation of high performance concrete. Cem Concr Res 32: 123–131.
17. Elinwa AU, Ejeh SP, Akpabio IO (2005) Using metakaolin to improve sawdust-ash concrete. Concrete International. 27: 49-52.
18. Yin C, Rosendahl LA, Kaer KSK (2008) Grate-firing of biomass for heat and power production. Progress in Energy and Combustion Science. 34: 725-754.
19. Campbell AG (1990) Disposing of wood ash. Tappi J 73: 141–143.

# Development and Characterization of Surface Porosity in Tannery Sludge Derived Adsorbent

**Geethakarthi A***

*Department of Civil Engineering, Kumaraguru College of Technology, Coimbatore, Tamil Nadu, India*

**Abstract**

In this work, activated carbons were prepared by a combination of physical and chemical activation from raw tannery sludge. The burn-off and the yield percentage of the acid activated carbons were evaluated for two particle size namely 600 µm (TSC600) and 300 µm (TSC300). The developed activated carbons were characterized and analysed for adsorption isotherm under nitrogen atmosphere at -196.86°C. The Scanning Electron Microscopy (SEM), Thermo Gravimetric Analysis (TGA) and the functional groups were also identified for the tannery sludge activated carbons. The thermal stability for the activated carbons TSC600 and TSC300 were confirmed by their exothermic peaks at 971.1°C and 897.4°C respectively. The higher surface area of the sludge carbons was obtained by the BET isotherm plot, suggesting the application of the produced carbon as a potential adsorbent. The surface area of TSC600 and TSC300 at the activating condition was 188.205 $m^2/g$ and 212.32 $m^2/g$. The pore size distribution of the sludge developed was obtained by BET isotherm under the BJH method (19.77 A°<200 A°) and HK method (5.25 A°<200 A°). The developed carbons were dominated by micropores and mesopores.

**Keywords:** Activation; Tannery sludge; SEM images; Yield percentage; Nitrogen isotherm; Micro pores

## Introduction

Adsorption is an efficient method for the removal of dyes from wastewater and activated carbon is one of the most studied adsorptive materials [1-4]. The adsorption capacity depends on the textural properties of the carbons such as porosity, surface area, nature of the adsorbate and the adsorbate concentration [5,6]. The surface characteristics and the internal pore structure of the activated carbon play an important role in adsorption processes. These parameters depend on both, the precursor used and the method of preparation. Therefore, an adequate characterization of the adsorbent is crucial to the adsorption and separation processes. Recently, the surface chemistry of activated carbons is focused to interpret results of dye adsorption experiments [7]. The use of activated carbons derived from expensive starting materials is not satisfactory for most pollution control applications. Research works on the usage of non-conventional adsorbents (natural materials, bio sorbents and waste materials from industry and agriculture) with lower cost and high efficiency [8-10] are being focussed. Several recent publications [11-13] indicate the interest in developing sustainable and environmentally safe recycling processes to produce value products such as activated carbon from wastes. The activation of carbon can be carried out by chemical or physical means. In this work, the development of tannery sludge carbon by acid activation combined with pyrolysis is carried out. The characterisation of the activated sludge carbon is supported by various analytical methods.

## Experimental

### Materials

Preliminary tannery sludge was collected from a common tannery effluent treatment plant, Ranipet, India. The chemical reagents used for the analysis of the activated carbon samples were of AR grade. Deionised water was used in preparing solutions throughout the investigations.

### Preparation of activated carbon

The dried tannery raw sludge was sieved to a particle size of 600 µm (TSC600) and 300 µm (TSC300). The sieved sludge particles were subjected to a combination of chemical and physical activation. Chemical activation was done by immersing the dried sludge in 5:1 sulphuric acid for an impregnation period of 12 hours with the impregnation ratio 2. The chemically activated samples were washed several times with deionised water to remove the effect of excess acid and dried in oven for 4 hours. The dried sample after desiccation was used for the activation in a horizontal tubular furnace for pyrolysis in the presence of nitrogen atmosphere at 650°C for 3 hours. These activated sludge carbons were used for the further analysis.

### Analytical methods

**Instrumental analysis:** Netzsch STA 409 (Simultaneous Thermal Analyzer) was used for the simultaneous measurements of TG–DTA (Thermogravimetric–Differential Thermal Analyser). The temperature of SiC furnace ranged from room temperature to 1600°C. The heating rate of the sludge sample/activated carbon was maintained at 10°C/min in an inert nitrogen atmosphere with a flow rate of 100 mL/min. The SEM analysis was carried out using JEOL JSM–6360 scanning electron microscope. The resolution of the microscope was 4.0 nm with an accelerating voltage of 0.5 to 30 keV. The magnification of the instrument was from 50X to 10000X. ASAP 2020 (Micromeritics) was used for the specific surface area analysis, pore volume and pore size distribution of the sludge sample/carbons using multipoint adsorption in the $P/P_0$ range 0.05 to 0.33, N2 isotherms at 77 K and Brunauer Emmett and Teller Equation. Barrett, Joyner and Halenda (BJH) method and the Horvath-Kawazoe (HK) method are the simple and popular methods used for the evaluating pore size distrubution of the mesoporous and microporous adsorbents. The pH of the sample was measured in a digital pH meter model 111E/01E with resolution 0.01 pH; 1 mV and an accuracy of pH ± 0.01.

**Boehm titration:** Procedures for the analysis of oxygen functional group for the activated carbons TSC600 and TSC300 were estimated

***Corresponding author:** Geethakarthi A, Department of Civil Engineering, Kumaraguru College of Technology, Coimbatore, Tamil Nadu, India
E-mail: ageethakarthi@yahoo.com

by Boehm [14]. Activated carbon samples were placed in an oven and dried at 105°C for 24 hours. 25 mL of an alkali solution (0.25 N $NaHCO_3$, $Na_2CO_3$, and NaOH) was added in Erlenmeyer flasks containing a given amount of activated carbon samples (1 g). The samples were constantly mixed in an orbital shaker at 25°C for 24 hours. 5 mL of the mixed sample were filtered and back titrated with 0.25 N HCl solution. The concentration of various functional groups was calculated by measured residual bases.

## Results and Discussion

### Characterization of tannery sludge activated carbon

The developed tannery sludge carbons TSC600 and TSC300 were characterized for its proximate and ultimate analysis. The burn-off percentage and yield percentage of the sludge carbon samples were also analyzed. The characterization of the sludge activated carbons are shown in Table 1.

### Analytical methods

**Thermal analysis:** TG-DTG spectrum of the sludge carbons are illustrated in Figures 1 and 2. The initial weight loss observed was due to

| Carbons | Proximate analysis (%) | | | | Ultimate analysis (%) | | | Burn off % | Yield % |
|---|---|---|---|---|---|---|---|---|---|
| | Moisture content | Volatile content | Ash content | Fixed content | C | H | N | | |
| TSC600 | 3.60 ± 0.02 | 32.50 ± 0.03 | 28.36 ± 0.02 | 35.54 ± 0.03 | 36.77 | 3.79 | 3.54 | 29 | 71 |
| TSC300 | 2.10 ± 0.00 | 32.15 ± 0.05 | 14.25 ± 0.02 | 49.40 ± 0.03 | 41.9 | 1.24 | 1.08 | 26 | 74 |

**Table 1:** Characterization of tannery sludge activated carbons.

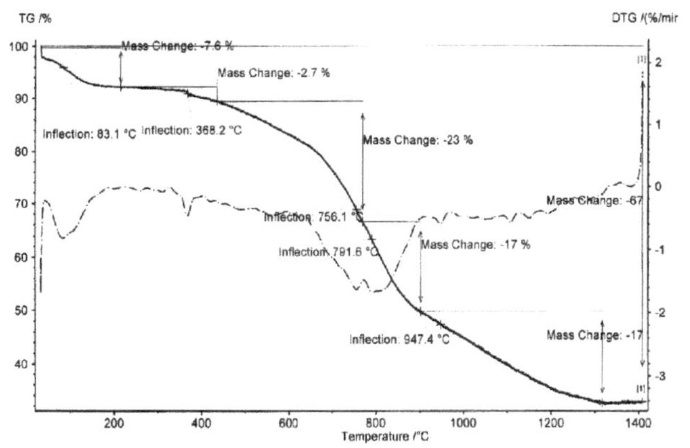


**Figure 1:** TG curves of tannery sludge activated carbon (SC600).

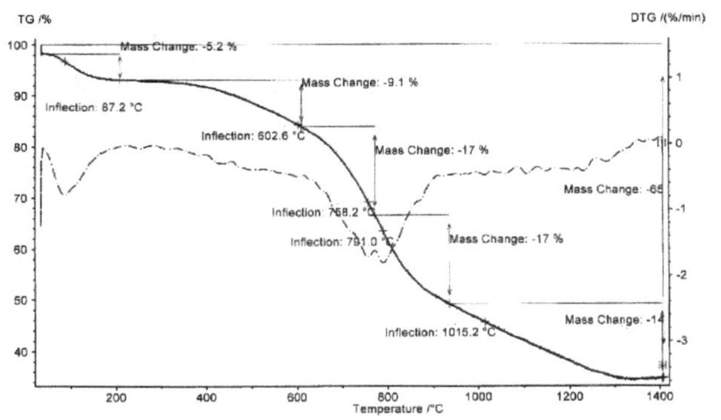


**Figure 2:** TG curves of tannery sludge activated carbon (SC300).

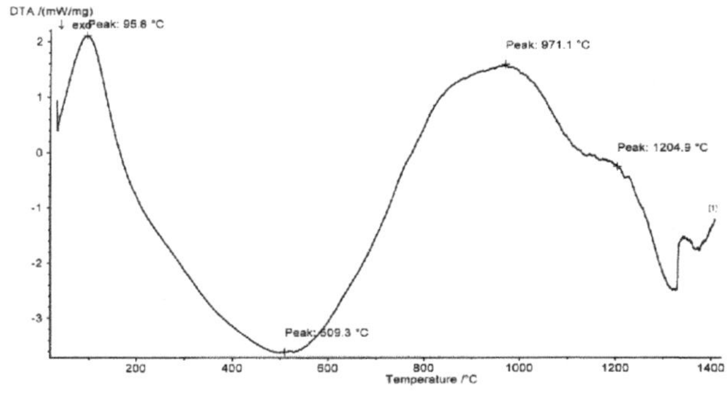


**Figure 3:** DT curves of tannery sludge activated carbon (SC600).

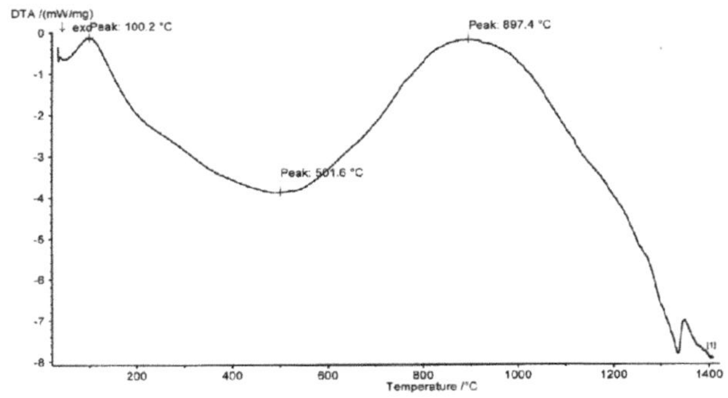


**Figure 4:** DT curves of tannery sludge activated carbon (SC300).

**Figure 5a:** SEM images of tannery sludge carbon TSC600 activated at 650°C respectively.

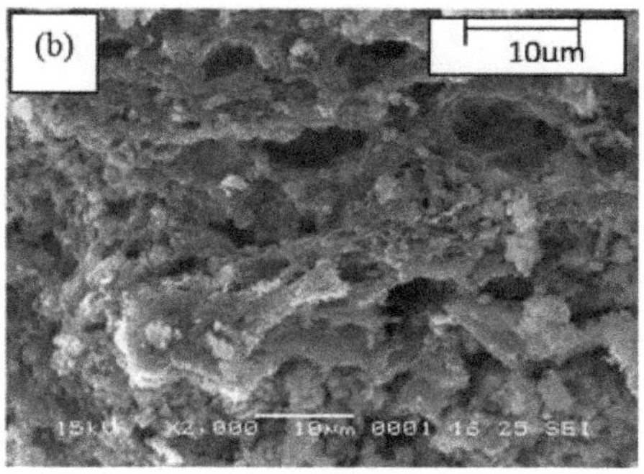


**Figure 5b:** SEM images of tannery sludge carbon TSC300 activated at 650°C respectively.

evaporation of free moisture content. The major weight loss of 23% and 17% was observed from the illustrations. This loss was due to the release of the volatile matter with increase in temperature. The exothermic peaks observed from the DT curves for TSC600 and TSC300 are shown in Figures 3 and 4. The thermal stability for sludge carbons were confirmed by its exothermic peak at 971.1°C and 894.4°C for TSC600 and TSC300 respectively.

**SEM analysis:** The presence of mesopores and micropores on the external surface of the developed tannery sludge carbon after the combined chemical and physical activation are revealed from the SEM micrographs of the tannery sludge activated carbons (TSC600 and TSC300) and shown in Figures 5a and 5b. The pores were formed due to the release of volatile matters and the reaction between the sulphuric acid and the carbon atom in the precursor (raw tannery sludge). The pore width was determined from the image with a magnification of 5,000X by using the appropriate scale of 10 µm represented in the image.

| Activated carbon | Surface area ($m^2$/g) | Pore volume ($cm^3$/g) | Pore width (A°) | |
|---|---|---|---|---|
| | | | BJH method | HK method |
| TSC600 | 188.25 | 0.134239 | 19.77 | 5.15 |
| TSC300 | 212.32 | 0.1325 | 19.73 | 5.07 |

**Table 2:** Porosity characterisation of the tannery sludge activated carbons by BET isotherm.

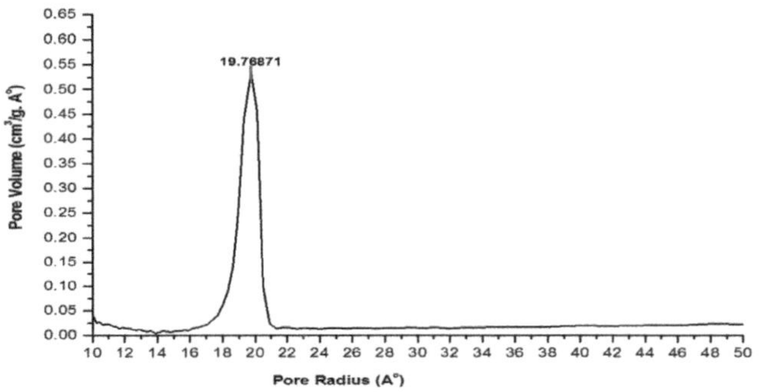


**Figure 6a:** BJH Desorption Pore Distribution of tannery sludge activated carbon TSC600.

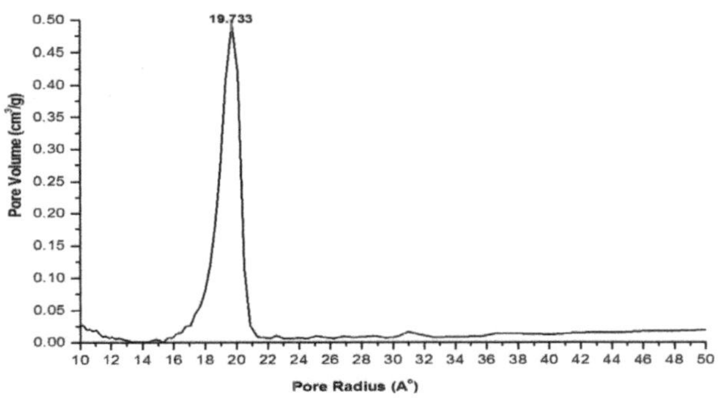


**Figure 6b:** BJH Desorption Pore Distribution of tannery sludge activated carbon TSC300.

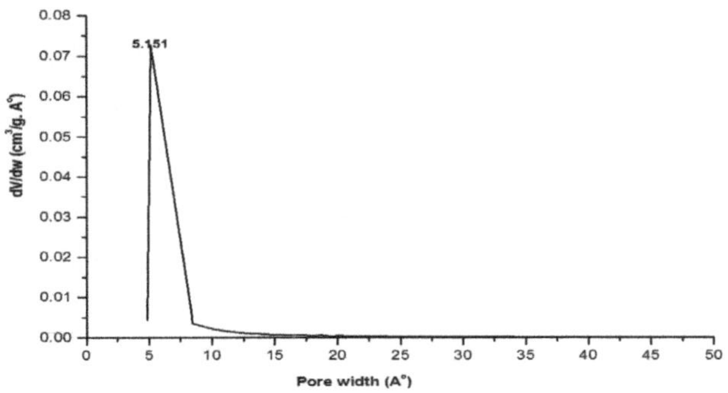


**Figure 7a:** HK Desorption Pore Distribution of tannery sludge activated carbon TSC600.

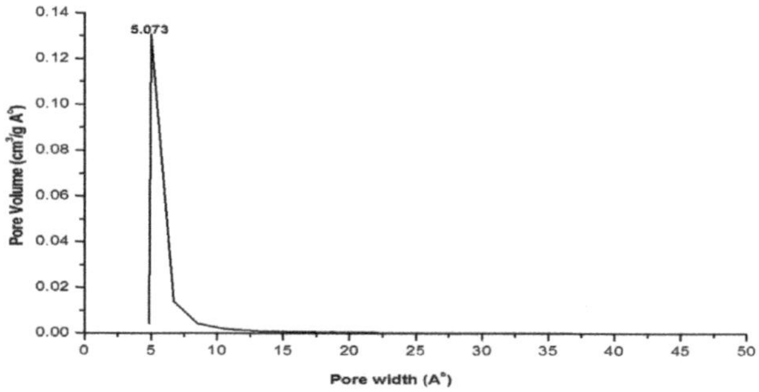


**Figure 7b:** HK Desorption Pore Distribution of tannery sludge activated carbon TSC300.

| Adsorbent | Carboxylic | Lactonic | Phenolic | Basic | Acidic | All |
|---|---|---|---|---|---|---|
| SC600 | 0.002 | 0.012 | 0.012 | 0.006 | 0.026 | 0.032 |
| SC300 | 0.006 | 0.01 | 0.011 | 0.007 | 0.027 | 0.034 |

**Table 3:** Results of Boehm titration in mmol/g for various activated carbons.

**BET isotherms:** The adsorptive capacity of activated carbon is related to its specific surface area, pore volume and pore size distribution. The porosity characterisation of the tannery sludge activated carbons are shown in Table 2. The surface area of the activated carbon decreased with the particle size of the activated sludge carbon. The pore size distribution of tannery sludge carbons was studied by BJH method and HK method of adsorption isotherm. The surface area of TSC300 was higher than TSC600 and the pore size distribution were dominated and uniformly distributed with mesopores and micropores. From Figures 6a and 6b and Figures 7a and 7b it is clear that the activated carbon produced from the tannery sludge was in the mesopore and micropore range.

**Boehm titration:** Introduction of functional groups containing oxygen is the major purpose of oxidation. Boehm proposed titrations of carbons with bases of various strengths. According to Boehm titration, the numbers of carboxylic, lactonic, and phenolic groups can be identified. Table 3 presents the Boehm titration results. The number of groups was reported in terms of surface concentration (mmol/g). For both sludge activated carbons, the number of acidic groups significantly increased with a decrease in the number of basic groups after oxidation [15]. The Boehm titration results revealed that, with increase in the particle size of the sludge carbons, the concentration of acidic and basic groups decreased. Hence lesser the particle size, greater is its surface functional group and quicker is the rate of adsorption of dyes. The acidic functional groups of the carbons were greater than the basic functional groups due to the activation of the carbon using acid as the precursor.

## Conclusion

The combined physical and chemical activation at 650°C was found to be optimum in using sulphuric acid as the activating agent. The burn-off of TSC600 and TSC300 were 29% and 26% and the yield percentages of the two carbons were 71% and 74% respectively. The thermal stability for different activated carbons was confirmed by their exothermic peaks at 971.1°C and 897.4°C respectively. The SEM images of the activated sludge carbons depicted the pore formation. The combined activation on the sludge has led to the development of micropores and mesopores on the external surface of the carbons. The surface areas of the various carbons were in the order of $S_{SC300}$:212.32 $m^2$/g >$S_{SC600}$:188.205 $m^2$/g respectively. The pore size distributions were dominated with mesopores and micropore were also uniformly distributed. The micropore widths of the activated carbons were 5.15 A° and 5.07 A° and the mesopore widths were 19.77 A° and 19.73 A°, respectively. The cationic nature of the developed activated sludge carbons was confirmed by the Boehm titrations methods with higher acidic functional groups.

### References


1. Pereira MFR, Soares SF, Orfao JJM, Figueiredo JL (2003) Adsorption of dyes on activated carbons: Influence of surface chemical groups. Carbon 41: 811-821.
2. Singh KP, Mohan D, Sinha S, Tondon GS, Gosh D (2003) Color removal from wastewater using low-cost activated carbon derived from agricultural waste material. Ind Eng Chem Res 42: 1965-1976.
3. Valdes H, Sanchez-Polo M, Rivera-Utrilla J, Zaror CA (2002) Effect of ozone treatment on surface properties of activated carbon. Langmuir 18:2111-2116.
4. Pelekani C, Snoeyink VL (2001) A kinetic and equilibrium study of competitive adsorption between atrazine and congo red dye on activated carbon: The importance of pore size distribution. Carbon 39: 25-37.
5. Barbooti MM, Mohamed TJ, Hussein AA, Abas FO (2004) Optimization of pyrolysis conditions of scrap tires under inert gas atmosphere. J Anal Appl Pyrolysis 72: 165-170.
6. Zabaniotou AA, Stavropoulos G (2003) Pyrolysis of used automobile tires and residual char utilization. J Anal Appl Pyrolysis 70: 71-72.
7. Al-Degs Y, Khraisheh MAM, Allen SJ, Ahmad MN (2003) Effect of carbon surface chemistry on the removal of reactive dyes from textile effluent. Water Res 34: 927-935.
8. Kaushik P, Malik A (2009) Fungal dye decolourization: Recent advances and future potential. Environ Int 35: 127-141.
9. Suteu D, Bilba D, Mureşan R, Mureşan A (2008) Using fibrous materials for textile wastewaters treatment. Proceedings of 4th International Textile, Clothing and Design Conference- Magic World of Textiles, Dubrovnik, Croatia. pp. 1112-1117.
10. Crini G (2006) Non-conventional low-cost adsorbents for dye removal: A review. Bioresour Technol 97: 1061-1085.
11. Geethakarthi A, Phanikumar BR (2011) Adsorption of two Azo reactive dyes by tannery sludge activated carbon: Effect of temperature and pH, International conference on environmental science and development. 4: 154-157.
12. Murillo R, Navarro MV, Lopez JM, Garcia T, Callen MS, et al. (2004) Activation of pyrolytic tire char with $CO_2$: Kinetic study. J Anal Appl Pyrolysis 71: 945-957.
13. El-Sheikh AH, Newman AP, Al-Daffaee HK, Phull S, Cresswell N (2004) Characterization of activated carbon prepared from a single cultivar of Jordanian olive stone by chemical and physicochemical techniques. J Anal Appl Pyrolysis 71: 151-164.
14. Boehm HP (1966) Chemical identification of surface groups. Adv Catal 16: 179-274.
15. Salame II, Bandosz TJ (1999) Experimental study of water adsorption on adsorption on activated carbons. Langmuir 15: 587-593.

# 21

# Repair Method of Damaged Steel Framed Structures and Ultimate Seismic State of Repaired Steel Frames

**Takumi Ito* and Kenjiro Mori**

*Department of Architechture, Tokyo University of Science, Japan*

**Abstract**

Recently, a lot of building structures have been experienced severe natural disasters, and it was reported that various types of terrible collapse mode were occurred. And also, there are many discussions about repairability and recovery on damaged buildings over the world. In particular, there has been focused on new keyword ಯ Resilience ಠ in any field including structural engineering. In Japan, a technical guideline for repairing damaged buildings has been established. However, the applicability and feasibility of repair method and recovery has not been well reported. Herein, to investigate the recovery and ultimate seismic state of repaired steel framed structures, experimental and analytical studies are conducted. Herein, the actual repairing technique for steel framed structure is suggested. During experimental study, the damaged steel member is reproduced by loading tests with consideration of past reports of mega-earthquake disasters. The next, the damaged portion is repaired by proposed method by use of steel-cover plate technique. After this repair process, the loading test is done again. And the recovery of structural performance is estimated by comparison of original and repaired state. From the test results, it is confirmed that the strength and ductility are improved after repair. Furthermore, the analytical model and restoring force characteristics of repaired steel member are suggested by observation of ultimate behavior during loading test. Here, the purpose on seismic design is to guarantee the overall failure mode formation on frames. So, the structural demand of strength and rigidity and column-to-beam strength ratio on repaired state are discussing analytically.

**Keywords:** Restoration method; Repairability; Seismic ultimate state; Steel frame

## Introduction

In Japan, a lot of building and infrastructure have been experienced severe earthquake disaster since the dawn of history. Recently, there are many discussions about repairability and recovery for these damaged building structures, in particular, after 2011 Great East Japan Earthquake. There has been focused on the new keyword "Resilience", which presents the revival potential or function maintenance of damaged structures and Bruneau and Reinhorn [1] have mentioned the concept of structural resilience, and high resilience system is as follows: 1) reduced failure probabilities, 2) reduced consequences from failures, in terms of lives lost, damage, and negative economic and social consequences, 3) reduced time to recovery. The technical guideline [2] has been published to repair the various kinds of structures in Japan. Particularly for steel structures, it suggests a repairing method based on the past experimental researches [3], however the number of such past researches was restricted. And also, few past research on repairability and recovery of steel structures limit to Japan have been conducted. Furthermore, authors have done the experimental studies to investigate the repairability of repaired H-shaped steel members [4] and also, an analytical model of repaired H-shaped steel member has been proposed. Its analytical model can evaluate the fundamental performance and hysteresis behavior.

## General Description of Damaged Steel Frames and Restoration Method

### Failure mode of steel frames after severe earthquake motions

In Japan, the various kinds of failure mode on building structures have been observed and reported during past earthquake disasters in Japan, such as the Great Hanshin-Awaji Earthquake 1995, the Great East Japan Earthquake 2011. And then, the Japanese technical guideline [2] has been published, and which suggests the estimation method of damage and restoration method for damaged members. Herein, the restoration method of damaged steel framed structures is explained.

### Restoration method of steel members

**Outline of restoration method for damaged steel members:** When the great earthquake motions occur, a various kind of failure mode is generated on the steel framed structures as shown in Figure 1. After the plastic hinge or local buckling is formed on the members, it is desirable to repair this damaged portion. And the example of restoration method on Restoration Manual is summarized as shown in Figure 2.

(a) Example of damaged steel frames (the Great Hanshin-Awaji Earthquake, Japan, 1995)

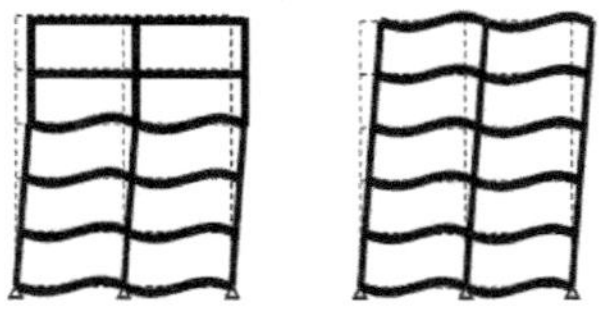

Local collapse mode    Overall collase mode

(b) Failure mode of steel frame

**Figure 1:** Failure mode of steel frames during earthquake.

***Corresponding author:** Takumi Ito, Department of Architechture, Tokyo University of Science, Japan, E-mail: t-ito@rs.kagu.tus.ac.jp

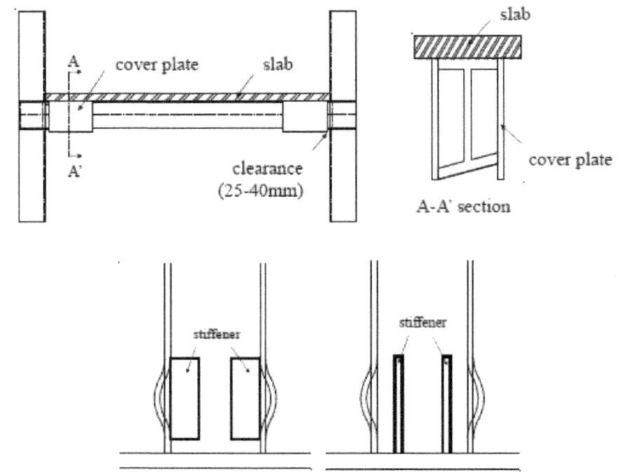


(a) H-shaped section steel members

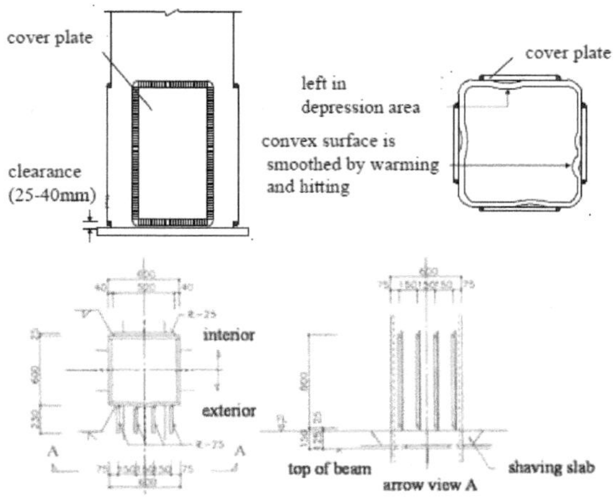


(b) Box-type steel members

**Figure 2:** Examples of restoration method of damaged steel members.

**Restoration method and repairing process on damaged H-shaped steel members:** From the Restoration Manual, some repair methods: for damaged H-shaped steel members are indicated as shown in Figure 2a. Herein, the diagram of actual repair method is illustrated in Figure 3. This method is as follows; the damaged portion is repaired by welding cover plates to the flanges with local buckling, and a clearance of 25 mm is opened between the end plate and the edge of the cover plates. The thickness of the cover plates is equal to that of the flange plate of the H-shaped steel members.

**Restoration method and repairing process on damaged Box-shaped steel members:** From the Restoration Manual, some repair methods for damaged box-type steel members are indicated as shown in Figure 2b. This method is as follows; the surface of damaged portion is corrugated, and this convex surface is smoothed by warming and hitting. After this smoothing, the cover plates are attached by welding, and a clearance of 25 mm is opened between the end plate and the edge of the cover plates. The thickness of the cover plates is equal to the box type steel member.

## Analytical Method for Repaired Steel Member and Frame

### Analytical model of steel member after repair

**H-shaped steel member:** The authors have conducted the experimental study to investigate the applicability and repairability of the restoration method for damaged H-shaped steel member as shown in Figure 3. From the test results, the ultimate states of test specimens were observed as shown in Figure 4. And the test results of moment-angle relation were obtained as shown in Figure 5, which compares the original state and repaired member. From the test results, the ultimate state behavior were observed as follows; on the repaired steel member, the plastic hinge were formed to the side of cover plates, and the repaired portion with cover plate behaved as rocking of rigid body. And also, the strength and ductility of repaired member were improved by

Comparison with original state and damaged member without repair. From these considerations, the analytical models of H-shaped steel member of original and repaired state were established as shown in Figure 6. And also, the hysteresis models during inelastic cyclic loading were applied with Skeleton Shift Model as shown in Figure 7. And the comparison of test results and analytical results are presented in Figure 8, and it can be said that the proposed analytical model can chase the test results well.

**Box-type steel member:** The authors have conducted the experimental study to investigate the applicability and repairability of the restoration method for damaged Box type steel member as shown

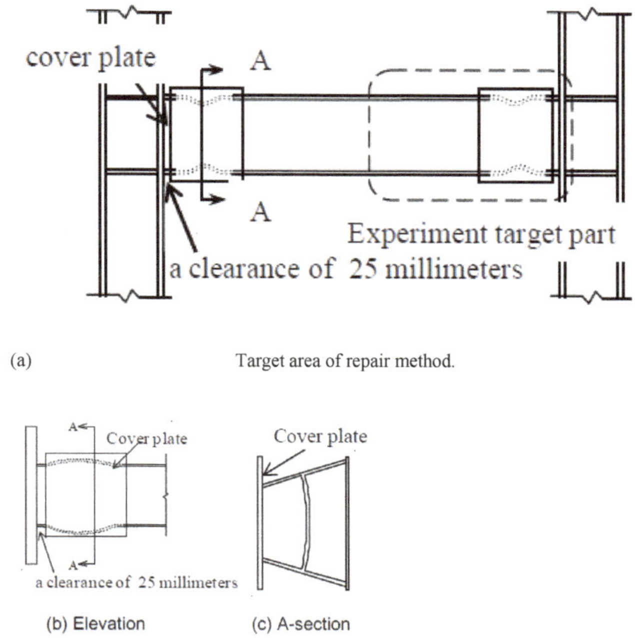


**Figure 3:** Diagram of restoration method of damaged H- shaped steel member.

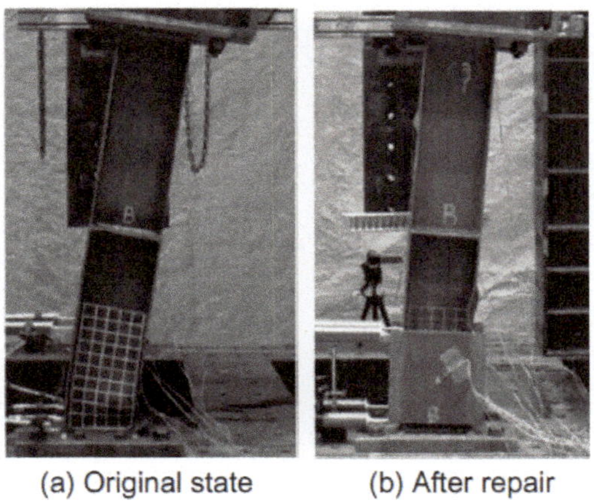


**Figure 4:** Ultimate state of test specimen.

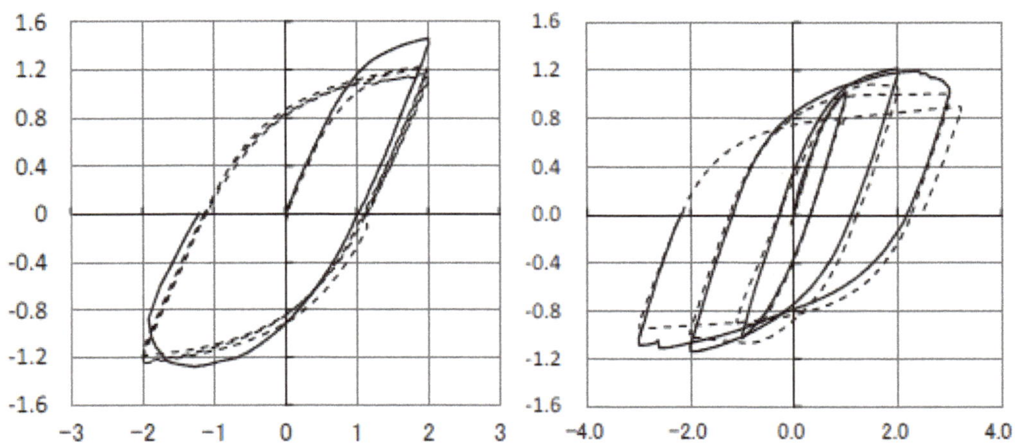


**Figure 5:** Comparison of test results.

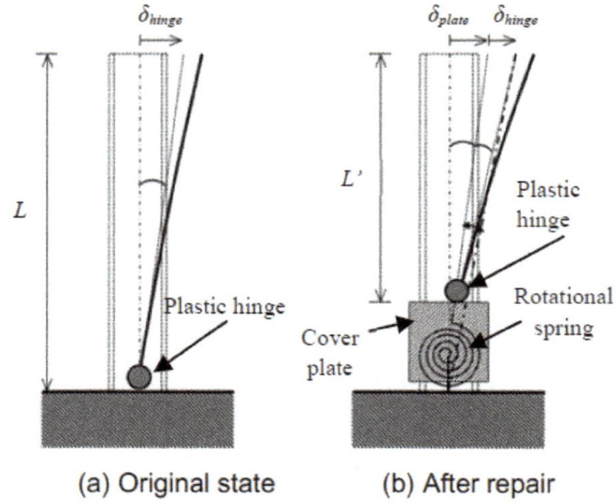


**Figure 6:** Assumption of analytical model.

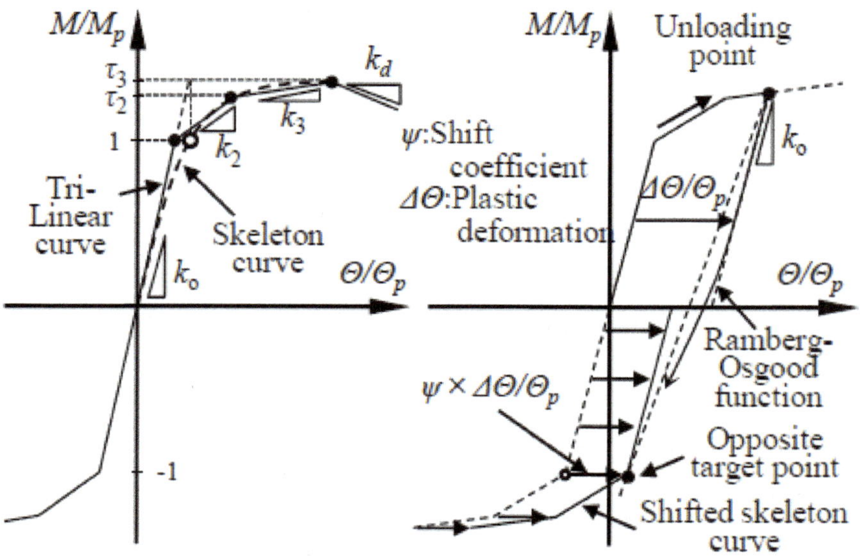


**Figure 7:** Hysteresis model during cyclic loading, called “Skeleton Shift Model”.

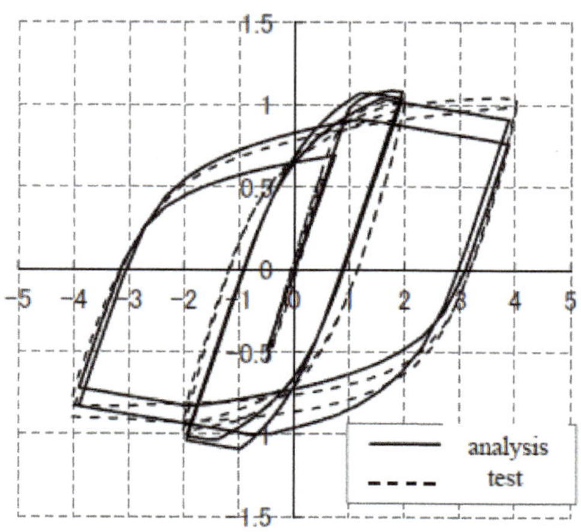


**Figure 8:** Comparison of test and analytical results.

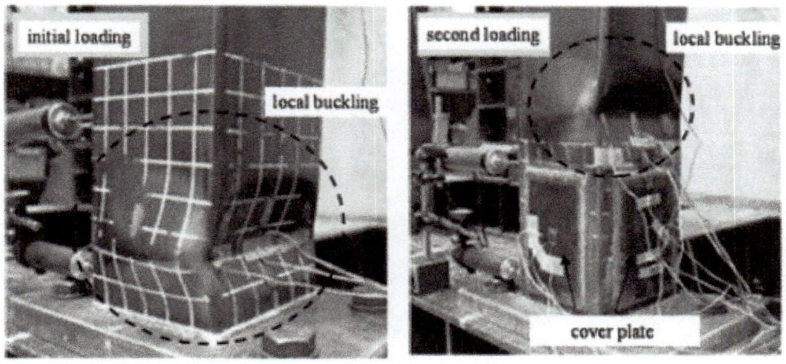


(a) Original state (b) After repair

**Figure 9:** Ultimate state of test specimen.

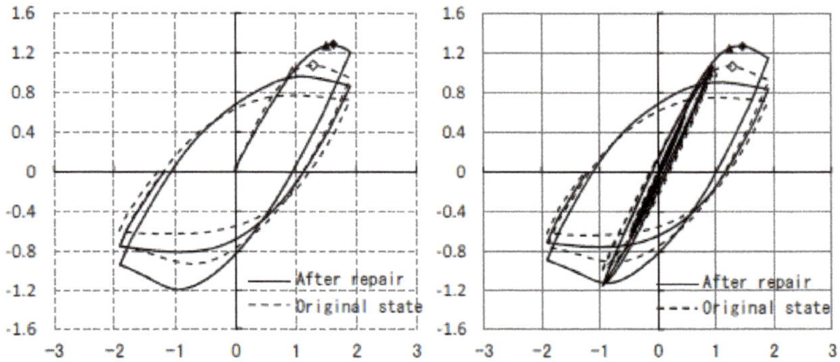


**Figure 10:** Comparison of test results.

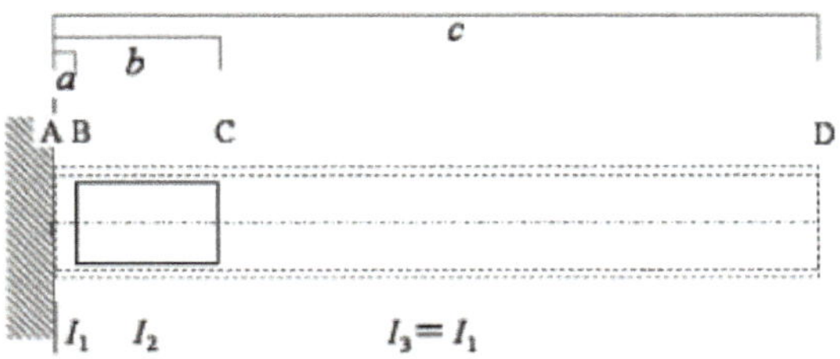


**Figure 11:** Assumption of analytical model.

in Figure 4. From the test results, the ultimate states of test specimens were observed as shown in Figures 8 and 9. And the test results of moment-angle relation were obtained as shown in Figure 10, which compares the original state and repaired member. From the test results, the ultimate state behavior were observed as follows; on the repaired steel member, the plastic hinge were formed to the side of cover plates, and the repaired portion with cover plate behaved as rocking of rigid body. And also, the strength and ductility of repaired member were improved by comparison with original state and damaged member without repair. From these considerations, the analytical model of box type steel member of original and repaired state was established as shown in Figure 11. And also, the hysteresis model during inelastic cyclic loading was applied with Skeleton Shift Model as shown in Figure 8 too. And the comparison of test results and analytical results are presented in Figure 12, and it can be said that the proposed analytical model can chase the test results well.

## Analytical model of repaired steel frame

Most steel moment resisting frame (rahmen frame) consists of H-shaped steel girder and Box type steel column. Generally, the failure mode of ultimate seismic state after large seismic input are formed as shown in Figure 13, in which the plastic hinges are generated on the end of most girder and few column end. The restoration method above mentioned (see 5.1) is adopted on this plastic hinge, the frame after repair is illustrated as shown in Figure 13a, and also the analytical mode of the frame can be given as Figure 13b. Furthermore, the restoring force characteristics model of this repaired member can be referred with model as shown in Figures 7, 8 and 11.

## Analytical study and considerations

Herein, by use of proposed analytical model above mentioned, the ultimate seismic behavior and seismic resistant performance of repaired steel frame are discussed. A 3-story 2-bay steel moment-resisting frame shown in Figure 14 is studied hereafter. The member properties are as follows; moment of inertia, beam =9.28 x 108 mm$^4$, column = 7.56 x 108 mm$^4$, plastic moment capacity, beam = 1,182 kN m, column = 682 kN m, and floor weights are 0.29M (2nd, 3rd), 0.42

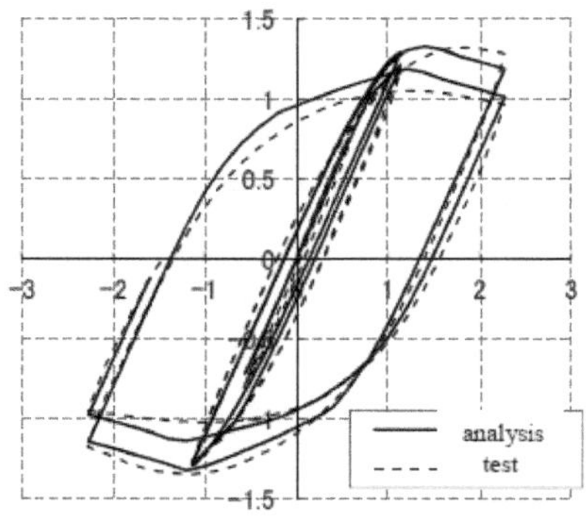


**Figure 12:** Comparison of test and analytical results.

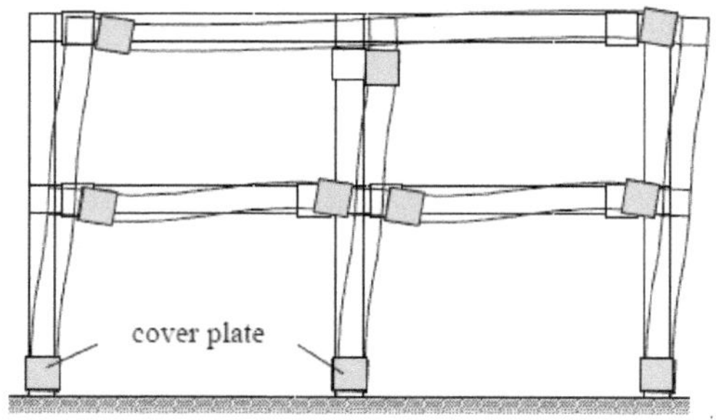


(a) Steel frame after repair.

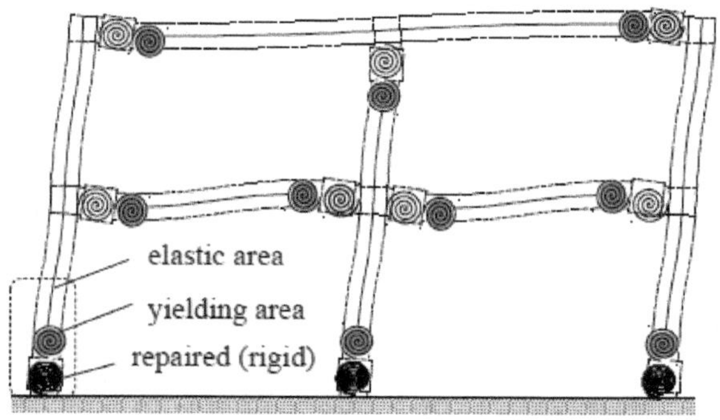


(b) Analytical model of repaired frame

**Figure 13:** General diagram of steel frame after repair.

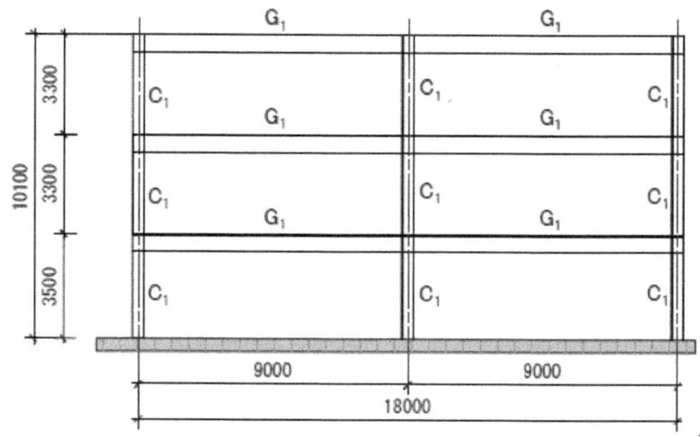


**Figure 14:** Analytical model of 3-story 2-bay steel frame.

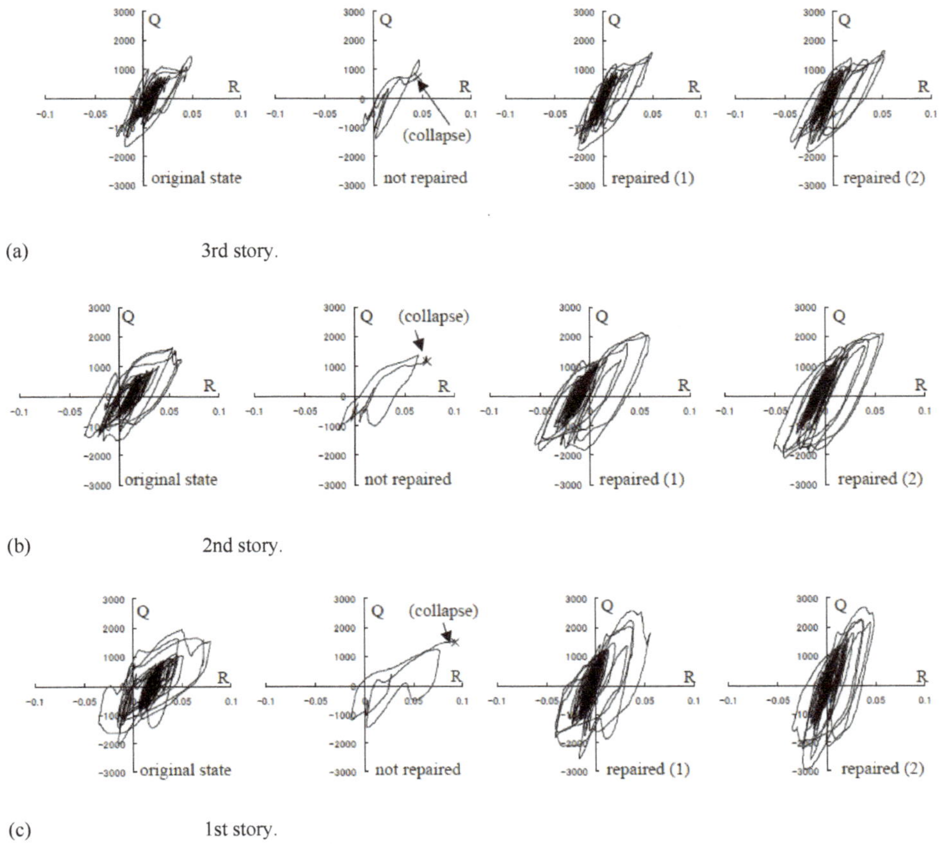


**Figure 15:** Hysteresis loops of story shear – story drift (vertical axis = shear Q (kN), horizontal axis = story drift R (rad)).

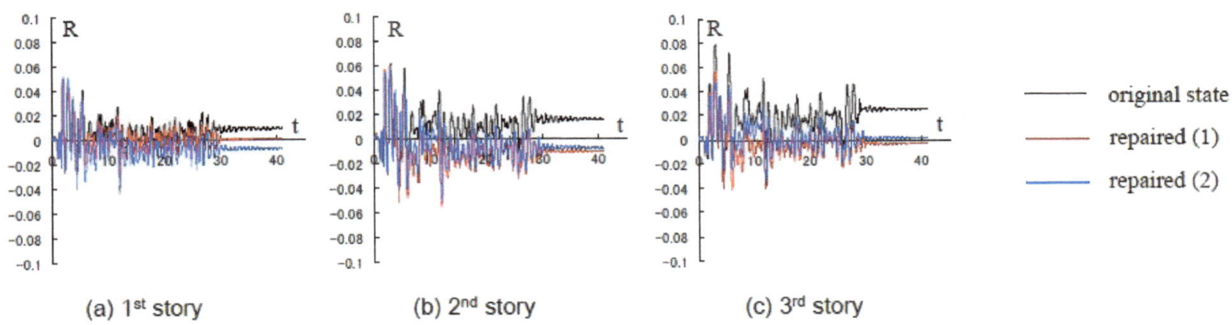


**Figure 16:** Time history of story drift (vertical axis = story drift R (rad), horizontal axis = time t (sec)).

M (roof), here M is total mass of frame. A constant modal damping ratio of 2% is considered. The elastic vibration periods are 0.866sec (1st mode), 0.225 sec (2nd mode), and 0.110 sec (3rd mode). In this paper, the effect of slab or interior/exterior material is not considered. This frame is analyzed for the ground acceleration records of El Centro 1940 NS (Imperial Valley Earthquake, here, PGV = 111 m/sec). The time history response analysis is performed on original state, and the results of story shear – story drift are presented in Figure 15, time history of story drifts are shown in Figure 16.

In this paper, the damaged member of plastic hinge formation is repaired by use of repairing method as shown in Figures 3 and 4. Furthermore, the strategy for repairing to damaged member is assumed as follows, strategy 1) all damaged member of plastic hinge formation is repaired, and its member is fully strengthened, strategy 2) to prevent the excessive increment of strength in specific member around beam-column-connection, the non-damaged member which remains within elastic is repaired alternatively. And then, the time history response analysis is performed on repaired state again, and the results of story shear – story drift and time history of story drift are presented in Figures 16 too. Furthermore, the frame without repair after failure mode formation is analyzed.

From the response analysis results,

1. On the frame which is not repaired after failure mode formation on original state, it is observed that the response of story drift becomes large, and the strength and rigidity become small. Finally, the frame is collapsed completely. It means that the damage frame does not have enough seismic resistant performance against aftershock,

2. However, the repaired frames have high strength comparing with original state. And also, the response of story drifts becomes smaller than original state. It means that the repaired frame has enough seismic resistant performance; furthermore, the seismic reinforcement effect is obtained.

## Conclusion

In this paper, the restoration method and its repairability for damaged steel frames are focused on. In Japan, the restoration guideline has been published, and it was actually referred and availed during past earthquake disasters. And also, the various kinds of restoration method are developed and suggested. This paper summarizes the outline of restoration method for H-shaped steel member and box type steel member. And also, the past experimental studies are explained, furthermore, the proposed analytical model are presented. By use of these proposed analytical model, the time history response analysis are performed on low-rise steel frame. Herein, to investigate it's effectively and effect to damage frame, a various strategy of repairing portion in frame has been assumed. From the response analysis results, the frame which is not repaired after failure mode formation on original state, the frame is collapsed completely. It means that the damage frame does not have enough seismic resistant performance against aftershock. However, the repaired frames survive against aftershock. And also, the repaired frame has enough seismic resistant performance; furthermore, the seismic reinforcement effect is obtained.

### References

1. Bruneau M, Reinhorn A (2006) Overview of the resilience concept. Proceedings of the 8th U.S. National Conference on Earthquake Eng.
2. The Japan Building Disaster Prevention Association (2001) Guideline for post-earthquake damage evaluation and rehabilitation technique (in Japanese).
3. Tanaka A, Izumi M, Narihara H (1990) Experimental study on the statistical characteristics of repaired or reinforced steel structures. J Struct Eng 36 B: 377- 384.
4. Mori K, Ito T, Sato H, Munemura H, Matsumoto T (2015) Repair ability performance and restoring force characteristics of damaged H-shaped steel members after repair. Int J High-Rise Build 4: 57-64.

# An Exploration of CO2 Emission Models for Horizontal Curves on Freeways

**Boya You, Fengxiang Qiao* and Lei Yu**

*Innovative Transportation Research Institute, Texas Southern University, Houston, Texas 77004, USA*

**Abstract**

Global climate change attributed by $CO_2$ emissions could seriously impact people's daily lives. According to the statistics, road transport is one of the largest contributors to greenhouse gases. Simultaneously, the increasing number of vehicles, especially vehicle mileages traveled, will continue to induce more $CO_2$ emissions. It is believed that the driving behavior is one of the most important factors that impact the $CO_2$ emissions. However, some other factors may also contribute to the $CO_2$ emissions, for instance, the geometric elements of a roadway. This paper explores the relationship between the horizontal curve radius and $CO_2$ emissions on freeways from field tests under real-world driving conditions through the development of an Emission Factor Model (EFM). The prediction results were also compared with the measured $CO_2$ emissions based on the default emission rates generated from Motor Vehicle Emission Simulator (MOVES). The modeling results indicate that the EFM has better performance for the estimation of $CO_2$ emissions factors for horizontal curves on freeways under free flow conditions with a predicting error of 5.5%, which is much smaller than the one from a method directly estimated from MOVES (16.8%).

**Keywords:** $CO_2$ emissions; Emission factor model; Freeway; Horizontal curve radius; MOVES

## Introduction

### Background

Climate change attributed by Greenhouse Gas (GHG) emissions could impact human's daily lives in a variety of ways. For example, the climate change enhances glacier melting and extends the length of the melt season, which increase river runoff and discharge peaks. The increased runoff could affect the quantity and quality of fresh water sources, thus affecting agriculture, industry, transportation, urbanization, and many other aspects. This would eventually affect the public health [1-3]. Carbon dioxide ($CO_2$) is the primary greenhouse gas [4], and mostly produced by human activities, such as industries, agriculture, and transportation. In 2012, transportation sector contributed 28% of US total GHG emissions. Within this sector, light-duty vehicles including passenger cars and light-duty trucks were the largest contributors with a percentage of 62% [1]. Meanwhile, many studies have demonstrated that vehicle emissions are sensitive to traffic operations [5] and drivers' driving behaviors [6,7]. Some strategies have been applied to address the vehicle emissions such as charging for $CO_2$ emissions and promoting public transportation, applying intelligent transportation systems (ITS) to improve drivers' driving behaviors, and so on [8-10]. However, there are always some restrictions to implement those strategies in practice. For example, urban sprawl may weaken the promotion of transit development and application; and it still take time before all vehicles on roads are equipped with ITS for emission reduction purpose. Charging emission by raising gasoline prices may only lead to a small decline in vehicle mileages traveled (VMT) [2]. To effectively reduce the amount of $CO_2$ emissions, the main focus of attention should be paid to the strategies dealing with the source of emissions. Facts released by the US Environmental Protection Agency's (EPA) show that, road transport is one of the largest contributors to GHG emissions. To estimate vehicle emissions, many mathematical models at macroscopic and microscopic levels have been developed, such as the MOBILE series and MOVES models developed by US EPA, which could be used to estimate the emissions at national, county, and levels. There are two main challenges for emission model development. One is the difficulty in obtaining accurate emission data. Another one is the identification of the structure and parameters of emission models [11-13]. Most emission models, especially the earlier developed models, focused on the linkages with vehicle speed and acceleration rate [12]. Nevertheless, other factors may also contribute to the vehicle emissions such as the roughness of pavement [14], and the geometric design of roadway [15]. Horizontal curves, which are the segments that provide a gradual change in direction, are important elements in roadway design. Due to the difficulty in judging the sight distance, drivers may apply the brakes or tap the gas pedal frequently when entering the winding segments, which may generate high-frequency accelerations and decelerations associated with extra vehicle emissions. According to the Geometric Design of Highways and Streets [16], also known as the Green Book, the current roadway alignment design method adopted by United States Federal Highway Administration (FHWA), is based on the concept of design speed, with no direct consideration of the environmental impacts [17].

### Objectives

The main objective of this study is to develop an Emission Factor Model (EFM) that could reflect the relationship between $CO_2$ emissions on freeway and horizontal curve radius using aggregated data from field tests in Houston, Texas, USA. The aggregated data used in the proposed exponential model aims to represent the overall condition for each single curve. This method views each single curve as a unit to reflect the relationship between $CO_2$ emissions and other independent variables. As a comparison, MOVES was used to also estimate the $CO_2$ emissions under the same curves based on the recorded instantaneous vehicle speed and acceleration/deceleration rates. The prediction results between the EFM and MOVES method would be finally compared.

***Corresponding author:** Fengxiang Qiao, Associate Professor And Co-Director, Innovative Transportation Research Institute, Texas Southern University, Houston, TX 77004, USA, E-mail: qiao_fg@tsu.edu

### Existing emission models and modeling methodologies

To effectively estimate the vehicle emissions, many emission models have been developed. The EPA released new vehicle emissions modeling tool MOVES replaces the previous series of models MOBILE. MOVES estimates emissions based on a bunch of default values and inputs, such as vehicle types, road types, VMT, vehicle age distribution, mileage accumulation rate provided by users. The 'default' database included in this model reflects emission information for the whole country. Also, MOVES can be used to estimate the vehicle emissions at state, county, and project levels. While MOVES could only be used for estimating the vehicle emission at macro and medium levels, the default value, for example, the emission rates and operation mode bin table from MOVES could be extracted and help researchers to calculated the emissions manually [16,18]. A study conducted by Ko generated the speed file for horizontal curves using the speed prediction model and calculated the vehicle emissions based on the emission rates generated from MOVES [16]. The results from this study indicated that several factors such as the curve radius, the 85$^{th}$ percentile tangent speed, and the instantaneous speed at the middle of the curve could affect the emissions on horizontal curves. Other emission models such as EMFAC focus on a macro level based on mean speed and VMT [19,20]. One drawback of these models is that the estimation may not reflect the real-world emission sufficiently without considering the driving dynamics [13,21]. These cycle-based models cannot be used to evaluate the microscope vehicle emissions [22,23]. In order to estimate the vehicle emissions at a micro level, modal-emissions models have been developed. According to Frey et al., there are basically two ways to develop the modal-emissions models [13]: one is to develop a speed-acceleration matrix to classify the operating mode of vehicles [24]; another way is to develop the modal models based on engine mapping [25]. The Mobile Emission Assessment System (MEASURE) developed by the Georgia Institute of Technology is able to estimate CO, $NO_x$, and $VOC_s$ using aggregated modal modeling approach [26]. Another modal model developed by Barth et al., could also predict the vehicle emissions for light-duty vehicles based on a series of vehicle operating modes [27]. In terms of the data sources, some emission models use field test data [14,15,28], which also known as the 'real-world' emissions, while other models were developed based on the dynamometer testing [29,30]. Also, some micro simulation software such as INTEGRATION and CORSIM, could also be used for the estimation of the vehicle emissions based on speed profiles or traffic conditions obtained from the field tests [31].

### Existing on-board emission data collection and data processing

The most common way to collect the emission data is to measure the vehicle emissions through laboratory tests using dynamometer [13,32]. The vehicles used for the laboratory tests are normally new vehicles or under good conditions, which is different from the real world traffic where contains composed of vehicles with varying conditions [33]. To develop an emission model to reflect the driving dynamics at micro level, the analysis of real-world emissions and special measurement techniques are needed. The portable emissions measurement system (PEMS) uses such techniques and could be used to measure the second-by-second tailpipe emissions with the record of vehicle activity information at the same time [33]. Some statistical issues of data processing for the on-board emissions data have been indicated in a previous study such as the representativeness of test vehicles [32]. Several recommendations on data synchronization and the identification of missing data for data quality and data analysis can be found in [5].

### Extraction of geometric data

In order to ascertain the relationship between geometric features and vehicle emissions, one challenge is how to obtain the geometric information of roadway alignment because the availability of original drawing files has been the Achilles' heel. Some researchers used a GPS-based method to collect data to compute curve radius and the results showed that GPS was able to provide the alignment data accurately [34,35]. Imran et al., conducted a field test with GPS to collect and calculate the geometric features for horizontal curves [35]. Pratt et al., also used a GPS-based method to collect data every 25 feet to calculate the radius and recommended posted speeds for horizontal curves. The GPS-based field test turned out to be time-consuming even it could produce data accurately [34]. Sanders used a GPS/GIS method to extract the geometric features for horizontal curves, where GPS was used to produce the centerlines of roadways and then the curve data were extracted through GIS. Results showed that the GPS/GIS methodology could provide even higher quality data than field-collected tests [36]. Another report in 2013 also adopted the GIS method to extract the geometric features using the embedded tool in GIS called ArcGIS COGO tool [37]. The horizontal curve types can be identified as simple curves, reverse curves, and spiral transition using satellite maps [38,39]. A previous study conducted by Kim et al., successfully extracted the lateral profiles and superelevation through laser scanning technology [40]. In fact, these cross section elements, for instance, the superelevation, are difficult in determining without the drawing files or field measurements. The simplest and most cost effective option is to identify the geometric information using digital maps. So far many approaches have been developed to identify the geometric features for horizontal curves from digital maps. For example, the Google Earth Pro developed by GOOGLE Company could be used for identifying the horizontal curves by fitting the curves with a radius-changeable circle [41]. Most of the applications use the semi-automatic data-extraction algorithm; no available full automatic method has been found. Curve Extension, Curve Finder, and Curve Calculator are all GIS based curve-extraction tools. The Curve Extension and Curve Calculator could only be used for identifying some geometric elements for a single circular curve, while the Curve Finder could be used to extract several curves on a designated route at one time [42]. Compared with the traditional data sources, GIS maps have advantages such as zero-cost, complete roadway data and short data preparation time [43]. One drawback of the GIS method for curve data extraction is the inclusion of tangent sections when the radii are very large. Khan suggested that curves with radius less than 10,000 ft. and greater than 200 ft. could be selected for tests when using the GIS method [44].

## Material and Methods

### Portable emission measurement system (PEMS)

PEMS can be used as a measurement tool to quantify the real-world emissions and fuel consumption [33]. The whole system has five main components: the dual gas analyzer, the GPS system, the on-board computer, the PM monitor and the engine data obtained [34]. It is able to provide second by second tailpipe emissions and vehicle activity. In the field test, two operators were involved in the installation of PEMS, which took about 20 minutes. To control the data quality from the start, some recommendations for equipment installation, data collection and data analysis for PEMS from a previous study were utilized [5].

### Test plan

This study focused on two 'test circles' around downtown Houston. Figure 1A shows the test sites in this study. Site 1 represents the small circle while site 3 represents the big one (I-610). Each 'circle' was tested for two rounds. Figure 1B indicates the trajectory of tests. The start/end point was at Texas Southern University. The total distance of the test routes is about 140 km. The testing vehicle was a 2004 Subaru Forester and the fuel type was gasoline. The field test was conducted during the time period of 4:30 p.m. - 8:30 p.m., which cover the peak hour and non-peak hour. The driver was instructed to follow the traffic flow in order to obtain natural data of driving behavior.

## Site selection

A series of criteria for site choice were defined. As for now, most of the previous studies related to geometric design, especially the horizontal curves, focused on two-lane rural highways [45-47]. This is because the two-lane rural highway has relatively lower traffic volume and less signalized intersections. Further, it could provide a variety of geometric features [48]. However, the freeways could also provide a variety of geometric features. In the meantime, it is able to naturally rule out the effects of intersections by providing continuous driving data. The roughness of pavement may somehow contribute to the vehicle emission, and the vibration of the testing equipment might also affect the accuracy of data [5]. The pavement of freeway might be more even compared to local roads, which could help reduce the impacts of pavement type. Therefore, this paper focuses on freeways in order to provide purer field-testing data, respectively. Considered that vehicle speed would be affected by the amount of traffic on the road, the data also need to cover different traffic conditions including congested and free flow for the variety of speed. The horizontal curve radius is the only geometric feature considered in this model. To identify the relationship between curve radius and emissions, other geometric elements, for

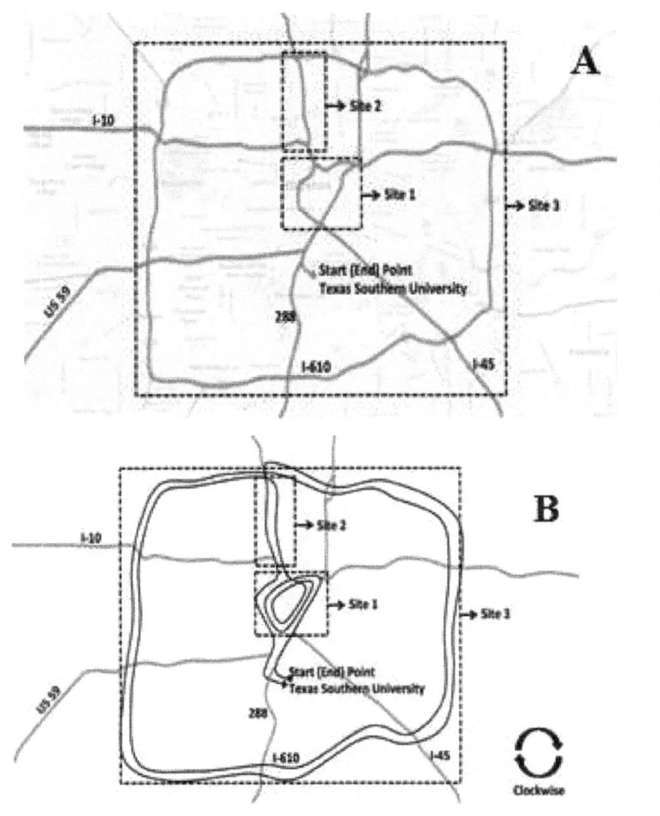


**Figure 1:** Field test routes in Houston area. (A) Test site 1, site 2 and site 3. (B) Start/end point and test routes.

example, lane width should be constant. Also, the roadway longitudinal grade was assumed to be zero.

- Freeway (e.g., I-610 Loop)
- A variety of geometric characteristics (radii)
- Constant geometric elements (lane width, shoulder width, etc.)
- Same advisory speed (60 mph)
- Data availability

## Data collection

Vehicle speed, acceleration/deceleration, longitude, latitude, and vehicle emissions ($CO_2$) were recorded using PEMS, which produced the records for each second. A Google Earth/Micro Station method was used to extract the geometric features for horizontal curves. First, the Google Earth will be used to generate the digital maps of roadways. Second, the maps produced by GPS will be imported into Micro Station. Third, Micro Station will be used to identify the geometric features. This methodology is composed of four steps: (1) identify the tangent segment; (2) identify the P.I. for curves; (3) identify the curves; and (4) measure the curve radius. Figure 2A shows the identification for geometric features for horizontal curves in Micro Station. The curve radius, degree of curvature, and curve type could easily be identified in the digital maps. The spiral transition curve is the segment between the straight section and a curve, which was designed to prevent sudden changes in lateral acceleration Figure 2B. As the transition segment, the spiral curve is hard to be identified and separated from the digital maps when the drawing files are not available. Due to this reason it is assumed that, all curves are simple curves with no spiral transition. In the modeling part, among the geometric features, only the curve radius will be used as the geometric related independent variables (Figure 2).

## Data preparation

PEMS provided a bunch of emission-related data including: instantaneous vehicle speed, acceleration/deceleration, latitude, longitude, and $CO_2$ emissions. The horizontal curve radii need to be determined manually using Google Earth/Micro Station method. After data collection, the first thing is to match two different data sets based on geo-locations. Due to the connection problem, some invalid data may occasionally be produced by PEMS. To ensure the data quality, data screening was conducted to identify errors and remove invalid data. Also, some outliers and unreasonable test data should be removed from the data set (e.g., an acceleration rate higher than 10 m/s2.) In this study, the curve based aggregated data were needed for modeling purposes. Therefore, the substantial amount of instant data generated by PEMS need to be aggregated based on each single curve. After aggregation, there will be a set of unique elements (e.g., the average speed for each single curve; and emission factors for a specific curve) assigned to each curve with different radius. There is one thing should be noticed here is that the aggregate process will neglect the detailed driving behavior.

## Modeling

After data preparation, the Emission Factor Model was developed. The EFM viewed each single curve as a unit. The emission factor for tangent will be viewed as a baseline in this model. The data including vehicle speed, acceleration/deceleration, and emission factors for $CO_2$ were aggregated for each single curve. SPSS was used to run this model and a self-compiled MATLAB program was used for validation.

$$EF_C = EF_T^{\sum \beta_i x_i} \quad (1)$$

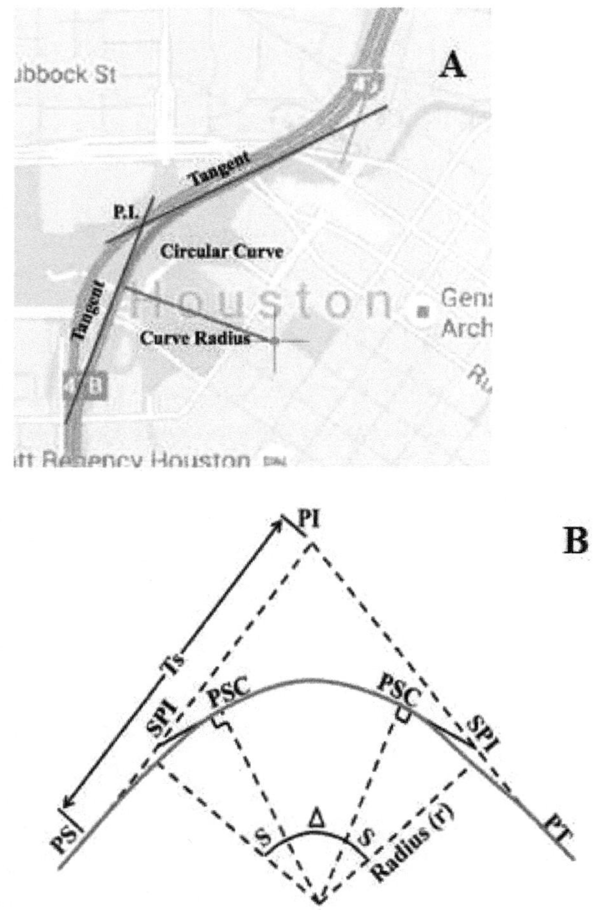


**Figure 2:** Identifying the horizontal curve radius from digital maps. (A) Curves extraction using Micro Station. (B) Sketch of a single curve (with spiral transition).

$$EF_C = EF_T^{\beta_1 \cdot v + \beta_2 \cdot a + \beta_3 \cdot r} \quad (2)$$

$$\log EF_C = (\beta_1 \cdot v + \beta_2 \cdot a + \beta_3 \cdot r) \cdot \log EF_T \quad (3)$$

$$\frac{\log EF_C}{\log EF_T} = \beta_1 \cdot v + \beta_2 \cdot a + \beta_3 \cdot r \quad (4)$$

Where,

$EF_C$: emission factor for horizontal curves ($g/m$)

$EF_T$: emission factor for tangent segments ($g/m$)

$v$: Average vehicle speed for each horizontal curve (km/hr)

$a$: Average absolute acceleration/deceleration for horizontal curve (m/s²), and

$r$: Horizontal curve radius (m)

## Simulation process using MOVES

In order to compare between the predicting results from EFM and the simulation results from MOVES, the second by second speed file and acceleration/deceleration information were used for calculating the $CO_2$ based on the default emission rates provided by MOVES. A previous research also estimated the vehicle emissions for horizontal curves of two-lane rural highways using this methodology [16]. First of all, the vehicle specific power (VSP) was calculated based on instantaneous speed, acceleration/deceleration. The equation for VSP calculation:

$$VSP = v*1.1*a + 9.81*grade(\%) + 0.132v + 0.000302*v^3 \quad (5)$$

Where $v$ is the instantaneous vehicle speed in the unit of m/s, $a$ is the second-by-second acceleration/deceleration in the unit of m/s2. The grade is the roadway longitudinal grade, which could be determined by field measurement or from the roadway profile. In this study, the grade was assumed to be zero. After generating the VSP, the data for each second was assign to a specific operating mode bin provided by MOVES. After matching the operating mode bin for each set of data, the default emission rates were extracted from MOVES's database based on the vehicle type (passenger car), fuel type (gasoline) and the age of the testing vehicle (10 years), and then the accumulated value of $CO_2$ emissions for each single curve was calculated. The aggregated results ($CO_2$ emission factor for each single curve) were transformed to Log(EFc)/Log($EF_T$) in order to compare with the results from EFM. The $EF_C$ is the emission factor for each single curve and the $EF_T$ is the average emission factor for tangent segments.

## Results and Discussion

Compared to the $CO_2$ emissions calculated based on the emission rates provided by MOVES, the emission factor model has a better performance to reflect the overall emissions for a single curve using aggregated data. In this session, the results of the EFM will be discussed in detail.

## The x-y plot of curve radius and emission factor

After data aggregation, there was one unique set of data for each single curve. For example, when the curve radius is 738.19 m, the average speed for this curve is 67.29 km/hr; the average absolute value of acceleration is 0.376 m/s2; the emission factor of this curve is 2.55 g/m. Figure 3 shows the x-y plotting results. The x-axis represents for the curve radius while the y-axis represents for the value of log ($EF_C$)/ log ($EF_T$). The plotting results show that when the curve radius is larger, the value of log ($EF_C$)/log ($EF_T$) is higher. When the curve radius is approaching 3,000 (estimate), the y-value is approaching 1, which means that the emission factor for curves is close to the emission factor for tangent segment. To explain this tendency, a function was used to fit these data samples:

$$y = \frac{\log EF_C}{\log EF_T} = 1 - e^{\frac{r}{690}} \quad (6)$$

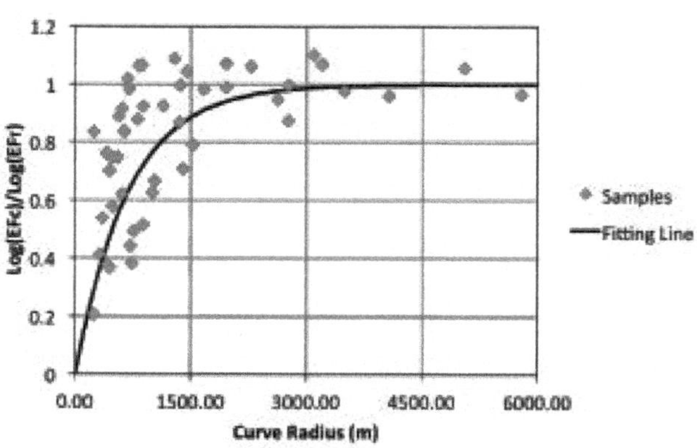


**Figure 3:** X-y plot of emission factor and curve radius.

Where,

$EF_C$: emission factor for horizontal curves (*g/m*)

$EF_T$: emission factor for tangent segments (*g/m*), and

$r$: Horizontal curve radius (*m*)

When r is zero, y equals to zero. In this case, no roadway exists, then no vehicle emissions. When r goes to infinity, y is 1. Assume that the 'radius' of tangent segment is infinity. When the curve radius goes to infinity, the emission factor for curves will be very close to the emission factor for tangent segment. Also, the yield point of this fitting line is when r equals to 2,070 (2,070 = 690×3, where 690 is a time constant in Equation (6), the 3 times of which means the termination of termination periods of an exponential curve like the one in Figure 3). This means that, when the curve radius is larger than 2,070 m, there is no big difference between emission factors for curves and tangent segments. In order to validate the relationship between curve radius and emission factors, the emission factor models should be developed.

## Bubble chart for curve radius, vehicle speed and $CO_2$ emissions

Figure 4 is the bubble chart showing the relationship between curve radius, vehicle speed, and $CO_2$ emission. The x-axis represents for the curve radius; the y-axis represents for the vehicle speed and the size of the bubbles represent for the emissions. The bigger the bubble is, the higher the emission is. This chart indicates that the average speed is higher for horizontal curves with larger radius, and also, the emission of $CO_2$ is higher when the average speed is higher.

## Modeling results

The x-y plotting results show that there might be a positive relationship between horizontal curve radius and $CO_2$ emissions. In order to verify this, a mathematical model in the form of Equation (1) was developed. Table 1 lists the modeling results for the first run in SPSS. The small p-value means the null hypotheses (coefficients=0) should be rejected, which also means that these two independent variables (Curve radius and vehicle speed) have significant impacts on dependent variable ($CO_2$). Also, the constant is not equal to 0. The

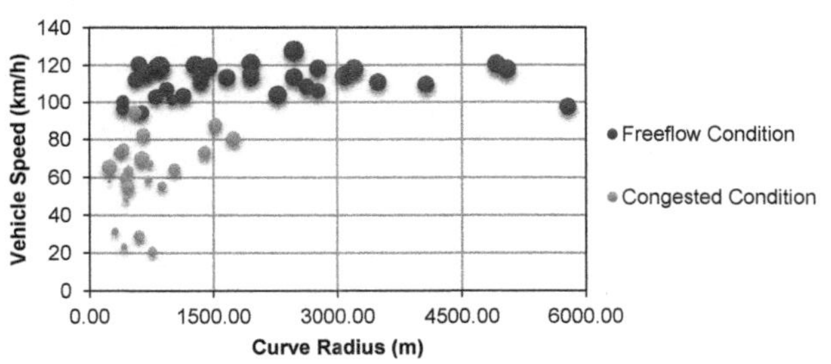


**Figure 4:** Bubble chart for curve radius, vehicle speed and CO2 emissions.

| Elements | Unstandardized Coefficients 2 | | t-value | p-value |
|---|---|---|---|---|
| | B | Std.Error | | |
| (Constant) | 0.276 | 0.082 | 2.653 | 0.011 |
| Curve Radius | 2.774E-05 | 0.000 | 8.452 | 7.6465E-11 |
| Vehicle Speed | 0.006 | 0.001 | 2.039 | 0.047 |
| Acceleration/ Deceleration | -0.069 | 0.127 | -0.493 | **0.625** |

**Table 1:** Coefficients for the first run.

p-value for acceleration/deceleration is 0.625 (larger than 0.05), which means the coefficient for the variable of acceleration/deceleration should equal to 0.

Before the second run, the modeling error for each sample was checked in order to screen some outliers. Some extreme sample with a high modeling error was re-examined. For example, the second sample with a curve radius of 241.26 m with a modeling error of 192% was studied and the driving behavior was found to be abnormal within this curve. Driver frequently slammed on the brakes, which may cause the vibration of the equipment and produce some invalid data. The linear regression rerun after deleted the variable of acceleration/deceleration and removed those outliers; the analysis results show in Tables 2 and 3. Table 2 summarized the goodness of fit of this statistical model. The R square is 0.872, which means the linear model fit the data well.

Table 3 shows the final results. Two variables were kept in this model after stepwise regression. The small p-value for Curve Radius (p-value = 0.037 < 0.05) and Vehicle Speed (p-value = 1.006E-13 < 0.05) means that these two variables have significant impacts on $CO_2$. Also, the rest data samples were used for validation using a MATLAB program. Figure 5A is the plotting for 56 samples. The green line represents for the calculation results from EFM. The black line represents for the real-world $CO_2$ emissions. The fitting results show that the EFM could well estimate the $CO_2$ emissions on freeway for horizontal curves. Most points on the blue line, which represents for the calculation results from MOVES, indicate that the MOVES may overestimate the $CO_2$ emissions on freeway for single curves with different radius. Figure 5B is the estimation error for these two models. The field test data of real-world emission were viewed as baseline with an error of 0. The error-curve radius plotting shows that for both of the model, the error for estimating the $CO_2$ emission factor is larger when the curve radius is smaller than 1,000 m. This Emission Factor Model covers all the driving conditions including congested conditions and free flow conditions. In order to look into the modeling error, the data under free flow condition and congested condition were separated and analyzed Figures 5C and 5D. The comparison results indicate that this model will produce a smaller error under free flow conditions. The modeling error for EFM under free flow condition was found to be 5.5% while the modeling error under congested condition for EFM was 28%. At the meantime, the modeling error from MOVES was also calculated and the results show that MOVES could estimate the $CO_2$ emission factor for horizontal curve under free flow condition with an error of 16.8%. The error under congested condition for MOVES is 58.7%. Both of the models have better performance under free flow condition. Figure 6 shows the validation results. The black line represents for the tested value from PEMS. The green line represents for the predicted value calculated by the equation. The blue line represents for the predicting results from MOVES. 10 samples (20%) were used for validation in this study and the prediction error is 4.63% for EMF, which is much smaller than the prediction error produced by MOVES.

The corresponding function is:

$$y = \frac{\log EF_C}{\log EF_T} = 0.006v + (3.213\mathrm{E}-5)r + 0 \cdot a + 0.246 \tag{7}$$

$$\mathrm{EF_C} = \mathrm{EF_T}^{0.006v+3.213\mathrm{E}-5r+0.246} \tag{8}$$

where, EFc is the predicted emission factor for horizontal curves with different radius, v (km/hr) is the average speed for this single curve; a (m/s2) is the average absolute value of acceleration/deceleration; r (m) is the circular curve radius. The coefficient 3.213E-5 means that, larger

| Std. Error of the Estimation | Sample Size | R Square | Independent Variables | Dependent Variable |
|---|---|---|---|---|
| 0.117246012 | 62 | 0.872 | Curve Radius (m) Vehicle Speed (km/h) | Emission Factor for $CO_2$ (g/m) |

**Table 2:** Summary of modeling results.

| Elements | Unstandardized Coefficients 2 | | *t*-value | *p*-value |
|---|---|---|---|---|
| | B | Std.Error | | |
| (Constant) | 0.246 | 0.055 | 3.784 | 0.000445 |
| Curve Radius | 3.213E-5 | 0.000 | 2.148 | 0.037 |
| Vehicle Speed | 0.006 | 0.001 | 9.689 | 1.006E-13 |

**Table 3:** Coefficients for the second run.

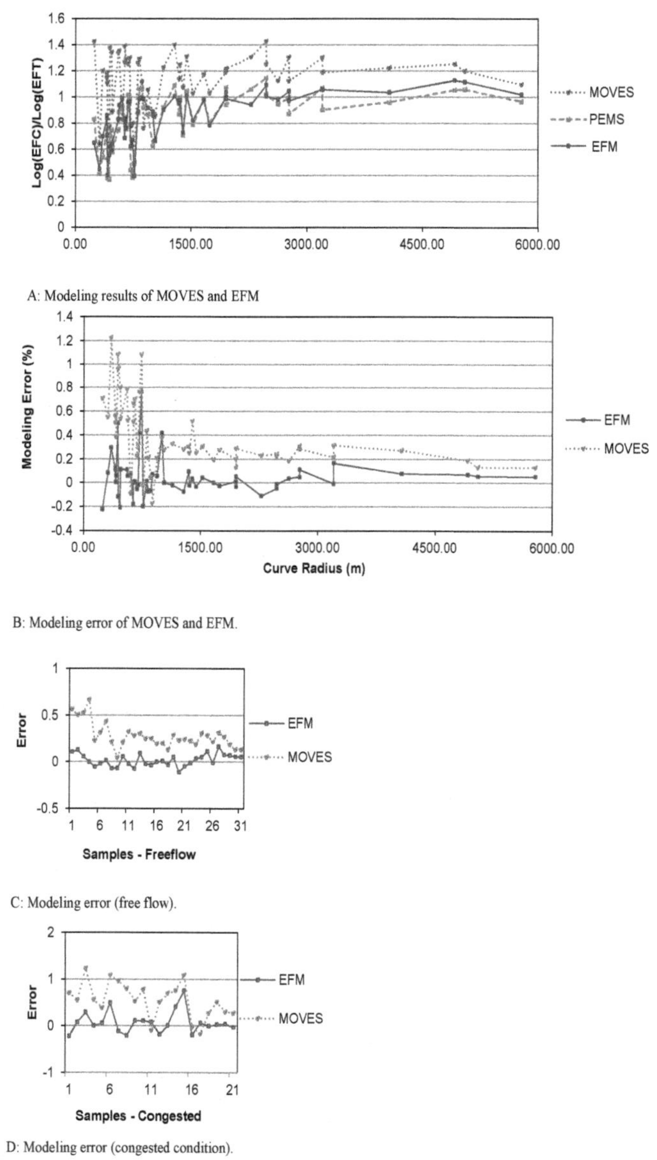


A: Modeling results of MOVES and EFM

B: Modeling error of MOVES and EFM.

C: Modeling error (free flow).

D: Modeling error (congested condition).

**Figure 5:** Comparison of prediction results and test results (56 samples).

curve radius will produce more $CO_2$. The coefficient 0.006 means that the higher the vehicle speeds is, the larger the amount of $CO_2$ will be.

## Conclusion

The modeling results indicate that the curve radius could contribute to the $CO_2$ emissions. For example, if the average speed for a single curve is constant and equal to 80 km/hr, when the curve radius is 2000 m, the emission factor for curve will be 6.915 g/m. When the curve radius is 4000 m, the emission factor for curve will be 8.099 g/m. Under the free flow condition, according to Ko, the fuel consumption and $CO_2$ emissions might be reduced due to the speed reduction on sharp curves while larger emissions are related to greater curve radius with tangent speed between 90 and 110 km/hr [16]. Also, the results from previous study also show that the $CO_2$ emission was minimum within the speed range of 70 to 80 km/hr [49]. The modeling error analysis shows that the EFM has better performance for estimating the $CO_2$ emission factor for horizontal curves under free flow condition. The predicting error of EFM for free flow condition is 5.5% while the predicting error for MOVES is 16.8%. It is possible that the error under congested condition may be caused by the complex and uncertain driving behavior. Also, the modeling error was smaller when the curve radius is larger than 1000 m. There are mainly two explanations for the error for smaller curve radius: one is that congested traffic flow tends to happen on sharp curves while the model has better performance for free flow; the second explanation is that the error may be generated during the process of extraction for the geometric features from the digital maps. A previous study also suggested that a specific range of curve radius might has better accuracy could be obtained when using the manual extraction methods for horizontal curves from digital maps [44].

The statistical analysis results show that the average speed has bigger impacts on the emission factor for curves with a coefficient of 0.006. The results also indicate that the acceleration and deceleration have no significant impacts on emission factors for horizontal curves on freeway. Under free flow condition, the average speed on freeway is relative high and the drivers tend to apply the brakes or tap the paddles slightly for safety concerns. This may explain why the variable of acceleration has no impact on emission of $CO_2$ on freeways. Previous studies also stated that the acceleration/deceleration has significant impacts on other vehicle emissions such as CO and $NO_X$ compared to $CO_2$ [16].

## Limitations and Future Work

In this study, the average speed and curve radius were included in the emission factor model, even though the statistical results show that the acceleration/deceleration behavior has no impact on the emission factor for horizontal curves on freeways. The average speed neglected the detailed driving behavior for horizontal curves because of data aggregation. The trips on horizontal curves with different

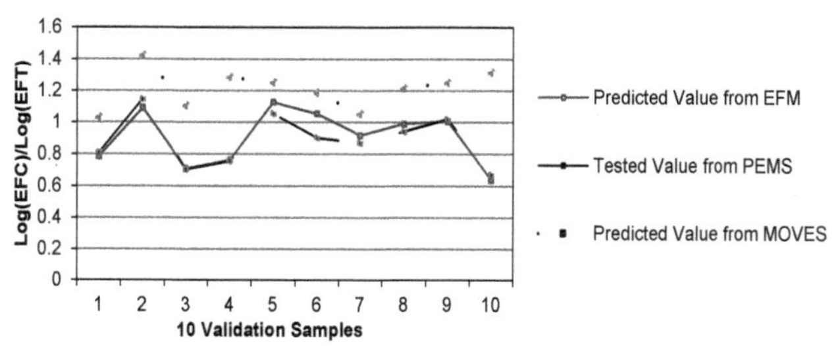


**Figure 6:** Comparison of prediction results and tested results (10 for validation).

driving behavior may also produce the same average speed. This is one of the major drawbacks in the EFM. After the data aggregation and data screening, the modeling sample size (80% for fitting and 20% for validation) is very limited. All the data from this study were collected from the freeways in Houston area. The EFM could be applied to the freeways with the same pavement type and traffic conditions. Also, in this study, a light-duty vehicle was used for data collection. For estimating the overall $CO_2$ emission, a factor should be used for the adjustments. There are mainly two assumptions for the horizontal curves in this study: one is that all the curves are simple curves without spiral transition; another assumption is that each curve was viewed as a unit, which neglected the combination of curves. For example, the reverse curves may result in different driving behavior and produce different speed files. As for the future work, more curve-related variables, for instance, the entered-tangent speed will be included in the emission factor model. The modeling results will be linked to the detailed roadway design process.

## Acknowledgments

The authors acknowledge that this research is supported in part by the United States Tier 1 University Transportation Center TranLIVE # DTRT12GUTC17/KLK900-SB-003, and the Natonal Science Foundation (NSF) under grants #1137732. The opinions, findings, and conclusions or recommendations expressed in this material are those of the author(s) and do not necessarily reflect the views of the funding agencies. The authors also appreciate Wu Ying and Pengfei Liu for their help in data collection.

## References

1. Environmental Protection Agency (EPA) – US Transportation Sector Greenhouse Gas Emission 1990-2012.
2. The Congressional Budget Office (2008) Climate-Change Policy and $CO_2$ Emissions from Passenger Vehicle.
3. Li Q, Qiao F, Yu L (2015) Will Vehicle and Roadside Communications Reduce Emitted Air Pollution? Int J Science Technol 5: 17-23.
4. Li Q, Qiao F, Yu L (2016) Vehicle Emission Implications of Drivers Smart Advisory System for Traffic Operations in Work Zones. J Air and Waste Manage.
5. Li Q, Qiao F, Yu L (2015) Implications of wireless communication system for traffic operations on Vehicle emissions. Publication in the proceedings of the 107th Air and Waste Management Association (AWMA), June 22-25, 2015. Raleigh, North Carolina, USA.
6. Munni J, Qiao F, Li Q, Yu L (2015) Driving behavior and emission analysis at yellow interval with advanced warning message under foggy weather condition: A simulator test. Publication in the proceedings of the 56th Annual Transportation Research Forum in Atlanta, Georgia, USA.
7. Rahman R, Qiao F, Li Q, Yu L, Kuo PH (2015) Smart phone based forward collision warning message in work zones to enhance safety and reduce emissions. Publication in the proceedings of the 94th Transportation Research Board Annual Meeting, National Academy of Sciences, Washington, DC 15: 0648.
8. Li Q, Qiao F, Yu L (2014) Impacts of vehicles to infrastructure communication technologies on vehicle emissions. Publication in the Proceedings of the International Conference on Environmental Science and Technology, American Academy of Sciences, Houston, TX, June 9-13.
9. Qiao F, Li Q, Yu L (2014) Testing impacts of work zone x2v communication system on safety and air quality in driving simulator, Publication in the proceedings of the 21st ITS World Congress, Detroit, USA, September 7-11.
10. Li Q, Qiao F, Wang X, Yu L (2013) Impacts of P2V wireless communication on safety and environment in work zones through driving simulator tests (paper # 26-179). Publication in the proceedings of the 26th Annual Conference of the International Chinese Transportation Professionals Association (ICTPA), Tampa, Florida, May 24-26.
11. Environmental Protection Agency (EPA) (2010) Mortor Vehicle Emissions Simulator (MOVES): User Guide for MOVES2010a. Publication. EPA-420-B-10-036.
12. Frey HC, Rasdorf WJ, Pang S, Kim K (2007) Methodology for activity, fuel use, and emissions data collection and analysis for non-road construction equipment. Proceedings of the Air and Waste Management Association Annual Conference. Pittsburgh, PA: AWMA.
13. Frey HC, Unal A, Chen J (2002) Recommended Strategy for On-Board Emission Data Analysis and Collection for the New Generation Model. Prepared for EPA.
14. Li Q, Qiao F, Yu L (2016) Calibrating emission factors for highways considering pavement roughness information. Paper #80. Proceedings of the 2016 Air Quality Measurement Methods and Technology. March 15-17, 2016, Chapel Hill, North Carolina.
15. Li Q, Qiao F, Yu L (2015) Clustering pavement roughness based on the impacts on vehicle emissions and public health. J Ergonomics. 5: 1000146. doi:10.4172/2165-7556.1000146. Online access.
16. Ko MH (2011) Incorporating vehicle emission models into the highway design process. Diss. Texas AandM University.
17. American Association of State Highway and Transportation Officials (2010) Highway Safety Manual, (1stedn) AASHTO, Washington, D.C.
18. Frey HC, Liu B (2013) Development and evaluation of a simplified version of MOVES for coupling with a traffic simulation model. Transportation Research Board. Washington, D.C.
19. Claggett M, Houk J (2008) Comparing MOBILE 6.2 and Emfac2007 emissions factors for mobile source air toxics. Presented at 87th Annual Meeting of the Transportation Research Board, Washington, D.C.
20. California Air Resources Board EMFAC (2007) Users Guide. State of California. Sacramento, CA.
21. Kelly NA, Groblicki PJ (1993) Real-World Emissions from a Modern Production Vehicle Driven in Los Angeles. J Air Waste Manage Assoc 43: 1351-1357.
22. NRC (2000) Modeling Mobile-Source Emissions; National Academy Press; National Research Council; Washington, D.C.
23. Barth M, An F, Norbeck J, Ross M (1996) Modal Emissions Modeling: A physical approach. Transportation Res Record 1520: 81-88.
24. West BH, McGill RN, Hodgson JW, Sluder CS, Smith DE (1997) Development of data-based light-duty modal emissions and fuel consumption models. J Soc of Automobile Engineers 2910: 1274-1280.
25. Shih R, Fable S, Sawyer RF (1997) effects of driving behavior on automobile emissions. Proceedings of the Seventh CRC On-Road Vehicle Emissions Workshop, Coordinating Research Council, Atlanta, GA, 6: 115-124.
26. Fomunung I, Washington S, Guensler R, Bachman W (2000) Validation of the MEASURE Automobile Emissions Model: A Statistical Analysis. J Transportation and Stat.
27. Barth M, Norbeck J (1997) NCHRP Project 25-11: The development of a comprehensive modal emission model. Proceedings of the Seventh CRC On-Road Vehicle Emissions Workshop, Coordinating Research Council, Atlanta, GA, 6: 53-71.
28. Li Q, Qiao F, Yu L (2016) Estimating vehicle idle emissions based on on-board diagnostic II data. Paper #70. Publication in the proceedings of the 2016 Air Quality measurement methods and technology. March 15-17, 2016, Chapel Hill, North Carolina.
29. Barth M, An F, Younglove T, Scora G, Levine C et al. (2000) Comprehensive Modal Emissions Model (CMEM), Version 2.0: User's Guide.
30. Toth-Nagy C, Conley JJ, Jarrett RP, Clark NN (2006) Further validation of artificial neural network-Based emissions simulation models for conventional and hybrid electric vehicles. J Air and Waste Manage Association 56: 898-910.
31. INTEGRATION Release 2.30 for WINDOWS (2000) User's Guide – Volume I: Fundamental model features.
32. Wenzel T, Singer BC, Slott R (2000) Some Issues in the statistical analysis of vehicle emissions. J of Transportation and Stat.
33. Yang F, Yu L (2007) A microscopic emission model for the light-duty vehicles based on PEMS Data. International Conference on Transportation Engineering.
34. Pratt M, Jeffrey DM, James AB (2009) Workshops on Using the GPS Method to determine curve advisory speeds. College Station, Texas: Texas Transportation Institute. Texas A and M University System. Report No. FHWA/TX-10/5-5439-01-1.
35. Imran M, Hassan Y, Patterson D (2006) GPS–GIS-Based Procedure for tracking vehicle path on horizontal alignments. Computer-Aided Civil and Infrastructure Eng 21: 383-94.

36. Sanders B (2005) Updating horizontal curve data using GPS centerlines. Kentucky Transportation Cabinet. Division of Planning. Presentation at the FHWA Highway Performance Monitoring System (HPMS) Data Collection Workshop. Salt Lake City, Utah. 8-10 Mar 2005.

37. Larsen KA (2013) Use of highway ROW for high-speed intercity passenger rail and dedicated freight transportation systems.

38. Easa SM, Dong H, Li J (2007) Use of satellite imagery for establishing road horizontal 24 alignments. In J Surveying Eng 133: 29-35.

39. Dong H, Easa SM, Li J (2007) Approximate extraction of spiraled horizontal curves from 26 satellite imagery. In J Surveying Eng 133: 36-40.

40. Kim JS, Lee JC, Kang IJ, Cha SY, Choi H et al. (2008) Extraction of geometric information on highway using terrestrial laser scanning technology. The International Archives of the Photogrammetry, Remote Sensing and Spatial Information Sciences. Vol. XXXVII. Part B5, Beijing, China, 2008: 539–544.

41. Google Earth Pro. Earth Help – Measuring Tool Features and Options. 2015.

42. Osei-Asamoah A, Jackson E (2015) Automated horizontal curve classification and parameter calculation Methods using vehicle collected road inventory data. Transportation Research Board. Paper Number: 15-4700.

43. Li Z, Chitturi MV, Bill AR, Noyce DA (2012) Automated identification and extraction of horizontal curve information from geographic information system roadway maps. Transportation Research Record, No. 2291, Transportation Research Board of the National Academies, Washington, D.C., 2012: 80–92.

44. Khan G, Bill AR, Chitturi M, Noyce DA (2013) Safety evaluation of horizontal curves on rural undivided roads. Prepared for the 92nd Annual meeting of the Transportation Research Board, Washington, D.C.

45. Voigt A, Krammes R (1995) An operational and safety evaluation of alternative horizontal curve design approaches on rural two-lane highways. International Symposium on Highway Geometric Design Practices.

46. Fitzpatrick K (2000) Speed prediction for two-lane rural highways. Publication FHWA- RD-99-171. FHWA, US Department of Transportation.

47. Hassan Y (2011) Modeling operating speed: Synthesis Report, Transportation Research Circular E-C151, Washington, D.C.

48. Choi PJ, Stephen GR, Cheol OH (2013) Methodology for generating individual vehicle speed profile for estimating freeway emissions. Transportation Research Board 92nd Annual Meeting. No. 13-5138.

49. Boriboonsomsin K, Barth M (2009) Impacts of road grade on fuel consumption and carbondioxide emissions evidenced by use of advanced navigation systems. J Transportation Res Board 2139: 21-30.

# Repair Method of Damaged Steel Braced Structure and Restring Force Characteristics after Repair State

**Ryotaro Arai*, Hanako Satoh, Kenta Aoki, Takumi Ito and Changhoon Choi**

*Department of Architecture, Tokyo University of Science, Japan*

**Abstract**

A lot of steel structures have experienced severe earthquake disasters in Japan, and its various kinds of failure mode have been reported. Especially on steel braced structure, the failure modes are generally categorized into two modes; buckling and fracture of brace member, and yielding of joints and fasteners. After a lot of earthquake disasters, the technical manual has been published to repair and recover the damaged building structures in Japan [1]. The manual suggests the actual repairing method for damaged steel braced structures; however, it is guessed that a few of these methods have some difficulties related to construction technique and estimation of recovery after repair. In this paper, new repairing method for joints on damaged steel brace structure has been proposed, and its applicability and feasibility are investigated experimentally. Herein, the horizontally loading test on diagonal steel brace are performed with parameters as follows; thickness of gusset plate, layout of bolts, slenderness ratio of brace and the method of joint. At first, the damaged state of gusset plate is reproduced by horizontally loading test. Next, the damaged gusset plate is repaired. Finally, the loading test is done to the repaired specimen again. From the test results, the strength is reinforced after repair, and the rigidity is almost same. The change of failure mode and slenderness ratio is presented, and which is related to eccentric distance and torsional parameters. Furthermore, the evaluation method of the failure mode and buckling strength are proposed, and this estimation method shows good utility.

**Keywords:** Steel braced structure; Ultimate seismic state; Repair method; Gusset plate; Evaluation method

## Introduction

Recently a lot of buildings are damaged variously after huge earthquake disasters, and it is necessary to decide whether they are rebuilt or reuse. The steel frame structures are classified roughly into a rahmen and braced structure, and serious damages are reported to both of them. Receiving that, the word “resilience” is focused on. From the viewpoint of reconstruction and restoration of city function promptly, it is desirable to repair and reuse the public buildings which become the base of recover and reconstruction in the stricken area. Also, these public buildings are mainly braced structure. There are many reports that the various types of failure modes of steel braced structure, such as buckling, fracture of brace member, and yielding of joints and fasteners. In Japan, the corresponding to these failure modes is discussed in the technical guideline). The guideline suggests some actual repair methods, however, it is not necessary the best method considering the state of stricken area. In this study, in case of the gusset plate is damaged, the new method which takes place the conventional method is suggested and the horizontally loading tests are conducted on diagonal steel bracing. And this paper aims to propose an evaluation method of repaired state.

## Materials and Methods

### Repair methods of damaged steel brace joint

This study focuses on the failure mode of joint which out of plane deformation is occurred during ultimate seismic state. It is premised that the damaged brace and fasteners are replaced and the gusset plate is repaired. When the new brace is attached on the damaged gusset plate, a clearance occurs between gusset plate and brace. And the tapered steel filler is inserted on this clearance. And to restrain the deformation of gusset plate, the side stiffener is attached. The outline of repair method is as follows: First, the tapered steel filler is welded on damaged gusset plate. Next, the side stiffener is welded on the side of gusset plate. The details of proposed repair methods are shown in Figure 1.

### Inspection of repair effect by horizontally loading test

**The outline of horizontally loading test:** In this loading test, the test specimen and test setup are illustrated in Figure 2. The flow of the loading test is as follows: First, the test frame (Figure 2) is horizontally loaded and a brace member and gusset plates are damaged (herein after referred to as “Initial loading”). Second, the gusset plates which damaged by initial loading is repaired. Finally, the loading test is done to the repair specimen again (herein after referred to as “Second loading”).

**Experimental parameters:** Experimental parameters are as follows: the layout of bolts on joint, the thickness of gusset plate, repair or not repair after initial loading, the slenderness ratio of the brace and the usage of tapered steel filler. The layout of bolts is determined by the length of brace. The thickness of gusset plate is 4.5 mm, 6 mm, 16 mm. The slenderness ratio of the brace is 90, 100,150. And also, the loading program is monotonic and cyclic loading. On the monotonic loading, the brace member or the gusset plate become the ultimate states. During the cyclic loading, maximum story deformation angle is 1/30 radian. Test specimens were instrumented with sensors for measuring below items: horizontal displacement, in and out of plane deformation of brace, out of plane of the gusset plate. Strain of brace is measured at the 200 mm, 400 mm, 600 mm, and 800 mm from the end of brace. Mechanical properties of materials and specimens list are summarized in Tables 1 and 2.

## Test results

### Initial loading test results

During loading test, three kinds of failure mode were observed, and it is summarized in Figure 3. Also, on the occasion of the buckling

***Corresponding author:** Ryotaro Arai, Department of Architecture, Tokyo University of Science, Japan, E-mail: 4115603@ed.tus.ac.jp

occurred, two kinds of bending mode were observed as showed in Figure 4.

**L150 series:** During the monotonic loading, it was observed that the brace remained within elastic when the buckling was occurred. The gusset plate and the brace members deformed to out of plane in one unit until the buckling is occurred. After buckling was occurred, the plastic hinge was formed at the middle of the brace and the connection between brace and gusset plate (herein after referred to as "failure mode C"). And also, the out of plane of the brace was remarkable. In case of large slenderness ratio, the bending moment to the gusset plate does not become so large. Therefore, out of plane of gusset plate wasn't so large. Furthermore, the connection between brace and gusset plate was twisted. During the cyclic loading, the connection was twisted and the plastic hinge was formed at the middle of brace and connection.

**L100-6 series:** From the results of strain during monotonic and cyclic loading, the occurrence of intermediate of elastic and plastic buckling was considered. A plastic hinge was generated on middle of the brace (herein after referred to as "failure mode A") on the all specimens. The out of plane deformation is observed on gusset plate.

**L100-16 series:** The occurrence of intermediate of elastic and plastic buckling was considered. The plastic hinge was generated on middle and both end of brace, because the gusset plate didn't deform. (herein after referred to as "failure mode B").

**L90 series:** In case of small slenderness ratio, axial load was larger than any other series and bending moment to gusset plate exerted enough. Therefore, the out of plane of gusset plate was the largest. Also the failure mode A was observed on all specimens.

## The damage situation and the repair method

Exclude the L100-16 series; all gusset plates have plastic deformation. The result of residual deformation was summarized in Table 3. In case of the L150 series, the tapered steel filler wasn't

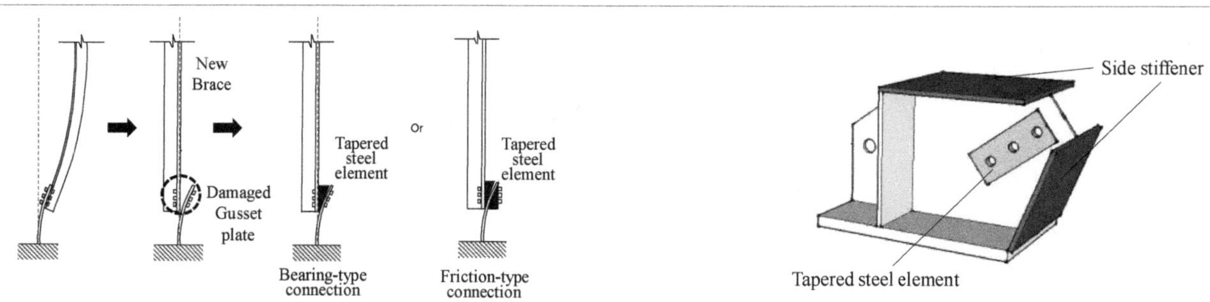


(a)Outline of repair method (b)Conception model

**Figure 1:** Repair method.

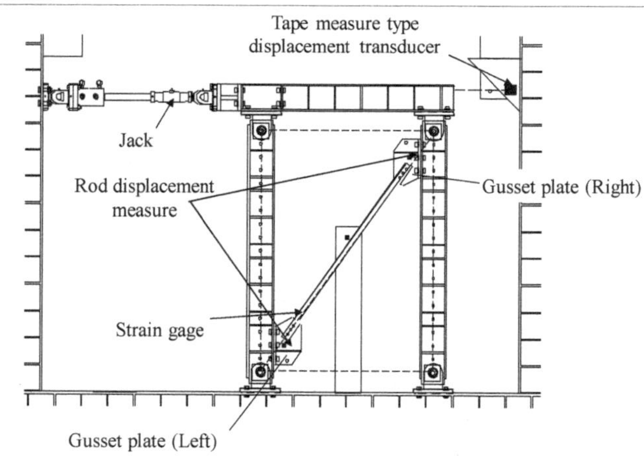


**Figure 2:** Test set up.

| | Rot | Young's modulus (GPa) | Yield strain (μ) | Yield strength (N/mm²) | Tensile strength (N/mm²) |
|---|---|---|---|---|---|
| Gusset plate | a | 205 | 1657 | 274 | 437 |
| | b | 218 | 1882 | 357 | 444 |
| | c | 203 | 1531 | 311 | 459 |
| | d | 201 | 1860 | 351 | 433 |
| | e | 207 | 2260 | 416 | 538 |
| Brace | a | 209 | 1558 | 337 | 472 |
| | b | 212 | 1598 | 338 | 454 |
| | c | 188 | 1422 | 299 | 398 |
| | d | 206 | 1510 | 285 | 396 |

**Table 1:** Mechanical properties of steel materials.

<table>
<tr><th rowspan="2">Specimen name</th><th rowspan="2">Slenderness ratio</th><th rowspan="2">Thickness (mm)</th><th rowspan="2">Bolt layout (mm)</th><th rowspan="2">Friction</th><th rowspan="2">Repair</th><th rowspan="2">Loading pattern</th><th colspan="2">Lot</th></tr>
<tr><th>Gusset plate</th><th>Brace</th></tr>
<tr><td>L90-4.5-0-F-R-M1</td><td rowspan="4">90</td><td rowspan="16">4.5</td><td rowspan="24">0</td><td rowspan="16">Friction</td><td rowspan="8">Repair</td><td rowspan="2">Monotonic</td><td rowspan="16">a</td><td rowspan="4">a</td></tr>
<tr><td>L90-4.5-0-F-R-M2</td></tr>
<tr><td>L90-4.5-0-F-R-C1</td><td rowspan="2">Cyclic</td></tr>
<tr><td>L90-4.5-0-F-R-C2</td></tr>
<tr><td>L150-4.5-0-F-R-M1</td><td rowspan="4">150</td><td rowspan="2">Monotonic</td><td rowspan="4">b</td></tr>
<tr><td>L150-4.5-0-F-R-M2</td></tr>
<tr><td>L150-4.5-0-F-R-C1</td><td rowspan="2">Cyclic</td></tr>
<tr><td>L150-4.5-0-F-R-C2</td></tr>
<tr><td>L90-4.5-0-F-N-M1</td><td rowspan="4">90</td><td rowspan="12">Non-repair</td><td rowspan="2">Monotonic</td><td rowspan="4">a</td></tr>
<tr><td>L90-4.5-0-F-N-M2</td></tr>
<tr><td>L90-4.5-0-F-N-C1</td><td rowspan="2">Cyclic</td></tr>
<tr><td>L90-4.5-0-F-N-C2</td></tr>
<tr><td>L150-4.5-0-F-N-M1</td><td rowspan="4">150</td><td rowspan="2">Monotonic</td><td rowspan="4">b</td></tr>
<tr><td>L150-4.5-0-F-N-M2</td></tr>
<tr><td>L150-4.5-0-F-N-C1</td><td rowspan="2">Cyclic</td></tr>
<tr><td>L150-4.5-0-F-N-C2</td></tr>
<tr><td>L100-6-0-NF-N-M1</td><td rowspan="18">100</td><td rowspan="12">6</td><td rowspan="8">Non- friction</td><td rowspan="2">Monotonic</td><td rowspan="4">d</td><td rowspan="4">d</td></tr>
<tr><td>L100-6-0-NF-N-M2</td></tr>
<tr><td>L100-6-0-NF-N-C1</td><td rowspan="2">Cyclic</td></tr>
<tr><td>L100-6-0-NF-N-C2</td></tr>
<tr><td>L100-6-0-NF-R-M1</td><td rowspan="8">Repair</td><td rowspan="2">Monotonic</td><td rowspan="8">b</td><td rowspan="8">c</td></tr>
<tr><td>L100-6-0-NF-R-M2</td></tr>
<tr><td>L100-6-0-NF-R-C1</td><td rowspan="2">Cyclic</td></tr>
<tr><td>L100-6-0-NF-R-C2</td></tr>
<tr><td>L100-6-40-F-R-M1</td><td rowspan="4">40</td><td>Friction</td><td rowspan="2">Monotonic</td></tr>
<tr><td>L100-6-40-NF-R-M2</td><td rowspan="3">Non- friction</td></tr>
<tr><td>L100-6-40-NF-R-C1</td><td rowspan="2">Cyclic</td></tr>
<tr><td>L100-6-40-NF-R-C2</td></tr>
<tr><td>L100-16-0-F-N-M1</td><td rowspan="6">16</td><td rowspan="4">0</td><td rowspan="6">Friction</td><td rowspan="2">Non- repair</td><td>Monotonic</td><td rowspan="6">e</td><td rowspan="2">d</td></tr>
<tr><td>L100-16-0-F-N-C1</td><td>Cyclic</td></tr>
<tr><td>L100-16-0-F-R-M1</td><td rowspan="4">Repair</td><td>Monotonic</td><td rowspan="4">c</td></tr>
<tr><td>L100-16-0-F-R-C1</td><td>Cyclic</td></tr>
<tr><td>L100-16-40-F-R-M1</td><td rowspan="2">40</td><td>Monotonic</td></tr>
<tr><td>L100-16-40-F-R-C1</td><td>Cyclic</td></tr>
</table>

**Table 2:** Specimens list.

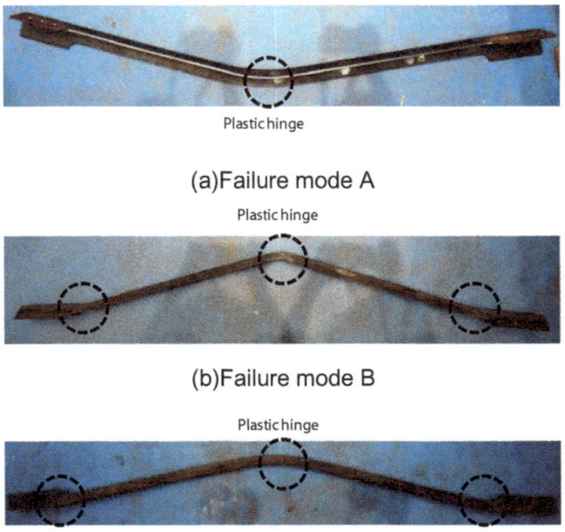


(c)Failure mode C

**Figure 3:** Failure modes on ultimate state.

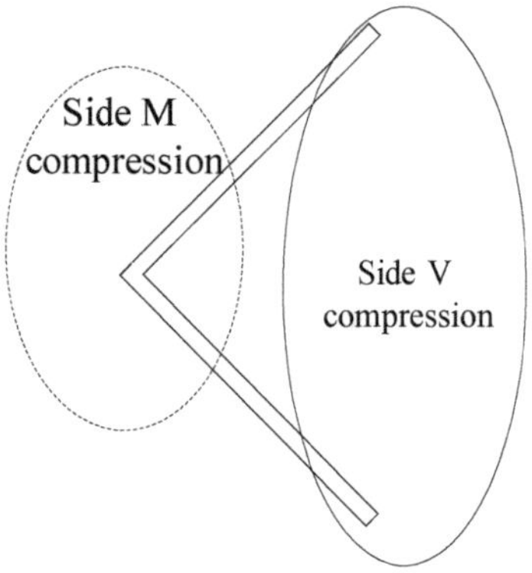


**Figure 4:** Compression side.

| Specimen | side | Residual deformation |
|---|---|---|
| L90-4.5-0-F-R-M | L | 13.9 |
| | R | 13.2 |
| L90-4.5-0-F-R-C | L | 11.2 |
| | R | 17.8 |
| L150-4.5-0-F-R-M | L | 5.1 |
| | R | 8.8 |
| L150-4.5-0-F-R-C | L | 5 |
| | R | 1.9 |
| L100-6-0-F-R-M | L | 3.1 |
| | R | 4.2 |
| L100-6-0-F-R-C | L | 8.1 |
| | R | 3.7 |
| L100-6-40-F-R-M | L | 38 |
| | R | 37.8 |
| L100-6-40-F-R-C | L | 11.2 |
| | R | 20.5 |

**Table 3:** Residual deformation (mm).

| Name of specimen | Axial rigidity (kN/mm$^2$) | Ratio of rigidity | Buckling load (kN) | Ratio of buckling load | Buckling length (mm) | Failure mode* | Compression side |
|---|---|---|---|---|---|---|---|
| L90-4.5-0-F-R-M1 | 10.5 | 1.0 | 57.0 | 1.5 | 1360 | B→TB | M |
| L90-4.5-0-F-R-M2 | 10.7 | | 85.1 | | 800 | TB | V |
| L90-4.5-0-F-R-C1 | 13.2 | 0.7 | 56.4 | 1.0 | 1440 | B→TB | M |
| L90-4.5-0-F-R-C2 | 9.3 | | 56.5 | | 880 | TB | V |
| L150-4.5-0-F-R-M1 | 8.4 | 1.0 | 20.3 | 1.1 | 1760 | B→TB | M |
| L150-4.5-0-F-R-M2 | 8.4 | | 21.7 | | 800 | B | V |
| L150-4.5-0-F-R-C1 | 8.4 | 1.0 | 23.2 | 1.0 | 1200 | B→TB | M |
| L150-4.5-0-F-R-C2 | 8.1 | | 23.2 | | 800 | B | V |
| L90-4.5-0-F-N-M1 | 12.0 | 0.8 | 57 | 1.2 | 1520 | B | M |
| L90-4.5-0-F-N-M2 | 10.0 | | 70.7 | | 1200 | B | M |
| L90-4.5-0-F-N-C1 | 12.4 | 0.8 | 57.9 | 1.2 | 1600 | B | M |
| L90-4.5-0-F-N-C2 | 9.7 | | 67.1 | | 1200 | B | M |
| L150-4.5-0-F-N-M1 | 7.0 | 0.9 | 19.1 | 0.9 | 960 | B→TB | M |
| L150-4.5-0-F-N-M2 | 6.1 | | 17.1 | | 800 | TB | V |
| L150-4.5-0-F-N-C1 | 6.6 | 0.8 | 18.6 | 1.0 | 1440 | B→TB | M |
| L150-4.5-0-F-N-C2 | 5.4 | | 18.3 | | 800 | TB | V |
| L100-6-0-NF-N-M1 | 15.2 | 0.9 | 38.1 | 1.2 | 1150 | B | M |
| L100-6-0-NF-N-M2 | 13.2 | | 47.5 | | - | B | M |
| L100-6-0-NF-N-C1 | 11.1 | 0.9 | 35.9 | 1.4 | | B | M |
| L100-6-0-NF-N-C2 | 10.5 | | 49.7 | | | B→TB | M |
| L100-6-0-NF-R-M1 | 15.4 | 0.6 | 37.8 | 1.1 | 1340 | B | M |
| L100-6-0-NF-R-M2 | 9.1 | | 41.4 | | 950 | B→TB | V |

| | | | | | | | |
|---|---|---|---|---|---|---|---|
| L100-6-0-NF-R-C1 | 12 | 1.0 | 39.8 | 1.2 | | B | M |
| L100-6-0-NF-R-C2 | 11.5 | | 48.5 | | | B | V |
| L100-6-40-F-R-M1 | 12.3 | 1.1 | 42.1 | 1.0 | 1320 | B | M |
| L100-6-40-NF-R-M2 | 13.3 | | 40.7 | | 1050 | B | V |
| L100-6-40-NF-R-C1 | 15.5 | 0.6 | 37.9 | 1.5 | | B | M |
| L100-6-40-NF-R-C2 | 9.4 | | 55.2 | | | B→TB | V |
| L100-16-0-F-N-M1 | 15.5 | | 63.4 | | 880 | B→TB | M |
| L100-16-0-F-N-C1 | 18.1 | | 61.2 | | | B→TB | M |
| L100-16-0-F-R-M1 | 14.6 | | 61 | | 800 | B | V |
| L100-16-0-F-R-C1 | 14.5 | | 57.9 | | - | B | V |
| L100-16-40-F-R-M1 | 14.6 | | 49.7 | | | B | V |
| L100-16-40-F-R-C1 | 12.8 | | 47.2 | | | B | V |
| *B = Bending, TB = Torsional bending | | | | | | | |

**Table 4:** Comparison of test results.

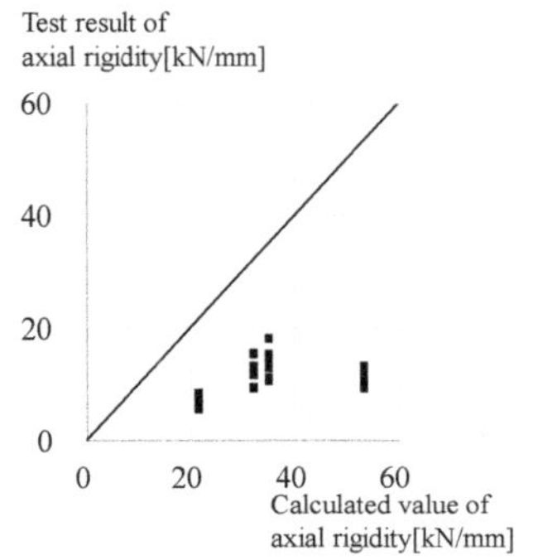


**Figure 5:** Comparison of axial.

attached. According to the "Guidebook on Design and Fabrication of High Strength Bolted Connections [2]. When the clearance is less than 1 mm, filler isn't necessary. The reason is that the gusset plate wasn't repaired by tapered steel filler. The details of repair materials are bellow. As mentioned in chapter 2, the tapered steel filler is welded on damaged gusset plate. Note that the repair method of connection is one of the experimental parameters, the specimen by using friction joint is soaked rust by medicine. And also, when the brace is fastened with the gusset plate by the bolts normally, and then the same shape of tapered steel filler is adopted on both side of brace test specimen. The usage of tapered steel filler is shown in Figure 1.

## Second loading test results

**L150-N series:** On the damaged state, a clearance between gusset plate and brace were existing. And this clearance was disappeared little after the bolts were fastened. From the test results, the elastic buckling was occurred on all specimens, and side V is presented as the state of compression. Since the initial deformation exist on the brace, not enough bending moment exerted to the gusset plate.

**L100-N series:** In a similar manner of L150 series, there was a clearance between the gusset plate and the brace. From the test results, with comparison to the initial loading, the maximum load (buckling load) becomes larger. Differ from the L150 series; bending moment exerted enough to return the gusset plate. Therefore, during second loading, the maximum load became larger compared with initial loading.

**L90-N series:** The clearance between gusset plate and the brace was the largest of all series. In a similar manner of the L100 series, the maximum load was larger than initial loading. However, the bolts slipped frequently.

**L150-R series:** Differ from the non-repair specimens, the clearance became less than 1 mm by welding the tapered steel filler on the gusset plate. And also, the side V is presented as the state of compression.

**L100-R series:** Since the bolts weren't vertically fastened on the gusset plate, they were stiffened as vertically as possible. And also all specimens slipped frequently. So the critical damage was confirmed around the bolt hole of tapered steel element and the gusset plate.

**L90-R series:** The tapered steel filler was welded on gusset plate. Since the residual deformation was large, measurement error and processing precision was not precisely. Therefore, clearance existed between brace and tapered steel filler. This clearance was disappeared little after the bolts stiffened. And also, initial deformation occurred on brace.

## A verification of repair

To make a comparison between initial loading and second loading is shown in Table 4. Expect for L150-4.5- 0-F-N-M, ratio of buckling load is 1.0 or more. From that, the effectiveness of this repair method is confirmed.

## Study on the strength and ductility

**Initial stiffness:** Figure 5 compares the rigidity obtained from test and theoretical result. On all specimens, test result underestimates the rigidity. This is assumed that the frictional force function not sufficiently because of slip.

**Buckling length:** The buckling length of repaired specimens become shorter than before repair. Judging from that, the bearing situation is enhanced and come close to the fix support.

**Deformed state:** About the L100-16 and repaired specimens, deformation of brace advances in plane, whereas other specimens (before repaired specimens) advance out of plane. By this, in case of the bearing situation come close to fix, direction of deformation proceeds to in plane. In addition, as showed in Table 4, repaired specimens (in brief, in the case of bearing situation of joint is enhanced) tend to be compressed "side V".

**Curvature distribution:** As a representative example of specimens, the curvature description of L100-6-0-F-R-M is described in Figure 6. Focusing on 600 mm in Figure 5a, as the ductility factor rises, the curvature rises too. However, focusing on 600 mm in Figure 5b, there is no difference in curvature. By this, side stiffener restrains the deformation not only gusset plate but brace.

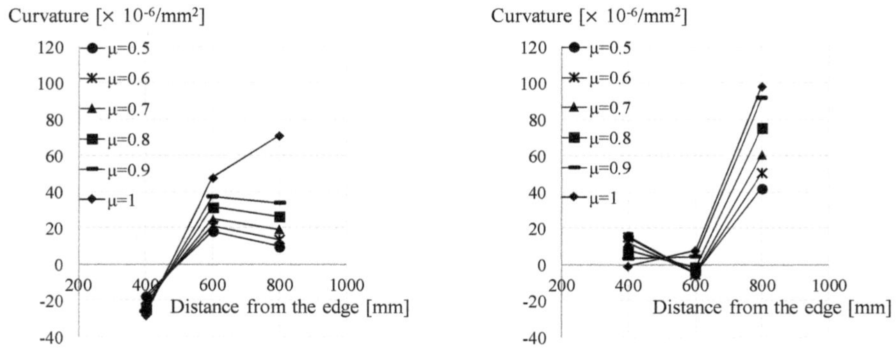


(a) Before repair (b) After repaired

**Figure 6:** Comparison of curvature distribution.

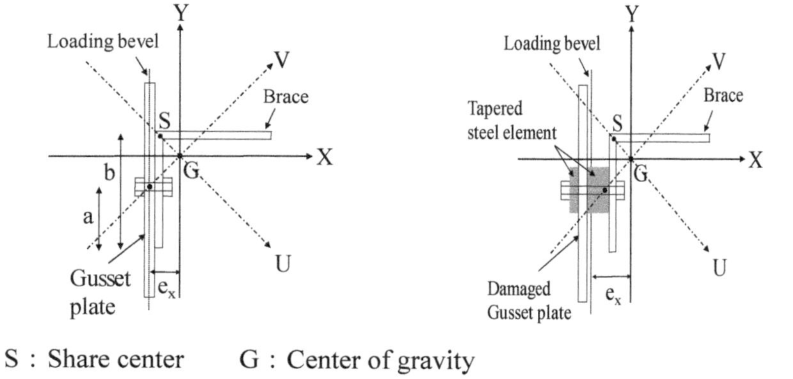


(a)Before repair (b)After repaired

**Figure 7:** Section of joint.

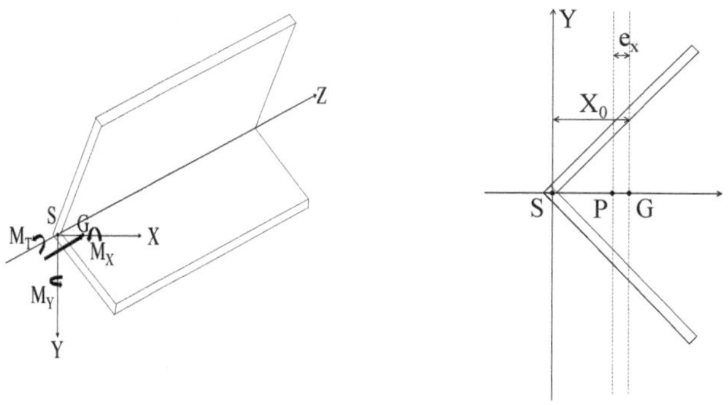

**Figure 8:** Definition of the variable.

## Proposal of evaluation method of buckling load

### General description of evaluation method

By a foregoing chapter, in case of the bearing situation is enhanced, torsional buckling is likely to occur. And also, the buckling load reverted to original state by repairing. Therefore, the repairing method in this study is effective for damaged bracing structure. In this chapter, the evaluation method of buckling load for original state and repaired state is discussed.

### Variable of evaluation method

**Bearing situation:** Considering the circumstances of connection, bearing situation is calculated. Furthermore, by attaching side stiffener, the rigidity of gusset plate is ten times as large as original state. So in this method, the rigidity of repaired gusset plate calculated the same way.

**Eccentric distance:** A cross section of the repaired gusset plate is shown in Figure 7. The eccentric distance of original state is calculated by equation (1), and after repaired is calculated by equation (2). Note that, the tsp1 and the tsp2 is thickness of tapered steel element.

$$e_x = \left( g + \frac{t_g}{\sqrt{2}} - \frac{(b-a) - \frac{t_g}{2}}{\sqrt{2}} - t_b \right)$$

$$e_x = \left( g + \frac{t_g + t_{spl} + t_{sp2}}{\sqrt{2}} \right) - \frac{(b-a) + \frac{t_g + t_{spl} + t_{sp2}}{2} - t_b}{\sqrt{2}}$$

### Evaluation method

The buckling state is classified into the bending buckling or the lateral torsional buckling by using equation of equilibrium as bending situation and eccentric distance are variables, and buckling load is evaluated. Considering the test situation, boundary conditions are as follows: Mx = Mt = 0, My = pex, P = P, y0 = 0. Equation (3) shows the buckling load in elastic region. The variables of equation (3) are shown in Figure 8. By these equations, the buckling load is called by eccentric distance and bearing situation.

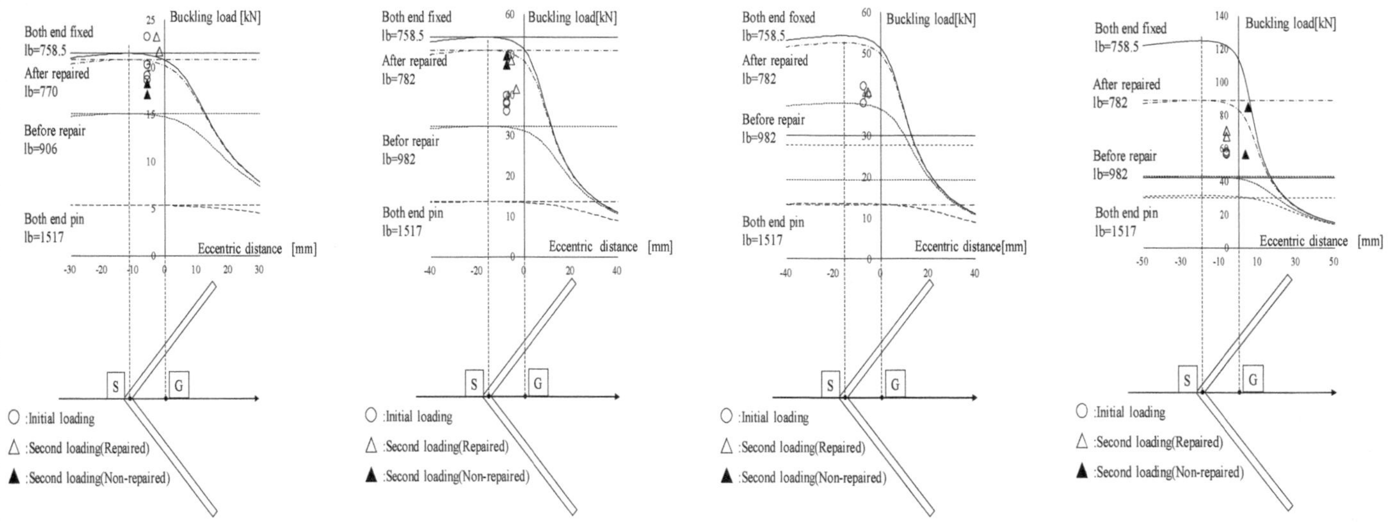


(a)L150series (b)L100-6-0series (c)L100-6-40series (d)L90series

**Figure 9:** Comparison of test results and evaluation method.

$$P=\frac{(P_w+\alpha P_{EX})-\sqrt{(P_w-\alpha P_{EX})^2+4P_wP_{EX}(x_0/\rho_0)^2}}{2\{\alpha-(X_0/\rho_0)^2\}}$$

$$\rho_0^2=\int(x^2+y^2)\,dA/A \qquad (3)$$

$$\alpha=1-\frac{e_x\left\{\int x(x^2+y^2)\,dA/A\right\}}{\rho_0^2 I_y}$$

## Inspection of evaluation method

**L150 series:** Relation of eccentric distance and buckling load is shown in Figure 9a. Straight line indicates a bending buckling load; curve indicates a lateral torsional buckling load. In all specimens, the test results exceed the evaluation value. Especially, initial loading test results greatly exceed the evaluation value; it's assumed that the buckling is occurred around the strong axis.

**L100-6 series:** Relation of eccentric distance and buckling load is shown in Figures 9b and 9c. On the loading test, some specimen became the plastic buckling. Therefore, the plastic buckling load was also described in Figures 9b and 9c. Some repaired specimen, the buckling load is less than the evaluating value. In this case, axial rigidity became around 60%. Therefore, compression stress is not transmitted enough to the brace. Except this specimen, the test results exceed the evaluation value.

**L90 series:** Relation of eccentric distance and buckling load is shown in Figure 9d. The plastic buckling load is also described in Figure 9d. Except some repaired specimen, test results exceed the evaluating value. In repaired specimens which are falling below the evaluating value, they became a plastic lateral torsional buckling. So this evaluation method is not enough to such situation. However, considering the Young's modules decline to around 65%, this evaluation method can apply to this series.

## Conclusion

This paper focused on repair methods for damaged steel brace members. The main conclusions are shown below:

1. To investigate the mechanical property, loading tests was performed. From the test results, it is confirmed the buckling load and rigidity were almost recovered to original state.

2. By using the equation of equilibrium, the evaluation method was proposed and comparison with test results. About specimens which became elastic buckling, evaluating value underestimate the buckling load. However, for other specimens, this method gives good result.

### Acknowledgement

This research was supported by Grants-in-Aid for Scientific Research(C) (Head: Takumi Ito).

### References

1. The Japan Building Disaster Prevention Association (2001) Technical Manual.
2. Architectural Institute of Japan (2011) Guidebook on Design and fabrication of high strength bolted connection.

# A Study on Ultimate Seismic States of Multi-story Horizontally Mixed Structural System and Feasibility on Connection Joints

**Kento Mori*, Takehiro Takahashi and Takumi Ito**

*Department of Architecture, Tokyo University of Science, Japan*

**Abstract**

This study is concerned with horizontally mixed structure system that an existing old R/C structure is seismically retrofitted by addition of exterior steel frame. This system is suggested as seismic retrofitting method in Japan. And some previous studies related to its seismic response mitigation effects were conducted analytically. This study focuses on ultimate seismic state on its multi-degree of freedom system, and the feasibility study of connection joint between exterior steel frame and existing R/C building is investigated. To determine the design point which estimates the ultimate seismic performance, monolithic load patterns considering seismic load effect are adopted on pushover analysis under structural design procedure. Herein, this composite system consists of two various structural systems, and it is assumed that the complicated elasto-plasticity behaviors are presented during inelastic response. So this study suggests the envelope curve model which approximates the inelastic seismic response domain, and the load pattern is obtained by reference of this model. From comparison of proposed model and seismic response analysis, the predicted design points are corresponded each other. And also, it is observed that the predominant failure mode is changed on original and retrofitted state. So this paper suggests the retrofitting strategy which overall failure mode formation is guaranteed. Moreover, the actual connection method between steel frame and R/C building has been suggested. Herein, the required strength is calculated by reference of design points, and ultimate strength of this proposed connection compares with the requirement. From comparisons, it is confirmed that this joint shows sufficient strength.

**Keywords:** Column-to-beam strength ratios; Connection joint; Design point; Envelope curve model; Horizontally mixed structural system; Load distribution

## Introduction

It has been discussed that old R/C institutional buildings, such as government offices or school buildings, have poor seismic resistant performance. In Japan, "An Act on Promotion of Seismic Retrofitting of Buildings" was enacted in 1995.

In these trends, "the horizontally mixed structural system" that an existing old R/C structure is strengthen by new steel frames attached is suggested as one of the seismic rehabilitation methods by The Japan Iron and Steel Federation (Figure 1). It aims that the retrofitted composite structure can expand its floor space and acquires architectural changeability and flexibility (Figure 2). So in terms of the addition of flexibility and conversion for existing buildings, the concern over the horizontally mixed structural system has risen increasingly in future. Iwabuchi and Ito et al. and Chiyo and Fujii et al. [1] conducted analytical researches on the response mitigation effect of this system. These researches showed that the connection joint between the existing old R/C structure and the steel frames needs to possess sufficient rigidity and strength under an earthquake motion (Figure 3). However, it has not investigated the ultimate seismic states and the retrofitting effect of this system in elasto-plasticity behaviors. Then this paper investigates ultimate seismic states of the multi-story horizontally mixed structural system and suggests the load pattern adopted in pushover analysis. Moreover, this paper investigates whether the connection joint is effective in the ultimate seismic state [2].

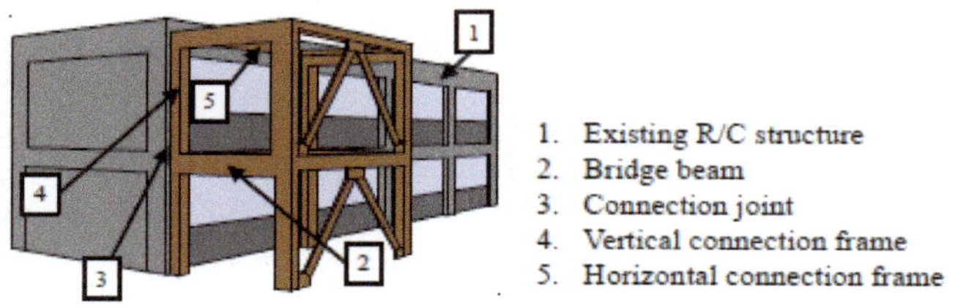


Figure 1: Horizontally mixed structural system.

Figure 2: Architectural changeability.

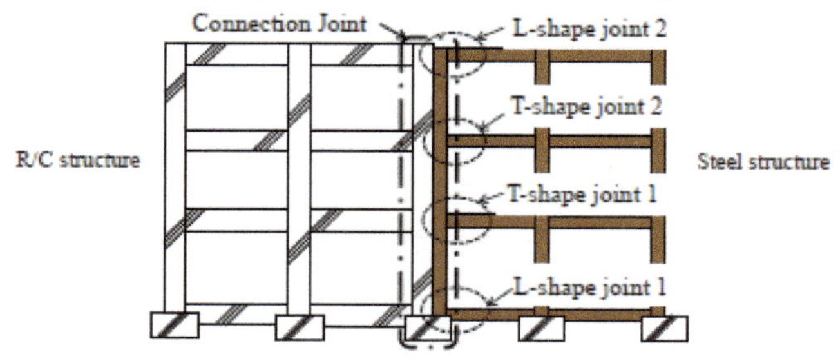


Figure 3: Connection joint.

*Corresponding author: Kento Mori, Department of Architecture, Tokyo University of Science, Japan, E-mail: pasuterukunsan@gmail.com

## Envelope Curve Model

### Design point and load distribution

This study adopts modal load coordinate in order to investigate the design point and the load distribution in elasto-plasticity response behaviors (Figure 4). Forces in normal coordinate are transformed into forces in modal load coordinate by Equation 1.

$$\{f\}=\left[\Phi^{T}\right]^{-1}\{r\} \tag{1}$$

Where,

{f }: restoring force vector in normal coordinate.

[Φ]: participation matrix.

{r}: restoring force vector in modal load coordinate.

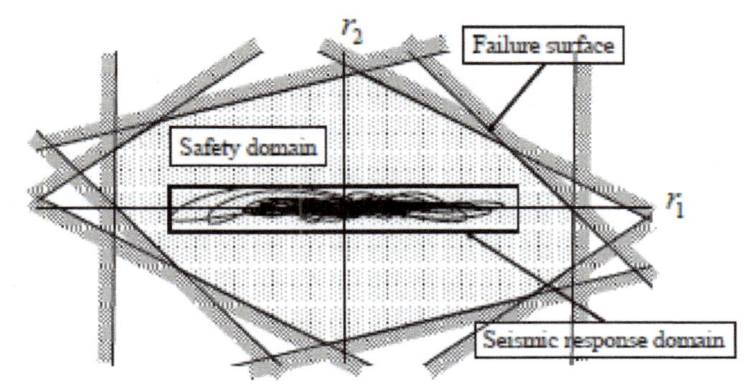


**Figure 4:** Failure mode in modal load coordinate.

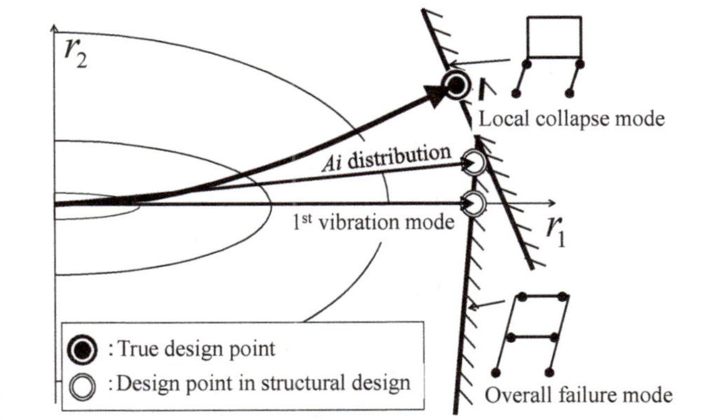


**Figure 5:** Conceptual diagram of design points.

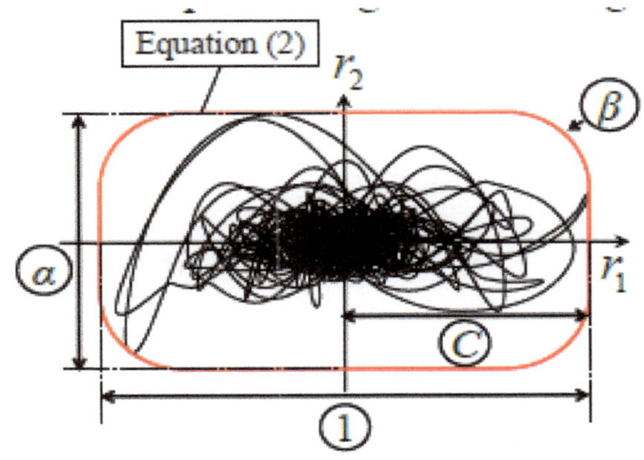


**Figure 6:** Conceptual diagram of envelope curve model.

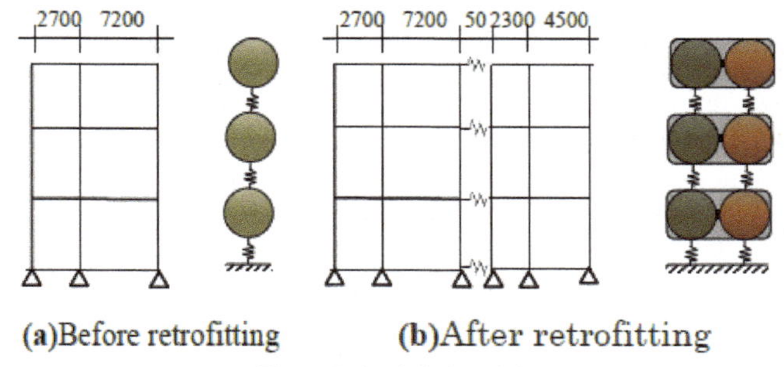


**Figure 7:** Analytical model.

In normal structural seismic design, the pushover analysis is often performed, and a monolithic load pattern such as Ai distribution is adopted. However, in the horizontally mixed structural system, it is assumed that the complicated inelastic behavior is presented because both the hysteresis characteristics of R/C and steel structure exist simultaneously. So it is likely that this effect is considered on load pattern adopted on limit state analysis during limit state design procedure (Figure 5) [3].

### Definition of envelope curve model

This study suggests the envelope curve model which approximates the inelastic seismic response domain. This model is described by Equation 2.

$$r_1^{\beta}+\left(\frac{r_2}{\alpha}\right)^{\beta}=C^{\beta} \tag{2}$$

Where,

α: ratio of 2nd mode component to 1st mode component.

β: constant representing roundness of the curve.

C: length of major axis.

The input level of seismic waves is assumed gradually from a range of elastic to a range of forming failure mechanisms. The values of α and β are determined so that the envelope curve model for each seismic wave can describe (Figure 6).

### Outline of seismic response analysis

The analytical model is the 3-story horizontally mixed structural system composed of the R/C structure and exterior steel frame, assuming 3-degree of freedom system (Figure 7). Here, the connection joint possesses sufficient rigidity and strength, and R/C structure and connection frame behave in a unified fashion. And it is assumed that the mass of each floor is sum of R/C structure and steel frame. The restoring force characteristics of R/C structure and steel structure are Takeda model and perfect elasto-plasticity model respectively (Figure 8). Considering aging deterioration, this study regards concrete strength as small value: $F_c$ = 18 N/mm². The weight and vibration characteristics of this model building and the seismic waves adopted in this analysis are summarized in Table 1. The sum of 1st mode effective mass and 2nd mode effective mass in the analytical model is more than 90% of the total, so this study excludes 3rd mode component from consideration [4-6].

### Analytical results

The six failure modes are expected in the analytical model frame as shown in Figure 9. The transitions of the value of α and β in each input wave level are shown in Figures 10 and 11.

Figure 12 shows an example of the load path obtained from seismic response analysis and the plastic surface in modal load coordinate. Plastic surfaces in Figure 12 and failure modes in Figure 9 correspond respectively, and the seismic response domain under seismic response reaches plastic surface (Figure 13).

This study adopts large input level (120 kine) that was observed at The South Hyogo Prefecture Earthquake in 1995. From the results of Figures 10a and 11a, $\alpha$ tends to rise as input level becomes large and converge at about 0.25 in all kinds of seismic waves. Comparing the result of the R/C structure before retrofitting and the horizontally mixed structural system under same input level, there is little difference between two values of $\alpha$. Therefore, it is confirmed that there is little effect of higher vibration mode on retrofitting state. From the results of Figures 10b and 11b, the relation $\beta$ under the elastic response is not confirmed; however, $\beta$ under the elasto-plasticity response takes averagely about 3.0. From the above, the seismic response domain of the horizontally mixed structural system in the ultimate seismic state can be described by the envelope curve model using $\alpha$ =0.25, $\beta$ =3.0, and the design point is identified by determination of value of C.

## Investigation of Design Point

### Comparison of design point and failure mode

Figure 14a shows the load path during seismic response in case of El Centro NS (100 kine) on original state (Figure 7a). Figure 14b shows the load path during seismic response in case of same input wave

**Members and weight of building**

| Structure | Members | Sectional shape | Material | Weight of the building [t] 1st story | 2nd story | 3rd story |
|---|---|---|---|---|---|---|
| R/C | Column | 550×550 (12-D19) | Reinforcing bar: SD345 | 33.9 | 36.5 | 36.5 |
| | Beam | 300×600 (10-D19) | | | | |
| | Footing Beam | 300×1200 (12-D19) | | | | |
| Steel | Column | □-300×300 | BCP235 | 4.2 | 9.3 | 9.3 |
| | Connection frame | H-300×300×12×15 | SN400 | | | |
| | Beam | H-400×200×7×11 | | | | |
| | Footing Beam | H-600×200×11×17 | | | | |

**Vibration characteristics**

| Mode | Before retrofitting | | Horizontally mixed structural system | |
|---|---|---|---|---|
| | Effective mass [t] | Natural period [s] | Effective mass [t] | Natural period [s] |
| 1 | 91.3 | 0.289 | 111.4 | 0.282 |
| 2 | 11.3 | 0.111 | 13.5 | 0.109 |
| 3 | 4.2 | 0.078 | 4.9 | 0.078 |

**Input of seismic waves**

| No | Seismic waves | Maximum acceleration [gal] | Duration time [s] |
|---|---|---|---|
| 1 | El Centro 1940 NS | 314.7 | 30.0 |
| 2 | El Centro 1940 EW | 210.1 | |
| 3 | Taft 1952 NS | 152.7 | |
| 4 | Taft 1952 EW | 175.9 | |
| 5 | Tohoku 1978 NS | 258.2 | |
| 6 | Tohoku 1978 EW | 202.5 | |
| 7 | JMA Kobe 1995 NS | 818 | |
| 8 | Fukiai 1995 NS | 804.6 | |
| 9 | Hachinohe 1968 NS | 225 | |
| 10 | Hachinohe 1968 EW | 182.9 | |

**Table 1:** Detail of analysis.

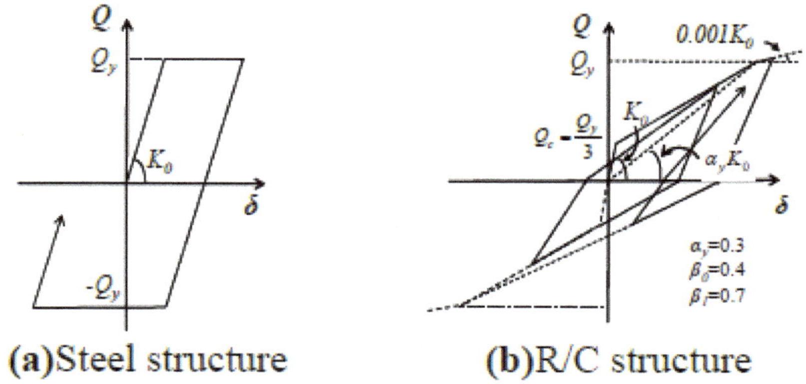


**Figure 8:** Restoring force characteristics.

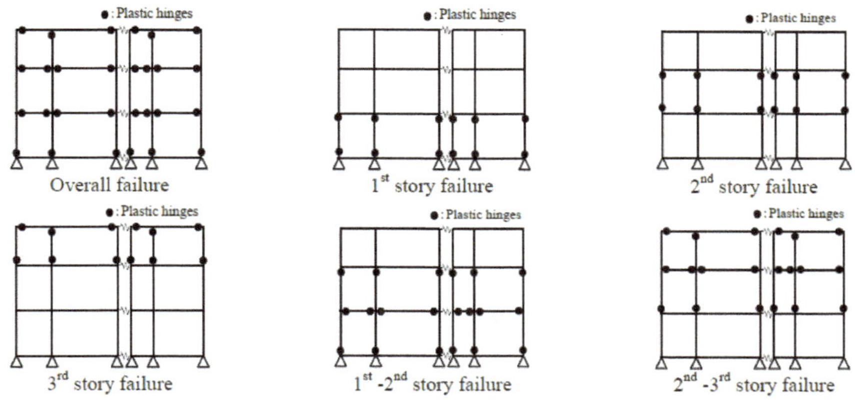


**Figure 9:** Failure mode.

**(a)**Value of $\alpha$. **(b)** Value of $\beta$

**Figure 10:** Transition of constants before retrofitting.

**Figure 11:** Transition of constants after retrofitting.

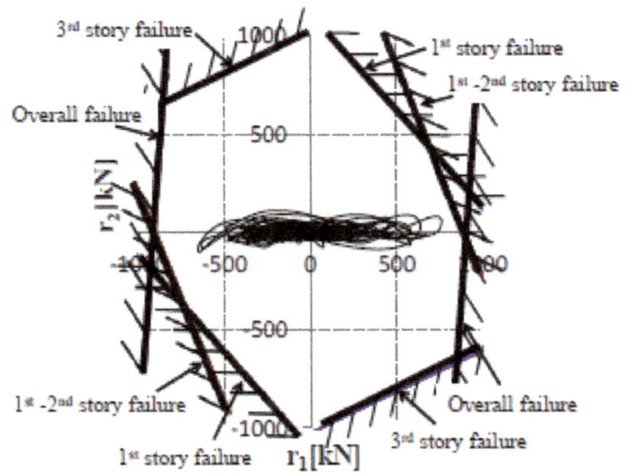


**Figure 12:** Load path and failure surface (40 kine).

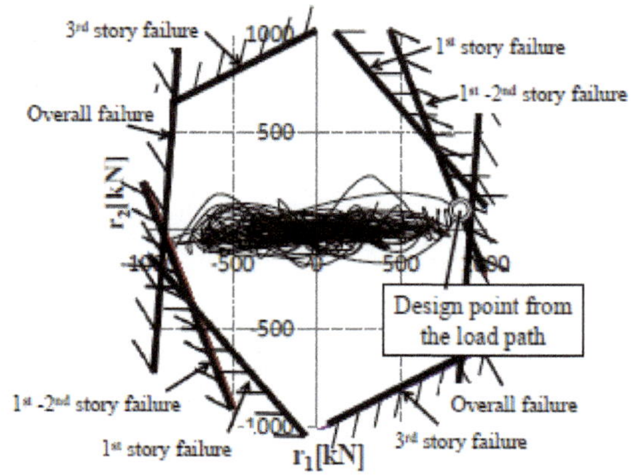


Figure 13: Load path and failure surface (150 kine).

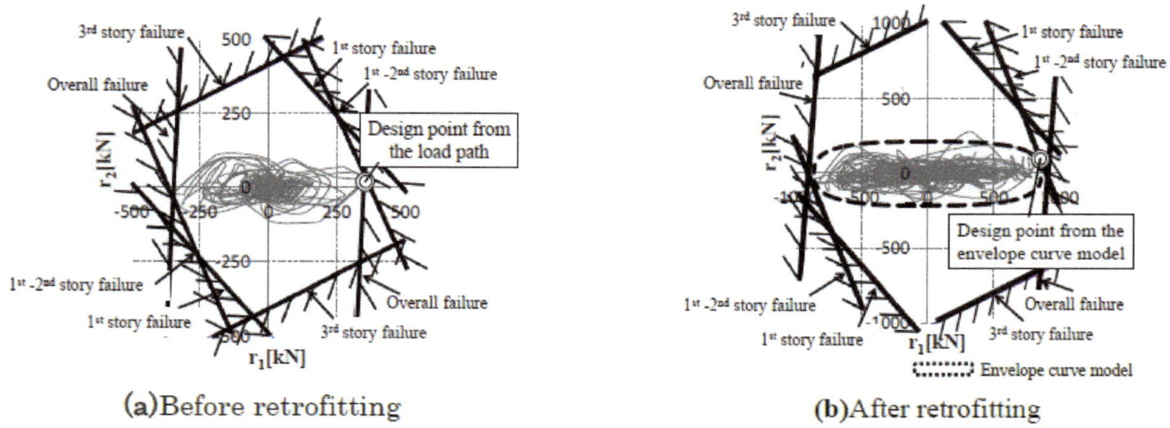


Figure 14: Change of design point and failure mode.

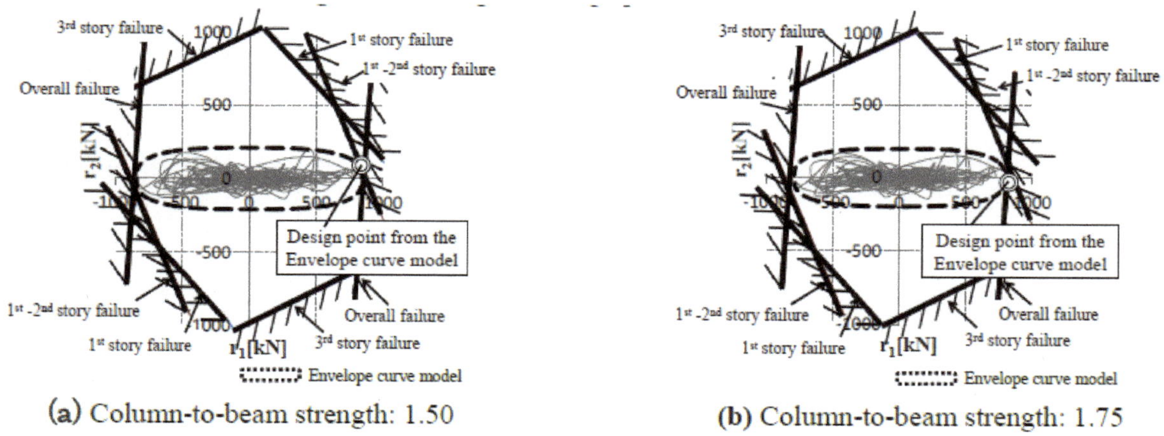


Figure 15: Relationship between column-to-beam strength ratios and design points.

for the horizontally mixed structural system. Comparing two cases, it is confirmed that these design point does not correspond with each other. And also, in case of other seismic input, these transitions of failure mode are confirmed. From these results, it can be said that the predominant failure mode moves by retrofitting with the horizontally mixed structural system.

## Retrofitting plan of steel structure

Now in the seismic design, it is desirable that buildings form overall failure mechanism in the ultimate seismic state. Then the required performance of the exterior steel structure for forming the overall failure mechanism after retrofitting is investigated.

This paper focuses on the relationship between the design points and the plastic surfaces when the column-to-beam strength ratio of steel structure changes. This investigation sets two column-to-beam strength ratios: 1.50 and 1.75 by changing only the beam's strength. Figure 15 shows the result of the seismic response analysis in case of column-to-beam strength ratios 1.50 and 1.75. From these results, in

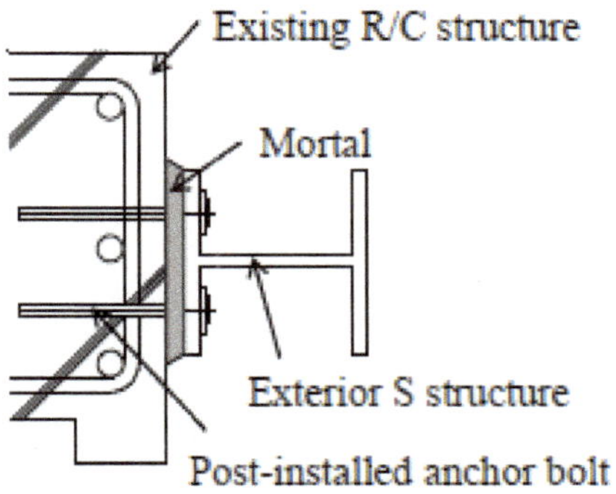


**Figure 16:** Direct-connection technique.

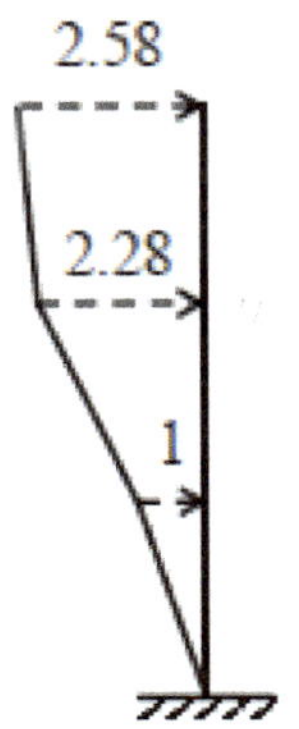


**Figure 17:** Load distribution.

| Parameter | | T-shape connection Joint | L-shape connection Joint |
|---|---|---|---|
| Tensile force on post installed anchor bolts [kN] | | 63.1 | 118.2 |
| Tensile strength [kN] | Axial-yield of anchor bolts | 197.7 | |
| | Cone-type failure | 548.4 | |
| | Adhesion section failure | 208.6 | |
| Joint coefficient *αc* | | 3.13 | 1.67 |

**Table 2:** Value of tensile forces on connection joint.

case of column-to-beam strength ratios 1.50, the design point is on the 1st -2nd story failure. On the other hands, in case of column-to-beam strength ratios 1.75, the design point is on the overall failure.

From the above, it can be said that adjusting the column-to-beam strength ratios of steel structure can control the failure mechanism of the horizontally mixed structural system in the ultimate seismic state.

## Feasibility Study of Connection Joint in Ultimate Seismic States

### Outline of investigation of connection joint

Mitomi et al. suggested the concrete connection joint techniques, and the effectiveness of this method is shown experimentally. However, it is likely that the stress on the connection in the ultimate seismic state is very complicated. Therefore, pushover analysis is performed for investigating whether the connection joint is effective in the ultimate seismic state. Herein, the load distribution is calculated from the design point by the envelope curve model.

### Outline of the connection joint

In this analysis, H-300×150×7×11 (SS400) is used for the beams in the exterior steel structure; the other members are the same as the analytical model (Figure 7b). The direct-connection technique (Figure 16) is adopted for the connection joint between the R/C structure and the steel structure, and D19-SD345 is adopted for post-installed anchor bolts.

The seismic response analysis is performed for this analytical model, and the collapse load and the load distribution is gained as shown in Figure 17. The pushover analysis for the investigation of the connection joint is adopted in these conditions of the ratios and design points.

### Result of the pushover analysis

From the results of the pushover analysis, the maximum tensile force at the each bridge beam (Figure 2) is calculated as shown Table 2, following the evaluation method suggested by Mitomi et al. The tensile strength of the three collapse mode with post-installed anchor bolts is also shown in Table 2. It is confirmed that the joint coefficient αc is more than 1.5 in each joint. Therefore, the connection joint has sufficient strength.

## Conclusions

This study investigates the design point and the load distribution of the multi-story horizontally mixed structural system, and the feasibility of the connection joint in the ultimate seismic state. The main conclusions are shown below:

1. The envelope curve model that approximates the inelastic seismic response domain in the horizontally mixed structural system is suggested. Moreover, by identifying the design in the ultimate seismic state from this model, the failure mechanism and the load distribution are investigated.

2. It is shown that the failure in the ultimate seismic state mode can move after retrofitting. Adopting the relevant column-to-beam strength ratios of the steel structure, the failure mechanism can be controlled.

3. The feasibility of the connection joint in the ultimate seismic state. Herein, adopting the load distribution at the design point, pushover analysis is done. Then it is shown that the connection joint technique suggested by Mitomi et al. can be effective in the ultimate seismic state.

### References

1. Fujii K, Chiyo Y (2013) Evaluation of seismic response of low-rise existing reinforced concrete buildings strengthened by addition of steel frames. J Struct Eng 59B: 55-64.

2. Mitomi R, Matsumoto T, Ito T, Fujii K (2013) Development of connection joint for horizontally mixed structural system by experimental study and frame analysis. Proceedings of the Pacific Struct Steel Conference p: 757-762.

3. Mitomi R, Takahashi K, Ito T, Kambe W, Fujii K (2012) An experimental study on connection joint of composite mixed structures consisting of existing R/C buildings retrofitted with exterior steel frames. Proceedings of World Conference on Earthquake Eng.

4. Mitomi R, Takahashi K, Ito T, Kambe W, Fujii K (2012) A feasibility study of direct-connection by use of post-installed anchor bolts for horizontally mixed structural system. Proceedings of International Symposium on Architectural Interchanges in Asia.

5. Takeda T, Sozen MA, Nielsen NN (1970) Reinforced concrete response to simulated earthquakes. The 3rd Japan Earthquake Eng Symposium p: 357-364.

6. Ito T, Ohi K, Mori Y, Fang P (2003) Hybrid design point search on collapse mechanism formation of framed structures. J of Struct and Construct Eng 567: 125-132.

25

# Prediction Method of Maximum Seismic Response for Vibration Control Steel Frames With Scaling-Frame Device Installed


**Akinobu Kano[1]*, Hideaki Azu[1], Takumi Ito[1], Donghung Wu[2], Takashi Nagumo[3] and Haruhiko Hirata[3]**

*[1]Department of Architechture, Tokyo University of Science, Japan*
*[2]Wu Office, P.R. Japan*
*[3]Hory Corporation, Japan*



## Abstract

The innovated vibration control device called as scaling frame ωSFď structure has been proposed by the authors, and the SF structures are applied to low-rise wooden structures already. This paper aims to investigate the applicability of SF structure on steel frames. Herein, SF structure consists of beam-column frame, diagonal bracing, and SF device ωSFDď installed, and SFD is made of Steel or aluminum. And vibration energy is absorbed by the plastic behavior of the diagonal deformation of SF device. In previous study, the experimental study on SFD and steel frame specimen with SFD installed were conducted to clarify the seismic response and seismic mitigation effect. And also, from observation of test results, analytical method has been suggested. From the comparison of test results, the proposed method shows good agreements with test results. Furthermore, the enormous past studies have suggested prediction method of maximum seismic response of vibration control frames. So this study aim to develop the prediction method of seismic response on vibration controlled steel frames with SFD installed, and the simple design procedure can be provided. The first, the simplified restoring force characteristics which can chase the test results is applied on prediction method procedure. So the prediction method is reformulated to adapt the bilinear model and strain hardening rule. The time history response analysis is performed to investigate the applicability and effectiveness. From the comparison of analysis results and predictions, it is confirmed that the proposed prediction method shows enough accuracy.



**Keywords:** Prediction method of maximum response; Steel framework; Seismic vibration control system; Time history response analysis


## Introduction

Recently, a various type of vibration control device and damper has been developed on country, and most devices are suggested and adopted to actual building structure. For example, oil damper and high damping rubber device are suggested. However, they have problems of durability and stability velocity. In addition, they consist of special device and material, so there are some difficulties of getting material and construction technics. In contrast, the innovated vibration control device called as scaling frame (abbreviated as "SF") structure has been proposed by the authors, and it is expected that SF structure can solve these difficulties. SF structure consists of beam-column frame, diagonal bracing, and SF device (abbreviated as "SFD") installed. SF as shown in Figure 1. From the points of mechanical characteristics and materials, SFD are categorized as shown Table 1 with damping system, resistant mechanics and materials. SFD is made of Steel or aluminum. And vibration energy is absorbed by the plastic behavior of the diagonal deformation of SFD. SFD has some advantage such as workability, transportability and containment because of small size, light weight, durability and sustainability of metal materials. And SF structures are applied to low-rise wooden structures already. This paper aims to

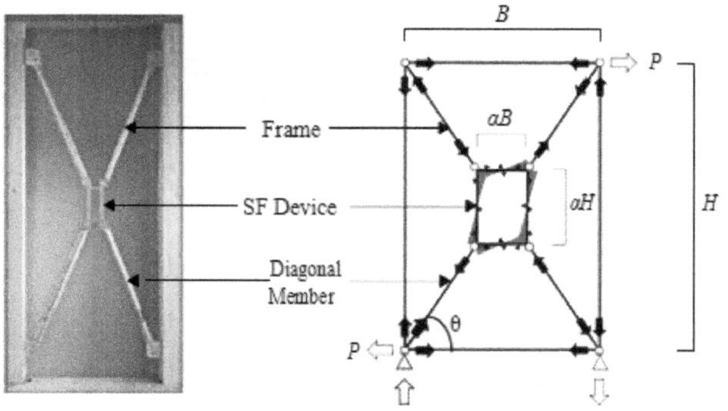


**Figure 1:** Concept of SF structure.

| Mechanics | System | Materials |
|---|---|---|
| Bending | Viscous | Metal |
| Shear | Hysteresis | Oil |
| Axial | Friction | Rubber |

**Table 1:** The mechanical characteristics and materials of SFD.

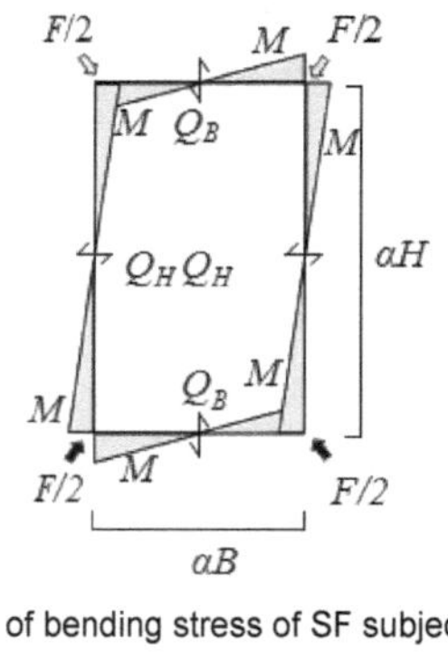


**Figure 2:** Diagram of bending stress of SF subjected to lateral force.

investigate the applicability of SF structure on steel frames. (Figure 2) This study aim to develop the prediction method of seismic response on vibration controlled steel frames with SFD installed, and the simple design procedure can be provided. To investigate the simplified restoring force characteristics, the horizontal static cyclic loading test is conducted on half-scaled frame test [1,2]. And, the prediction method is reformulated to adapt the bilinear model and strain hardening


***Corresponding author:** Akinobu Kano, Department of Architechture, Tokyo University of Science, Japan, E-mail: 4116617@ed.tus.ac.jp

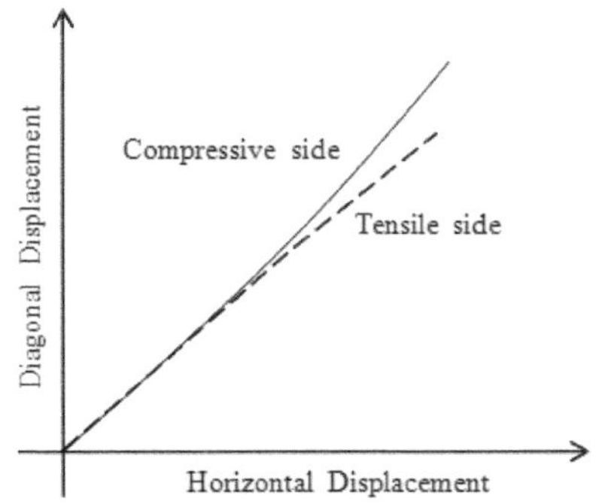


**Figure 3:** Comparison of compressive and tensile deformation.

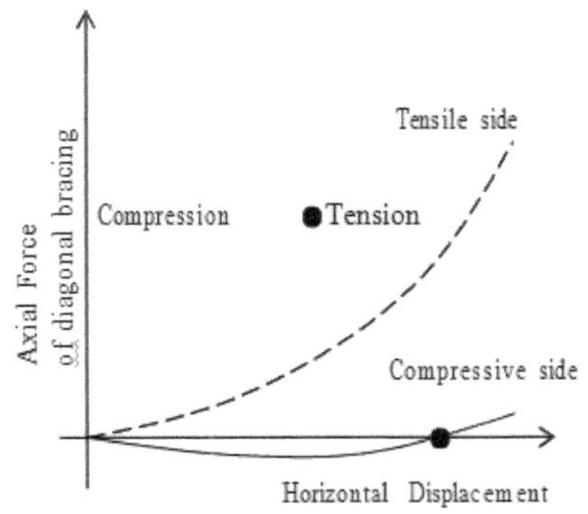


**Figure 4:** Relation of axial force and lateral displacement.

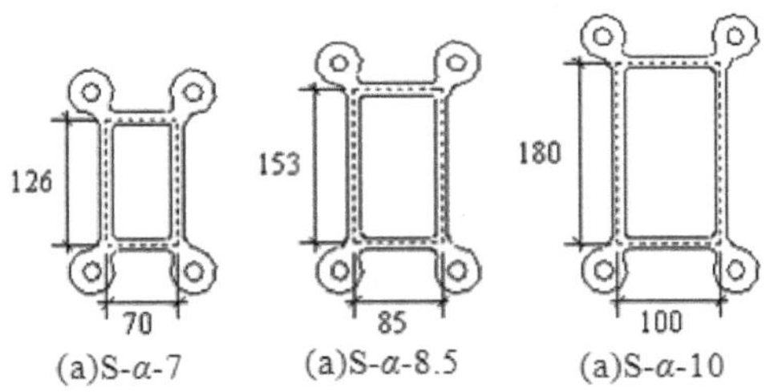


**Figure 5:** The test specimen of SFD.

| Member | Reduction ratio | JIS grade | σy (N/mm²) | σu (N/mm²) | E (kN/mm²) |
|---|---|---|---|---|---|
| S-α-7 | 7 | SS400 | | | |
| S-α-8.5 | 8.5 | SS400 | 285 | 433 | 210 |
| S-α-10 | 10 | SS400 | | | |
| Bracing | - | SS400 | 342 | 472 | 202 |

**Table 2:** Name of test specimen and test parameters and mechanical properties of steel.

rule. To investigate the applicability and effectiveness of the method, the time history response analysis is performed geometrically. From Figure 3, it can be said that the compressive displacement grows larger than the diagonal displacement of extension. Thus, the diagonal member on tensile side will resist increasingly, on the other hands, the diagonal member on compressive side will be reduced as shown in Figure 4. So the lateral buckling of diagonal member on compressive side is prevented, and bending reflection of the SF is restricted in large deformation range, and rigidity in the axial direction of the SFD becomes dominant (Figure 5).

## General Description of SF Structure

Herein, the general description of resistant mechanism and characteristics of SF structure is explained. As a definition of SF structure, the relation of size of out-frames and SFD is defined as reduction rate α as shown in Figure 1. Figure 2 shows the resistant mechanism of SF structure. From the principal of vertical work, the rigidity KSF and yield strength Py are obtained as follows:

$$K_{SF} = \frac{12EI_{SF}L^2\cos^2\theta}{\propto^3 B^2H^2(B+H)} \quad (1)$$

$$P_y = \frac{4f_b Z_{SF}}{\propto H} \quad (2)$$

E: The young's modulus of the SFD

ISF: Moment of inertia of area of the SFD

ZSF: Elastic modulus of SFD

B: The length of the beam

H: The length of the column

L: The diagonal length of beam-column frame

fb: Allowable bending stress of the SFD

From the above equations, it can be predicted that KSF is inverse proportion to a cube of reduction rate α, and the strength is inverse proportion to reduction rate α. Therefore, SF structure with small reduction ratio shows the large rigidity, strength and high energy absorption. However, in case of very small reduction rate of SFD, the fracture would be occurred, so it is desirable that the limitation of reduction ratio is decided. The next, the tensile and compressive deformation in the diagonal displacement of SFD subjected to lateral force are dominated, it means that the diagonal bracing system is appeared. So the tensile and compressive displacements of each member are drawn in Figure 3 geometrically. From Figure 3, it can be said that the compressive displacement grows larger than the diagonal displacement of extension. Thus, the diagonal member on tensile side will resist increasingly, on the other hands, the diagonal member on compressive side will be reduced as shown in Figure 4. So the lateral buckling of diagonal member on compressive side is prevented, and bending reflection of the SF is restricted in large deformation range, and rigidity in the axial direction of the SFD becomes dominant.

## Summary of Horizontal Static Loading Test

### General description of loading test

This paper investigates the fundamental restoring force characteristics of SFD experimentally. Herein, the horizontal static loading tests are conducted on elementally test to investigate them. Figure 5 shows the test specimen of SFD. Herein, the reduction ratio of SFD is considered with parameters as 7%, 8.5%, and 10%. SFD (thickness 19 mm) are made of steel. Table 2 shows member of test specimen and test parameters on elementally test and the mechanical properties of steel used for the test specimen. Herein, α is reduction rate. σy is yield strength and σu is tensile strength. And, E is Young's modulus. SS400 present the grade of JIS (Japanese Industrial Standards). This paper investigates the fundamental restoring force characteristics of SFD experimentally. Herein, the horizontal static loading tests are conducted on elementally test to investigate them. Figure 5 shows the test specimen of SFD. Herein, the reduction ratio of SFD is considered with parameters as 7%, 8.5%, and 10%. SFD (thickness 19 mm) are made of steel. Table 2 shows member of test specimen and test parameters on From the above equations, it can be predicted that KSF is inverse proportion to a cube of reduction rate α, and the strength is inverse proportion to reduction rate α. Therefore, SF structure with small reduction ratio shows the large rigidity, strength and high energy absorption. However, in case of very small reduction rate of SFD, the fracture would be occurred, so it is desirable that the limitation of reduction ratio is decided. The next, the tensile and compressive deformation in the diagonal displacement of SFD subjected to lateral force are dominated, it means that the diagonal bracing system is appeared. So the tensile and compressive displacements of each member are drawn in Figure 3 elementally test and the mechanical properties of steel used for the test specimen. Herein, α is reduction rate. σy is yield strength and σu is tensile strength. And, E is Young's modulus. SS400 present the grade of JIS (Japanese Industrial Standards) [3,4].

## Test setup and loading test method

Figure 6 shows a loading test setup. In this paper, the length of column is 1800 mm, and that of beam is 1000 mm on half-scaled frame. SFD connected with column jig through gusset plate and diagonal bracing. Columns are supported to the beam by pin joint.

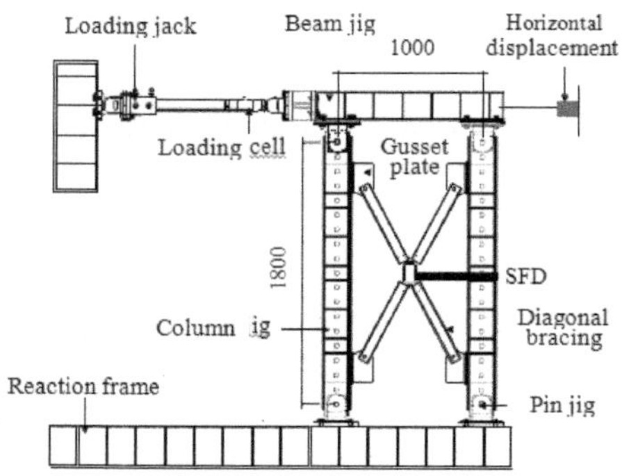


**Figure 6:** Elevation of test setup and location of sensors.

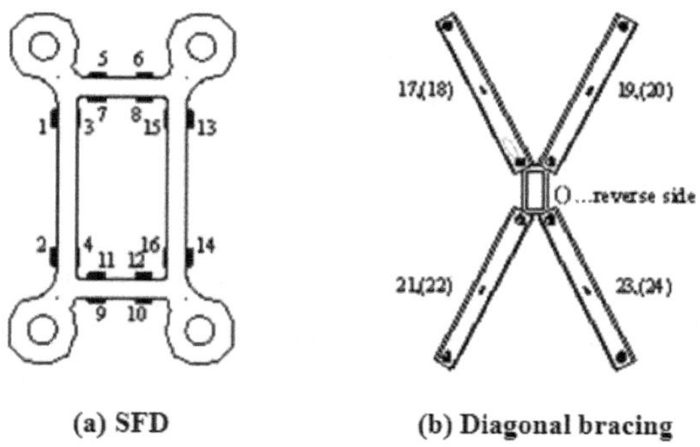


**Figure 7:** Location of strain gauges.

| No | Input Wave | PGA (gal) | Duration (s) | Dh |
|---|---|---|---|---|
| 1 | El Centro NS | 341.7 | 53.76 | 25 |
| 2 | Hachinohe NS | 229.65 | 50.98 | 25 |
| 3 | Taft NS | 152.7 | 54.38 | 30 |
| 4 | Fukiai NS | 804.58 | 30 | 15 |
| 5 | Tottori NS | 714 | 60 | 25 |
| 6 | Miyagi NS | 1093.66 | 60 | 25 |
| 7 | Niigata NS | 863.06 | 60 | 15 |
| 8 | Takatori EW | 605.55 | 60 | 20 |
| 9 | JMA-Kobe NS | 818 | 39.33 | 15 |
| 10 | BCJ-L2 | 355.66 | 120 | 75 |
| 11 | Ocean EQ(GPV100 kine) | 196.55 | 81.93 | 75 |
| 12 | Epicentral EQ (GPV100 kine) | 671.15 | 81.93 | 7 |

**Table 3:** List of input seismic wave.

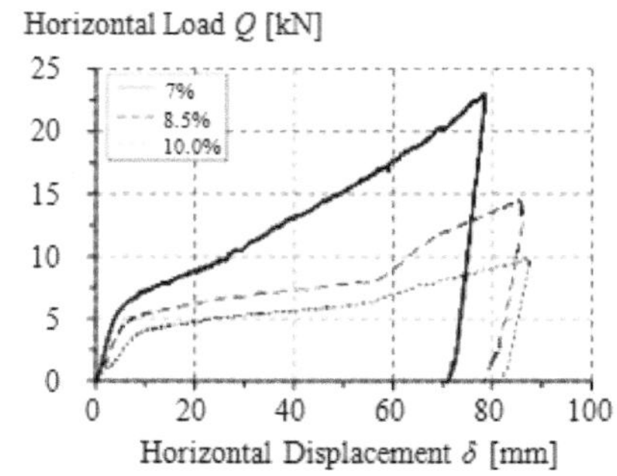


**Figure 8:** Monotonic loading test results of SFD.

The horizontal force is measured from the load cell built into the loading jack. The horizontal displacement at the top of the column is measured by using a tape measure type displacement transducer. The strain gauges are placed on the SFD and the diagonal bracing as shown in Figure 7. Loading pattern is monotonic or cyclic. The loading program is arranged by the reference of the target displacement, and the maximum angles of the column at each loop are 1/200, 1/100, 1/75, 1/50, and 1/30 rad. It is gradually increased and the same two angles of the column for each cycle are repeated (Table 3).

## Result of the Test Study

### Test result of monotonic loading

Figure 8 shows the relation of horizontal load - horizontal displacement of monotonic loading test result. From the test results of Figure 8, it is confirmed that the rigidity and strength of the SFD become large in case of small reduction rate α. And also, the strain hardening is appeared during ultimate state. Moreover, strength and rigidity of SFD is raised strongly during ultimate states. From the analysis of axial and curvature of SFD obtained from strain gauges, the curvature is dominated during early stage, however the axial deformation is advanced during ultimate states. According to the above results, it can be said that resistance state of the SFD translates from the bending mechanism to axial force mechanism with progress of the deformation [1,2].

### Test result of cyclic loading

Figure 9 shows the relation of horizontal load - horizontal displacement of monotonic loading test result. From the results of Figure 9, it is confirmed that the stable inelastic behaviors are presented.

## Proposal of Restoring Force Characteristics

From the test results of Figure 9, the skeleton curve of SFD is modeled as bi-linear curve as shown in Figure 10. A break point of bi-linear curve is determined by the yield strength and the rigidity of SFD. It is the theoretical value based on Eq.(1),(2). And the second rigidity is determined from monotonic loading test of SFD and material testing. The ratio of initial rigidity and second rigidity are 0.04 in the steel SFD and 0.015 in the aluminum SFD. Herein, the hysteresis behavior model is assumed with the strain hardening model as shown in Figure 11. It's confirmed that this model is the static state by several cycles after it passed through hardening phenomena repeatedly when repetitive stress was given under the constant distorted amplitude. In this paper, the analytical coefficient β related to hardening rule is decided to correspond with test result of SFD (herein, β=0.5). When β is 0.5; this model is the static state by two cycles. Herein, the hysteresis behavior model is the elasto-plastic force-displacement relation model, when the ratio of initial rigidity and second rigidity are 0. Figure 12 shows the comparison of horizontal load - horizontal displacement curves of the cyclic loading test and analytical results. Figure 13 shows the comparison of horizontal load - cumulative displacement curves of the cyclic loading test and analytical results. From the results of Figures 12 and 13, the analysis results can chase test results well.

## Prediction Method of Maximum Seismic Response Based on Equivalent Linearization

In this paper, the strain hardening model of hysteresis rule is

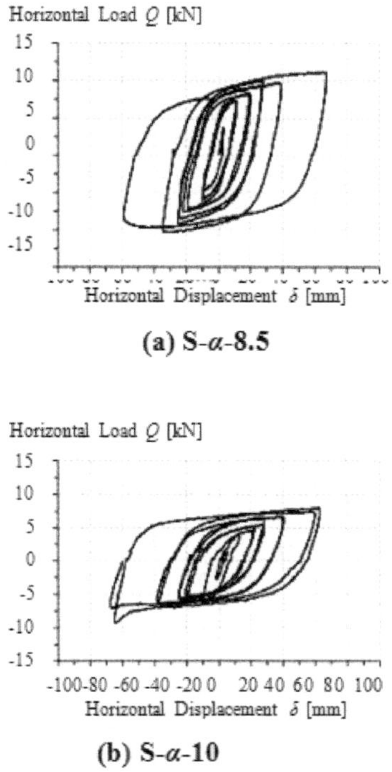

**Figure 9:** Cyclic loading test results of SFD.

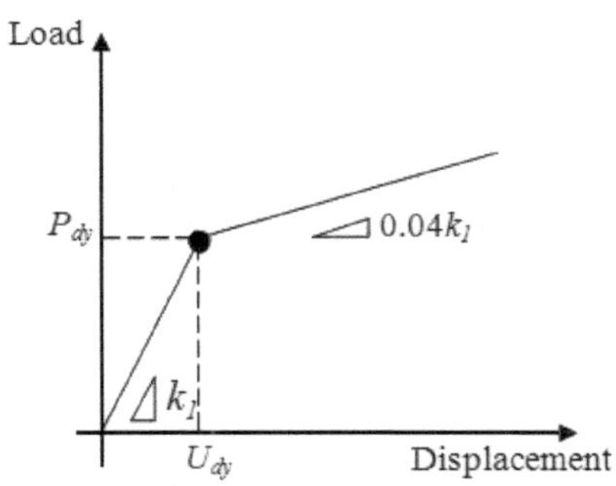

**Figure 10:** Skeleton curve of SFD.

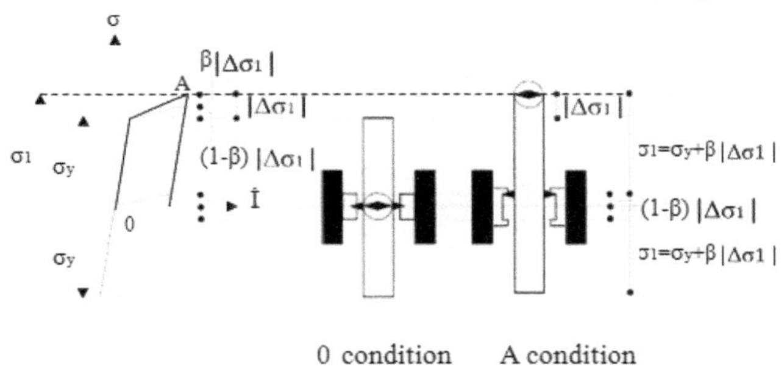

**Figure 11:** Strain hardening model of SFD.

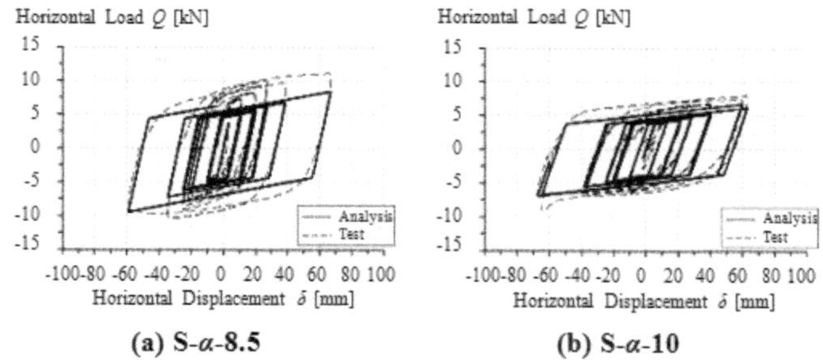

**Figure 12:** Hysteresis curve.

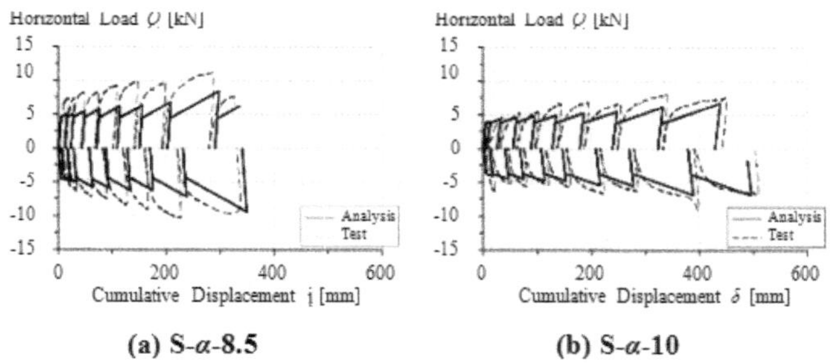

**Figure 13:** Cumulative deformation curve.

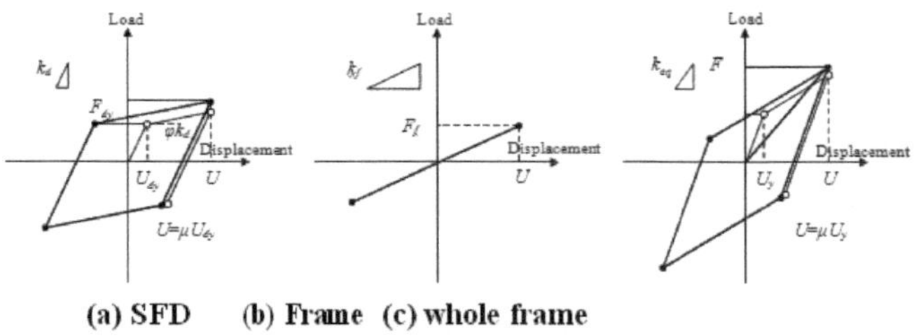

**Figure 14:** Analytical frame model.

assumed as shown in Figure 14. The revised equivalent natural period Teq of vibration controlled frame is as follow:

$$T_{eq} = T_f\sqrt{\mu/[\varphi r[2\mu - \varphi\{2\mu - \varphi(\mu-1)+2\} - 2] + r + \mu} \quad (3)$$

Where, Tf is natural period of steel frame, r is ratio of elastic stiffness ( = kd/kf), kd is elastic stiffness of SFD, kf is elastic stiffness of frame, kb is elastic stiffness of bracing member, φ is second rigidity ratio of SFD, μ is maximum ductility. Then, the equivalent natural period of φ = 0 is as follow:

$$T_{eq} = T_f\sqrt{(1+\alpha)\mu/(\alpha+\mu)} \quad (4)$$

And the hysteresis loop presents the perfectly elasto-plastic. Furthermore, by use of Newmark-Rosenblueth

method [3], the equivalent damping heq is as follow:

$$D_h = \sqrt{(1+bh_0)}/(1+bh_{eq})$$
$$U = D_h(T_{eq}/2\pi)\bar{S}_{pv}(T_{eq},h_0) \quad h_{eq} = h_0 + 1/\mu\int_1^{\mu} h_{eq}(x)dx \quad (5)$$
$$F = MU(2\pi/T_{eq})^2$$

Herein, larger maximum ductility μ leads to larger $T_{eq}$ and to a certain extent, larger $h_{eq}$.

Peak response of SFD will be obtained from common linear response spectrum as well as $T_{eq}$ and heq explained above. Ductility by each cycle is various and the period by each cycle change between $T_0$

and $T_{eq}$ under seismic response included a various volume of amplitude. Then, maximum velocity by use average of velocity response spectrum from $T_0$ to $T_{eq}$ is as follow:

$$\bar{S}_{pv}(T_{eq},h_0)=\frac{1}{T_{eq}-T_0}\int_{T_0}^{T_{eq}} S_{pv}(T,h_0)\,dT \tag{6}$$

The effect of damping increase is treated by the Dh. Dh represents the average reduction Sd, Spv and Spa for an ensemble of earthquakes over the entire range of vibration periods. It is given as follow [4]:

$$D_h=\sqrt{(1+bh_0)/(1+bh_{eq})} \tag{7}$$

The coefficient b is related to earthquake as shown in Table 3.

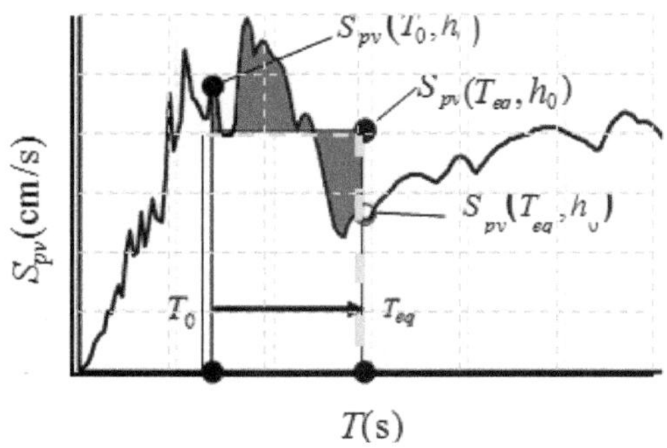


**Figure 15:** Concept of equivalent acceleration response spectrum.

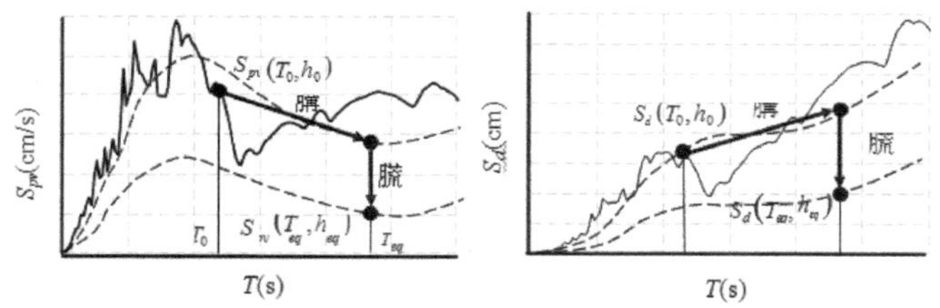


**Figure 16:** Outline of prediction method.

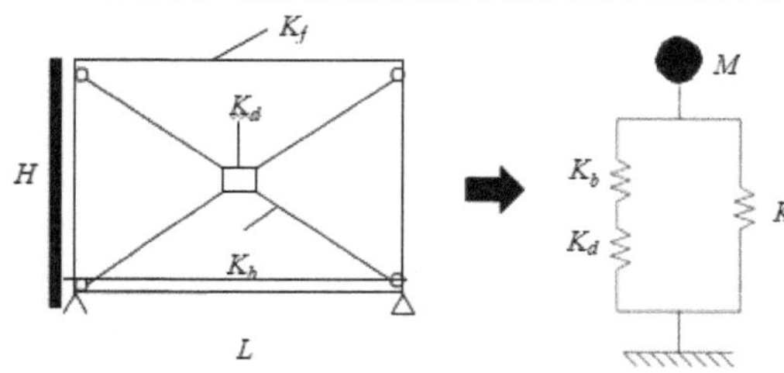


**Figure 17:** Analytical frame model.

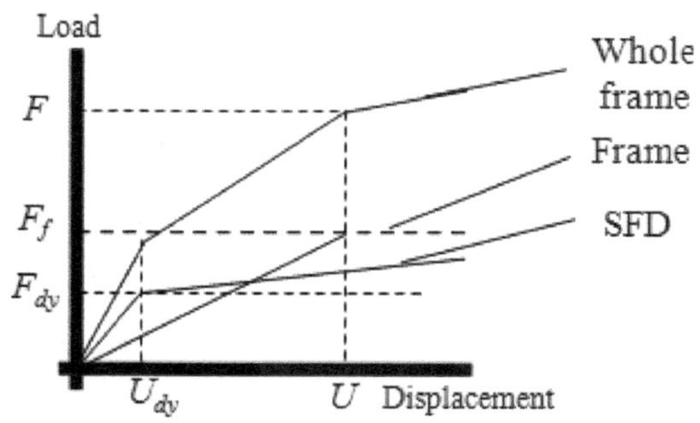


**Figure 18:** The skeleton curve of whole frame.

Moreover, pseudo-displacement response spectrum and pseudo-acceleration response spectrum is as follow:

$$S_d(T_{eq},h_{eq})=\frac{T_{eq}}{2\pi}S_{pv}(T_{eq},h_{eq})=\left(\frac{T_{eq}}{2\pi}\right)^2 S_{pa}(T_{eq},h_{eq}) \tag{8}$$

Equivalent acceleration response spectrum is calculated by smoothing by an integration average of response spectrum and the extension of the period to maximum ductility μ as shown in Figures 15 and 16 (arrow (1)). Velocity response spectrum is predicted using the response reduction rate by the response is reduced by increase of damping by the extension of the period (arrow (2)). Therefore, maximum response displacement and maximum shear force are following as:

$$U=D_h(T_{eq}/2\pi)\bar{S}_{pv}(T_{eq},h_0) \tag{9}$$

$$F=MU(2\pi/T_{eq})^2 \tag{10}$$

## Verification of Accuracy of Prediction method

### Analytical frame model

At first, a single-story model is adopted as fundamental study as shown in Figure 17. The analytical frame model is pinned support at column base, and it is assumed that a column (H-300 × 300 × 10 × 15, SS400) and a beam (H-300 × 150 × 6.5 × 9) behave within elastic, so then SFD show the inelastic behavior. It is assumed that the inertial mass is 9000kg and damping factor is 2% of this frame. Full plastic moment of the column is larger than that of the beam. A natural period of framework is determined from the inertial mass and initial stiffness of the whole framework. The P-δ Effect is not considered herein. In addition, Figure 18 shows the skeleton curve of whole frame.

### Seismic response analysis

The time history response analysis is performed to investigate the applicability and effectiveness. The seismic response analysis is performed subjected to 12 waves which include the past records during earthquake disaster and artificial seismic waves. The damping effect coefficients Dh are summarized on Table 3. Ratio of elastic stiffness r is considered with parameters as 1, 4 and 9. Yield deformation ration y is considered with parameters as 3, 4, 5 and 6.

## The Comparison of Analysis, Results and Predictions

Figure 19 shows the comparison of Prediction values by use prediction method and results of the time history response analysis. Figure 19 shows the comparison of maximum response displacement and maximum shear force. Furthermore, the mean value and the standard deviation of result of comparison are shown the same figure. From the comparison of analysis results and predictions, it is confirmed that the proposed prediction method shows enough accuracy. Moreover, it is confirmed that damping evaluation show approximate accuracy when the hysteresis behavior model is assumed with the strain hardening model.

## Conclusions

1) The prediction method is reformulated to adapt the bilinear model and strain hardening rule.

2) Prediction method of maximum seismic response based on equivalent linearization technique shows enough accuracy.

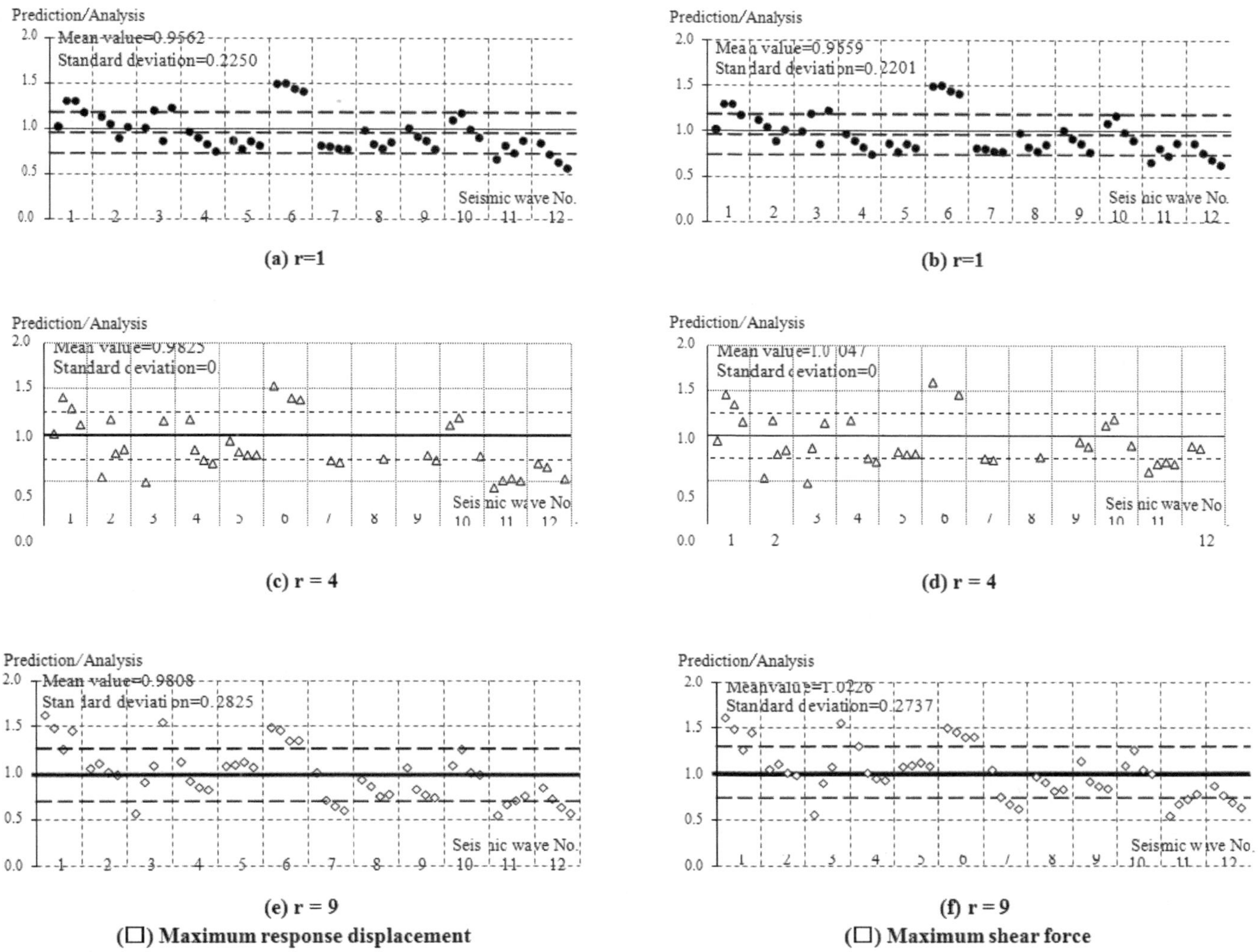


**Figure 19:** Comparison of analysis results and predictions.

## References


1. Tani S, Shimada S, Ito T, Wu D, Nagumo T, et al. (2014) An experimental study on inelastic behavior and restoring force characteristics of vibration control device as steel scaling-frame (ʊ), The 5th Asian Conference on Earthquake Engineering.
2. Azu H, Shimada S, Kurokawa K, Ito T, Wu D, et al. (2015) A study on inelastic behavior and resistant mechanism of steel framed structure with vibration control devices called as scaling-frame (ʊ), The 5th Structural Engineers World Congress.
3. Newmark NM, Rosenblueth E (1971) Fundamentals of earthquake engineering. Prentice-Hall Inc.
4. Kasai K, Ito H, Watanabe A (2003) Peak response prediction rule for a sdof elasto-plastic system based on equivalent linearization technique. J Struct and Construction Eng 571: 53–62.

# Establishment of a Composite Safety Index for Pavement Management

**Stephen A Arhin* and Asteway Ribbiso**
*Department of Civil and Environmental Engineering, Howard University, Washington, DC, 20059, USA*

**Abstract**

One of the goals of local transportation agencies is to improve the quality of life for citizens and visitors by ensuring the efficient and safe movement of people and goods through the roadway system. Maintenance and rehabilitation of pavements are necessary to ensure that roadway networks continue to perform at their optimum. Currently, maintenance and rehabilitation of roadway networks depend on several factors including pavement condition indices, funding availability, among others. Previous studies have established relationships between crash frequency and pavement condition indices. However, the combined influence of speed, volume, and crash frequency on pavement indices, and thereby pavement management efforts has not been thoroughly examined. In this paper, a multinomial logistic regression was employed for 193 arterial segments to establish a new categorical variable: Composite Safety Index (CSI). The CSI values or ratings were based on pavement indices, crash frequency, traffic volumes and vehicular speeds to help categorize pavement sections for either maintenance or rehabilitation. The results indicated that the selected independent variables were statistically reliable in ranking pavement sections for rehabilitation or maintenance based on their CSI values.

**Keywords:** Composite safety index; Pavement management; Traffic safety data; Traffic operations

## Introduction

A vibrant transportation infrastructure is essential for sustaining the social and economic growth of a city. When roadway networks are efficient, they provide economic and social opportunities and benefits which result in better access to the markets, and employment for citizens. Surface transportation infrastructure in the United States of America (USA) is aging according to a report prepared by Department of Homeland Security in 2014 [1]. The USA was ranked 16th in the overall quality of roads in a report developed by World Economic Forum despite a 3rd place ranking in an overall competitiveness of the economy among the rest of the world [2]. The American Society of Civil Engineers (ASCE) indicated in another study that it would take up to $225 billion to upgrade the United States' aging transportation infrastructure to meet current demands and future growth. There are two significant concerns regarding the aging transportation infrastructure: the failing existing conditions, and how to prioritize and allocate funding for their Maintenance and Rehabilitation (M&R). Hence, the need for the United States Department of Transportation (USDOT) and local jurisdictions to develop programs and efficient tools to improve and rehabilitate roadway systems. These programs and tools can help local agencies to allocate fund to enhance and preserve transportation networks appropriately.

Annual pavement condition assessments are conducted by transportation agencies to determine the physical states of roadway networks. Maintenance and rehabilitation (M&R) of road segments are scheduled based on certain factors and funding availability in most jurisdictions. The notable factors include International Roughness Index (IRI) and Pavement Condition Index (PCI). However, most of these M&R programs do not include safety variables and other traffic operations measurements in the decision-making processes. The goal of this research is to incorporate safety and operational characteristics in the prioritizing of M&R of roadway networks.

Safety on roadway networks is an essential component for both economic success and quality of life for all stakeholders. Maintenance and rehabilitation is necessary to ensure that roadway networks may operate safely and efficiently. Even though the primary objective of local transportation agencies is to improve the roadway system within their jurisdictions, there are limited studies that focused on investigating the impact of safety and traffic operational issues on pavement conditions and the utilization of additional variables to improve pavement M&R strategies. Currently, several states only consider pavement condition indices such as IRI and PCI, along with budget constraint as the primary factors for selecting roadway segments for M&R. Even though this is a methodological way of selecting road segments for M&R, it does not consider the safety component and traffic operational factors of the roadway segment selected.

## Objectives

The goal of this research is to develop a safety-incorporated pavement logistic regression model that would aid engineers and local agencies in prioritizing and selecting roadways for M&R. This is achieved by developing a composite safety index (CSI) based on traffic and safety data, and pavement condition by employing multinomial logistic regression (MLNR) analysis. This evaluates a segment based on its CSI value on the basis of which the road segment may be selected for M&R.

The objectives of this research are:

1. To define a composite safety index based on crash frequency, crash rate, fatalities, injuries, and PDOs on each segment.

2. Use a multinomial logistic regression to predict the impact(s) of crash frequency, *IRI, PCI*, average speed, and AADT on *CSI.*

Ultimately, this research is aimed at developing a CSI classification model that can be used to aid in the selection of roadway segments for M&R while taking critical safety and traffic operational measures into consideration.

***Corresponding author:** Stephen A. Arhin, Assistant Professor, Department of Civil and Environmental Engineering, Howard University, Washington, DC, 20059, USA, E-mail: saarhin@howard.edu

## Literature Review

The Asset Management Primer, published in 1999 by the Federal Highway Administration (FHWA), establishes the process and decision-making framework which utilizes engineering and economic principles as a base to measure asset performance [3]. The Primer was developed to improve the allocation of funds and plan repairs as well as rehabilitation of pavements. Pavement management is different from all other infrastructure maintenance. A Pavement Management System (PMS) is a tool that is used to collect and monitor the physical state of pavement sections, and forecast future conditions. This system only focuses on the distress and irregularities on the roadway systems.

Pavement distress can be expressed in terms of surface irregularities that can potentially affect the ride quality and safety of road users [4]. Many models have been established to predict the effect that pavement distress has on ride quality, but the potential of crash occurrence is only limited to road characteristics, environmental factors, and traffic conditions. Several factors can influence crashes on roadway networks such as pavement distress or roughness, speed, and volume. Pavement roughness is typically measured by automation and is quantified by an International Roughness Index (IRI). The National Cooperative Highway Research Program (NCHRP) Report 228 (Transportation Record Board) first defined IRI in the late 70's. IRI is commonly obtained from the longitudinal profile of a roadway segment measured in/mi or m/km [5]. It has become an accepted measure of smoothness/roughness since its introduction in 1986 [6]. Data obtained from the profile of a road is used in computer algorithms to calculate the pavement smoothness or roughness; the resultant value is called the *IRI*. *IRI* values range from 50 to 1,000. Lower scores are an indication of good quality roads (i.e., smooth), and values greater than 170 indicate lower quality (i.e., rough) road segments [4]. Figure 1 presents an image of a typical survey van used for *IRI* data collection.

In addition, distress photographs obtained from roadway profiles are reviewed and scored by trained pavement engineers and technicians for classification by severity type. The resulting score is known as the Pavement Condition Index (PCI). The United States Army Corps of Engineers [7] created PCI based on visual survey review of the number and types of distresses in the pavement segment. The outcome of the evaluation, PCI, is a value between 0 and 100, with 100 representing pavements with no observed distress and 0 representing pavement with extensive distresses.

**Figure 1:** Survey van used for IRI data Collection Florida DOT.

Speed is an important variable used for design and as a performance measure for the evaluation of roadway networks [8]. In an article published by the World Health Organization (2004), speed was identified as a primary risk factor in traffic crashes influencing both potential of accident and severity of injuries. Excessive speeding is defined as speed exceeding the 85th percentile speed of the segment. Each jurisdiction is responsible for setting speed limits based on roadway characteristics and geometry. Speed has been a major factor contributing to roadway crashes [9]. According to the World Health Organization's road safety facts, an increase in average operating speed of 1 km/hr typically results in a 3% higher risk of injury crashes, and a 4 - 5% increase in fatalities on all roadways. Speeding characteristics of a roadway network may be impacted by pavement conditions.

Traffic volumes could potentially contribute to the congestion and crashes on road segments. Average annual daily traffic of a roadway segment is defined as the average number of vehicles that pass a specific point on the road for 365 days expressed in vehicles per day (vpd). Average annual daily traffic is also a variable used in the classification of a roadway segments. Furthermore, engineers typically examine crash rates when comparing roadways with different traffic volumes and crashes using the following expression:

$$Crashrate(R) = \frac{100,000,000 * C}{365 * N * V * L}$$

where C is number of crashes, N is number of years, V is volume of segment, and L is length of the roadway (Federal Highway Administration 2015).

Shan et al. [10] conducted a study to explore the possibility of using a predictive model to improve pavement management in Indiana. The study, conducted in 2001, used average annual daily traffic, age, and IRI in a simple regression to evaluate the performance of arterials and collectors road segments over time. IRI was used as the dependent variable, with age of pavement and average annual daily traffic representing the independent variables. Roadway segments were randomly selected. The analysis resulted in a statistically significant prediction model. Three of the six regression models had a $R^2$ values higher than 0.50 which were recommended for use to the Indiana Department of Transportation (IDOT). The models are as follows:

$$\text{IRI} = 43+1.8 \times \text{AGE}+0.0004 \times \text{AADT},\ R^2 = 0.70$$

$$\text{IRI} = 65+1.9 \times \text{AGE}+0.0003 \times \text{AADT},\ R^2 = 0.50$$

$$\text{IRI} = 65+8.1 \times \text{AGE}+0.0009 \times \text{AADT},\ R^2 = 0.90$$

In another study, Zeng et al. [11] explored the influence of pavement condition on safety to prioritize highway maintenance. Crash and pavement condition indices (IRI, PCI, and rut depth) data were collected from Virginia Department of Transportation (VDOT) for 2007 through 2011. The study evaluated the performance of good pavement conditions versus deficient pavement conditions on rural, two-lane undivided highways in Virginia. The research employed the empirical Bayes method to explore the relationship between pavement conditions and crash frequency and crash severity. The results indicated that good pavement segments could reduce fatal and injury crashes by 26% compared to deficient pavements. According to the author, improvement of pavement conditions from deficient to good, offers a significant safety benefit in reducing crash severity, but not crash frequency. The resulting model was significant with an $R^2$ value of 0.75.

Currently, for the District Department of Transportation (DDOT) in the District of Columbia, M&R of pavements are typically scoped during the winter season for implementation during spring and summer. In addition to budget constraints, DDOT considers pavement PCI data for its M&R programs. IRI data is submitted to FHWA, but is not utilized for pavement M&R. DDOT uses a software program called PAVER to manage its pavement M&R projects. After field evaluation of roadway segments, the resulting PCI is used as an input variable in PAVER which prioritizes the selection of roadway segments based on acceptable PCI value and/or financial constraints. This allows the agency to allocate funds for the most critical roadway segments without taking into consideration traffic operations or safety characteristics of the segments.

According to crash statistics for urban cities in the United States in 2013, major arterials experience the most crashes, followed by local roads as shown in Figure 2.

Other factors including traffic volume and speed may potentially affect the reliability and safety of a roadway segment, which may ultimately influence the need for maintenance and rehabilitation. Previous studies have established relationships between crash frequency and pavement conditions. However, the combined effect of pavement conditions, speed, volume, and crash frequency on the deterioration of pavement has not been examined thoroughly. This paper is aimed at developing a composite safety index that can be used to make decisions in selecting of roadway segments, while taking safety into consideration.

## Research Methodology

In this research, a composite safety index ($CSI_1$, $CSI_2$, and $CSI_3$) was established using the following variables: crash rate (*CR*), fatalities (*F*), property damage only (*PDOs*) and injuries (*I*). The index ranged from one (1) to three (3), with three (3) representing a poor safety condition, and one (1) representing an excellent safety condition. These four crash characteristics were combined to establish the composite safety index using the following equation:

$$CSI = 0.25*CR + 0.30*I + 0.35*F + 0.10*PDOs$$

where *CR* is crash rate, *I* is injuries, *F* is fatalities, and PDOs is property damage only. The weights were computed using average percentage cost of impact of each variable. Each roadway segment information was entered in the equation to determine the crash composite safety index. The four factors were weighted using values of 0.25 for crash rate, 0.30 for injuries, 0.35 for fatalities, and 0.10 for property damage only, as shown in the equation. The segments were then sorted in descending order based on the CSI values. The composite safety indices were classified as either 1, 2, or 3 based cumulative curve. The first 50% of the segments were assigned to $CSI_1$, 51% to 75% to $CSI_2$, and those greater than 75% were assigned $CSI_3$.

Five independent variables were identified for the regression analysis based on data availability and the pertinent variables from the literature review. Table 1 presents the variables and their descriptions.

International Roughness Index (IRI), Pavement Condition Index (PCI), and the average speed data for 2014 were provided by the District Department of Transportation (DDOT). The International Roughness Index (IRI) and Pavement Condition Index (PCI) data sets included various types of pavement information such as pavement type, segment length, and functional classification. Crash data for the District of Columbia was accessed through DDOT's Traffic Accident Reporting and Analysis System (TARAS). AADT data was obtained from the Metropolitan Washington Council of Government, which oversees regional transportation data, and from DDOT. The most recent AADT data available was for 2013. The data was projected to 2014 using a 5% growth rate. The growth factor was based on a three-year moving average of AADT data from 2006 to 2013.

Multinomial Logistic Regression (MNLR) is typically used to model relationships between dichotomous dependent or response variable,

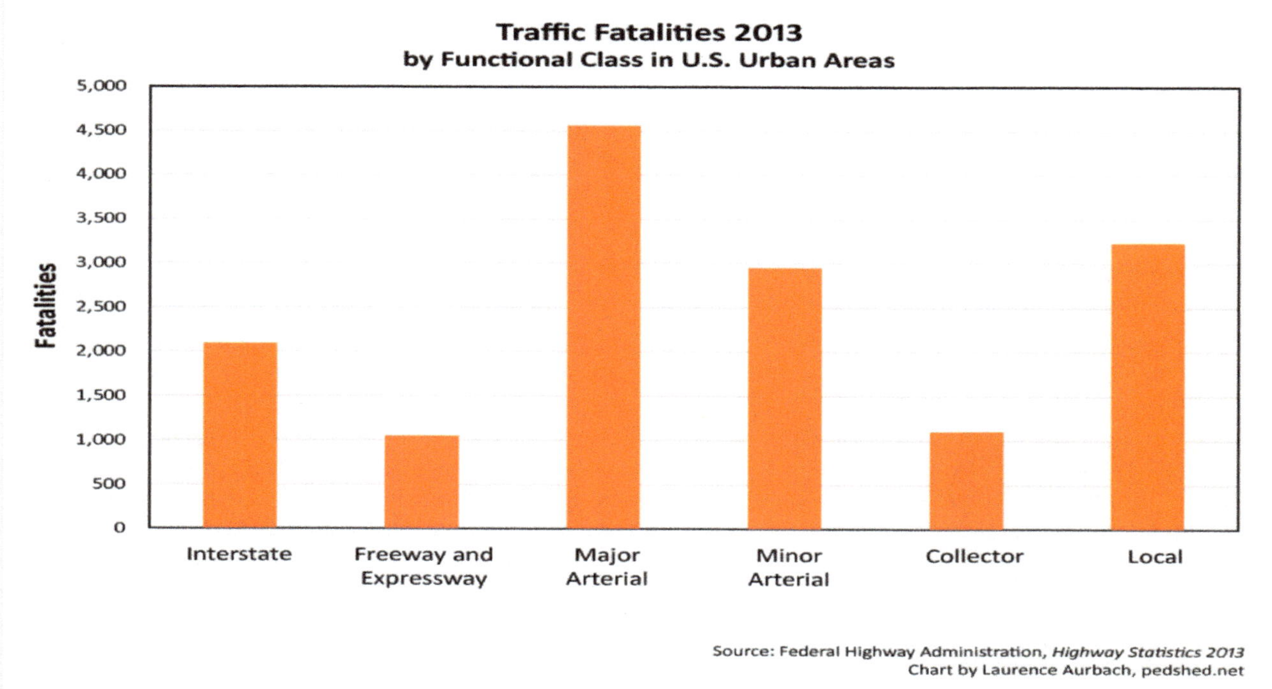


**Figure 2:** Fatal crashes based on functional classification [3].

| Independent Variables | Variables Description | Value of Measurement |
|---|---|---|
| International Roughness Index (IRI) | An index for condition of pavement | Index (in/mi) |
| Pavement Condition Index (PCI) | An index which evaluates distress on pavement surface | Index value 0 to 100 |
| Crash Frequency (CF) | Number of crashes from 2012-2014 | Number of Crashes |
| Average Vehicular Speed (AVS) | Average segment speed | Average vehicular speed (mph) |
| Average Annual Daily Traffic (AADT) | Annual average daily traffic | Number of Vehicles |

**Table 1:** Composite safety index rating.

and a set of independent variables [12]. Multinomial logistic regression is applicable since the dependent variable is categorical and not in any order. For multinomial logistic regression analysis, it is required that a comparison between two categories be made with one category used as a reference. The reference category for this analysis was $CSI_3$, which represents a poor safety condition, was compared to $CSI_1$ and $CSI_2$. The general equation is as follows:

$$\ln\left(\frac{P}{1-P}\right) = \beta_0 + \beta_1 x_1 + \beta_2 x_2 + \dots + \beta_p x_p$$

In this analysis, the dependent variables have three levels: $CSI_1$, $CSI_2$, and $CSI_3$. The general equation can be rearranged to express dependent variable with three levels. As such, the dependent variable is divided into j categories and one category was be selected as the reference. In this case, the probability equation can be denoted as:

$$P_j(Y_j) = \frac{\exp(\beta_0 + \beta_{1j} X_{1i} + \dots + \beta_{pj} X_p)}{1 + \sum_{j=1}^{k-1} \exp(\beta_0 + \beta_{1j} X_{1i} + \dots + \beta_{pj} X_p)}$$

where j = 1, 2, …, k-1 and i = 1, 2, …, n. The regression analysis compared $CSI_1$, relative to $CSI_3$ as well as $CSI_2$ relative to $CSI_3$. The logistic equation for each comparison group of categories can denote by these two logistic models:

$$\ln\left(\frac{P(Y=1)}{P(Y=3)}\right) = \beta_0 + \beta_1 x_1 + \beta_2 x_2 + \dots + \beta_p x_p$$

$$\ln\left(\frac{P(Y=2)}{P(Y=3)}\right) = \beta_0 + \beta_1 x_1 + \beta_2 x_2 + \dots + \beta_p x_p$$

where P denotes the number of dependent variable for the binary response Y. β is the coefficient of the equations and X represents the value of each independent variable associated with each coefficient of the equation.

For most regression analysis, it is assumed that the observed data are independent and unbiased. The dependent variable in a MLNR is assumed to follow a multinomial distribution. The probabilities are linked to the independent variable by a logit link function which is assumed to be linear. The logistic regression analyzed 193 segments in the District.

This regression employed the maximum likelihood and the likelihood ratio tests to evaluate the "goodness of fit." For MLNR, the overall fit is tested using the difference between the constant and final model. The constant model uses the marginal probabilities to predict categories of the dependent variable, while the final model incorporates the influence of all the independent variables to predict the resulting categories. The chi-square test was used to determine the resulting model's significance and is defined as the difference between the final and constant models. The difference between these two values is referred to as chi-square (χ2) for goodness of fit and is computed by:

$$X^2 = -2LL_{Constant} - 2LL_{Final}$$

The regression output from SPSS shows how well the multinomial regression predicts each CSI category. The "enter" method in SPSS was used to evaluate all the independent variables in the regression analysis and included all the variables at the same time.

In linear regression analysis, the $R^2$ explains the proportion of variance that can be accounted for by the predictor variable. A high $R^2$ indicates that the variation in the model and ranges between 0 to 1. However, for MLNR, computing $R^2$ is not the same as that for a linear regression. The output of the multinomial regression provides a pseudo $R^2$ values as an approximation of the regression. One of the pseudo $R^2$ values in a multinomial regression is called the Nagelkerke $R^2$. The Nagelkerke $R^2$ is a modified Cox and Snell $R^2$. It ranges between 0 to 1. A high Nagelkerke $R^2$ value indicates a strong relationship in the logistic regression model.

## Results

In the first step of the analysis, pearson correlation analyses between the dependent and each independent variable were performed to identify high relationships between two different groups of data sets. Due to multicollinearity, only four of the five variables were used to predict CSI catego ries. and each of the independent variables.

A MLNR analysis was conducted to predict CSI categories for 193 arterial segments using IRI, AVS, CF, and AADT as the independent variables. A test of the final model against a constant only model was statistically significant, indicating that the independent variables as a set are reliably distinguished between CSI categories ($\chi^2_{(8,193)} = 113.42$, p < .000 with df = 8) as presented in Table 2.

Nagelkerke's $R^2$ of 0.537 indicates a moderately good relationship between the prediction and classification of roadway segments. The overall prediction rate was approximately 75%. The Wald criterion confirmed that only CF, IRI, and AADT were significant in contributing to the model while the intercept and AVS were not. The independent variables, AADT and CF, were statistically significant with $\chi^2_{(2,193)} = 34.7$, p = 0.000, and $\chi^2_{(2,193)} = 17.38$, p = 0.000, respectively when comparing $CSI_1$ to $CSI_3$. IRI was statistically significant with $\chi^2_{(2,193)} = 4.007$, p = 0.000 when comparing $CSI_2$ and $CSI_3$. However, AVS was determined not to be statistically significant as presented on Table 3.

The odds ratio is less than 1 for CF when comparing $CSI_1$ to $CSI_{3,}$ which means that the outcome is less likely to be in the $CSI_1$ category when the CF values are high. On the other hand, when comparing $CSI_2$ to $CSI_{3,}$ only IRI has the most statistically significance influence on the model ($p<0.05$). Thus, when IRI values are high, the category will more likely be in $CSI_{3.}$

### The refence category is $CSI_3$

Figure 3 presents a comparison of the actual CSI category versus the predicted CSI category from the resulting MLNR analyzed in the study.

From the graph, it can be concluded that the predicted CSI

| Model | Model Fitting Criteria | Likelihood Ratio Tests | | |
|---|---|---|---|---|
| | -2 Log Likelihood | Chi-Square | df | Sig. |
| Intercept Only | 339.375 | | | |
| Final | 225.934 | 113.441 | 8 | .00 |

**Table 2:** Regression "goodness of fit".

categories are closer to the actual. The CSI model was able to predict categories 1, 2, and 3 at respective rates of 95%, 27%, and 47%, resulting in an overall prediction rate of approximately 75%.

## Results and Discussion

A composite safety index (CSI) with a range from 1 to 3 was established using crash rate, fatalities, injuries, and property damage only. This paper used the CSI as the dependent variable with independent variables such as crash frequency (CF), international roughness index (IRI), average vehicular speed (AVS), and average annual daily traffic (AADT) in a multinomial logistic regression (MLNR) at 5% significance level.

The resulting model used arterial segments in a dense urban environment. The final sample size for the regression was 193. The analysis was performed using SPSS to predict CSI using four independent variables. Multinomial logistic regression analysis yielded statistically significant results. From the MLNR analysis, CR, IRI, AVS, and AADT were used as independent variables for the model. Due to the presence of multicollinearity, PCI was excluded from this analysis. The logistic regression model yielded an $R^2$ value of 0.537 and a statistically significant $\chi^2_{(8,193)} = 113.42$, $p < .000$ with df = 8 at 5% level of significance.

The Nagelkerke's $R^2$ of 0.537 (53.7%) represents a moderate relationship between prediction and grouping, indicating that a good prediction of each sample into the appropriate CSI category. The prediction success for the overall model was 74.6%. Thus, the CSI categories can help identify segments with different safety attributes for M&R considerations, in addition to funding or budget constraints.

| CSI | | B | Std. Error | Wald | Sig. |
|---|---|---|---|---|---|
| 1* | Intercept | 1.914 | 2.457 | 0.607 | 0.436 |
| | CF | -0.269 | 0.046 | 34.714 | 0.000 |
| | AADT | 0.000 | 0.000 | 17.384 | 0.000 |
| | IRI | -0.004 | 0.002 | 3.813 | 0.051 |
| | AVS | 0.066 | 0.093 | 0.494 | 0.482 |
| 2* | Intercept | 3.174 | 2.447 | 1.683 | 0.195 |
| | CF | -0.041 | 0.033 | 1.533 | 0.216 |
| | AADT | 0.000 | 0.000 | 2.263 | 0.133 |
| | IRI | -0.005 | 0.002 | 4.007 | 0.045 |
| | AVS | -0.066 | 0.092 | 0.518 | 0.472 |

**Table 3:** Results of regression analysis.

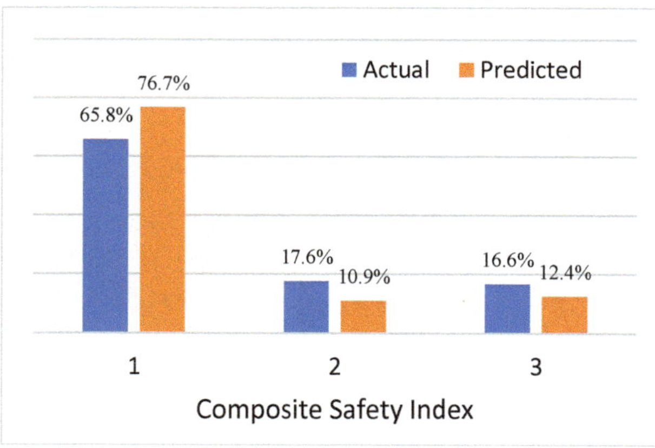


**Figure 3:** Actual vs. Predicted CSI.

## Conclusion

Over the past few years, there has been a significant effort for state agencies to develop to new methods to manage and improve their roadway network. Funding from the Federal government can be effectively applied for the implementation of pavement management programs based on data-driven models. The concept of adding safety and traffic operational factors to pavement condition indices and budget constraints may potentially improve the evaluation, selection, and prioritization process for roadway maintenance and/or rehabilitation. This research evaluated the potential of including four variables in a statistically significant logistic regression model in addition to pavement indices for pavement M&R for arterial roadway segments. The models were developed using data for an urban area (Washington DC) therefore they are not necessarily recommended for use in other jurisdictions.

Future research needs would include the following:

(i) Validate the model using the list of arterial roadway segments selected for maintenance or rehabilitation in 2014.

(ii) Develop and validate similar models for other roadway functional classifications for the District of Columbia.

### References

1. Department of Homeland Security (2010) Infrastructure: Issues, research, and technology. Buildings and Infrastructure Protection Series BIPS 01.
2. Schwab K (2015) The Global Competitiveness Report 2015–2016, World Economic Forum, Geneva.
3. Federal Highway Administration Office of Asset Management (1999) Asset Management Primer.
4. Arhin S, Williams L, Ribbiso A, Anderson M (2015) Predicting pavement condition index using in a dense urban area. Journal of Civil Engineering Research. Volume 5-1:10-17.
5. Sayers MW, Karamihas SM (1998) Little book of profiling. University of Michigan Transportation Research Institute.
6. Sayers MW, Gillespie T, Paterson W (1986) Guidelines for the conduct and calibration of road roughness measurements. World Bank Technical Paper No. 46.
7. ASTM International (2007) Standard practice for roads and parking lots pavement condition index surveys. ASTM International.
8. Fitzpatrick K, Carlson P, Brewer M, Wooldridge M (2003) Design speed, operating speed, and posted speed practices. Report submitted to Washington, DC. NCHRP 504.
9. Peden, M, Richard S, David S (2004) World Report on Road Traffic Injury Prevention. World Health Organization Geneva.
10. Gulen S, Zhu K, Weaver J, Shan J, Flora W (2001) Development of improved pavement performance prediction models for the Indiana Pavement Management System. Publication FHWA/IN/JTRP-2001,1:69.
11. Zeng H, Smith BL (2014) Estimation of the safety effect of pavement condition on rural, two-lane highways transportation research record: Journal of the Transportation Research Board, No. 2435: 45–52.
12. Kutner MH, Nachtsheim C, Neter J, Li W (2005) Applied linear statistical model. (5th edn). New York, The McGraw-Hill Companies, Inc. NY.

# High Modulus Asphalt Concrete: a Long Life Asphalt Pavement

**Maria Inmaculada García Hernández***

*Uppsala University, Ångströmlab, Lägerhyddsvägen, Uppsala, Sweden*

**Abstract**

This study uses a layered elastic analysis and a computational model to evaluate the fatigue life of the base layer stabilized with conventional bitumen 60/70 compared with a base layer stabilized with high modulus bitumen 13/22 for a lifetime of 20 years. This study makes a comparative between conventional bitumen and high modulus asphalt concrete in terms of fatigue or cycles to failure of both asphalts at 15°C. The result of this study shows that HMAC 13/22 has much longer fatigue life compared with conventional bitumen 60/70, in some cases even four times longer; the lifetime of the pavement increases considerably with the use of HMAC in the base layer of pavement according to mechanical calculations. HMAC raise elastic modulus of middle layer. HMAC can also improve stress state of pavement structure, reduce shear strain and asphalt pavement rutting.

**Keywords:** Airports; Fatigue; High modulus asphalt concrete; Lifetime

## Introduction

High Modulus Asphalt Concrete (HMAC) originated in France in the early 1980s. HMAC is a mixture of asphalt concrete fabricated for its use in the base course of asphalt pavements. HMAC has closed structure with comparatively large content of bitumen. It has been shown [1] that hard bitumen increases the mixtures resistance to rutting. However large content of bitumen assure workability, fatigue durability and water resistance.

One study [2] showed that HMAC has higher rutting resistance and higher complex modulus compared with styrene-butadiene-styrene (SBS) modified binders (2.7–3.0 times higher) and even higher than conventional bitumen. It has also proved that [3] rutting resistance of HMAC is higher than that of the SBS modified asphalt mixtures or conventional mixtures. The thickness of the base course stabilized with HMAC is 1/4–1/3 times lower than the traditional base course resulting significant saving of asphalt binder and aggregates [4].

With these mentioned properties, the concept of a long life pavement came up initially from the HMAC [5] and is this concept the main idea of this study. The definition of the long life pavement is a pavement that can last at least more than 40 years without major structural strengthening. The basic idea of a long life pavement is to increase the layer moduli and/or thickness of asphalt base layer, and then reduce the potential of structural distresses by minimizing the tensile strain at the bottom of the AC layer and the compressive strain on top of the subgrade. In this way, one can confine all the major pavement distresses within the surface layer. Only periodical rehabilitation for the surface layer might be needed for the entire design life.

## Laboratory Test

### Materials and sample fabrication

In this study two types of asphalt binders were under investigation, the conventional bitumen 60/70 and the HMAC 13/22 (Table 1). The bitumen content of the mixture in the conventional bitumen sample was 5, 5% and the void air 4, 5%. The optimum asphalt content in the samples of HMAC at the 5, 1% of air void were 4, 5%. The compaction and estimate mixing was done in the test following the Marshall test and the Swedish normative in compaction for bitumen samples. The mixing temperature for conventional bitumen was 165°C and 185°C for HMAC. The test samples used in the laboratory to obtain the dynamic modulus of the mixtures had geometry of 150 mm of diameter and a height of 50 mm for both conventional bitumen and HMAC. The

| | B13/22 | | B60/70 | |
|---|---|---|---|---|
| | Min | Max | Min | Max |
| Penetration at 25°C, 0.1 mm | 13 | 22 | 150 | 200 |
| Softening point °C | 60 | 72 | 40 | 47 |
| Ductility at 25°C | 10 | - | 180 | - |
| Fraas temperature, °C | - | +1 | - | -16 |

**Table 1:** Main properties of the binders used in this research.

| Sieve (mm) | 25 | 20 | 12,5 | 8 | 4 | 2 | 0,5 | 0,125 | 0,25 | 0,063 |
|---|---|---|---|---|---|---|---|---|---|---|
| % Passing | 100-100 | 75-95 | 55-75 | 40-60 | 25-42 | 18-32 | 7-18 | 4-12 | 3-8 | 2-5 |

**Table 2:** Grading curve for conventional bitumen 60/70.

| Sieve (mm) | 25 | 20 | 12,5 | 10 | 5 | 2,5 | 0,63 | 0,32 | 0,16 | 0,08 |
|---|---|---|---|---|---|---|---|---|---|---|
| % Passing | 100-100 | 80-95 | 65-80 | 60-75 | 43-58 | 30-45 | 15-25 | 10-18 | 8-14 | 7-10 |

**Table 3:** Grading curve for high modulus asphalt concrete 13/22.

samples were compacted using a Marshall compactor. The grading curve of the asphalt mixes that were used in this research are the following:

For both, HMAC and conventional bitumen are these grading curves the usual ones that are used in Sweden for the base course in drainage curves. All the mixes has been compacted to the same level leading to close void contents (from 3,8 to 5,0%) for all the test samples (Tables 2 and 3).

### Dynamic modulus test

The linear visco-elastic properties of the asphalt mixes samples of this research were measured from the dynamic modulus test in the asphalt lab of a geotechnical company, GEOCISA. All the tests were done at local ambient temperature, about 15°C, for conventional bitumen B60/70 and for the HMAC. Every test was performed at three different

***Corresponding author:** Maria Inmaculada García Hernández, Uppsala University, Ångströmlab, Lägerhyddsvägen 1, SE-752 37, Uppsala, Sweden
E-mail: Maria.Espersson@angstrom.uu.se

frequencies (10, 0, 5 and 0, 1 Hz). The master curve was represented by obtaining the dynamic modulus data at different frequencies. The explanation to the much higher dynamic young modulus of HMAC at high temperatures can be due to the high boiling point binder what increases the stiffness of the HMAC and in this way increases the linear visco-elastic properties of the asphalt mixtures.

### Test results

The values that were obtained in the lab with the dynamic press can be seen in the Table 4.

## Methodology of the Study

### Fatigue test with a mechanical-computational model

The aim of this research work is to evaluate the fatigue in the base layer of the pavement depending on the CBR, annual departures and if the base layer is stabilized with conventional bitumen or with high modulus asphalt concrete [6-10]. For fatigue calculation this study uses a mechanical-computational model in order to carry out an iterative process that gives the following steps: calculation of stress at critical points in the pavement during the experimental time, calculation of the number of cycles to failure (Nf) during the experimental time, calculation of the damage factor (n/Nf) during the experimental time, sum of the factors of fatigue damage produced in the pavement, increasing or decreasing the thickness of the layer if D, total damage to the pavement, is not close to 1 and as final step determine the final calculation of the cross section. To perform all these calculations have been used the EVERSTRESS program. The calculations give the number of times (n) that the aircraft design, Airbus 380, can use the runway until the failure of the pavement [11]. The following diagram is a schema of the mechanical-computational model that has been used in this study (Figure 1).

| | Conventional Bitumen 60/70 | | High Modulus Bitumen 13/22 | |
|---|---|---|---|---|
| **Tested Dynamic Modulus (MPa)**<br>**Test at local ambient temperature** | 6021,13 | Mean Value: 6.172 | 13990,5 | Mean Value: 13.497 |
| | 6324,35 | | 14006,6 | |
| | --- | | 12495,4 | |

**Table 4:** Tested Dynamic Modulus at local ambient temperature, at 15°C, for 60/70 and 13/22 bitumen.

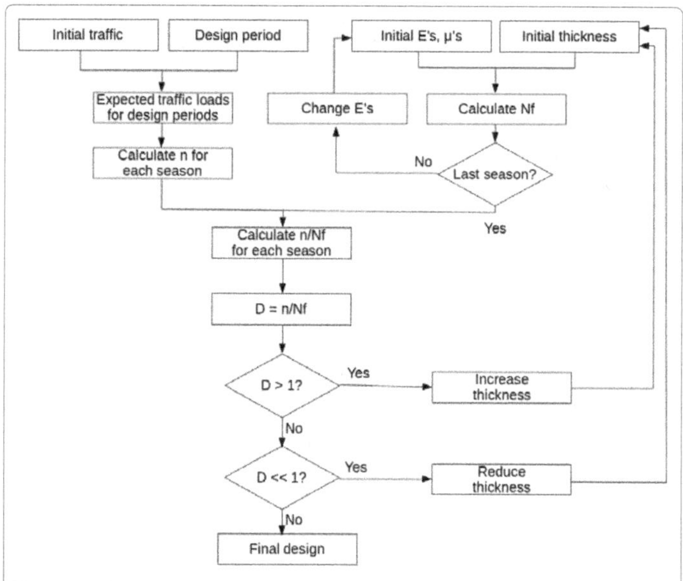


**Figure 1:** Flow Diagram of the Mechanistic-Empirical design procedure of the test.

### Calculation data

In this investigation has been set the thickness of the surface course in 60 mm, and the sub-base course in 250 mm, for calculating later the base course thickness for HMAC and conventional bitumen with the program LEDFAA by considering the dynamic young modulus of the Table 4. Initially the thickness was selected as the minimum thickness allowed according to the type of materials used for the pavement and the operating float. The critical stress in the pavement structure is determined at different positions that can be seen in the above figure and these positions have been used also in the program for the stress analyses. The number of load cycles to fatigue failure (Nf) were computed using failure criteria. In a multilayer model the mechanical characterization of materials is given as a function of the elasticity modulus (E) and Poisson's ratio (υ). Hence the accurate estimation of these parameters is one of the most important data to take into account. The Elasticity Modulus in bituminous mixture is a function of both the temperature (weather conditions) and the total thickness of the mixture. It has been estimated that the temperature gradient affects mainly to the first superior 15 cm pavement. A thermal hypothesis applies to: winter, spring and summer-autumn [12] (Tables 5-7).

### Strains and stresses

To calculate the stresses and strains in different parts of the pavement, the EVERSTRESS program has been used. The calculations were performed for different CBR and different number of annual departures. With this data and the fatigue formula 1 it is possible to calculate the number of passes on the pavement until failure. In that way the next step is calculating the life time of the pavement with both high modulus bitumen and conventional bitumen (Figure 2).

With the EVERSTRESS program was calculated the stresses and strains for CBR 5, 10, 15, 20 and 25.

An analysis was done to test the deformations in a base layer stabilized with conventional bitumen compared with high modulus asphalt concrete. From the result obtained in the test for the N values has been calculated the strains with the following fatigue formula 1.

$$\varepsilon_{xx,}\ \varepsilon_{yy} = 6{,}925.10^{-3} N^{-0{,}27243} \qquad (1)$$

With the test the following data for Nf was obtained, and this were the data used for calculating the strains (Formula 1) (Tables 8-13).

| Aircraft Operation | Total Aircraft Weight (tons) | Tire Inflation Pressure (MPa) | Speed (km/h) |
|---|---|---|---|
| **Take-off** | 560 | 1.33 | 140 |
| **Landing** | 386 | 1.33 | 120 |

**Table 5:** Design aircraft data for A380.

| Layer | Poisson's Ratio | Thickness (cm) | Moduli (MPa) |
|---|---|---|---|
| 1 | 0.33 | 13 | 6000 |
| 2 | 0.33 | 30 | 5000 (60/70); 12000 (13/22) |
| 3 | 0.35 | 25 | 500 |
| 4 | 0.35 | --- | 100 |

**Table 6:** Pavement Layer Data.

| | 1200 Annual Departures | 6000 Annual Departures | 15000 Annual Departures | 25000 Annual Departures |
|---|---|---|---|---|
| CBR 5 | 48,24 cm | 56,90 cm | 67,19 cm | 68,91 cm |
| CBR 10 | 20,7 cm | 24,59 cm | 29,38 cm | 28,70 cm |
| CBR 15 | 13,62 cm | 16,27 cm | 20,17 cm | 16,97 cm |
| CBR 20 | 10,99 cm | 11,91 cm | 16,61 cm | 14,09 cm |
| CBR 25 | --- | --- | 12,42 cm | 10,90 cm |

**Table 7:** Calculated base course thickness depending on annual departures and CBR.

## Analysis of the mechanical characteristics of the conventional bitumen compare with high modulus asphalt concrete

With the obtained data for the dimensions of the layers for the runway, specially the thickness of the base course that has been calculated with the LEEDFAA and after this, the obtained data from the EVERSTRESS for the number of load cycles to fatigue failure (Nf)

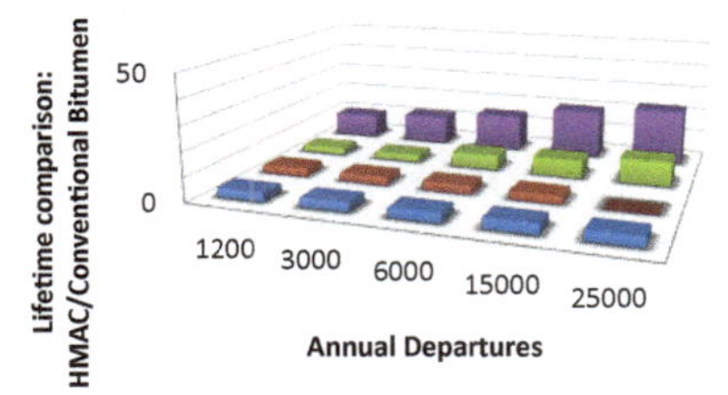


**Figure 2:** Life time comparison High Modulus/conventional bitumen.

| | 1200 | 3000 | 6000 | 15000 | 25000 |
|---|---|---|---|---|---|
| **CBR 5** | --- | --- | --- | --- | --- |
| **CBR 10** | 390 752 | 517 811 | 650 495 | 1 184 312 | 1 089 936 |
| **CBR 15** | 184 496 | 227 208 | 267 529 | --- | 382 770 |
| **CBR 20** | --- | 154 177 | 177 969 | 285 459 | 241 709 |
| **CBR 25** | --- | --- | --- | 215 733 | 162 157 |

**Table 8:** Number of passes in the runway with conventional bitumen until runway failure. (Table: CBR/Annual Departures).

| | 1200 | 3000 | 6000 | 15000 | 25000 |
|---|---|---|---|---|---|
| CBR 5 | --- | --- | --- | --- | --- |
| CBR 10 | 1 512 831 | 2 080 471 | 2 690 888 | 5 265 524 | 4 798 773 |
| CBR 15 | 628 795 | 797 138 | 960 383 | 1 762 058 | 1 442 878 |
| CBR 20 | --- | 507 313 | 597 118 | 1 076 466 | 845 713 |
| CBR 25 | --- | --- | --- | 692 061 | 137 217 |

**Table 9:** Number of passes in the runway with High Modulus Asphalt Concrete until runway failure. (Table: CBR/Annual Departures).

| | 1200 | 3000 | 6000 | 15000 | 25000 |
|---|---|---|---|---|---|
| CBR 5 | --- | --- | --- | --- | --- |
| CBR 10 | 3,78 | 4,02 | 4,14 | 4,45 | 4,40 |
| CBR 15 | 3,41 | 3,51 | 3,59 | --- | 3,77 |
| CBR 20 | --- | 3,29 | 3,36 | 3,77 | 3,50 |
| CBR 25 | --- | --- | --- | 3,21 | 0,85 |

**Table 10:** Comparison of number of passes in the runway until runway failure. High Modulus Asphalt Concrete vs conventional bitumen. (Table: CBR/Annual Departures).

| | 1200 | 3000 | 6000 | 15000 | 25000 |
|---|---|---|---|---|---|
| **CBR 5** | --- | --- | --- | --- | --- |
| **CBR 10** | 163 | 86 | 54 | 39 | 22 |
| **CBR 15** | 77 | 38 | 22 | --- | 8 |
| **CBR 20** | --- | 26 | 15 | 10 | 5 |
| **CBR 25** | --- | --- | --- | 7 | 3 |

**Table 11:** Life time for the pavement with a stabilized base with conventional bitumen. (Table: CBR/Annual Departures).

| | 1200 | 3000 | 6000 | 15000 | 25000 |
|---|---|---|---|---|---|
| **CBR 5** | --- | --- | --- | --- | --- |
| **CBR 10** | 630 | 347 | 224 | 176 | 96 |
| **CBR 15** | 262 | 133 | 80 | --- | 29 |
| **CBR 20** | --- | 85 | 50 | 36 | 17 |
| **CBR 25** | --- | --- | --- | 23 | 10 |

**Table 12:** Life time for the pavement with a stabilized base with High Modulus Asphalt Concrete. (Table: CBR/Annual Departures).

| | 1200 | 3000 | 6000 | 15000 | 25000 |
|---|---|---|---|---|---|
| **CBR 25** | 3,86 | 4,03 | 4,14 | 4,51 | 4,36 |
| **CBR 20** | 3,4 | 3,5 | 3,6 | | 3,6 |
| **CBR 15** | 3 | 3,2 | 6,33 | 7,6 | 10,4 |
| **CBR 10** | 10 | 12 | 15 | 20 | 23 |

**Table 13:** Life time comparison of High Modulus Bitumen vs Conventional Bitumen. (Table: CBR/Annual Departures).

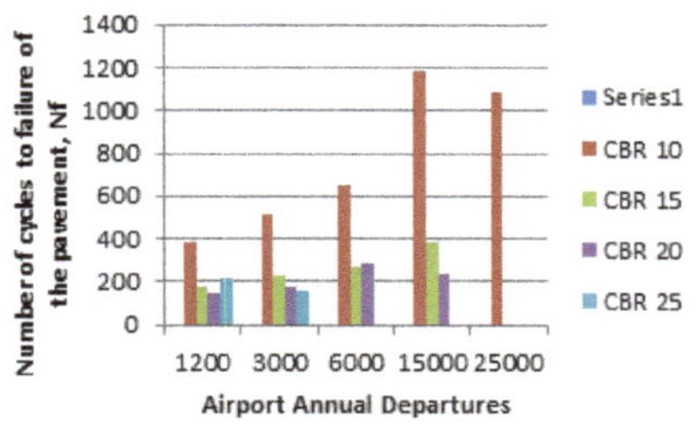


**Figure 3:** Number of cycles to failure for conventional bitumen.

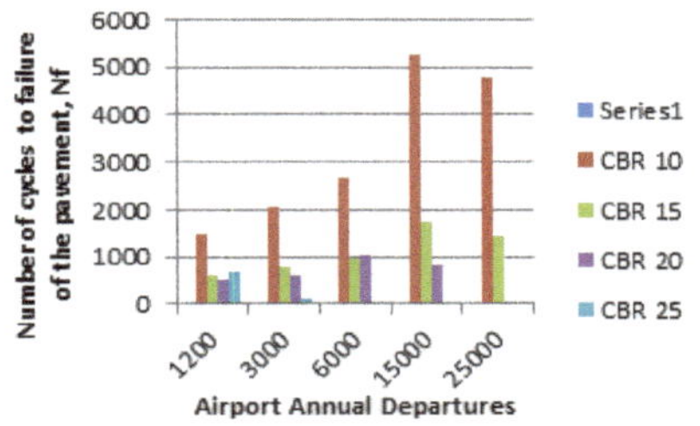


**Figure 4:** Number of cycles to failure for High Modulus bitumen.

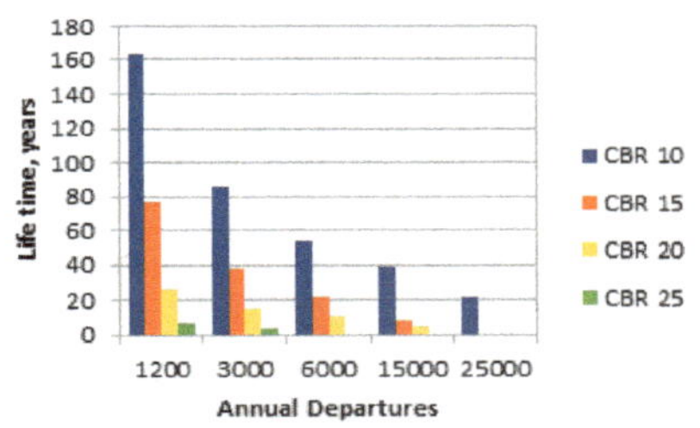


**Figure 5:** Life Time for a pavement with base stabilized with conventional bitumen.

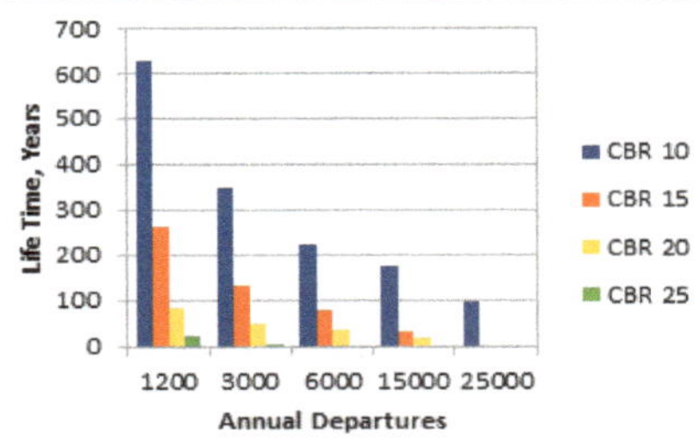


**Figure 6:** Life Time for a pavement with base stabilized with high modulus bitumen.

for conventional bitumen and high modulus bitumen an analysis and comparison in terms of fatigue and long life of both bitumen [13]. This data has been represented in the following graphs and an analysis of the data has been done (Figures 3-6).

## Conclusions

In this research two binders were tested, HMAC 13/22 and conventional bitumen 60/70. The results were compared and the binders were evaluated. The initial data were obtained from laboratory test, the thicknesses of the layers were calculated using LEDFAA, and computational calculations were performed with EVERESTRESS program. The analysis of the results allowed the following statements to be made:

1. This study compares the behaviour of the conventional bitumen 60/70 and the high modulus asphalt concrete 13/22 at a temperature of 15°C on the basis of a computational calculation and two variables: annual departures (1200, 3000, 6000, 15000, 25000) and CBR (10, 15, 20, 25). This study found that the life time of the high modulus asphalt concrete 13/22 is 4 times longer than the conventional bitumen 60/70 for soil CBR =10 and 2 times longer for soil = CBR 25. So there is a clear advantage in terms of life time of using High Modulus Asphalt Concrete specially for soil CBR = 10. The study shows also that the annual departures give the same results for both 60/70 and 13/22 bitumen.

2. In the computational model with EVERSTRESS, the longitudinal strains are generally greater than the transverse strains. This study shows that the calculated strains are systematically greater than the measured values at 15°C.

3. The study has found large differences in the strains between the sections of different thicknesses what indicate the significant effect that the thickness of the asphalt layer has taking into account the annual departures and CBR. This has naturally a high relation also with the number of cycles N at which the structure fails as it is shown in the Figures 3 and 4.

4. From an economical point of view, this study shows a clear advantage of using HMCA for the base course, in the case of CBR =10. Taking into account that the high modulus bitumen is 30% more expensive than the conventional bitumen and with the use of High Modulus Asphalt Concrete for the base layer of the pavement we obtain a reduction of 20% of the thickness of the base course, Figure 5 and 6 show that the use of High Modulus Asphalt Concrete is still very advantageous economically. According to the results obtained in this article the number of cycles to failure (Nf) increase considerably for the high modulus bitumen and increase in the same proportion the number of years that the pavement can be used with minor maintenance.

5. HMAC and conventional bitumen behaved differently in terms of stiffness and elasticity. HMAC in the study had higher complex modulus at all evaluated temperatures and viscosity was also higher than in conventional bitumen 60/70.

6. In this research has been observed that the HMAC´s dynamic young modulus at high temperatures is 50% higher than the conventional bitumen mixture's dynamic young modulus. Rutting resistance of HMAC is twice as high temperature compare to the conventional bitumen mixtures, and the fatigue resistance is 5–10 times higher. Besides it is interesting to add that the research of HMAC at high temperature, made by Jaczewski [14] in Poland, presented similar results that those that are shown in this research and it is that the usage of HMAC in asphalt pavement structure allows reduction in comparative deformations and slower deformation timing, at the same time increasing rutting resistance and durability.

## References

1. Sybilski D, Bankowski W, Krajewski M (2010) High Modulus Asphalt Concrete with Limestone Aggregate. Int J Pavement Res Technol 3: 96-101.
2. Han G, Cristian SC, Hussain UB (2013) Effects of High Modulus Asphalt Binders on Performance of Typical Asphalt Pavement Structures. Construct Build Mater 44: 207-213.
3. Corte JF (2001) Development and uses of hard-grade asphalt and of high-modulus asphalt mixes in France. Transport Res Circ 503: 12-31.
4. Espersson M (2014) Effect of the temperature in High Modulus asphalt bitumen. Elsevier Build Construct Mater 71: 638-643.
5. Nunn M (1997) Long-life Flexible Roads, The 8th International Conference on Asphalt Pavements 3-16.
6. Audrius V, Viktoras V (2013) Use of local aggregates in high modulus asphalt concrete layers Vaitkus. Građevinar 65: 353-360.
7. Čygas D, Laurinavičius A, Vaitkus A, Perveneckas Z, Motiejūnas A (2008) Research of Asphalt Pavement Structures on Lithuanian Roads (I). The Baltic J Road and Bridge Eng 3: 77-83.
8. Paliukaitė M, Vaitkus A (2011) Analysis of Temperature and Moisture Influence on Asphalt Pavement Strength, The 8th International Conference Environmental Engineering 1160-1165.
9. Vaitkus A, Laurinavičius A, Oginskas R, Motiejūnas A, Paliukaitė M (2012) The Road of Experimental Pavement Structures: Experience of Five Years Operation. The Baltic Journal of Road and Bridge Engineering 7: 220-227.
10. Wie O (2010) Research on Anti-Rutting Performance of High Modulus Asphalt Concrete Pavement. Advanced Mater Res 4474: 163-167.
11. David N (2001) Concepts of Perpetual Pavements. Transportation Research Circular Number ISSN 503: 0097-8515.
12. Motiejūnas A, Paliukaitė M, Vaitkus A, Čygas D, Laurinavičius A (2010) Research on The Dependence of Asphalt Pavements Stiffness Upon The Temperature of Pavement Layers. The Baltic J Road and Bridge Eng 5: 50-54.
13. Xiaohu L, Ulf I (2000) Artificial ageing of polymer modified bitumens. J Applied Polymer Sci 76: 1811-1824.
14. Jaczewski M (2010) Evaluation of High Temperature Behaviour of High Modulus Asphalt Mixes, Abstract for year.

28

# The Repair Method of Damaged H-shaped Steel Members and Experimental Study on Recovery after Repair

**Tomoe Onoda*, Takumi Ito, Kenjiro Mori, Hiromu Fuse and Changhoon Choi**

*Department of Architecture, Tokyo University of Science, Japan*

**Abstract**

Recently, there have been many discussions about repairability for damaged building structures after severe disasters. In japan, a technical manual for repairing damaged buildings has been established; however, its repair abilities have not been reported sufficiently. In this study, the Box-Shaped Repair Method which has been suggested on Japanese technical guideline is adopted for the damaged steel members, and its applicability and feasibility are investigated experimentally. Herein, the damaged portion on steel member such as plastic hinge or local buckling occurrence is covered with steel plate by welding. So the loading test is performed as parameters with section size of steel members, thickness of cover plate and welding size during repair process, and loading path. The procedure of this study is as follows; the first, the damaged test specimens of H-shaped steel member with local buckling are reproduced by initial loading test. Next, the damaged specimens with plastic residual deformation are returned to the original position. Finally, the loading test is done after repair. From test results, it is confirmed that the fundamental structural performance such as the rigidity, maximum strength, and absorbed energy after repair are improved by comparison of original state. And also, this performance can be controlled by adjusting the thickness of cover plate and welding condition. Furthermore, from the observation of test results, the analytical model of repaired steel member is suggested. From the comparison of test results, it can be said that the proposed model shows good agreements both test results and theoretical figure.

**Keywords:** Box shaped repair method; H-shaped steel members; Local buckling; Repair ability; Seismic limit state; Ultimate

## Introduction

Recently, there have been many discussions about repairability for damaged building structures after severe disasters. From these trends, new keyword "Resilience" which means revival potential or function maintenance of damaged structure is noticed. And in recent years, there are global environmental and resources problems. Considering these problems, the establishment and maintenance of technique and design method considering with the concept of resilience and repairability becomes increasingly important.

In Japan, a technical guideline for repairing damaged buildings has been established in the technical guideline for restoration by the Japan Building Disaster Prevention Association (2001, herein after referred to as "the restoration guideline"). But its repairabilities have not been sufficiently reported in past researches. Munemura H et al. [1] investigated the post-repair structural performance of H-shaped steel members, which the technical guidelines have suggested. The Moment-Rotation angle curve (M-Θ curve) is shown in Figure 1. Furthermore, the definition of the moment of member end M and rotation angle Θ

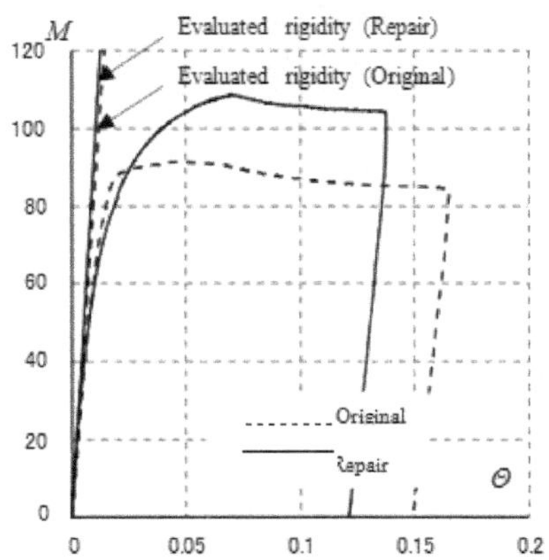


Figure 1: Structural performance of repaired member [1].

| Mechanics | System | Materials |
|---|---|---|
| Bending | Viscous | Metal |
| Shear | Hysteresis | Oil |
| Axial | Friction | Rubber |

**Table 1:** The mechanical characteristics and materials of SFD.

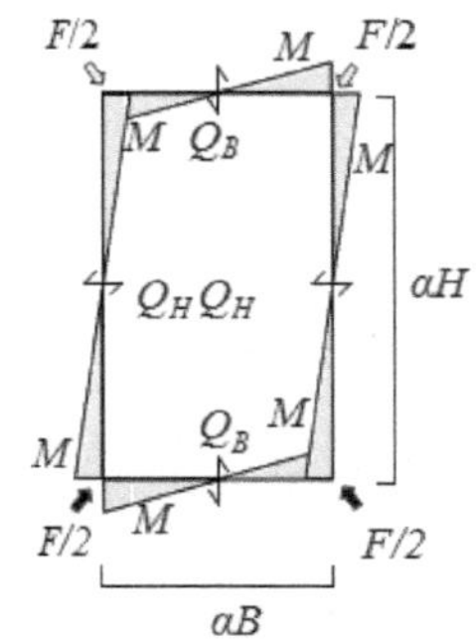


**Figure 2:** Diagram of bending stress of SF subjected to lateral force.

investigate the applicability of SF structure on steel frames. (Figure 2) This study aim to develop the prediction method of seismic response on vibration controlled steel frames with SFD installed, and the simple design procedure can be provided. To investigate the simplified restoring force characteristics, the horizontal static cyclic loading test is conducted on half-scaled frame test [1,2]. And, the prediction method is reformulated to adapt the bilinear model and strain hardening

***Corresponding author:** Tomoe Onoda, Department of Architecture, Tokyo University of Science, Japan, E-mail: 4116614@ed.tus.ac.jp

## Detail of Repair Method for Local Buckling

### Based on the restoration guideline

The restoration guideline has suggested a repair method mentioned that the damaged area around the flange local buckling of H-shaped steel members is repaired by cover plate (herein after referred to as "Box- shaped repair method") as shown in Figure 3. Before repaired by this method, residual displacement of damaged H-shaped members should be almost zero as same as original position. The damaged area around the flange local buckling is repaired by welding cover plate, and a clearance of 25 mm is provided between the end plate and the edge of the cover plate. The thickness of the cover plates is equal to that of the flange of the H- shaped steel member.

### Proposal of repair method

Herein, based on original Box-shaped repair method, this study suggests a modified repair method to adjust the strength and rigidity of repaired members. That is, it is assumed that the strength and rigidity of repaired members can be adjusted by changing either of the following:

1. The welding condition on repaired area
2. The thickness of cover plate

Authors have studied the affectivity and feasibility of the former repair method experimentally. Herein, the effects of welding condition as parameters with the number of welding and length are investigated (Figure 4). Table 1 shows the list of test specimens of monotonic loading test. The number of welding and length is based on Matsumoto T et al. [4]. The number of welding is 2 or 3 point. In case the number of welding is 2 point, the welding length (*l*) is 20 mm, 40 mm or 60 mm. On the other hand, in case the number of welding is 3 point, the welding length (*l*) is 20 mm, and the medium welding length (*l'*) is 20 mm or 40 mm. Figure 5 shows an elevation view of the test specimens. Test specimens are set as cantilever. The ultimate state of the normalized Moment-Rotation angle curve ($M$-$\Theta$ curve) and the test specimens with loading pattern are shown in Figures 6 and 7. Where $Mp$ is full plastic moment of the specimen, $\Theta p$ is the rotation angle when the load reaches $Mp$ on monotonic curve. Moreover, Table 2 shows the comparison of the maximum strength ($Mm$) and the energy absorption capacity when load reaches maximum strength ($c\eta m$). From the test results, the strength and the energy absorption capacity after repair show almost same with original state; however, the ultimate behavior is unstable. From the comparison of test results, it can be said that the fundamental structural performance such as the strength and the energy absorption capacity after repair depend on the damage condition of welding between cover plate and the damaged area around the flange local buckling. This study aims to provide the adjustment method of strength and rigidity of repaired member by changing the thickness of cover plate experimentally. The test specimens are repaired by cover plate which is half or 3/4 thick of the flange.

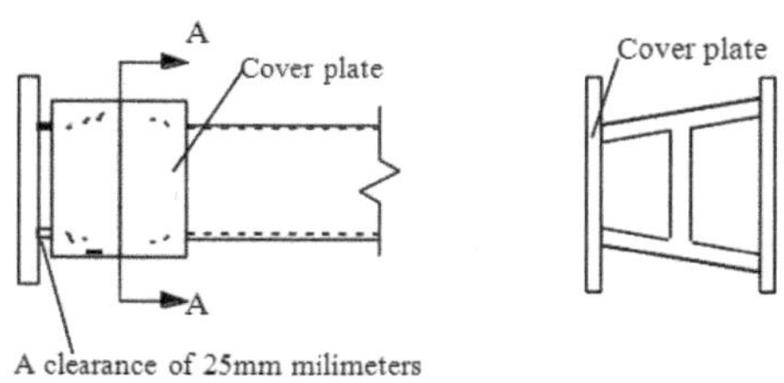

(b) Elevation (c) A-Section

**Figure 3:** Detail of box-shaped repair method.

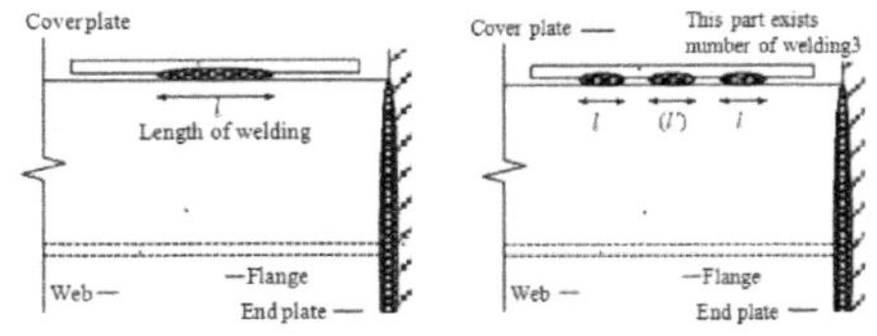

(a) Number of welding: 1 (b) Number of welding: 2, 3

**Figure 4:** Method of welding on cover plate.

| Name | Steel grades and section | Thickness of cover plate (mm) | Welding condition (*l-l'*) |
|---|---|---|---|
| NS150-NR-M | - | - | - |
| NS150-IR1-M | - | - | 20 |
| NS150-IR2-M | - | - | 40 |
| NS150-IR3-M | - | - | 60 |
| NS150-IR4-M | SM490 180×150×6×6 | - | 20-20 |
| NS150-IR5-M | - | 6 | 20-40 |

**Table 1:** Test specimen list (based on Matsumoto T et al. [4]).

## Outlines of Experimental Study

### General description

At first, test specimens which are damaged local buckling, and that residual deformation are reproduced by loading tests. Next, the damaged test specimens are returned to the original position (herein after referred as "Returning"). Then, the damaged area around the local buckling is repaired by the method described in chapter 2. Finally, the loading tests are done after repair, which the loading programs are same as initial loading tests. The evaluation and comparison with performance of the original states and that of the repaired states are examined.

### Test setup and loading test methods

Figure 5 shows an elevation view of the test specimens. This is as same as previous study which examined by changing the number of welding and welding length on repaired area. Test specimens are set as cantilever. The top of the test specimen was laterally loaded to produce the bending moment at the base of the H-shaped steel members. Test specimens were instrumented with sensors for measuring below items: lateral displacement at the top of the test specimens, strain of flange and cover plate.

### Test specimens

In this paper, the name of test specimens with non-scallop are called as NS150 and these specimens are composed by welding of steel plates. Herein, the configuration of test specimen is illustrated in Figure 8. The name of test specimens and loading patterns are summarized in Table 3. NS150-6C-M is the non-scallop test specimen repaired by the cover plate which thickness is equal to that of the flange of the H-shaped steel members, and investigated in Munemura H et al. [1]. In this study, 3.2 mm and 4.5 mm thick of cover plate means half or 3/4 thick of the flange respectively.

### Mechanical properties of materials

Mechanical properties of steel are summarized in Table 4. According to Japanese seismic resistant codes, NS150 specimens are classified as FC-rank. This is as same as previous study which examined by changing the number of welding and welding length on repaired area.

## Results of Test Study

### Test results of original state and after repair

The ultimate state of the test specimens and the normalized Moment-Rotation angle curve ($M$-$\Theta$curve) with loading pattern are shown in Figures 8 and 9 respectively. Table 3 shows the comparison of the maximum strength ($Mm$), the energy absorption capacity when load reaches maximum strength ($c\eta m$), and failure mode of each specimen. The results in case of 6.0 mm thick of cover plate are also shown in the same table [2].

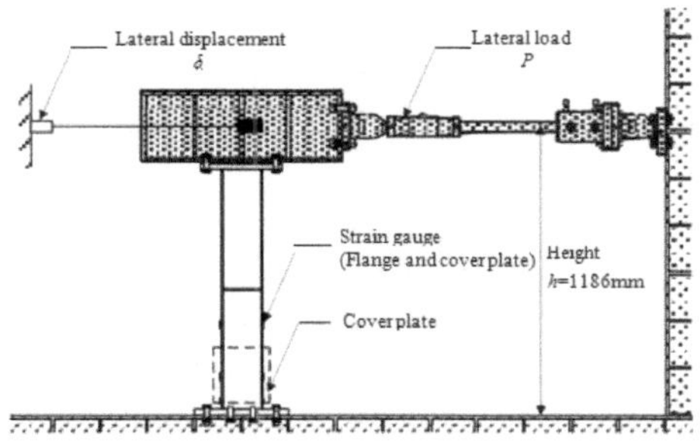


**Figure 5:** Elevation of test set up and location of sensors.

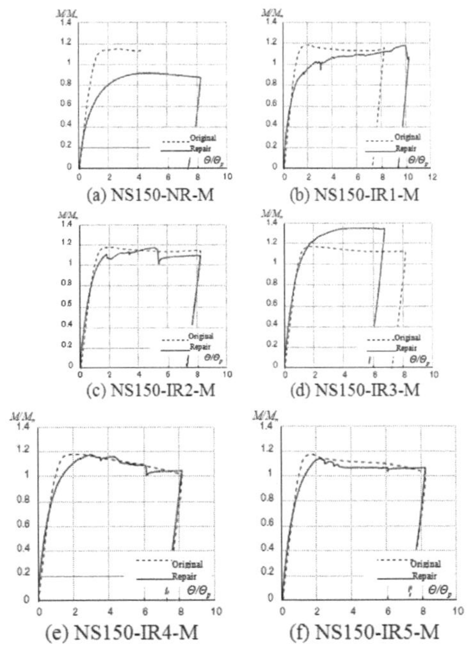


**Figure 6:** Test results of normalized M-Θ curves (based on Matsumoto T et al. [4]).

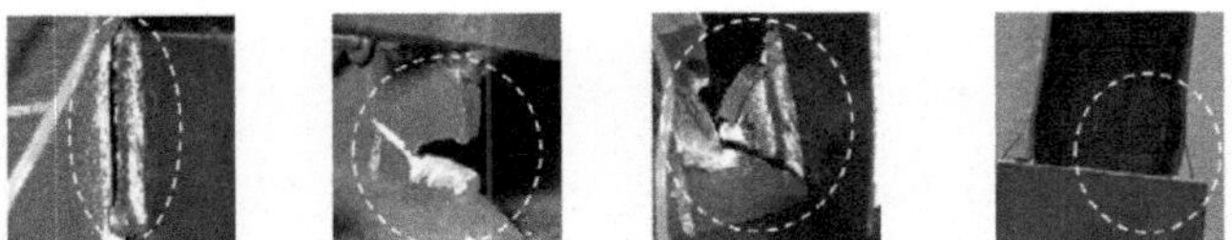

(a) Cracks on the repaired area of welding (b) Cracks on member end (c) Cracks on flange (d) New local buckling

**Figure 7:** Photographic observations of damage condition (based on Matsumoto T et al. [4]).

**Test results of monotonic loading:** The normalized $M$-$\Theta$ curves of monotonic loading tests are shown in Figure 8. On the loading tests before damaged (first loading), the local buckling was observed on the flange at the compression side after reaching $Mp$. After reached the maximum strength, the local buckling of the web progressed and the strength was deteriorated. On the second loading test of NS150-NR-M specimens (Non-repair) after Returning, rigidity lowered by the existence local buckling. Furthermore, the maximum strength was deteriorated, and the maximum strength stopped at 0.88 $Mp$. In case of NS150-3.2C-M specimen (repaired by 3.2 mm thick cover plate), initial rigidity after repair became almost equal to initial rigidity before damaged. The maximum strength improved compared with test specimen before damaged. The damage process is described below. When the moment reached around 1.75$\Theta p$, the welding on the repaired area was cracked and the rigidity lowered. Then, the crack around the repaired area on member end was observed when the moment reached around 1.11 $Mp$. After new local buckling is observed in the upper cover plate at the compression side, the maximum strength was reached.

**Test results of cyclic loading:** The normalized $M$-$\Theta$ curves of cyclic loading tests are shown in Figure 8. During the first loading, the local buckling was observed on the flange at the compression side after reaching $Mp$. After the local buckling of the flange progressed, the maximum strength was reached and the strength was deteriorated. This result is almost same with the results of monotonic loading. On the second loading test of NS150-3.2 C-C specimen, after the shape of cover plate is changed, the welding on the repaired area was cracked. After the maximum strength, the local buckling was observed in the upper cover plate at the compression side. Then, the crack around the repaired area on member end was observed. In case of NS150-4.5C-C specimen, after the welding on the repaired area was cracked around 1.0$\Theta p$, the maximum strength was reached and the crack around the repaired area on member end was observed. Then, new local buckling is observed in the upper cover plate at the compression side [3,4].

### Examination of the test results

**The maximum strength:** The recovery of performance is compared by the difference of the thickness of cover plate. The results are shown in Figure 10. The maximum strength of every repaired test specimen is experimentally improved. However, it is assumed that the reason why the maximum strength of NS150-3.2C-M was larger than that of NS150-4.5 C-M is due to the early cracks around the repaired area on member end. From the observation of test results, the analytical model

| | First Loading | | | Second Loading | | | Ratio (after/before) | |
|---|---|---|---|---|---|---|---|---|
| Name | $Mm$ | $c\eta m$ | The determinant of maximum strength | $Mm$ | $c\eta m$ | The determinant of maximum strength | $Mm$ | $c\eta m$ |
| NS150-NR-M | 1.15 | 1.88 | - | 0.92 | 3.11 | The existence Local Buckling | 0.8 | 1.66 |
| NS150-IR2-M | 1.18 | 1.45 | - | 1.17 | 4.42 | Cracks on the repaired area of welding | 1 | 3.06 |
| NS150-IR3-M | 1.17 | 1.35 | - | 1.36 | 5.87 | Cracks on flange | 1.16 | 4.33 |
| NS150-IR4-M | 1.17 | 1.32 | - | 1.17 | 2.17 | Cracks on the repaired area of welding | 1 | 1.64 |
| NS150-IR5-M | 1.17 | 0.84 | Local Buckling | 1.14 | 1.23 | Cracks on the repaired area of welding | 0.97 | 1.46 |

**Table 2:** Comparison of test results (based on Matsumoto T et al. [4]).

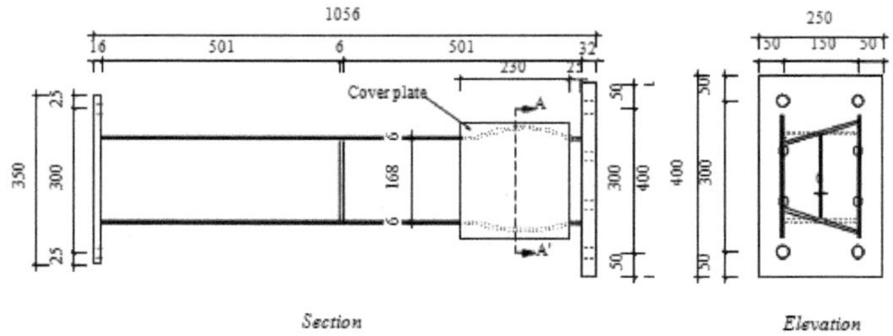


**Figure 8:** Configuration of test specimen after repair.

| Name | Steel grades and section | Thickness of cover plate(mm) | Welding condition | Loading pattern |
|---|---|---|---|---|
| NS150- NR-M | SM490 180×150×6×6 | - | - | Monotonic |
| NS150- 3.2C-M | | 3.2 | Continuous | Monotonic |
| NS150- 4.5C-M | | 4.5 | | |
| NS150- 3.2C-C | | 3.2 | | Cyclic |
| NS150- 4.5C-C | | 4.5 | | |
| NS150- 6C-M | | 6.0 | | Monotonic |

**Table 3:** Test specimen list.

| | Yield strength (N/mm2) | Tensile strength (N/mm2) | Young's modules (GPa) | Elongati on (%) |
|---|---|---|---|---|
| SM490 (t=6.0 mm) | 436 | 573 | 206 | 32 |
| SS400 (t=3.2 mm) | 354 | 438 | 198 | 31 |
| SS400 (t=4.5 mm) | 298 | 445 | 208 | 35 |

**Table 4:** Mechanical properties of material.

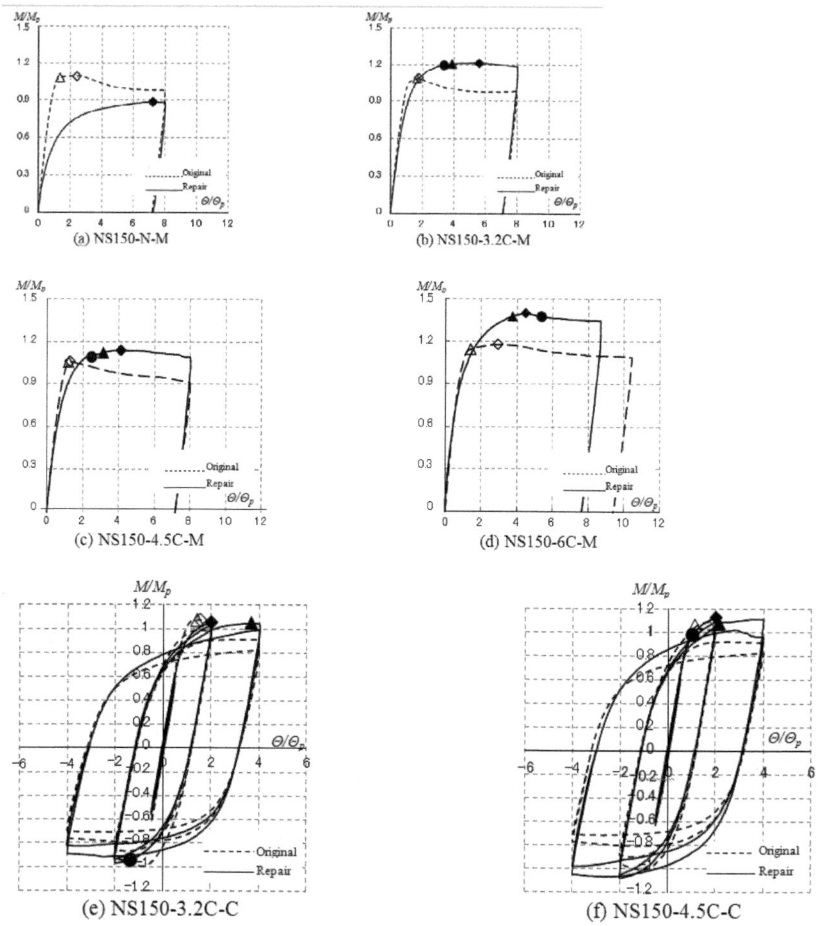


**Figure 9:** Comparison of test results of normalized M-Θ curves on original state and repaired state.

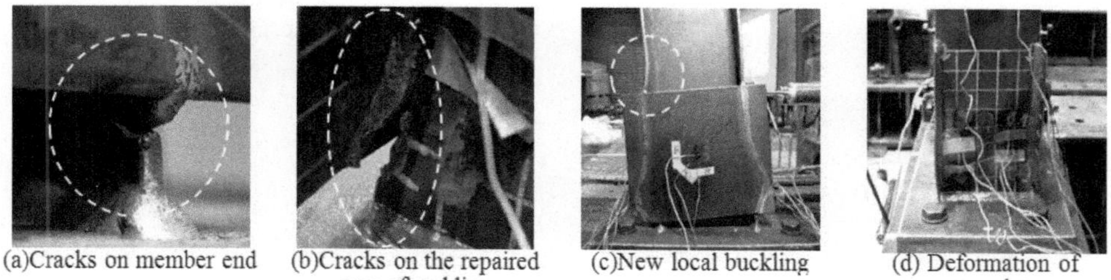

(a)Cracks on member end (b)Cracks on the repaired area of welding (c)New local buckling (d) Deformation of cover plate

**Figure 10:** Photographic observations of damage condition.

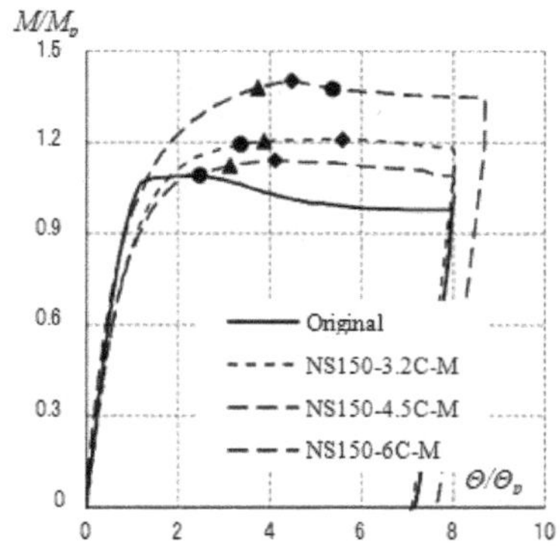


**Figure 11:** Comparison of test results of monotonic loading.

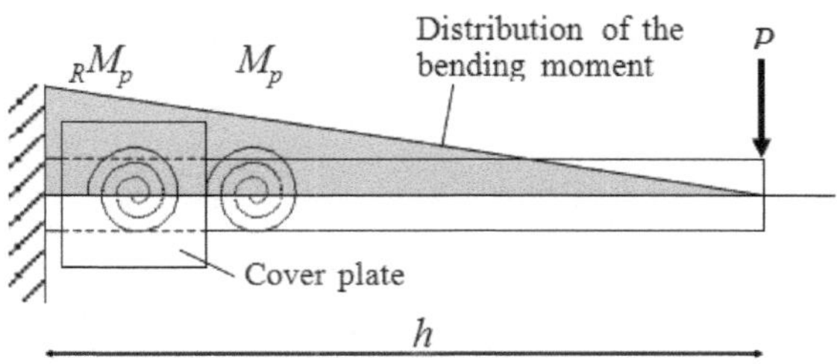


**Figure 12:** The analytical model of repaired steel.

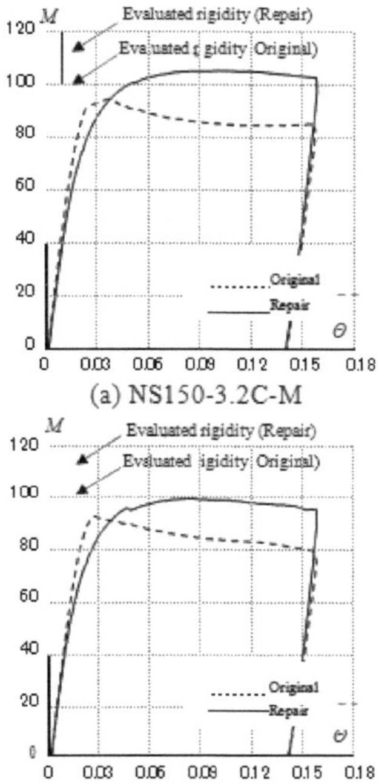


(a) NS150-3.2C-M

(b) NS150-4.5C-M

**Figure 13:** Comparison of the results of theoretical initial rigidity and M-θ curves.

| | First Loading | | | Second Loading | | | Ratio (after/before) | |
|---|---|---|---|---|---|---|---|---|
| Name | *Mm* | *cηm* | The determinant of maximum strength | *Mm* | *cηm* | The determinant of maximum strength | *Mm* | *cηm* |
| NS150-NR-M | 1.09 | 1.61 | | 0.88 | -4.99 | The existence Local Buckling | 0.81 | -3.09 |
| NS150-3.2C-M | 1.09 | 0.91 | | 1.21 | 5.21 | Cracks on the repaired area of welding | 1.11 | 5.73 |
| NS150-4.5C-M | 1.06 | 0.39 | | 1.14 | 3.44 | The crack on member end | 1.08 | 8.74 |
| NS150-3.2C-C | 1.07 | 0.82 | | 1.05 | 1.24 | Cracks on the repaired area of welding | 0.98 | 1.5 |
| NS150-4.5C-C | 1.09 | 1.28 | | 1.13 | 1.31 | the crack on member end | 1.04 | 1.02 |
| NS150-6C-M | 1.18 | 2.4 | Local Buckling | 1.4 | 4.51 | New flange local Buckling | 1.19 | 1.88 |

**Table 5:** Comparison of test results.

of repaired steel member composed of $_RMp$ (repaired area occurred plastic deformation or rigidity rotation) and $Mp$ (the elastoplastic element occurred local buckling) could be suggested. The analytical model of steel member is shown in Figure 11. From the theoretical results, it is predicted that the cracks on the repaired area of welding and the crack around the repaired area on member end was observed in the specimens repaired by 3.2 mm thick and 4.5 mm thick cover plate respectively. And also, it is also predicted that the crack around the repaired area on member end was observed in the specimens of 6.0 mm thick of cover plate. However, the crack around the repaired area on member end was not observed in case of repaired by 3.2 mm thick cover plate. This is due to the influence area of the welding heat. From the above, it was predicted that the proposed model shows good agreements both test results and theoretical figure.

**The rigidity:** The experimental and theoretical results of initial rigidity in the specimens of 3.2 mm thick and 4.5 mm thick of cover plate are shown in Figures 12 and 13. From these results, it is confirmed that initial rigidity could be controlled by changing the thickness of cover plate. Furthermore, initial rigidity after repair in this study became almost equal to initial rigidity before damaged.

**The absorbed energy:** From test results shown in Table 5, it is confirmed that the absorbed energy after repair was improved by comparison of original state.

## Conclusion

In this study, the experimental and theoretical study about recovery of H-shaped steel members is performed. The main conclusions are shown below:

1. It is confirmed that the fundamental structural performance such as the rigidity, maximum strength, and absorbed energy after repair are improved by comparison of original state.

2. These performances can be controlled by adjusting the thickness of cover plate and welding condition.

3. From the observation of test results, the analytical model of repaired steel member is suggested.

4. From the comparison of test results, it can be said in Figure 10 that comparison of test results of monotonic loading that the proposed model shows good agreements both test results and theoretical figure.

### Acknowledgement

The experiment in this study is supported by a specific study subsidy of Tokyo University of Science in 2015. I would like to express our sincere appreciation from the heart.

### References

1. Munemura H, Mori K, Matsumoto T, Sato H, Ito T, et al. (2014) Experimental study on repairability and recovery of structural performance for damaged steel members. Proceedings of Second European Conference on Earthquake Engineering and Seismology, Istanbul Aug. 25-29.
2. The Japan Building Disaster Prevention Association (2001) Guideline for post-earthquake damage evaluation and rehabilitation technique (in Japanese).
3. Sato H, Mori K, Munemura H, Matsumoto T, Ito T, et al. (2014) Transition of failure mechanism of damaged multi-story moment resisting frame after repair. The 5th Asia Conference on Earthquake Engineering, October 16-18.
4. Matsumoto T, Munemura H, Sato H, Mori K, Ito T, et al. (2014) Examination of repair methods for damaged steel members by finite element analysis. The 5th Asian Conference on Earthquake Engineering, October 16th-18th 2014.

# Effect of Adding Silicon Carbide and Titanium Carbide Nanoparticles on the Performance of the Cement Pastes

**Aamir Mahawish[1]*, Sarmed Imad Ibrahim[2], Ali Hussein Jawad[2] and Ferhad Mohamed Othman[2]**

[1]*Department of Highway and Transportation Engineering, University of Al-Mustansiriyah, Iraq*
[2]*Department of Materials Engineering, University of Technology, Baghdad, 00964, Iraq*

**Abstract**

The nanoparticle is used into cement pastes to improve its engineering properties. In this paper, silicon carbide (SiC) and titanium carbide (TiC) nanoparticles with a particle size (20 nm) were added at different percentages 1%, 2%, 3%, 4%, 5% and 6%) to cement pastes. The cement pastes were tested for mechanical and physical properties at different curing ages (7, 14, 21, and 28 days). The results of mechanical and physical tests showed a significant improvement with all mixtures of nanoparticles, and the maximum values of the mechanical and physical tests were associated with the cement paste samples with 5% TiC at curing time 28 days. SEM results showed the control samples of cement paste had needle shapes to structure Ca(OH)2 with the of large pores. However, adding of nanoparticles resulted in absence of needle shape structures, pore spaces and produced a homogeneous structure of compounds (C-SH).

**Keywords:** Nano particles; Cement paste; Strength; Scanning electron microscopy

## Introduction

In the recent decade, the nano-materials have shown a great development since its impact on construction materials [1,2]. The use of nanotechnology products especially nano-powder has increased. Nano-powder can improve the mechanical properties of cement pastes due to accelerating the hydration, the formation of small-sized crystals (such as Ca(OH) and Atomic Force Microscopy (AFM)) and the uniform clusters of C-S-H [1-5]. Nano-powder can also improve the durability of cement-based composites due to filling very small pores. The size of nano-powder is important as it has to be compatible with the pore structures of cement [1-5]. Mehta stated the cement particle size is important for its reaction with water as more fine cement, the more rapid reactions with water [6]. Many researches have been carried out experimental works over the recent years with the view to understand the effect of fine and ultra-fine materials (nanoparticles) on the cement properties [2,7-9].

Nanoparticles are used either to replace part of cement, inducing ecologic profile concrete [7] or as an admixture in the cement pastes [8,9]. In both cases, the addition of nano-scale particles improves the performance of cement; for instance, an improvement of rheological properties has been observed in the fresh cement mixtures [10], whereas, the compressive strength was increased in the hardened state [11]. The nano-scale observations revealed the nano-crystallized structure of C–S–H cement phase [12,13], and the nanoparticles work as a nucleation site for the cement phase promoting a cement hydration [14,15]. The most common nanoparticles in cement products are $SiO_2$, $TiO_2$, $Al_2O_3$ and carbon nanotubes [16-18]. Nano-$SiO_2$ has played a significant role according to Sobolev and Gutiérrez to modify cement mixtures [19].

Nano-$SiO_2$ particles lead to the chemical reaction between $SiO_2$ and $Ca(OH)_2$ which is released during cement hydration. The previous mechanism can cause to reduce of $SiO_2$ grain size contributing to increasing the strength and reduce permeability. "The great reactivity of nanoparticles is attributed to a high purity and a specific area in relation to its volume" [20]. The effect of nanoparticles sizes has been studied by numerous studies as the nanoparticle cement mixtures tend to increase the demanding for water for the same workability [9,20]. In addition, nanoparticles tend to agglomerate, which can be avoided by using dispersing additives or applying different techniques during mixing [15,16]. Li recorded a dense, homogeneous structure in nanoparticles modified mortars and a compact transition zone between aggregates and paste [8]. Sobolev et al. [20] reported that the nano-modified mixtures can reduce the initial strength with the curing age. Li tested cement products with fly ash and concluded that the pozzolanic reaction of fly ash is improved by adding nano-$SiO_2$ to the mixture [8].

The present study focuses on the structural changes of cement pastes when adding (SiC and TiC) nanoparticles as admixtures in different percentages. Scanning Electron Microscopy and Energy Dispersive Spectroscopy have been employed in identifying the structural changes in nanoparticles modified pastes. The mechanical and physical tests have been conducted at different curing ages.

## Materials and Methods

### Materials

Ordinary Portland cement was used throughout the current study. Nanoparticles powder was supplied by Nanostructured and Amorphous Material Inc., USA. The chemical compositions of cement are presented in Table 1, and the properties of nanoparticles (SiC and TiC) powder are given in Table 2.

### Sample preparation

A Portland cement was replaced with nanoparticles with (1%, 2%, 3%, 4%, 5% and 6%) by weight of cement. The nanoparticles were mixed with cement in a dry condition, and the dry mixtures were sieved several times for a proper dispersion of the nanoparticles. The water-binder ratio (W/B) used was 0.45. Cement paste cubes of 50 mm size were cast and vibrated in a table vibrator. The specimens were

***Corresponding author:** Aamir Mahawish, Department of Highway and Transportation Engineering, University of Al-Mustansiriyah, Iraq
E-mail: aamir.mahawish@monash.edu

| Oxides | Content (%) |
|---|---|
| $SiO_2$ | 20.3 |
| $Al_2O_3$ | 4.4 |
| $Fe_2O_3$ | 2.6 |
| CaO | 62.5 |
| MgO | 2.5 |
| $SO_3$ | 2.1 |
| LOI | 1 |

**Table 1:** Chemical composition of cement.

| Constituents | Purity | Average particle size | Specific Surface area | Color | TRUE density |
|---|---|---|---|---|---|
| Nano-SiC | 99.50% | ≤ 20 nm | 195 $m^2/g$ | Greyish white | 3.216 $g/cm^3$ |
| Nano-TiC | 99.50% | ≤ 20 nm | 210 $m^2/g$ | Black | 4.93 $g/cm^3$ |

**Table 2:** Properties of nanoparticles.

then demolded and cured in tap water until testing. Scanning Electron Microscopy and Energy Dispersive Spectroscopy of hydrated cubes were conducted on the samples taken from the specimen that induced the best mechanical and physical results.

## Test methods

**Compressive strength:** The compressive strength test was determined by two testing machines:

1. ***Portable mechanical tester:*** The compressive strength test was determined according to ASTM C (780-002). The compressive strength of the specimen was calculated by dividing the maximum load carried by the specimen during the test by the average cross-sectional area of the specimen. The load was applied perpendicularly and continuously on the specimen by using hydraulic testing compression machine with a maximum load (15 ton).

2. ***Electrical digital compression tester (ELE International Accu-Tek 250):*** An electrical digital compression tester can provide a consistent testing on a wide range of specimens through a hydraulic pump and a digital readout system.

**Splitting tensile strength:** Splitting tensile test was performed according to ASTM C (496-04). The determination of the splitting tensile strength was done by conducting the splitting tensile test on the cylindrical specimens.

The cement paste specimens were placed horizontally between plates of the testing machine. In order to support the testing specimens, narrow strips of plywood were required to interpose between the specimens and the plates. The strips were usually a 3mm thick and a 25mm width as per ASTM C (496-04).

**Flexural strength:** The flexural strength tests were conducted on 40*40*160 mm prism specimens in conformity with ASTM C (348-02). The prism specimens were subjected to two-point loading. The flexural strength of the cement paste was tested at ages of (7, 14, 21, and 28 days). The average of three specimens was recorded for each testing age. Flexural strength was calculated by using the following equation:

$$Sf = 0.0028P \quad (1)$$

Where: Sf: Flexural strength (MPa), P: Total maximum load (N).

**Water absorption:** Water absorption tests were conducted on the control cement paste specimens and the cement paste modified with the nanoparticles. The test was carried out in accordance to ASTM C (642-1997). In the test procedure, the specimens were dried in an oven at (100-110°C) for 24 hours, and then their masses were measured. Next, the specimens were fully immersed in a water tank for 24 hours, then they were taken out. Their surfaces were dried with a cloth and finally their weights were measured again. The test was conducted on the specimens with different curing ages (7, 14, 21, and 28 days), and the percentage of water absorption was calculated from the following equation:

$$\text{Water absorption (\%)} = (W2-W1)/W1 \times 100 \quad (2)$$

Where: W1: The average weight of dry sample (g), W2: The average weight of the wet sample (g).

**Scanning Electron Microscopy (SEM) and Energy Dispersive X-Ray (EDX):** The scanning electron microscope (SEM) is comparatively more useful in studies which investigate quantification of microstructural properties, such as micro-cracks, voids and to diagnose of the formed microstructural phases as studied by Shah in 2004. SEM is a powerful technique for analyzing the morphology of the prepared cement paste samples mixed with the nanoparticles and the control cement pastes. Scanning electron microscope (SEM) (Type TESCAN) was used in this study. The microstructure of the prepared samples was examined to determine the influence of adding the nanoparticles. Discs with a 20 mm thick were extracted using a diamond precision cutter from the samples that were moist cured for 28 days. The discs were dried in a vacuum oven at 50°C until a constant weight reached. The specimens were then impregnated with epoxy, polished and sputter coated. Scanning Electron Microscope (SEM) examinations were carried out in the secondary electron mode using an Apex Personal SEM. Energy dispersive X-ray was also used to identify the phases.

## Results and Discussion

### Compressive strength

The Compressive strength is the most important property of the concrete that is considered in the structural designs. The compressive strength results of the cement pastes with different nanoparticles (SiC and TiC) percentages (1%, 2%, 3%, 4%, 5% and 6%) at a curing time (7, 14, 21 and 28 days) are presented in Figures 1-6, respectfully. It can be seen from Figures 1-6 that all the specimens of the cement pastes modified by the nanoparticles were higher compressive strength than the control cement paste specimens' at all curing periods.

It is also seen from Figures 1-6 that a comparison between the cement paste specimens reinforced by SiC and TiC is presented along with the control samples at various curing periods. It is clearly seen from these figures that the compressive strength of control specimens was

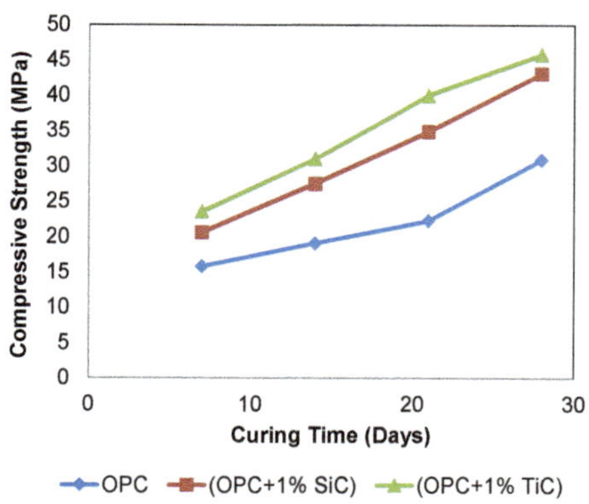


**Figure 1:** Effect of (1% SiC) and (1% TiC) nanoparticles on the compressive strength for the cement paste specimens at different curing time.

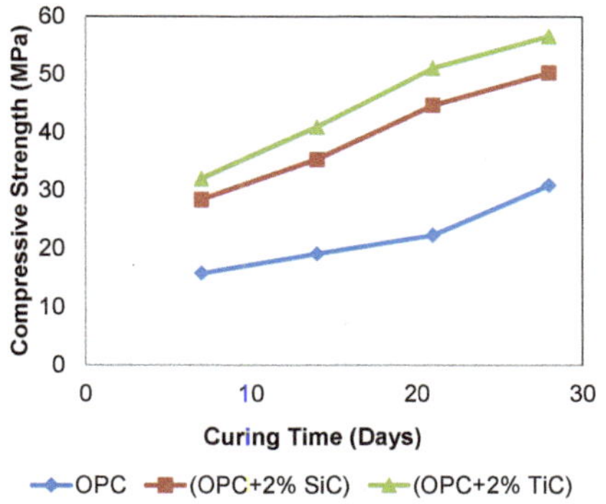


**Figure 2:** Effect of (2% SiC) and (2% TiC) nanoparticles on the compressive strength for the cement paste specimens at different curing time.

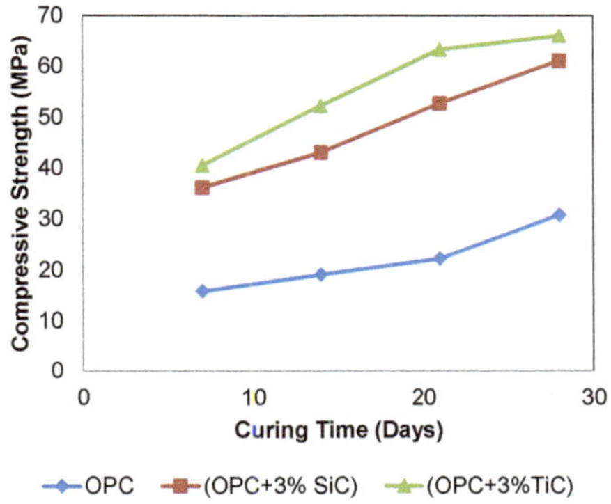


**Figure 3:** Effect of (3% SiC) and (3% TiC) nanoparticles on the compressive strength for the cement paste specimens at different curing time.

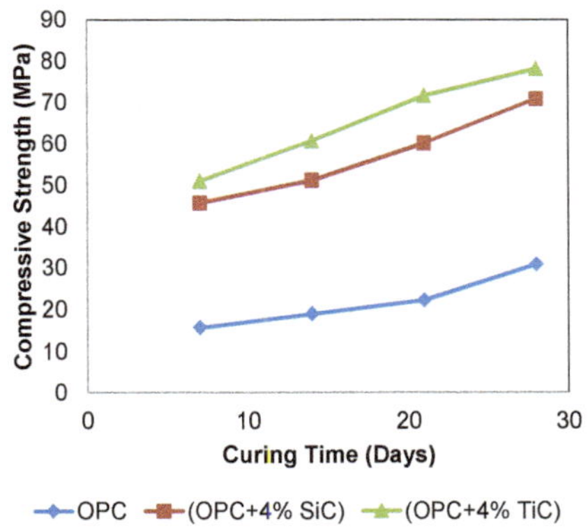


**Figure 4:** Effect of (4% SiC) and (4% TiC) nanoparticles on the compressive strength for the cement paste specimens at different curing time.

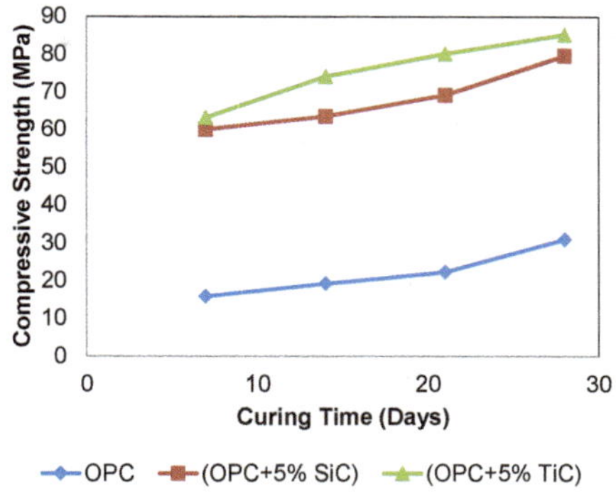


**Figure 5:** Effect of (5% SiC) and (5% TiC) nanoparticles on the compressive strength for the cement paste specimens at different curing time.

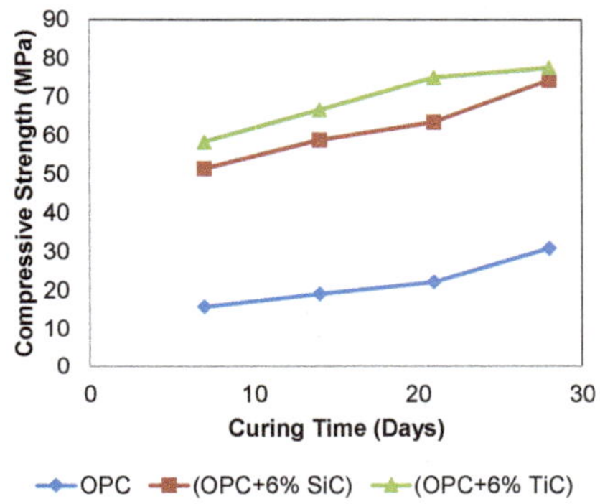


**Figure 6:** Effect of (6% SiC) and (6% TiC) nanoparticles on the compressive strength for the cement paste specimens at different curing time.

lower than the cement paste samples reinforced by the nanoparticles. This is because "nanoparticles packed into the cement pastes pores can increase its resistance to compressive forces and the tiny nanoparticles size can be efficient to fill the void of cement paste mixtures" [21]. Another important factor to note is that the nanoparticles may promote the pozzolanic reactions [21].

The compressive strength of cement paste reinforced by (TiC) nanoparticles was greater than cement paste reinforced by (SiC) nanoparticles in all testing conditions (i.e., percent of nanoparticles and curing periods) as shown in Figures 1-6. Also, these figures reveal that the maximum compressive strength of cement paste samples was achieved by adding 5% TiC and 5% SiC at curing time 28 days.

The effect of curing period on compressive strength for the control cement pastes samples and the cement pastes - nanoparticles mixtures is presented in Figures 1-6. The compressive strength increased with increasing the curing period for all cement paste mixtures. It can also be seen that the cement paste mixtures with (TiC) nanoparticles had the highest strength at all curing periods, and the compressive strength of the cement paste increased with increasing the percentage of nanoparticles up to (5%). The early effect of the nanoparticles may be due to its influence on accelerating the pozzolanic transformation of C3S, C2S and CH into the C-S-H gel which was responsible for the strength improvements of the cement pastes. The cement pastes mixtures with (5% TiC) nanoparticles (Figure 5) were the highest in gaining early strength at all curing periods due to it has small particles, which were superior in the filling the cement paste pores [22,23].

## Splitting tensile strength

The tensile strength results of the cement pastes and the cement pastes with different nanoparticles percentages are presented in Figures 7-12. It is clearly observed in Figures 7-12 that the tensile strength increased with an increase in the nanoparticles percentages to reach an optimum value (Figure 11). The tensile strength of the cement paste prepared with different percentages of nanoparticles (20 nm particle size) was higher than the tensile strength of the control cement pastes samples as presented in Figures 7-12. It can be clearly seen from these figures that the cement paste mixtures with 5% TiC and 5% SiC

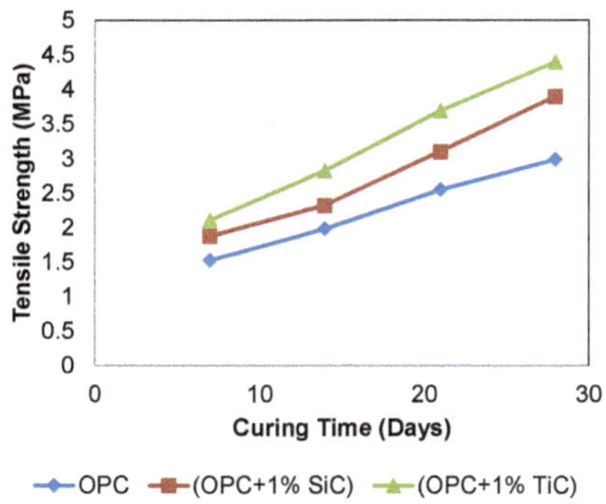


**Figure 7:** Effect of (1% SiC) and (1% TiC) nanoparticles on the tensile strength for the cement paste specimens at different curing time.

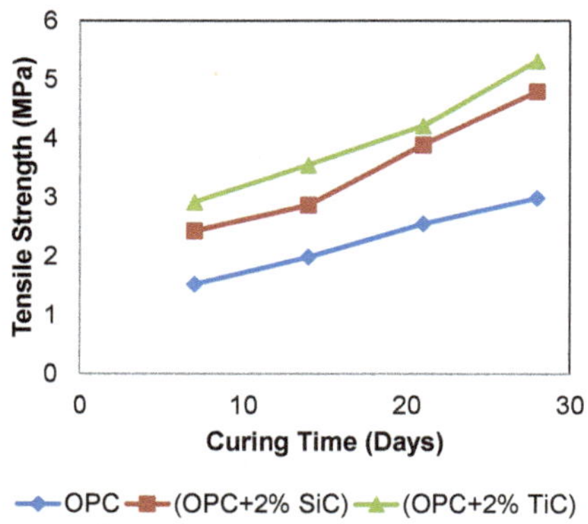


**Figure 8:** Effect of (2% SiC) and (2% TiC) nanoparticles on the tensile strength for the cement paste specimens at different curing time.

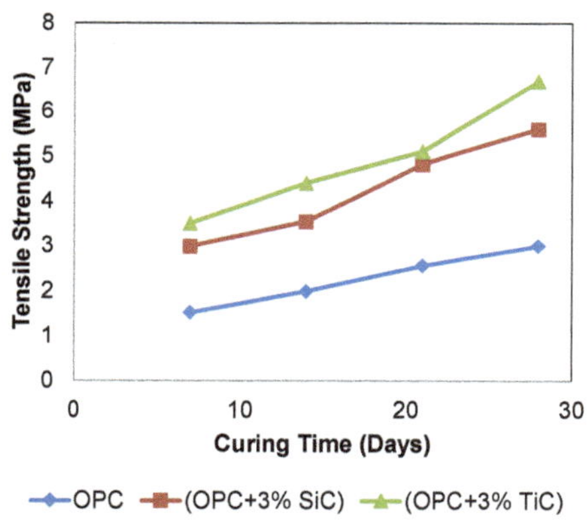


**Figure 9:** Effect of (3% SiC) and (3% TiC) nanoparticles on the tensile strength for the cement paste specimens at different curing time.

(Figure 11) showed a higher tensile strength than their counterparts prepared with low nanoparticles percentage at the curing age of 28 days. Another point to note from these figures is that adding SiC and TiC to cement paste increased the tensile strength of the specimens at inconsistent rates [8].

The effect of curing ages (7, 14, 21, and 28 days) on tensile strength of control mixtures, and the cement pastes with the different nanoparticles percentages (1%, 2%, 3%, 4%, 5% and 6%) mixtures is presented in Figures 7-12. It can be seen from these figures that the tensile strength consistently increased with the increase in the curing

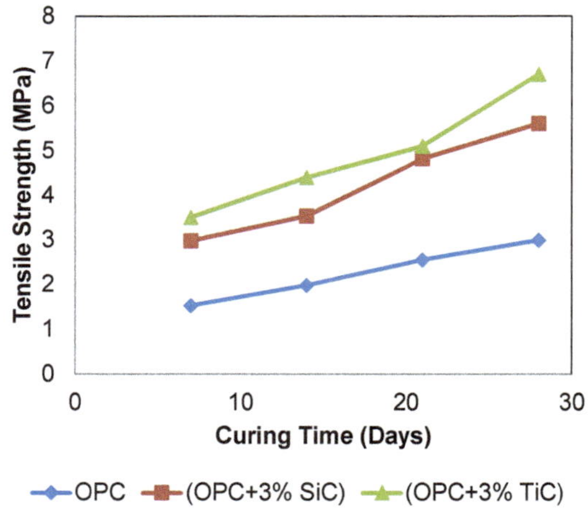


**Figure 10:** Effect of (4% SiC) and (4% TiC) nanoparticles on the tensile strength for the cement paste specimens at different curing time.

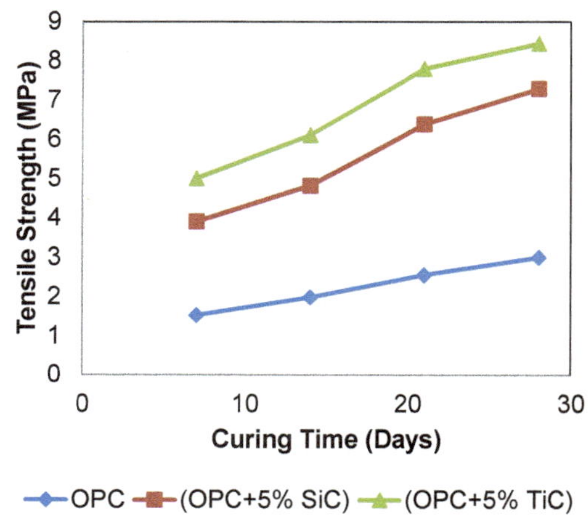


**Figure 11:** Effect of (5% SiC) and (5% TiC) nanoparticles on the tensile strength for the cement paste specimens at different curing time.

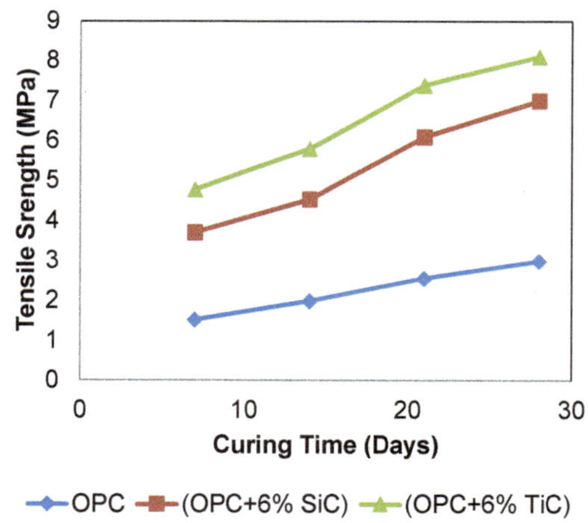


**Figure 12:** Effect of (6% SiC) and (6% TiC) nanoparticles on the tensile strength for the cement paste specimens at different curing time.

age for all cement paste mixtures. It can also be observed from these figures that the highest tensile strength was gained for cement paste mixtures with (TiC and SiC) nanoparticles at the curing age of 28 days. It could be concluded that adding nanoparticles to cement paste mixtures increased its tensile strength. The reason for this may be related to the large surface area of nanoparticles which promote the pozzolanic reaction to form C-S-H gel which in turn gives the mixture its strength [22,23].

## Flexural strength

The results of the flexural strength of the control cement pastes and the cement paste samples modified by various percentages (1%, 2%, 3%, 4%, 5% and 6%) of (SiC and TiC) nanoparticles at different curing time (7, 14, 21 and 28 days) are presented in Figures 13-18. It

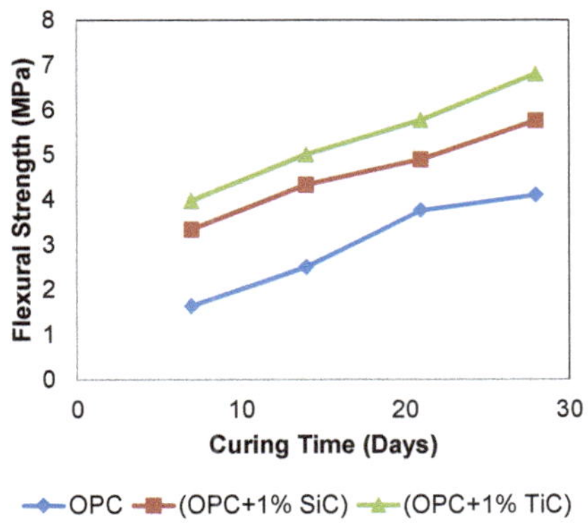


**Figure 13:** Effect of (1% SiC) and (1% TiC) nanoparticles on the flexural strength for the cement paste specimens at different curing time.

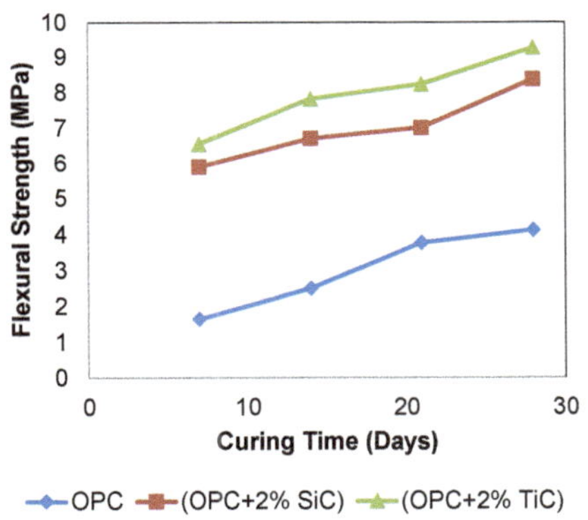


**Figure 14:** Effect of (2% SiC) and (2% TiC) nanoparticles on the flexural strength for the cement paste specimens at different curing time.

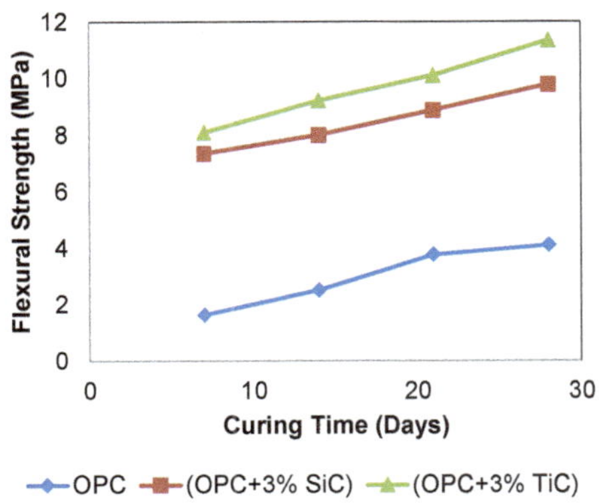


**Figure 15:** Effect of (3% SiC) and (3% TiC) nanoparticles on the flexural strength for the cement paste specimens at different curing time.

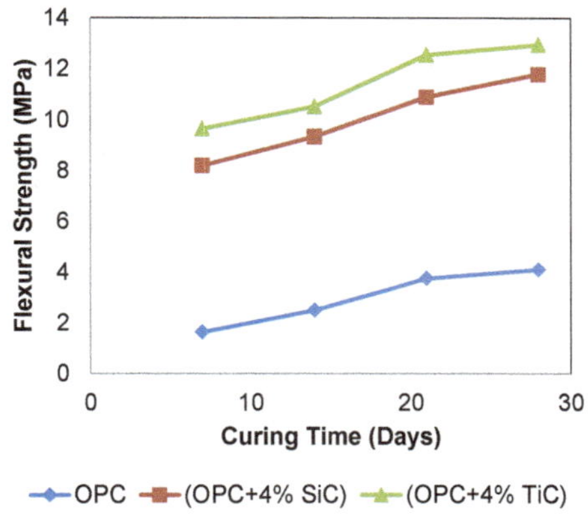


**Figure 16:** Effect of (4% SiC) and (4% TiC) nanoparticles on the flexural strength for the cement paste specimens at different curing time.

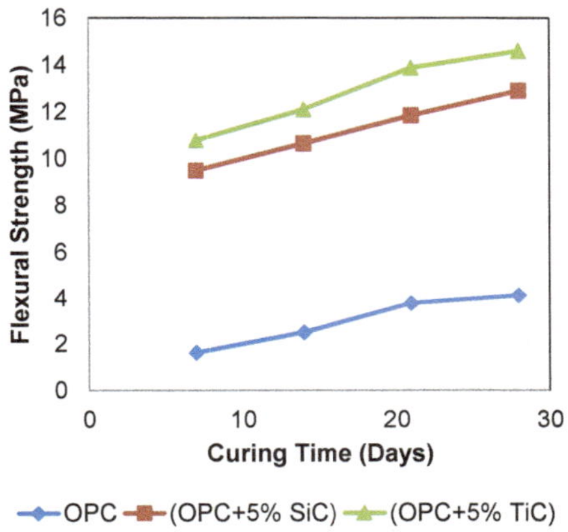


**Figure 17:** Effect of (5% SiC) and (5% TiC) nanoparticles on the flexural strength for the cement paste specimens at different curing time.

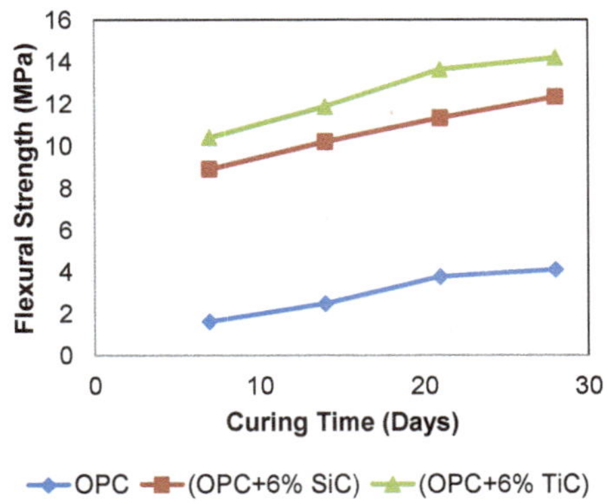


**Figure 18:** Effect of (6% SiC) and (6% TiC) nanoparticles on the flexural strength for the cement paste specimens at different curing time.

can be clearly seen from the figures that the flexural strength increased with increasing nanoparticles percentages and the curing time. This behaviour was attributed to the continuity of hydration process, which induced new hydration by products (calcium silicate hydrate (CSH)) within the cement paste [24,25].

It is also observed that the maximum flexural strength could be achieved when using a combination of 5% TiC and 5% SiC nanoparticles (Figure 17) with the cement pastes at curing time 28 days. The nanoparticles can lead to having a very dense structure, less porosity and reduce capillary voids of the cement pastes. This is because the nanoparticles can react with CH to induce CSH, fill the voids the cement pastes [24,25]. The flexural strength of the cement pastes specimens increased with an increase the percentages of nanoparticles up to (5%). The addition of (5% TiC) and (5% SiC) nanoparticles to the cement pastes specimens resulted in the higher flexural strength than the other experimental specimens.

However, it is interesting to note from the flexural strength results that a little reduction in the flexural strength was associated with adding 6% nanoparticles. This may be related to the fact that the quantity of (TiC) and (SiC) nanoparticles presented in the cement paste mixtures were higher than the required amount to combine with the liberated lime during the hydration process [26,27] Thus, leading to the leaching out of excess silica and causing a reduction in strength It may also occur in the poor dispersion of nanoparticles, which cause weak zones [26,27].

## Water absorption

The water absorption test was conducted in order to study the effect of mixing nanoparticles with the cement paste mixtures. Figures 19-24 show the water absorption test results of the control cement pastes mixtures and the cement paste mixtures modified by various (SiC and TiC) nanoparticles percentages (1%, 2%, 3%, 4%, 5% and 6%) respectively at different curing time (7, 14, 21 and 28 days). It is clearly seen from these figures that the water absorption for the modified cement pastes specimens was lower than the control cement pastes samples. These figures also reveal that the water absorption decreased

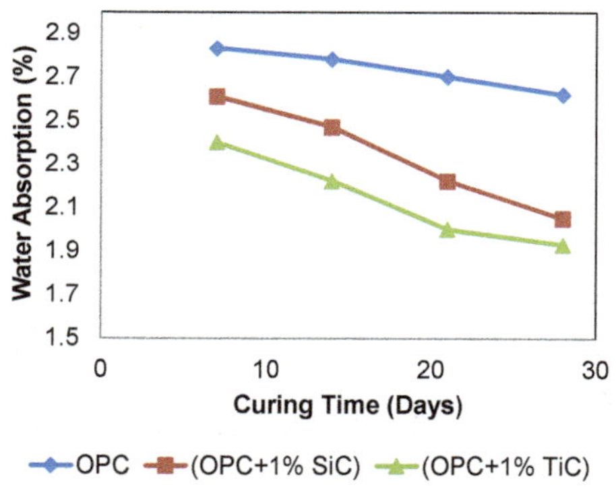


**Figure 19:** Effect of (1% SiC) and (1% TiC) nanoparticles on the water absorption for the cement paste specimens at different curing time.

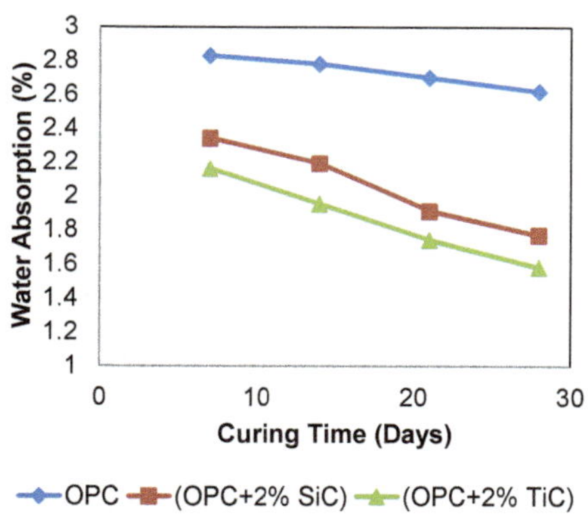


**Figure 20:** Effect of (2% SiC) and (2% TiC) nanoparticles on the water absorption for the cement paste specimens at different curing time.

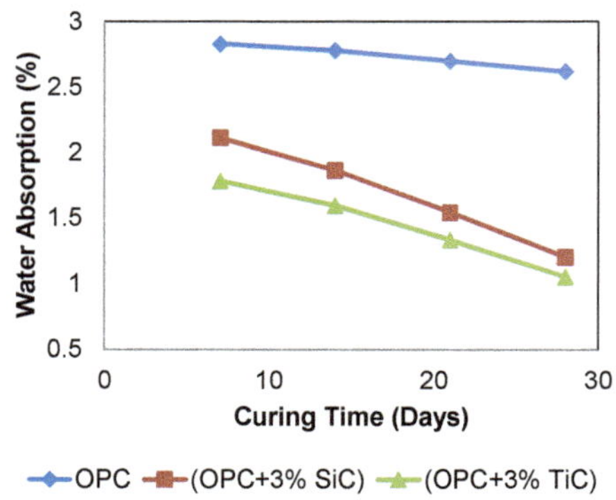


**Figure 21:** Effect of (3% SiC) and (3% TiC) nanoparticles on the water absorption for the cement paste specimens at different curing time.

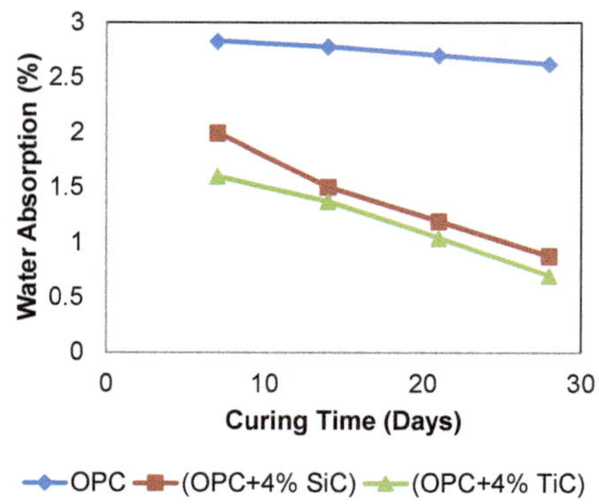


**Figure 22:** Effect of (4% SiC) and (4% TiC) nanoparticles on the water absorption for the cement paste specimens at different curing time.

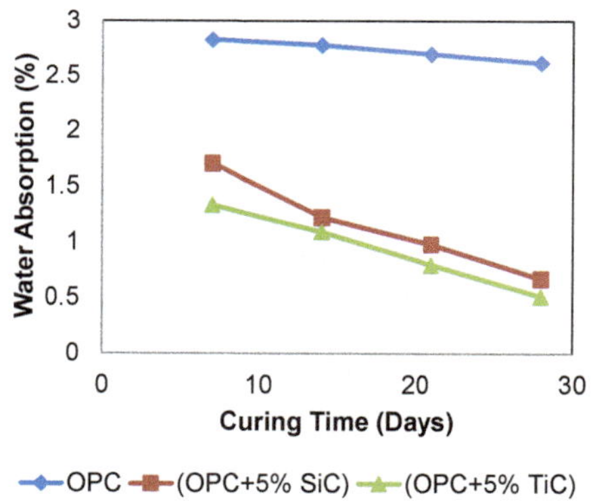


**Figure 23:** Effect of (5% SiC) and (5% TiC) nanoparticles on the water absorption for the cement paste specimens at different curing time.

with increasing curing time and the nanoparticles percentages up to (5%). The lowest water absorption of cement paste samples was associated with 5% TiC and 5% SiC at curing time 28 days. The nanoparticles have generally decreased the total water absorption particularly in the case of using (TiC) nanoparticles which substantially reduced the water absorption. This may be due to "the higher fineness of nanoparticles that led to fill the pores spaces and to disconnect the continuity of the capillary pores" [24].

The testing results also indicated that the absorption of the cement pastes modified by nanoparticles continuously decreased with

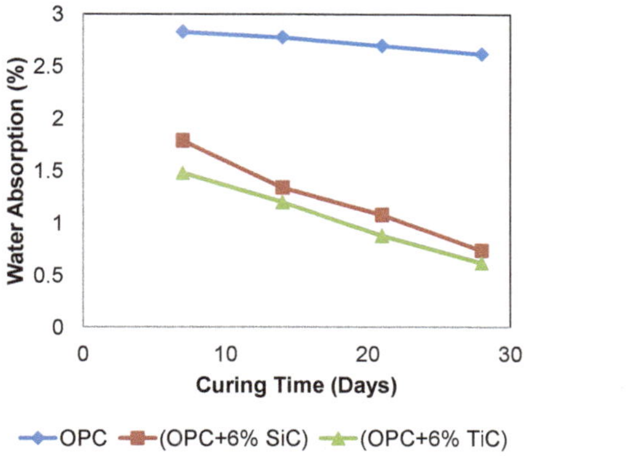


**Figure 24:** Effect of (6% SiC) and (6% TiC) nanoparticles on the water absorption for the cement paste specimens at different curing time.

increasing of the nanoparticles. This is because the partial filling of the cement pastes pores by the hydration by products, which reduce the capillary porosity [24]. The decreasing in water absorption may also be due to the enhancement of the cement paste porosity by both the physical and the chemical mechanisms of the nanoparticles. The nanoparticles can make the pore structure of cement paste more homogeneous by decreasing the number of large pores [26]. However, the absence of the nanoparticles can lead to "growing large crystals of CH that tend to be parallel to aggregate particles surface" [24]. The CH crystals cannot contribute to the strength of cement pastes due to the CH crystals can be easily cleaved [24,28].

## Scanning Electron Microscopy (SEM)

The analysis of structural, quality and morphology of (SiC and TiC) nanoparticles can be identified by SEM. The SEM images shown in Figures 25 and 26 illustrate the morphology and the distribution of SiC and TiC nanoparticle materials, respectively. It can be seen from these figures that some of SiC and TiC particulates have fused together. It is also shown in Figures 25 and 26 that SiC and TiC nanoparticles had various particle structures including spherical particles and irregular shape particles, as well as agglomerated and rounded particles. The tendency for agglomeration of particles may be related to Van-der-Waal forces that act between the individual particles (data not shown).

Figure 27 shows the SEM micrograph of the control cement paste mixture at the curing time of 28 days. It can be noticed from Figure 27 that the control cement pastes mixture had a porous structure that contained a full of large size pores and the presence of $Ca(OH)_2$. The existence of many CH crystals connected to the C-S-H gel indicated that the hydration process was not completed which explained the reduction of mechanical properties for the control cement pastes mixtures. It is also depicted from Fig. 27 that the concentration of the CH was higher than the C-S-H gel concentration, and the CH hydrate needles covered a large area in the control cement pastes specimens [29].

Figures 28 and 29 show SEM of the cement pastes samples with (5% SiC and 5% TiC) nanoparticles, respectively at curing time 28 days that resulted in the highest mechanical properties than the other specimens having different nanoparticles percentages. These Figures show that the cement pastes with (SiC and TiC) nanoparticles samples had lower densities, and were more uniform than the control cement pastes specimens. They also had a small number of $Ca(OH)_2$ crystals and small sized pores. It can also be noted from Figures 28 and 29 that the CH needles were invisible, and there was a compact structure with the absence of the un-hydrated crystals and voids. It is also shown

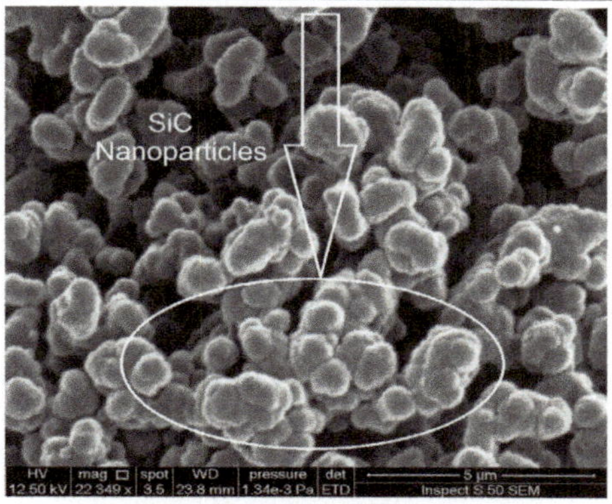


**Figure 25:** SEM image of (SiC) nanoparticles.

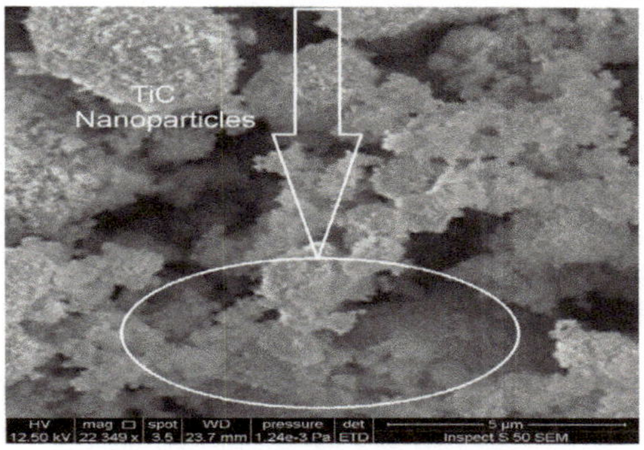


**Figure 26:** SEM image of (TiC) nanoparticles.

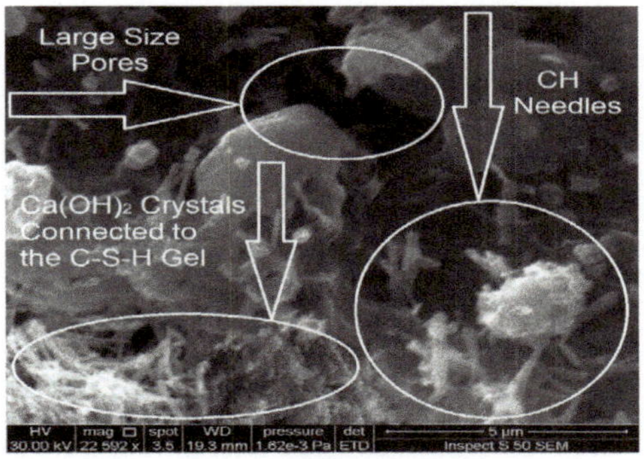


**Figure 27:** SEM image of the cement paste without nanoparticles at curing time 28 days.

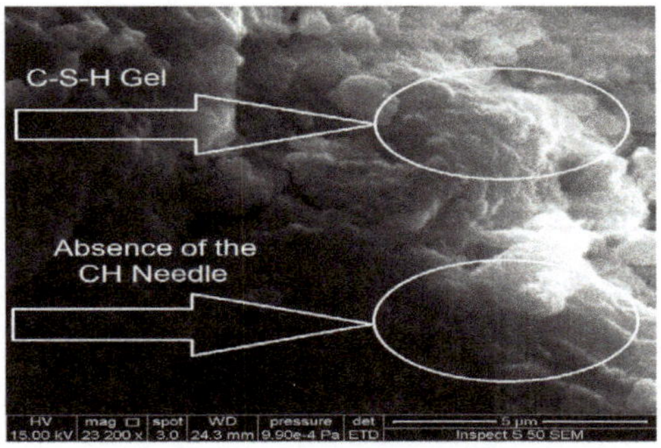


**Figure 28:** SEM image of cement paste with (5% SiC) nanoparticles at curing time 28 days.

from these figures that the cement pastes specimens with nanoparticles were more uniform and homogeneous than the control cement paste specimens. This could interpret the improvements of mechanical properties (flexural strength, splitting tensile strength and Compressive strength). The improvements of mechanical properties may be related to the high activity of many nanoparticles. The nanomaterial can consume "calcium hydroxide crystals, fill pores to increase the strength, reduce the size of the crystals at the interface zone, transmute the calcium hydroxide feeble crystals to the C-S-H crystals, and improve the interface zone and the cement paste structures" [29,30].

## Energy Dispersive X-ray Spectroscopy (EDS)

EDS composition analyses of the (SiC) nanoparticles are shown in Figure 30 and Table 3 The compositions mainly consisted of silicon and carbon, whereas Figure 31 and Table 4 show EDS compositions of (TiC) nanoparticles. The compositions mainly contained titanium and carbon. The EDX composition analysis for the control cement paste sample is shown in Figure 32 and Table 5. Figure 32 reveals that the compositions of the cement paste comprised of carbon, oxygen, silicon, and calcium.

Figures 33 and 34 and Tables 6 and 7 present EDX of the cement pastes modified by (5% SiC and 5% TiC) nanoparticles, respectively. The main reason for conducting EDX on these samples was because using (5% SiC and 5% TiC) resulted in the highest mechanical properties at a curing time 28 days than other specimens having different nanoparticles percentages. It can be seen from Figures 33 and 34 that the chemical compositions of the samples mainly contained carbon, oxygen, silicon, calcium, and nanoparticles. EDS analyses showed that the cement paste samples with nanoparticles had a more uniform distribution and a higher concentration of nanoparticles than the control cement pastes [31].

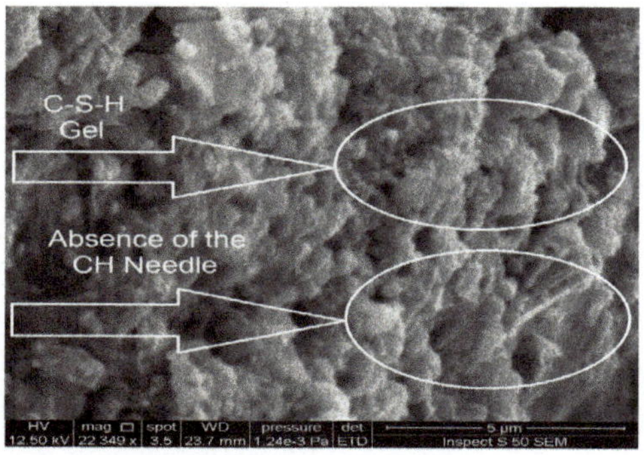


**Figure 29:** SEM image of cement paste with (5% TiC) nanoparticles at curing time 28 days.

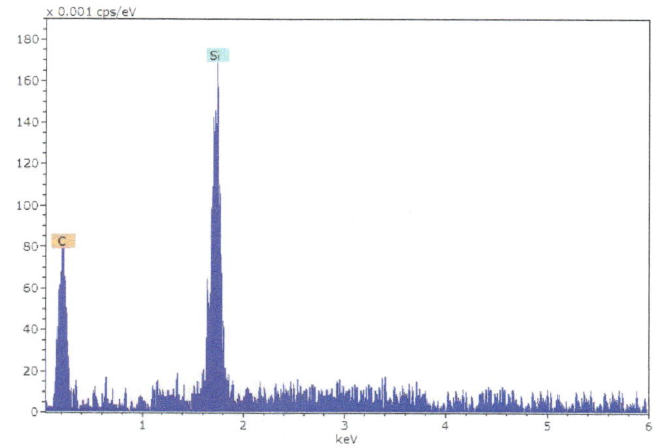


**Figure 30:** EDS analysis of (SiC) nanoparticles.

| Element | Weight % | Atomic % |
|---|---|---|
| C | 27.58 | 46.68 |
| Si | 72.42 | 53.32 |
| Totals | 100 | 100 |

**Table 3:** EDS results of SiC nanoparticles.

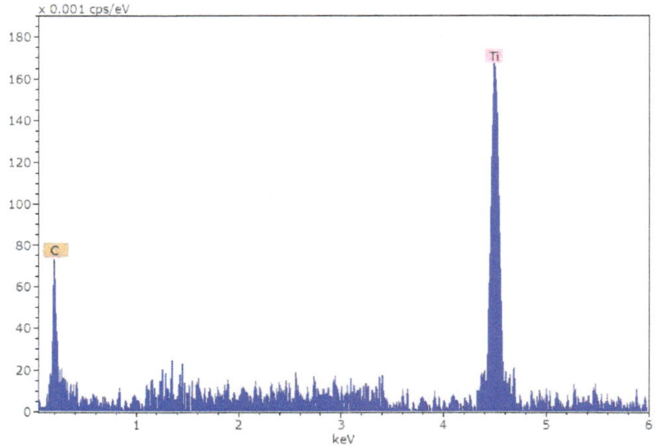


**Figure 31:** EDS analysis of (TiC) nanoparticles.

| Element | Weight % | Atomic % |
|---|---|---|
| C | 26.62 | 45.05 |
| Si | 73.42 | 54.95 |
| Totals | 100 | 100 |

**Table 4:** EDS results of TiC nanoparticles.

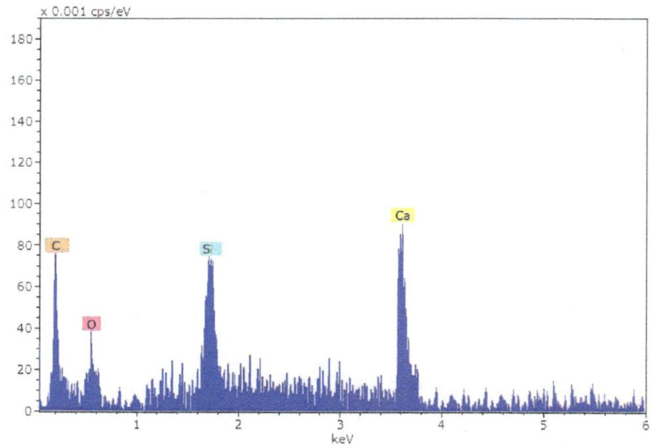


**Figure 32:** EDS analysis of cement paste without nanoparticles at curing time, 28 days.

| Element | Weight % | Atomic % |
|---|---|---|
| C | 26.81 | 45.65 |
| O | 14.64 | 18.72 |
| Si | 26.53 | 19.3 |
| Ca | 32.02 | 16.33 |
| Total | 100 | 100 |

**Table 5:** EDS results of the cement pastes.

## Conclusion

In recent years, different types of nanoparticles have been used in Portland cement in order to improve the pore structure of cement

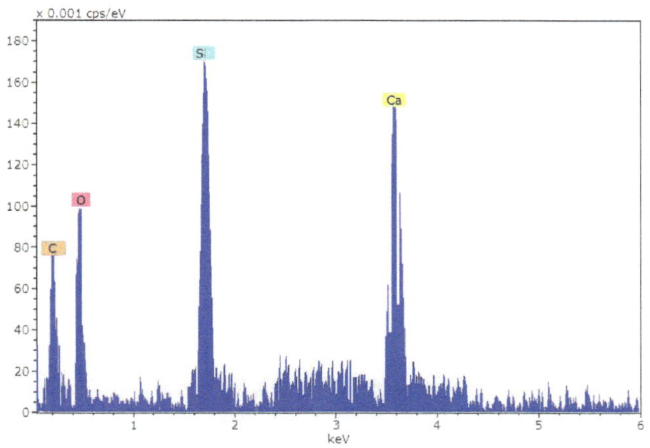


**Figure 33:** EDS analysis of cement paste with (5% SiC) nanoparticles at curing time, 28 days.

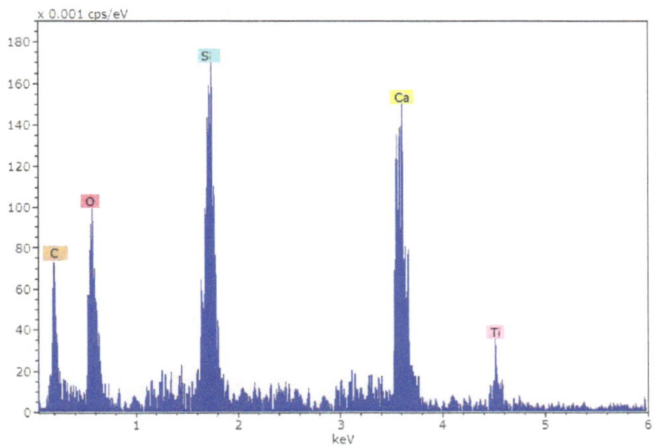


**Figure 34:** EDS analysis of cement paste with (5% TiC) nanoparticles at curing time, 28 days.

| Element | Weight % | Atomic % |
|---|---|---|
| C | 17.73 | 31.61 |
| O | 20.95 | 27.97 |
| Si | 33.69 | 25.59 |
| Ca | 27.63 | 14.83 |
| Total | 100 | 100 |

**Table 6:** EDS results of SiC nanoparticles modified the cement pastes.

| Element | Weight % | Atomic % |
|---|---|---|
| C | 16.07 | 29.24 |
| O | 21.31 | 29.23 |
| Si | 33.03 | 25.72 |
| Ca | 28.66 | 15.6 |
| Ti | 0.93 | 0.21 |
| Total | 100 | 100 |

**Table 7:** EDS results of TiC nanoparticles modified the cement pastes.

pastes. Most research has studied the use of $SiO_2$ nanoparticles. Few studies have incorporated other nanoparticles in cement pastes. In the present study, the influence of (SiC and TiC) nanoparticles on compressive strength, splitting tensile strength, flexural strength and water absorption of cement paste was investigated.

The results obtained can be summarized as follows:

1. Adding nanoparticles has improved the mechanical properties such the compressive strength, the splitting tensile strength and the flexural strength and the maximum values of the mechanical properties were associated with the cement pastes modified by (5% TiC) and (5% SiC) nanoparticles at a curing time 28 days.

2. The physical properties such as water absorption has enhanced by adding the nanoparticles, where the water absorption was substantially reduced. The lowest water absorption obtained from the samples modified with (5% TiC) and (5% SiC) nanoparticles at curing time 28 days.

3. SEM and EDS results indicate that the addition of nanoparticles led to a significant consume of Portlandite (CH) in the pozzolanic reaction, and resulted in a significant reduction of voids and gaps.

4. Increasing the percentages of nanoparticles up to (5%) of (SiC and TiC) enhanced the mechanical and physical properties. However, an adverse trend was detected when the percentage of (SiC and TiC) nanoparticles was increased to (6%) leading to slumped properties.

5. The (mechanical and physical) properties were improved with increasing the curing time (7, 14, 21, and 28 days).

6. The results revealed that the cement pastes treated by (TiC) nanoparticles had higher mechanical and physical properties than (SiC) nanoparticles.

## Acknowledgments

The authors acknowledge the support provided by the department of building and construction engineering of university of technology, Baghdad and the SEM center at university of technology.

## References

1. Sobolev K, Flores I, Hermosillo R, Torres-Martínez LM (2006) Nanomaterials and nanotechnology for high-performance cement composites, Proceedings of ACI Session on Nanotechnology of concrete: Recent developments and future perspectives, Denver, USA.
2. Sanchez F, Sobolev K (2010) Nanotechnology in concrete – A review. Constr Build Mater. 24: 2060-71.
3. Nazari A, Riahi S (2011) $TiO_2$ nanoparticles effects on physical, thermal and mechanical properties of self compacting concrete with ground granulated blast furnace slag as binder. Energ Buildings 43: 995-1002.
4. Zhang MH, Li H (2011) Pore structure and chloride permeability of concrete containing nano-particles for pavement. Constr Build Mater 25: 608-616.
5. Chen J, Kou SC, Poon CS (2012). Hydration and properties of nano-$TiO_2$ blended cement composites. Cement and Concrete Composites 34: 642-649.
6. Diamond S (1986) Concrete structure, properties and materials. Cement Concrete Res 16:790
7. Collepardi M, Collepardi S, Skarp U, Troli R (2004) Optimization of silica fume, fly ash and amorphous nano-silica in super plasticized high-performance concretes. Proceedings of 8th CANMET/ACI International Conference on Fly Ash, Silica Fume, Slag and Natural Pozzolans in Concrete, SP-221.
8. Li H, Xiao HG, Yuan J, Ou J (2004) Microstructure of cement mortar with nano-particles. Composites Part B 35:185-189.
9. Jo BW, Kim CH, Tae GH, Park JB (2007) Characteristics of cement mortar with nano-$SiO_2$ particles. Constr Build Mater 21:1351-1355.
10. Senff L, Hotza D, Repette WL, Ferreira VM, Labrincha JA (2010) Mortars with nano-$SiO_2$ and micro-$SiO_2$ investigated by experimental design. Constr Build Mater 24:1432-1437.
11. Qing Y, Zenan Z, Li S, Rongshen C (2006) A comparative study on the pozzolanic activity between nano-$SiO_2$ and silica fume. Journal of Wuhan University of Technology. Mater Sci Ed 21:153-157.
12. Ulm FJ, Vandamme M (2009) Probing Nano-structure of C-S-H by Micro-

mechanics Based Indentation Techniques. Nanotechnology in Construction 3: 43-53.

13. Zyganitidis I, Stefanidou M, Kalfagiannis N, Logothetidis (2011) Nanomechanical characterization of cement-based pastes enriched with $SiO_2$ nanoparticles. Mater Sci Eng B. 176: 1580-1584.

14. Corr D, Shah SP (2005) Concrete materials science at the Nano scale. Applications of Nanotechnology in Concrete Design 2-12.

15. Quercia G, Brouwers HJH (2010) Application of nano-silica (nS) in concrete mixtures. 8th fib International Ph. D. Symposium in Civil Engineering. Lyngby.

16. Porro A, Dolado JS, Campillo I, Erkizia E, De Miguel Y, et al. (2005) Effects of nano silica additions on cement pastes. Applications of Nanotechnology in Concrete Design 2: 87-96.

17. Metaxa Z, Konsta-Gdoutos MS, Shah SP (2009). Carbon nanotubes reinforced concrete. Special Publication 267: 11-20.

18. Lee BY, Kurtis KE (2010) Influence of $TiO_2$ nanoparticles on early C3S hydration. J Am Ceram Soc 93: 3399-3405.

19. Sobolev K, Gutiérrez MF (2005) How nanotechnology can change the concrete world. American Ceramic Society Bulletin. 84:14-18.

20. Sobolev K, Flores I, Torres-Martinez L, Valdez P, Zarazua E, et al. (2009) Engineering of $SiO_2$ nanoparticles for optimal performance in nano cement-based materials. Nanotechnology in Construction 3:139-148.

21. Said AM, Zeidan MS (2009) Enhancing the reactivity of normal and fly ash concrete using colloidal nano-silica. Special Publication 267: 75-86.

22. Hosseini P, Booshehrian A, Farshchi S (2010) Influence of nano-$SiO_2$ addition on microstructure and mechanical properties of cement mortars for ferrocement. Transp Res Rec 2141: 15-20.

23. Haruehansapong S, Pulngern T, Chucheepsakul S (2014) Effect of the particle size of nanosilica on the compressive strength and the optimum replacement content of cement mortar containing nano-$SiO_2$. Constr Build Mater 50: 471-477.

24. Jahren P (1983) Use of silica fume in concrete International Concrete Abstracts Portal, Special Publication 79: 625-642.

25. Gjørv OE, Sakai K, Banthia N (1998) Concrete under severe conditions 2: Environment and loading: Proceedings of the Second International Conference on concrete under severe conditions, CONSEC'98, Norway, CRC Press, USA.

26. Lee SJ, Kriven WM (2005). Synthesis and hydration study of Portland cement components prepared by the organic steric entrapment method. Mater Struct 38: 87-92.

27. Li H, Zhang MH, Ou JP (2007) Flexural fatigue performance of concrete containing nano-particles for pavement. Int J Fatigue 29: 1292-1301.

28. Neville AM (1995) Properties of concrete (5th edn), Pearson Education Limited, Edinburgh-Harlow, England.

29. Quercia G, Spiesz P, Hüsken G, Brouwers HJH (2014) SCC modification by use of amorphous nano-silica. Cem Concr Compos 45: 69-81.

30. Shah VS (2004) Detection of microcracks in concrete cured at elevated temperature. PhD dissertation, University of Florida, USA.

31. Li G (2004) Properties of high-volume fly ash concrete incorporating nano-$SiO_2$. Cem Concr Res 34: 1043-1049.

# Evaluation of the Effect of Aggregates Mineralogy and Geometry on Asphalt Mixture Friction

**Marco Magnoni*, Filippo Giustozzi, Emanuele Toraldo and Maurizio Crispino**

*Politecnico di Milano – Department of Civil and Environmental Engineering , Transportation Infrastructures, Piazza Leonardo da Vinci 32, Milan 20133, Italy*

**Abstract**

Road pavement friction is a main physical phenomenon of interaction between tires and the road surface; it provides skid resistance during accelerations and decelerations. The paper focuses on the effects, at laboratory scale, of geometry (in terms of fine aggregates - minimum/maximum size of 0.063/2 mm - content), mineralogy (basalt or metamorphic - minimum/maximum size of 8/16 mm) and shape (in terms on Flakiness and Shape index) on friction performance of asphalt wearing courses for road pavements, in terms of British Pendulum Number and Mean Texture Depth. Laboratory results were statistically treated and used to define predictive formulas which correlate friction characteristics and the explanatory variables considered by the study. The amount of fine particles in the mix and the shape of coarse aggregates demonstrated to affect friction properties and macro texture.

**Keywords:** Asphalt; Friction; Macro-texture; Mean texture depth; Micro-texture; Skid resistance; Road pavement

## Introduction

Road and airport pavements should guarantee high performance in terms of friction and bearing capacity to allow safety movements of vehicles and aircrafts; it is well-known that several accidents are indeed caused by a lack of friction. Road engineers commonly distinguish among four variables of the pavement surface depending on deviations from a planar surface: micro-texture, macro-texture, mega-texture, and irregularity [1]. Micro and macro-texture are particularly investigated due to their great influence on pavement friction. Micro-texture is provided by the roughness of the individual aggregate particle; it is measured with wavelengths smaller than 0.5 mm and it is mainly connected to the aggregate mineralogy.

Macro-texture, usually between 0.5 mm and 50 mm, is mainly attributed to the size, shape, angularity and distribution of the aggregates on the surface; inadequate macro-texture reduces water drainage effectiveness and, consequently, pavement friction, particularly at high speeds, while increases splash & spray and tire-pavement noise. Surface texture is strictly connected to skid resistance, which is divided in two major components: adhesion and hysteresis [2]. Adhesion, which is proportional to aggregates micro-texture, is computed by multiplying the interface shear strength, developed between tire and pavement surface, and the actual contact area.

When the tire and the asphalt pavement come into contact, molecular bonds, which need some energy to get broken, are thus established: the energy involved in the breakage is the adhesion force. On the other hand, hysteresis is connected to macro-texture: the energy produced by the deformation of the rubber of the tire, alternately compressed and expanded when it rolls over the pavement asperities, accentuates the grip and creates the hysteresis component. Gunaratne et al. [3] defined a correlation between wet and dry friction coefficients and the texture properties; they found a stronger dependency of the dry friction coefficient on adhesion (due to micro-texture), while the wet friction coefficient depended mostly on hysteresis force (related to macro-texture).

The friction of a road pavement is mainly affected by the following factors: void content, surface wear, polishing of surface aggregates, rutting due to post-compaction effects or lateral distortions, bleeding and flushing of the asphalt binder to the surface, contamination, and type of aggregate. Petrographic and mineralogical nature of the different type of aggregates and their relative percentage within the mix represent two essential parameters to define micro and macro-texture of the pavement. The present paper focuses on the effects, at laboratory scale, of geometry (in terms of fine aggregates - minimum/maximum size of 0.063/2 mm - content), mineralogy (basalt or metamorphic - minimum/maximum size of 8/16 mm) and shape (in terms on Flakiness and Shaped index) on the friction performance of slab-shaped asphalt samples, in terms of British Pendulum Number and Mean Texture Depth.

## Literature Review

Several studies evaluated the tire-pavement interaction in terms of BPN and MTD depending on aggregate characteristics, polishing, surface texture, and temperature. Choo [4] used the BPN to analyze granite and steel slag aggregates and the effect of aggregate gap width on the frictional resistance; it was found that the skid resistance was positively related to the magnitude of the sliding contact surface and the number of gaps within the sliding area. Other authors tried to establish a relationship between aggregate characteristics and friction: Kanitpong [5] considered the aggregate type as a significant factor which influenced the micro-texture and the macro-texture. Results from multiple linear regression analysis showed that the Polished Stone Value (PSV) affected the BPN, while fineness modulus influenced MTD values; furthermore, both the values were correlated with angularity and texture.

Shah et al. [6] evaluated the effect of the aggregate shape: angular aggregates showed greater skid resistance, followed by elongated and flaky aggregates, which exhibited the worst BPN. Davis [7] found a relationship between the Nominal Maximum Size (NMS) of aggregate and Mean Profile Depth (MPD), which can be correlated with the MTD too [8,9] and voids in the mixture and Skid Number: he found an increasing in Skid Number when voids in the mixture increase (more voids in the mixture result in more surface voids, which would allow

***Corresponding author:** Marco Magnoni, Politecnico di Milano – Department of Civil and Environmental Engineering, Transportation Infrastructures, Piazza Leonardo da Vinci 32, Milan 20133, Italy, E-mail: marco.magnoni@polimi.it

water to flow from the tire-pavement interface), and an increasing in MPD when NMS increases.

In a research led by Sengoz [10] basaltic aggregate exhibited higher angularity values compared to limestone aggregate: this was probably due to the mineral grain. During testing, the slabs prepared by using basaltic aggregate showed greater MTD and MPD values compared to limestone slabs. Also, dynamic friction tester provided greater friction values for basaltic. Several authors tried to predict friction evolution over time due to texture deterioration.

Eyad [11] did not find any significant difference in MTD values before and after the polishing of the slabs of asphalt mixture (some values increased, while other decreased), but mixes with different aggregates provided different MTD values according to the following principle: the smaller nominal aggregate size provided a reduced macro-texture and BPN values showed a decreasing trend with an increasing in the polishing.

Vaiana [8] monitored an experimental road section for 2 years and analyzed four different dense graded wearing courses (100% limestone aggregate, 85% limestone aggregate and 15% basaltic aggregate, 70% limestone aggregate and 30% basaltic aggregate, 82% limestone aggregate and 8% expanded clay); results indicated that aggregates with smaller PSV values – e.g.; limestone – had a lower resistance to stripping and became more exposed to traffic deterioration. Macro-texture analysis pointed out an initial reduction due to the migration phenomenon of the asphalt binder inside the voids of the mix, while, after 6 months, the migrated binder was removed by the traffic and MTD began to increase.

Chen et al. [12] evaluated the evolution of the aggregate-surface texture during polishing processes: the Wehner/Schulze polishing machine was adopted to analyze three types of aggregate (basaltic aggregate, greywacke aggregate, and gabbro aggregate). It was found that a) the basaltic aggregate provided the worst resistance to polishing if compared with gabbro and greywacke – which exhibited the best skid resistance, power spectral density curves confirmed the global height reduction, and b) the evolution of aggregate friction was governed by micro-texture changes due to differential removal of mineral components.

Significant contributions about the texture-friction relationship were made by Tsohos [13] who identified a correlation between skid resistance of a pavement surface and its fractal dimension; a pavement with a rough fractal interpolation function exhibited a greater skid value than a pavement with a smooth fractal interpolation function.

Empirical Mode Decomposition analysis, which is part of Hilbert-Huang Transformation, was used by Kane [14] to set basic profiles, called Intrinsic Mode Funtions (IMFs). Base Intrinsic Mode Functions (BIMFs), which are the sum of four IMFs, were then introduced, and they were characterized by the density and sharpness of the peaks. The correlation between texture and friction was analyzed through the Dynamic Friction Texter (DFT) and the Circular Texture Meter (CTM) to find a strong correlation between BIMFs, expressed by density per sharpness, and the measured friction: the increasing of BIMF pointed out an increasing of DFT (which was measured at three different speeds – 20 km/h, 40 km/h and 60 km/h).

Bogardi [15] applied the International Friction Index (IFI) to connect skid resistance and surface macro-texture in a probabilistic pavement performance modeling framework; IFI is useful to convert the device-specific friction and macro-texture measurements into an international scale and it was used to define threshold values to assess the adequacy of pavement friction.

Gunaratne [3] predicted tire wear rate and tire-pavement friction on smooth concrete pavements by using Fourier Transform technique and constructing power spectral density plots of texture: he found a good correlation among the dry friction coefficient, mostly based on adhesion, and pavement micro-texture; furthermore, it was seen that whenever the macro-texture increased, the contact between the tire and individual aggregate particles was reduced and, on the other hand, when micro-texture increased, the surface became rougher owing to the increased contact between the tire and individual elements.

Bazlamit [16] and Voussough [17] investigated the effect of temperature on friction properties. They found a strong dependence among the two elements: skid resistance decreased when the temperature increased following a linear relationship; for this reason, compensations should be made to account for temperature effects, as reported in the European standard EN 13036.

## Objective and Experimental Plan

The main objective of the study was to define, at laboratory scale, how friction performance of asphalt pavements, in terms of micro and macro texture, is influenced by:

- mineralogy, geometry and shape of lithic aggregates;
- Volumetric characteristics of the mixtures.

To this end, six types of aggregates currently used in Italy were selected to produce laboratory slab-shaped asphalt samples and measure their volumetric characteristics (void contents) and friction performance in terms of micro texture, by means the British Pendulum Number (BPN); macro texture was instead determined by the Mean Texture Depth (MTD).

Results were firstly treated by using the Cook's distance to identify and remove the outliers.

Then, Analysis of Variance (ANOVA) was used to validate laboratory results, and the possible correlations among the explanatory variables (mineralogy, geometry and shape of the aggregates and volumetric characteristics of the asphalt slabs) and friction performance of the corresponding slabs were computed with the Bravais-Pearson linear correlation method. Finally, prediction formulas according to the least square method and multiple linear regressions were identified.

A detailed flow-chart of the experimental plan is shown in Figure 1.

## Experimentation: Materials and Test Results

The aggregates considered in this study were, in terms of mineralogy (and geometry): crushed metamorphic (8/16 mm; Los Angeles Index - LA of 15%; Polished Stone Value - PSV of 42.8%), crushed basalt (8/16 mm; LA of 16%, PSV of 43.4%), crushed limestone (8/16 mm; LA of 14%; PSV of 51.4% - 4/8 mm; LA of 15%; PSV of 51.4%), sand (2/4 mm), fine aggregate (0.063/2 mm). Although some PSV values could be considered too low for some countries, they are perfectly suitable for the current Italian standards.

As far as the shape of the aggregates, flaky and elongated shapes were considered in terms of Shape Index (SI, EN 933-4) and Flakiness Index (FI, EN 933-3), as reported in Table 1.

Aggregates were then used to produce eighteen asphalt mixtures, including the reference mix, which was also adopted as wearing course for a newly constructed highway in the North of Italy.

50/70 pen bitumen (6% by weight of the total aggregates) and filler (11%) contents were maintained constant for all the mixtures.

The investigated mixtures were divided into two groups:

– Mix 1 to 9, Mix Z, Mix ZZ and reference mixture were obtained by modifying the amount of fine and basaltic aggregates (Table 2);

– Mix 10 to Mix 15 were manufactured with a specific sieve size distribution but varying particles geometry (SI and FI index), as shown in Table 3.

For each mixture, the groups of aggregate were mixed together to reestablish the same gradation of the reference mix and, in any case, within the dotted-line limit of the sieve size distribution shown in Figure 2.

Laboratory sample production included asphalt mixture preparation, using a common laboratory mixer ($T_{mix}$ = 160°C ± 5°C), and compaction ($T_{comp}$ = 150°C ± 5°C) using a roller compactor (EN 12697-33, Procedure C), obtaining two slabs for each mixture. The void content of the slabs (Figure 3) was computed according to EN 12697-8: bulk density of the slabs was measured (EN 12697-6 Procedure B) and the Maximum Theoretical Density (EN 12697-5) was determined on the corresponding loose mixes. For each mix it was computed the average value between six measures.

As mentioned above, friction-related characteristics of the slabs were evaluated by both the BPN according to EN 13036-4, which is an indicator of the micro-texture of a wearing course, and the MTD obtained by the Sand Patch Method Test (EN 13036-1), currently used to determine the macro-texture of a surface mixture. BPN was measured on each slab according to 5 alignments (Figures 4a and 4b) in order to obtain a satisfactory number of test results and avoid possible influence of edge effects and compaction methodology: three alignments in the longitudinal direction (equal to the compaction direction) and two in the transversal direction. Six measures were conducted on each alignment therefore resulting in 30 measurement points on each slab. The average friction value for each asphalt mixture was evaluated according to the measurements on two slabs (30 + 30 measurements on each mix). As reported in Figure 4c, only one measure of MTD was taken on each slab (1 + 1 measurements on each mixture).

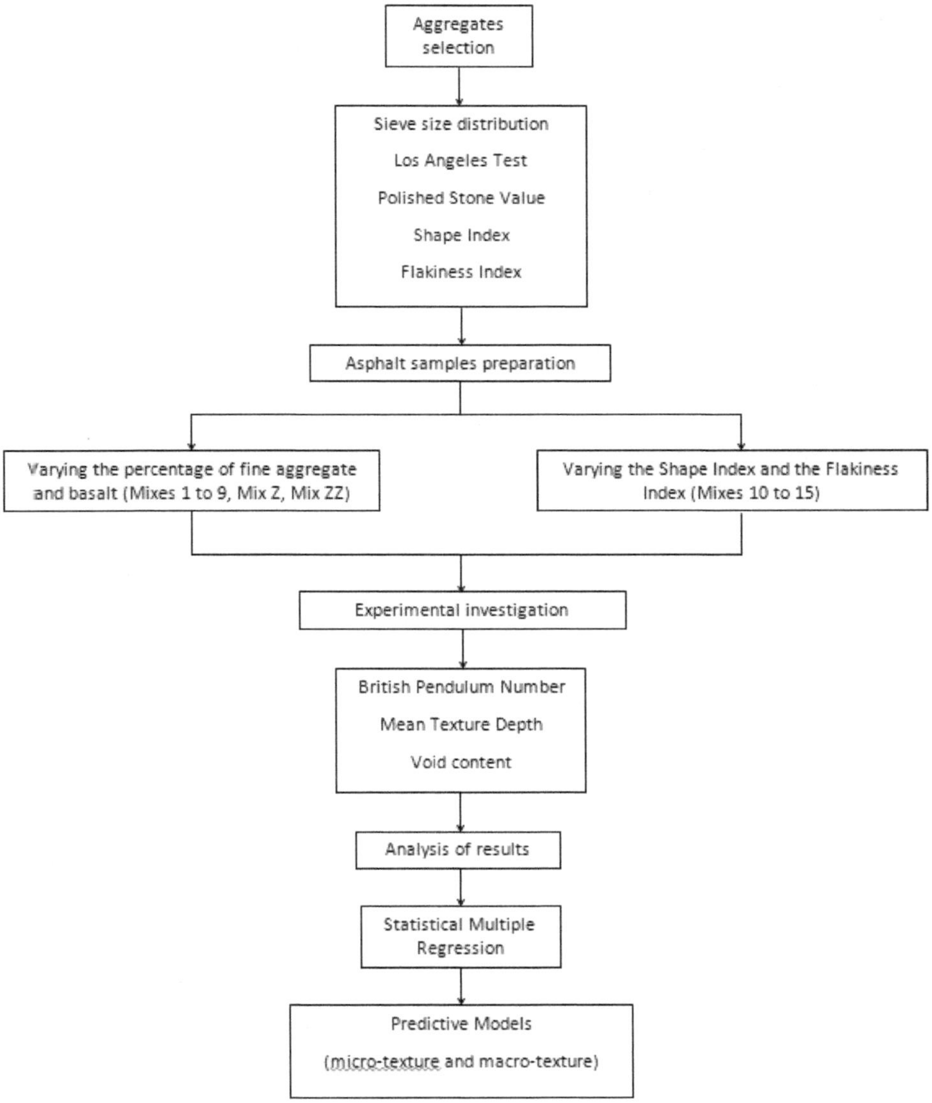


**Figure 1:** Experimental program.

| Type of aggregate | SI [%] | FI [%] |
|---|---|---|
| Metamorphic 8/16 mm | 7.9 | 8.9 |
| Basaltic 8/16 mm | 11.4 | 13.0 |
| Limestone 8/16 mm | 6.5 | 16.6 |
| Limestone 4/8 mm | 8.0 | 9.9 |

**Table 1:** Shape index and flakiness index of the aggregates.

| | Fine aggregate 7.1 % | | | Fine aggregate 11.9 % | | | Fine aggregate 14.2 % | | | Fine aggregate 16.6 % | | |
|---|---|---|---|---|---|---|---|---|---|---|---|---|
| Basaltic aggregate 0% | - | | | - | | | **Reference mix** | | | - | | |
| | | | | | | | sand | 2/4 mm | 3.80% | | | |
| | | | | | | | Limestone aggregate | 4/8 mm | 29% | | | |
| | | | | | | | | 8/16 mm | 30% | | | |
| | | | | | | | Metamorphic Aggregate | 8/16 mm | 12% | | | |
| Basaltic aggregate 9% | - | | | - | | | **Mix Z** | | | - | | |
| | | | | | | | sand | 2/4 mm | 3.80% | | | |
| | | | | | | | Limestone aggregate | 4/8 mm | 25% | | | |
| | | | | | | | | 8/16 mm | 37% | | | |
| Basaltic aggregate 19% | - | | | **Mix 1** | | | **Mix 2** | | | **Mix 3** | | |
| | | | | sand | 2/4 mm | 3.10% | sand | 2/4 mm | 3.80% | sand | 2/4 mm | 4.40% |
| | | | | Limestone aggregate | 4/8 mm | 34% | Limestone aggregate | 4/8 mm | 28% | Limestone aggregate | 4/8 mm | 26% |
| | | | | | 8/16 mm | 21% | | 8/16 mm | 24% | | 8/16 mm | 23% |
| Basaltic aggregate 26% | **Mix ZZ** | | | **Mix 4** | | | **Mix 5** | | | **Mix 6** | | |
| | sand | 2/4 mm | 1.90% | sand | 2/4 mm | 3.10% | sand | 2/4 mm | 3.80% | sand | 2/4 mm | 4.40% |
| | Limestone aggregate | 4/8 mm | 28% | Limestone aggregate | 4/8 mm | 28% | Limestone aggregate | 4/8 mm | 32% | Limestone aggregate | 4/8 mm | 29% |
| | | 8/16 mm | 26% | | 8/16 mm | 20% | | 8/16 mm | 13% | | 8/16 mm | 13% |
| Basaltic aggregate 34% | - | | | **Mix 7** | | | **Mix 8** | | | **Mix 9** | | |
| | | | | sand | 2/4 mm | 3.10% | sand | 2/4 mm | 3.80% | sand | 2/4 mm | 4.40% |
| | | | | Limestone aggregate | 4/8 mm | 32% | Limestone aggregate | 4/8 mm | 31% | Limestone aggregate | 4/8 mm | 28% |
| | | | | | 8/16 mm | 8% | | 8/16 mm | 6% | | 8/16 mm | 6% |

**Table 2:** Mix 1 to 9, mix Z, and mix ZZ.

| | SI = 8% | SI = 12% | SI = 17% |
|---|---|---|---|
| FI = 8% | mix 10 | - | - |
| FI = 12% | mix 11 | mix 12 | - |
| FI = 17% | mix 13 | mix 14 | mix 15 |

**Table 3:** Shape index and flakiness index matrix, mix 10 to 15.

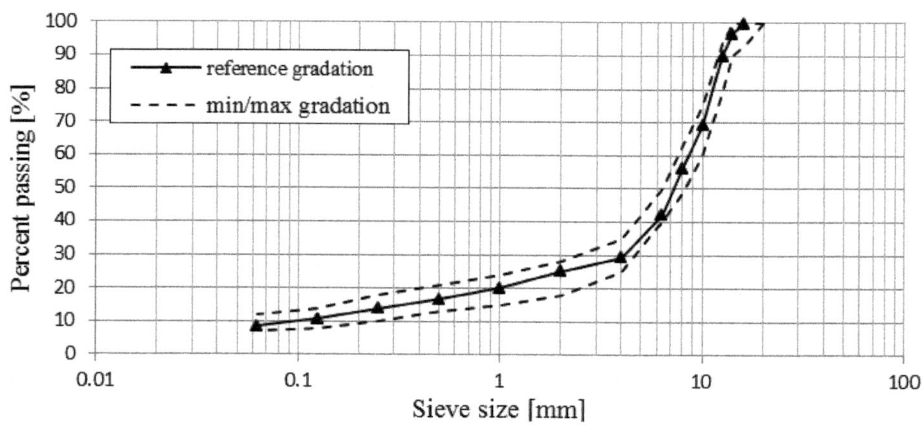


**Figure 2:** Sieve size gradation of the reference mix.

Texture characteristics of the mixtures are given in Figure 5, in which maximum and minimum results are shown as error bars.

## Analysis and Discussion of the Results

### Linear regression

The first step of the analysis was performed taking into account the effects of mineralogy (content of coarse basaltic aggregate), dimension (fine aggregate content) and shape (Flakiness and Shape index) of lithic particles on the friction performance (BPN and MTD) of asphalt mixes. To this end, a regression analysis was used. Prior to that, the Cook's Distance statistical method was used to identify and remove outliers from the friction results.

Graphs in Figures 6 and 7 show the correlation between friction performance (BPN and MTD) and both mineralogy (Figure 6) and dimension (Figure 7) of the aggregates (mixtures 1-9, Z, ZZ, and Reference mix). Volumetric characteristics (in terms of void content) were also considered, as shown in Figure 8.

Analyzing the results in Graphs A-C of Figures 6 and, a general increase of BPN (Figures 6 and 7) can be observed according to both fine and basaltic aggregates content. A slight decrease was instead detected for MTD values (Graphs D, E and F of Figures 6 and 7).

Considering the coefficient of determination ($R^2$), which define the quality of a correlation from a statistical point of view, results revealed that BPN (microtexture) had a good correlation with the investigated

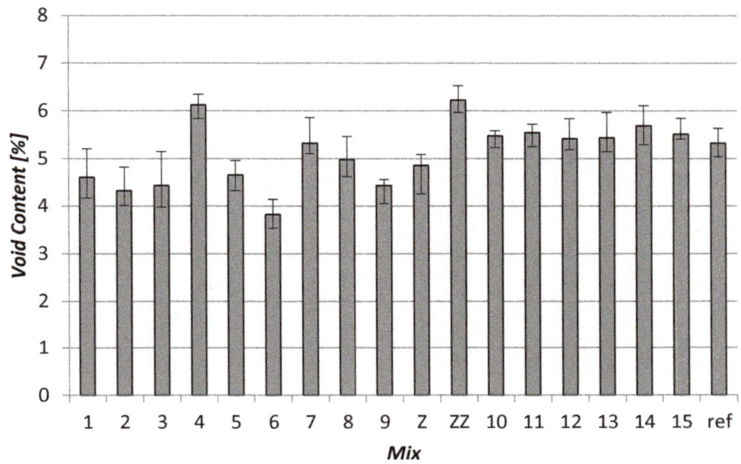


**Figure 3:** Average void content into the asphalt slabs.

**Figure 4:** (a) and (b): BPN alignments; (c) Sand patch method tests.

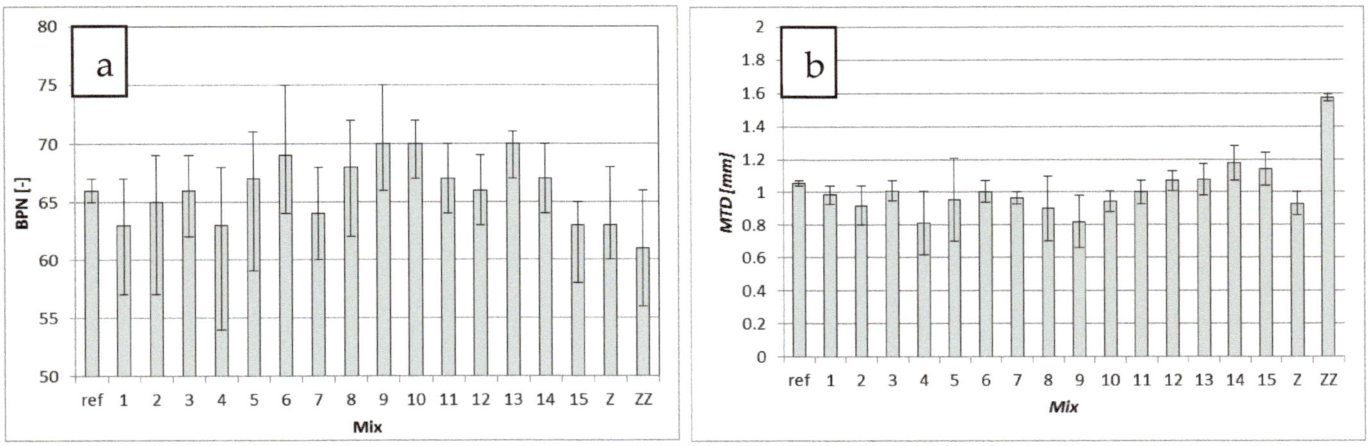


**Figure 5:** (a) Mix versus BPN (B) Mix versus MTD.

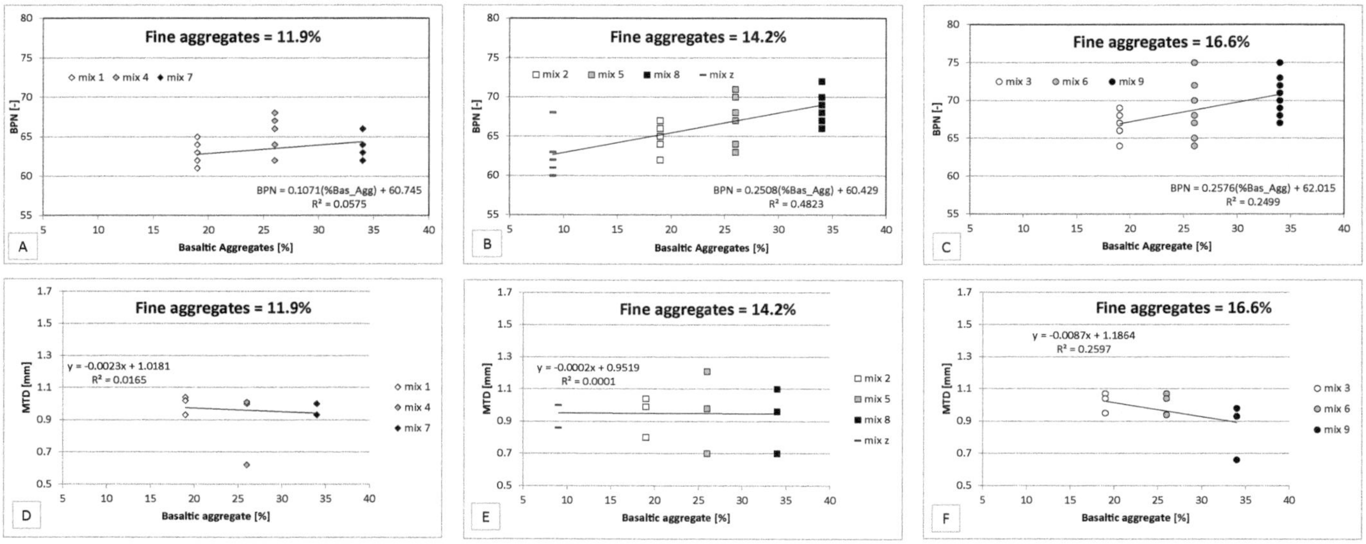


**Figure 6:** Basaltic coarse aggregate contents versus friction (BPN and MTD) results (Mixtures 1-9 Z, ZZ and Ref).

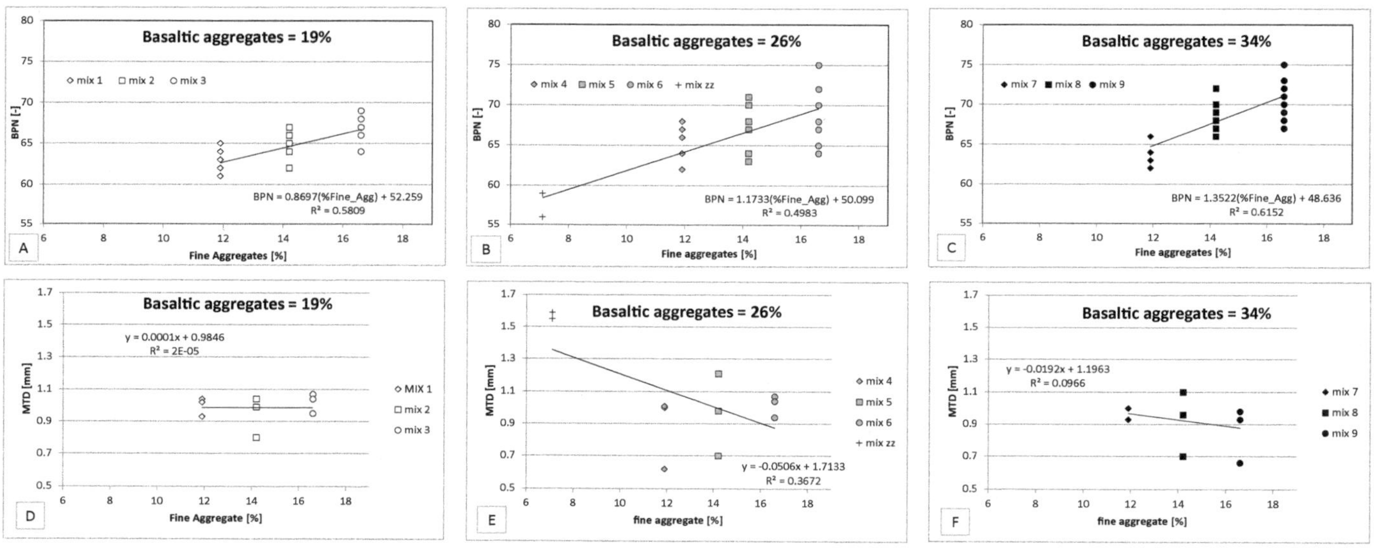


**Figure 7:** Fine aggregate contents versus friction (BPN and MTD) results (Mixtures 1-9 Z, ZZ and Ref).

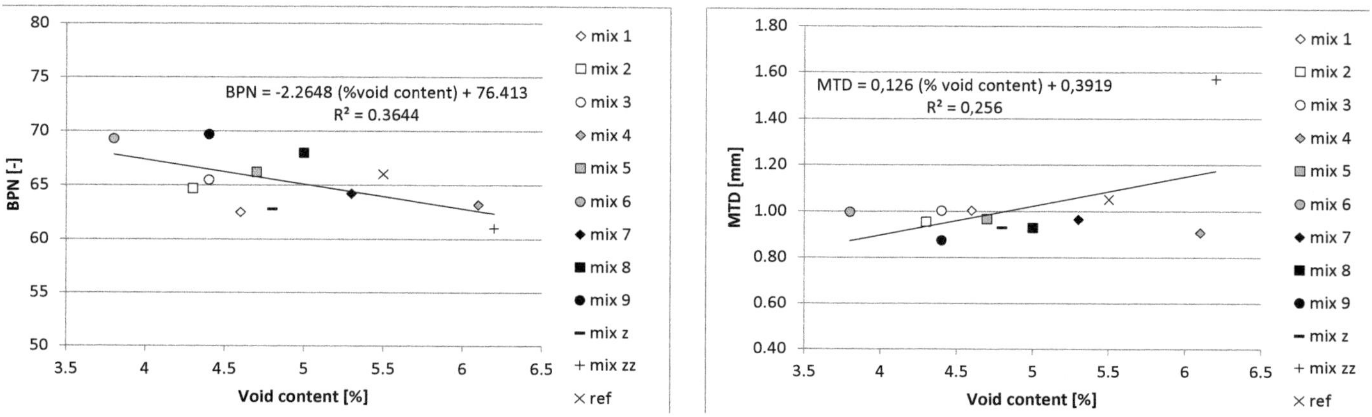


**Figure 8:** Void content versus friction (BPN and MTD) results (Mixtures 1-9, Z, ZZ and Ref).

variables, in particular when fine aggregate-mixes were investigated. On the contrary, unsatisfactory results were obtained in terms of macrotexture (MTD). As far as the compaction effects, results in Figure 8 indicate that BPN increases and MTD decreases for reduced void content. Thus, BPN decreases because the number of asperities per unit of surface decreases, diminishing the ability of the pavement to brake the pendulum during the tests. On the contrary, the same decrease of asperities increases the number of surface voids that manages the ability of the sand to penetrate into the surface during the Sand Patch test. Finally, the coefficient of determination ($R^2$) was found to be lower.

Graphs in Figure 9, shows the correlation between friction results (BPN and MTD) and shape of the aggregates (SI and FI), mixture 10 to 15. The comparison between friction parameters and volumetric characteristics was not performed since the asphalt slabs were compacted at the same content of voids (5.5%). Graphs A, B and C of Figure 9 indicate that BPN is more sensible to SI than to FI. In particular, a decrease of this friction parameter can be observed due to the increase of SI index. On the contrary, an increasing of MTD according to FI and SI was observed. However, in both cases the determination coefficient ($R^2$) is not very high, except for BPN versus SI (Graph A), and for both friction measures on slabs with SI=FI.

## Statistical validation of the results

For validation purposes, correlations discussed in the previous paragraph were computed with the Bravais-Pearson linear correlation method, which is an index defined as the covariance of two statistical variables divided by the product of the standard deviations [18]. The correlation index evaluates the relationship between two statistical variables and ranges between +1 and -1: if the coefficient is approximately +/- 1 it means that there is a strong direct/inverse correlation, while if it is close to 0 there is no correlation between the considered variables.

The Bravais-Pearson correlation indexes for mixes 1-9, Z, ZZ and Reference mix are given in Table 4. Table 5 reports the same indexes for mixes 10-15. Given the complexity of the goals, it is more useful to discuss the Bravais-Pearson indexes considering three ranges. Particularly, the range between 0 and ±0.3 can be defined as a poor level of correlation, the ranges +0.3/+0.7 and -0.3/- 0.7 can be defined as good, and finally over ±0.7 the level of correlation is high. As far as data given in Table 4, it is possible to claim that dimension of the aggregates are more important to their mineralogy; actually both coarse and fine aggregates have a good Bravias-Pearson Index with the surface texture parameters (BPN and MTD); on the contrary, basalt appears not to affect the above-mentioned parameters. Finally, as expected, voids content affects the surface characteristic of a wearing course; in particular, if void content increases, micro-texture is reduced, thus the BPN value decreases and the macro-texture is increased and, consequently, the MTD values rise.

Referring to the shape of the aggregates, results in Table 5 identify the Shape Index (SI) as the parameter that mainly influences the surface characteristics of an asphalt wearing course, especially in terms of micro-texture (BPN), even if also the FI index shows a good

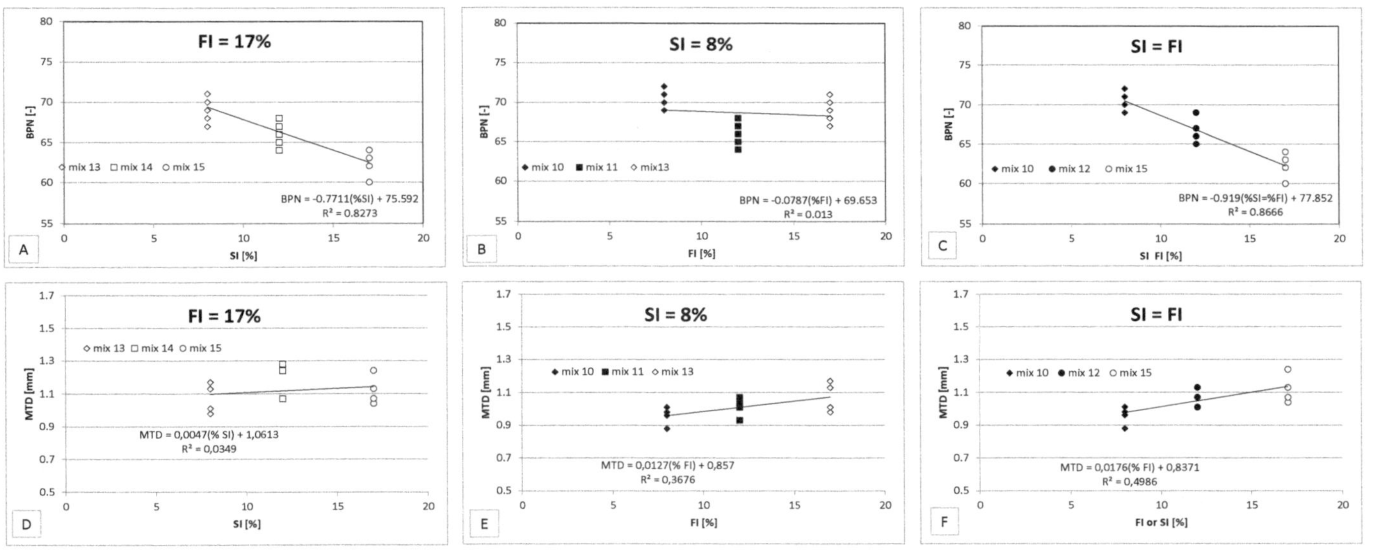


**Figure 9:** Shape of coarse aggregates versus friction (BPN and MTD) results (Mixtures 10-15).

| | BPN [-] | MTD [mm] |
|---|---|---|
| Fine aggregate (0.0063/2 mm) [%] | 0.46 | -0.40 |
| Coarse Aggregate (8/16 mm) [%] | -0.37 | 0.43 |
| Basaltic aggregate (8/16 mm) [%] | 0.19 | -0.21 |
| Voids content [%] | -0.35 | 0.45 |

**Table 4:** Bravais-Pearson correlation indexes for mixes 1-9, Z and ZZ.

| | BPN [-] | MTD [mm] |
|---|---|---|
| SI [%] | -0.71 | 0.30 |
| FI [%] | -0.34 | 0.22 |

**Table 5:** correlation coefficients between friction properties, volumetric features, and mix design (mixes 10 to 15).

correlation. On the contrary, MTD appeared not to be significantly affected by both aggregate shape indexes.

Finally, the Bravais-Pearson correlation index confirms the trend obtained using the linear regression method discussed in the paragraph 5.1.

## Pavement friction predictive formulas

The analysis of variance (ANOVA) was conducted to identify predictive formulas able to define the multiple linear relations (regression) among the friction characteristics (BPN and MTD) of the investigated asphalt wearing courses and the explanatory variables considered in the study. For this reason, six predictive formulas were identified as reported in Table 6, in which PF-1-BPN and PF-1-MTD are related to dimensions and mineralogy of the aggregates, PF-2-BPN and PF-2-MTD are associated to the shape of the aggregates (SI and FI), PF-3-BPN and PF-3-MTD describe the relation with the volumetric characteristics of the asphalt mixtures (void content).

The multiple linear regression allowed to estimate the two descriptive variables, BPN and MTD, using the least-square method and determining $R^2$ adjusted, which confirmed if the relationship between the dependent variables and the set of explanatory variables was significant (higher than 0.50) or not (less than 0.50). Moreover, a t-Student test was performed in order to define the t-Student values for each explanatory variable in the predictive formulas, which is to be compared with the t-Student critical value (equal to 1.66) as a test to validate the significance of the variable in the formulas.

Both coefficients of the predictive formulas and their t-Student achieved values are given in Table 7.

As a first general consideration, Table 7 suggests that BPN values are more sensible to the considered explanatory variables than MTD, at laboratory scale.

In fact, considering MTD predictive formulas, t-Student values were smaller than the critical |t| value (equal to 1.66) except for t-Student tests of Flakiness Index in PF-2-MTD and Voids content in PF-3-MTD, even if in both cases the predictive coefficients were close to zero. This showed that MTD is not affected by the investigated explanatory variables. Obviously, this is not believed to be true in general but the findings suggested that it is unlikely to study the effects of aggregates (mineralogy, dimension, and shape) on the macro-texture of a road wearing course using laboratory-produced asphalt slabs. This is mainly due to laboratory compaction procedures which are not able to correctly simulate the macro-texture obtained in the field by using asphalt pavers and rollers.

More interesting are instead the results deriving from the predictive formulas of BPN.

As far as the effects of aggregates dimension and mineralogy on BPN

| Dependent Variables | BPN | MTD |
|---|---|---|
| Predictive Formulas | PF-1-BPN | PF-1-MTD |
| Explanatory variables | Fine aggregates | |
| | Coarse aggregates | |
| | Basaltic aggregates | |
| Predictive Formulas | PF-2-BPN | PF-2-MTD |
| Explanatory variables | Shape Index | |
| | Flakiness Index | |
| Predictive Formulas | PF-3-BPN | PF-3-MTD |
| Explanatory variables | Void content | |

**Table 6:** Variables considered into the predictive formulas.

| Explanatory variables | PF-1-BPN | | PF-1-MTD | |
|---|---|---|---|---|
| | Coefficient | t-Student value | Coefficient | t-Student value |
| Intercept | 58.69 | 9.12 | 0.68 | 1.02 |
| Fine aggregates | 0.76 | 5.64 | -0.02 | -1.56 |
| Coarse aggregates | -0.20 | -1.75 | 0.02 | 1.35 |
| Basaltic aggregates | 0.18 | 5.04 | 0.00 | -0.11 |
| $R^2_{adj}$ | 0.57 | | 0.23 | |
| **Explanatory variables** | **PF-2-BPN** | | **PF-2-MTD** | |
| | Coefficient | t-Student value | Coefficient | t-Student value |
| Intercept | 73.98 | 74.72 | 0.80 | 11.14 |
| Shape Index | -0.72 | -10.11 | 0.01 | 1.20 |
| Flakiness Index | 0.06 | 0.92 | 0.01 | 2.48 |
| $R^2_{adj}$ | 0.71 | | 0.34 | |
| **Explanatory variables** | **PF-3-BPN** | | **PF-3-MTD** | |
| | Coefficient | t-Student value | Coefficient | t-Student value |
| Intercept | 78.99 | 27.06 | 0.65 | 7.16 |
| Voids content | -2.66 | 4.57 | 0.07 | 4.09 |
| $R^2_{adj}$ | 0.20 | | 0.20 | |

**Table 7:** Multiple linear regression - coefficients of predictive formulas and t-student.

(PF-1-BPN predictive formula), the |t| values of the coefficients were greater than the critical value, meaning that the accidental probability of correlation between dependent and explanatory variables was lower than 5%, which represented the actual critical value. In particular, the contribution of the coarse aggregate percentage was not as considerable as the other two coefficients (fine and basaltic aggregate) because of its smaller t-Student value. $R^2$adj for PF-1-BPN was equal to 0.57; this predictive formula therefore demonstrated to be adequately reliable and it well described the correlation between fine aggregate, coarse aggregate, basaltic aggregate and BPN.

PF-2-BPN aimed to define BPN values as a function of SI and FI. Unfortunately, t-Student value for FI coefficient (equal to 0.92) was smaller than the critical |t| value (equal to 1.66). Moreover, the prediction formula showed |t| value for SI coefficient equal to 10.11, largely greater than the critical |t| value. This result confirmed the trend already obtained in Figure 9a and the negative coefficient revealed that the Shape Index significantly affected the BPN values, which means that a decrease of BPN can be expected as a consequence of the increase of the quantity of elongated aggregates in the asphalt wearing course. Actually, elongated aggregates, usually lacking of well-defined surface asperities, cannot generate great micro-texture. Finally, it can be highlighted that the computed $R^2$adj obtained in PF-2-BPN was significant, thus SI variation is strictly correlated with BPN.

Coefficient of PF-3-BPN formula indicates that friction resistance (BPN) is negatively related to void content. Moreover, the value of t-Student for the void content coefficient is 4.57, considerably greater than the critical value (1.66). However, $R^2$adj was only equal to 0.20, which means the prediction formula did not show enough reliability and further analysis is necessary.

## Conclusions

The present study aimed at evaluating the friction properties of asphalt slab laboratory-prepared samples by changing mineralogy, dimension and shape of the aggregates, and considering the volumetric characteristics of the corresponding asphalt slabs. The experimental results were statistically analyzed to identify possible correlations between the variables. Results highlighted the followings.

- Mineralogy and geometry. Geometry, in terms of fine aggregate (nominally, 0.063-2 mm) content, appears to be the most significant parameter that influences microtexture; actually, an increase up to 10% of BPN was observed according to fine aggregate content. An increase of BPN was also observed according to the increase of basalt, even if less evident if compared to the effect of fine aggregate. On the contrary, MTD decreases according to fine and basalt content.
- Shape. The increase of non-polyhedral particles into the mixtures appears to be detrimental in terms of microtexture (BPN results) and plays a positive role in raising the macro texture (MTD). Even if in this last case, the correlation was small from a statistical point of view.
- Volumetric characteristics. Compaction characteristics (void contents) produce a general BPN increase and MTD decrease. This was possibly due to the increase of voids that reduces the number of asperities on the surface slowing down the pendulum during the BPN tests, and consequently increases the number of surface voids that govern the ability of the sand to penetrate into the surface.

### References

1. Descornet G (1989) A criterion for optimizing surface characteristics. Transportation Research Board, 173-177.
2. Smith KL, Hall JW, Titus-Glover L, Wambold JC, Yager TJ et al. (2009) Contractor's Final Report for National Cooperative Highway Research Program.
3. Gunaratne M, Bandara N, Medzorian J, Chawla M, Ulrich P (2013) Tribology: Friction, wear, and lubrication- a summer professional program.
4. Fwa TF, Choo YS, Liu Y (2003). Effect of aggregate spacing on skid resistance of asphalt pavement. J Transp Eng 129: 420-426.
5. Kangkhajitre C, Kanitpong K (2011) Effect of aggregate characteristics on texture and skid resistance of asphalt pavement surface. Eastern Asia Soc Transp Stud 8: 247.
6. Shah SMR, Abdullah ME (2010) Effect of aggregate shape on skid resistance of compacted Hot Mix Asphalt (HMA). 2nd International Conference on Computer and Network Technology, ICCNT 2010, 421-425.
7. Davis RM (2001) Comparison of surface characteristics of hot-mix asphalt pavement surfaces at the virginia smart road. Thesis submitted to the Faculty of Virginia Polytechnic Institute and State University.
8. Vaiana R, Capiluppi GF, Gallelli V, Iuele T, Minani V (2012) Pavement surface performances evolution: an experimental application. Procedia - Social and Behavioral Sci 53: 1149-1160.
9. Kevin K, McGhee, Gerardo WF (2003) High-speed texture measurement of pavements, Virginia Transportation Research Council.
10. Sengoz B, Onsori A, Topal A (2014) Effect of aggregate shape on the surface properties of flexible pavement. KSCE J Civil Eng 18: 1364-1371.
11. Eyad M, Arash R, Arif C, Pat H (2009) Predicting asphalt mixture skid resistance based on aggregate characteristics. Performed in cooperation with the Texas Department of Transportation and the Federal Highway Administration.
12. Chen XH, Wang, DW (2011) Fractal and spectral analysis of aggregate surface profile in polishing process. Wear 271: 2746-2750.
13. Kokkalis AG, Tsohos GH, Panagouli OK (2002) Consideration of fractals potential in pavement skid resistance evaluation. J Transportation Eng 128: 591-595.
14. Kane M, Zoltan R, Timmons A (2015) Exploring the texture-friction relationship: from texture empirical decomposition to pavement friction. Int J Pavement Eng 1-12.
15. Fulop IA, Bogardi I, Gulyas A, Csicsely-tarpay M (2000) Use of friction and texture in pavement performance modeling 126: 243-248.
16. Bazlamit SM, Reza F (2005). Changes in asphalt pavement friction components and adjustment of skid number for temperature. J Transportation Eng 131: 470-476.
17. Jahromi SG, Mortazavi SMR, Voussough S, Yingjian L (2011) Evaluation of pavement temperature on skid frictional of asphalt concrete surface. Int J Pavement Eng 12: 47-58.
18. Conover WJ, Iman RL (1981) Rank transformation as a bridge between parametric and nonparametric statistics. Am Statistician 35: 124-129.

# An Experimental Study of Embedded Connection for Lightweight Steel Square Tube Filled with Adhesives

**Itsuki Yagi*, Mana Kitahonoki and Takumi Ito**

*Department of Architecture, Tokyo University of Science, Japan*

**Abstract**

Recently, a various type of steel connection technique has been developed, which aims to possess the multi-demands on steel structures. In this study, a new fitting connection technique for steel square tube is suggested, and this joint consists as follows; lightweight steel square tube is just embedded to steel connector, and an adhesive is filled to the clearance between each member as filler, and a pin is inserted as a fail-safe mechanism against the uplift of the column. The adhesives adopted on this connection are widely used to adhere nonstructural members such as tiles and marbles in Japanese architectural building field. The advantage of this connection is not only improvement of workability due to needless of bolting and welding but also the adjustment of the bending strength and rigidity by filling methods and materials. This study investigates the resistant mechanism subjected to bending experimentally. Herein, the loading test is conducted as parameters with the existence of adhesives, pin and loading path. From test results, it is confirmed that the bending strength and rigidity are improved by the filling effect of adhesives, and the maximum bending strength is enhanced by the resistance of a pin during the ultimate state. Furthermore, the stress diagram subjected to bending is investigated from the test results, and a kinematic model of this connection considering the exfoliations of adhesives is suggested. It can almost evaluate the strength of the test results well.

**Keywords:** Adhesive; Bending resistant mechanism; Embedded connection; Lightweight steel square tube; Loading test; Steel structure; Stress distribution

## Introduction

Generally, bolting and welding are used as joint components in steel structures. In perspective, a connection of steel structures is divided into rigid connection, semi-rigid connection and pin connection by the fixation of the connection. However, the joint with various restoring force characteristics and structural performance are proposed and developed in recent years [1]. For example, the socket connection method not using bolting or welding is suggested, and it is expected to ease the inclination and displacement of jointed materials [2]. A fitting connection method on steel structures is sometimes adopted which aims to possess the multi-demands such as improvement of workability, disassemble, and any other purposes. In this study, a new fitting connection technique for steel square tubes is suggested as shown Figure 1. This proposed method is that a lightweight steel square tube column is just embedded to another member by not using bolting and welding. A clearance between each steel tube is filled with organic or inorganic fillers so as to prevent from the slipping and adjust the bending rigidity and strength for structural design. Herein, inorganic adhesive materials are adopted to fill the clearance as filler on this connection, which adhesives widely used to adhere nonstructural members such as tiles and marbles in Japanese architectural building field. This paper proposes a new type embedded connection method using a lightweight steel square tube, and also, an experimental study is conducted to clarify the resistant mechanism and inelastic behavior of about the connection method Figure 1a. The loading test is performed as parameters with the existence of adhesives, pin and loading path.

## Details of Embedded Connection

### General description

A construction process of the embedded connection as shown Figure 1b is described below. First of all, adhesives are coated to a connector, and fit a column (an outside steel tube) into the connector. Finally, a pin is inserted as a fail-safe mechanism against the uplift of members. The construction process of embedded connection is simple so that its workability is excellent because of unnecessary of bolting and welding.

### Composition of test specimens

Figure 2 shows a specimen of embedded connection. A specimen is consisted of below parts; a connector with spacers, an outside steel tube, a pin and adhesives. Herein, spacers are made spot welding to the connector in order to adjust and maintain a layer thickness of adhesives that are filled between the connector and the outside steel tube. The layout of spacers is arranged over the entire circumference in specimens. Moreover, epoxy resin adhesives are adopted as filler. Epoxy resin adhesives are composed of a based resin containing an epoxy resin and a curing agent containing a modified aliphatic polyamine. In this paper, epoxy resin adhesives are called adhesives.

## The Outline of Loading Test

### General description

In this study, the loading test was conducted to verify the resistant mechanism of the embedded steel connection. The mechanical properties of steel elements are summarized in Table 1, and adhesive elements are presented in Table 2. In addition, Table 3 summarizes the test specimens. According to Table 3, the existence of adhesives, pin and loading path are considered as parameters.

### Test set up and loading test methods

Figure 3 illustrates a loading test set up, and a lateral load was applied to the top of specimens to produce the bending moment at the connection. Test specimens were instrumented with sensors to measure lateral load and displacement at the top of test specimens, and strain of connectors and outside steel tubes.

***Corresponding author:** Itsuki Yagi, Department of Architecture, Tokyo University of Science, Japan, E-mail: 4116655@ed.tus.ac.jp

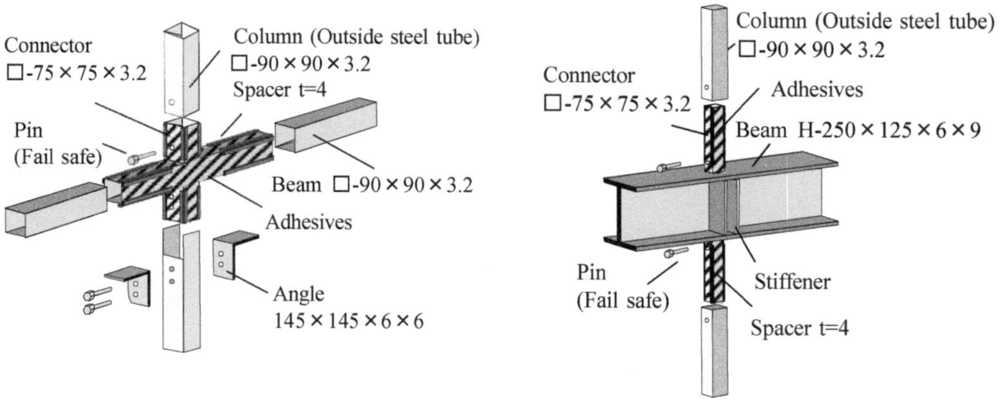


(a) Steel square tube beam (b) H-shaped beam

**Figure 1:** Conceptual diagram of embedded steel connection.

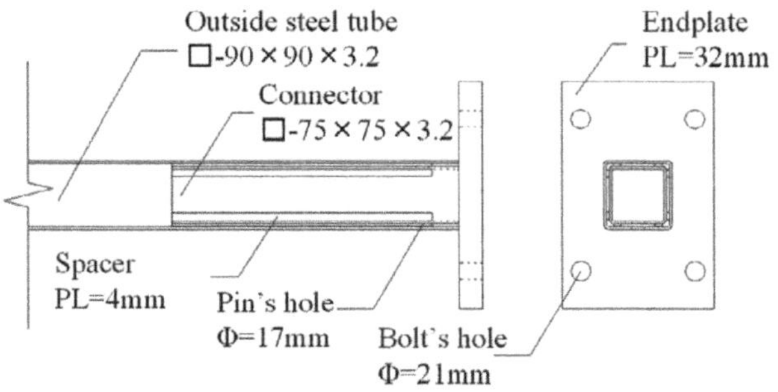


**Figure 2:** Diagram of test specimens (unit: mm).

| Parts | Steel grade (Grade of JIS) | Yield strength (N/mm²) | Tensile strength (N/mm²) | Young's modulus (kN/mm²) |
|---|---|---|---|---|
| Connecter | STKR400 | 333 | 416 | 212000 |
| Outside steel tube | STKR400 | 361 | 443 | 210000 |

**Table 1**: Mechanism properties of steel components.

| Thickness of adhesives (mm) | Maximum load (kN) | Shear strength of exfoliations of adhesives (N/mm²) |
|---|---|---|
| 4 | 2.48 | 2.02 |

**Table 2:** Mechanism properties of adhesives.

| Specimens | Thickness of adhesives | Pin | Loading path |
|---|---|---|---|
| A-P-m | 4 mm | Pin | Monotonic |
| A-P-c | | | Cyclic |
| A-nP-m | | non-Pin | Monotonic |
| A-nP-c | | | Cyclic |
| nA-P-m | non-Adhesives | Pin | Monotonic |
| nA-P-c | | | Cyclic |
| nA-nP-m | | non-Pin | Monotonic |
| nA-nP-c | | | Cyclic |

**Table 3:** List of test specimens.

## Loading pattern

Monotonic and cyclic loading tests were conducted, herein. The maximum rotational angle of the specimens on monotonic loading tests was 1/10 rad. The maximum rotational angle of the specimens on cyclic loading tests were ±1/300, ±1/200, ± 1/150, ± 1/100, ± 1/75, ± 1/50, ± 1/30 and ± 1/10 rad. (Figure 4).

## Results of Test Study

### Test result of monotonic loading

Figure 5 shows the moment-rotation angle curves of connection (M-θ curve) under monotonic loading tests. In addition, Figure 5 shows collapse modes and exfoliation of adhesives decided by the

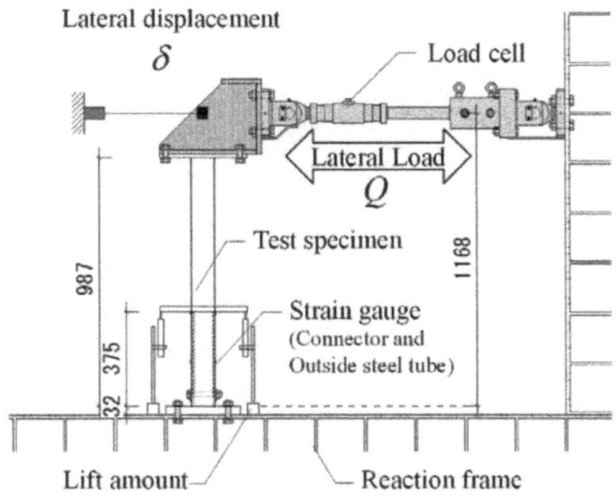


**Figure 3:** Set up of loading test.

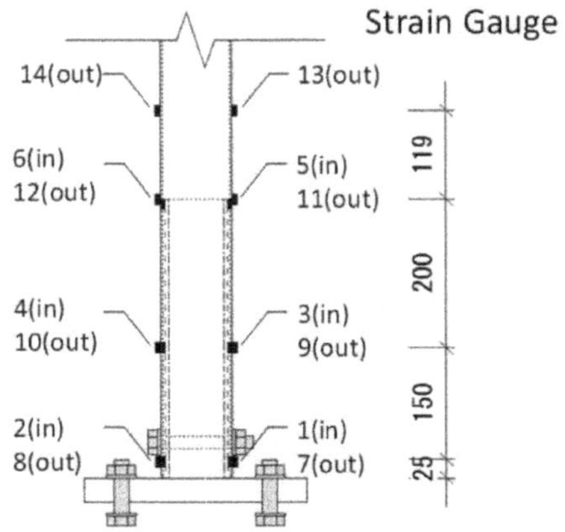


**Figure 4:** Position of strain gauges.

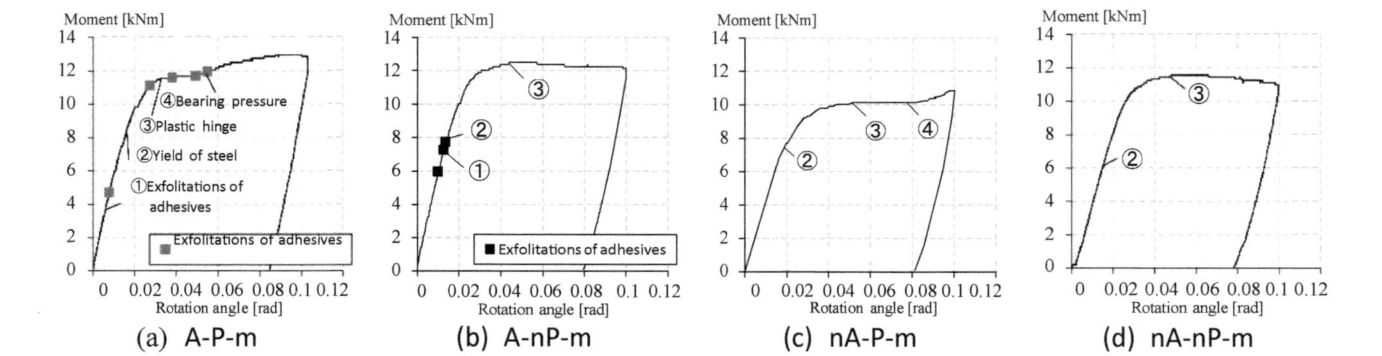


**Figure 5:** Test results of M-θ curve of monotonic loading.

| Specimens | Bending Strength (kN) | | | Bending Rigidity (kNm/rad) |
|---|---|---|---|---|
| | Exfoliations of Adhesives | Yield Strength | Maximum Strength | |
| A-P-m | 4.67 | 9.46 | -404 | 579 |
| A-nP-m | 5.95 | 7 | 12.5 | 579 |
| nA-P-m | 0 | 8.06 | 0 | 413 |
| nA-nP-m | - | 6.31 | 11.6 | 404 |

**Table 4:** Test results of monotonic loading test.

**Figure 6:** Section of test specimens.

noise, observation and behavior of strain during loading test. And also, Table 3 shows the test results (Exfoliations of adhesives, yield strength, maximum strength, and bending rigidity). Collapse modes of the embedded connection were observed as follows;

1. Exfoliation of adhesives
2. Yield of steel
3. Plastic hinge of connector
4. Bearing pressure of steel tube by pin.

Figure 6 shows a picture of section of specimens after monotonic loading test. From this figure, the exfoliations of adhesives were occurred in boundary layer between adhesives and an outside steel tube. According to Figure 5 and Table 4, in case of A-P-m and A-nP-m, the bending strength and rigidity were improved by the filling effect of adhesives. Figure 7 presents the test results of stress distribution of web. From this results in case of non-adhesive (metal touch), the stress was only transmitted little from connector to outside steel tube. By contrast, if adhesives were used as filler, the stress from connector to outside steel tube was increased by the filling effect. As a result, in case of adhesives filling, the bending strength and rigidity became high. From the result of Figure 5, it is confirmed that the bending rigidity shows almost same in both case of pin and non-pin, however, the maximum bending strength was enhanced during ultimate state in case of using pin. The pin's whole deformation was observed after the loading test, thus the specimen is resisted bending moment by bearing pressure of a pin.

## Test result of cyclic loading

Figure 8 shows the moment-rotation angle hysteresis curves (M-θ curve) under cyclic loading tests. And also, Table 5 shows the test results (Yield strength, Maximum strength, Bending rigidity). From the results of Figure 8 and Table 5, it is confirmed that adhesives prevent the slipping, and it improves the bending strength and rigidity at moment 0 in case of A-P-c and A- nP-c. Furthermore, in case of A-P-c and nA-P-c, the bending rigidity shows almost same; however the bending strength was enhanced at ultimate (Figure 9).

## The bending moment distribution of embedded connection

From the results of Figure 10, it is confirmed that the outside steel tube resisted the bending moment at height from 375 mm to 1200 mm. The bending moment was decreased toward column base (height 0mm), on the other hands, the bending moment of connector was increased. Consequently, it is confirmed that the bending moment of connector showed the maximum value at height 0mm. At height 375 mm, the stress of outside steel tube became large because the outside steel tube was contacted with the connector and received reaction force. Moreover, in case of A-P-m and nA-P-m, the sum of bending moment of connector and outside steel tube corresponds with the bending moment of specimen regardless of adhesives. However, in case of A-nP-m and nA-nP-m, the bending moment of connector was corresponded with the bending moment of specimen. Therefore, the connector and the outside steel tube collaborated to resist the bending moment when a pin was used. On the other hand, in case of non-pin, the connector mainly resisted the bending moment. Furthermore,

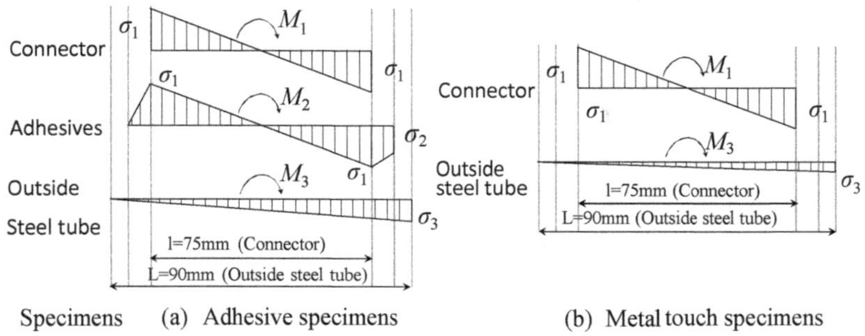


**Figure7:** Test results of stress distribution of web.

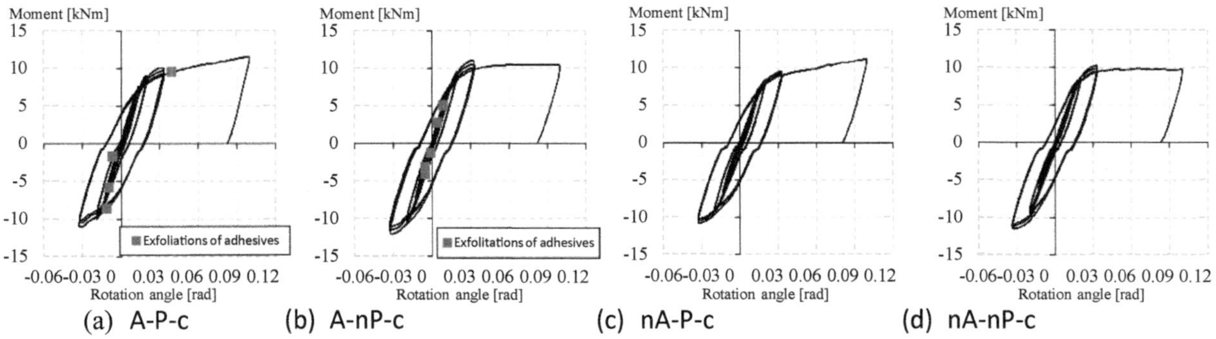


**Figure 8:** Test results of M-θ hysteresis curve under cyclic loading.

| Specimens | Bending strength [kNm] | | Bending rigidity [kNm/rad] |
|---|---|---|---|
| | Yield strength | Maximum strength | |
| A-P-c | -6.66 | W | 541 |
| A-Np-c | 4.79 | 11.0 | 540 |
| nA-P-c | -8.99 | W | 474 |
| nA-Np-c | 3.62 | -11.6 | 481 |

**Table 5:** Test results of cyclic loading test.

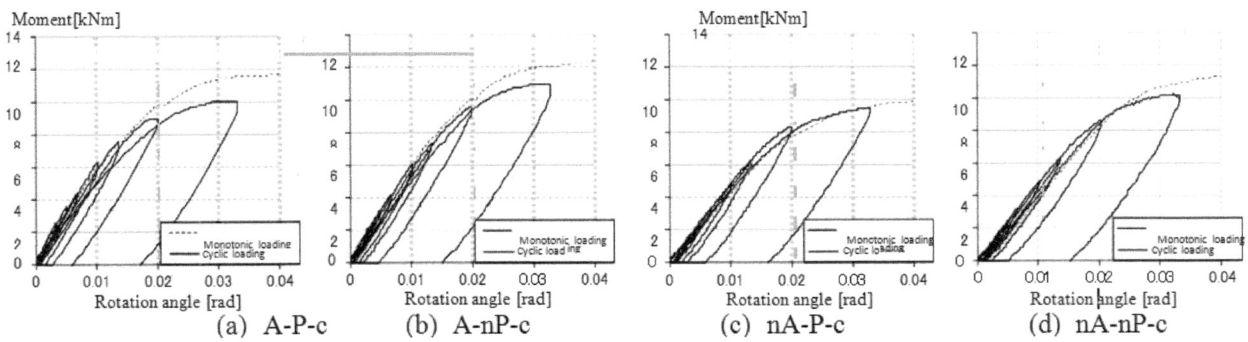


**Figure 9:** Comparison of envelope curves and M-θ curve of monotonic loading.

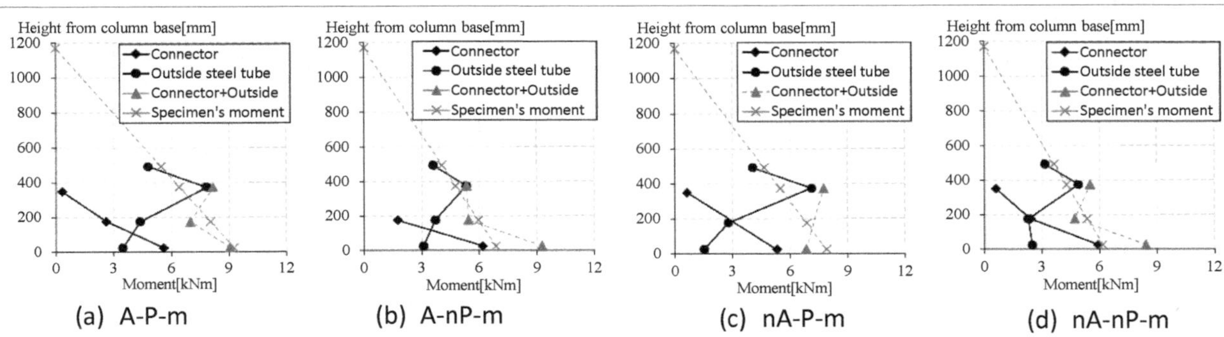


**Figure 10:** Bending moment distribution of embedded connection.

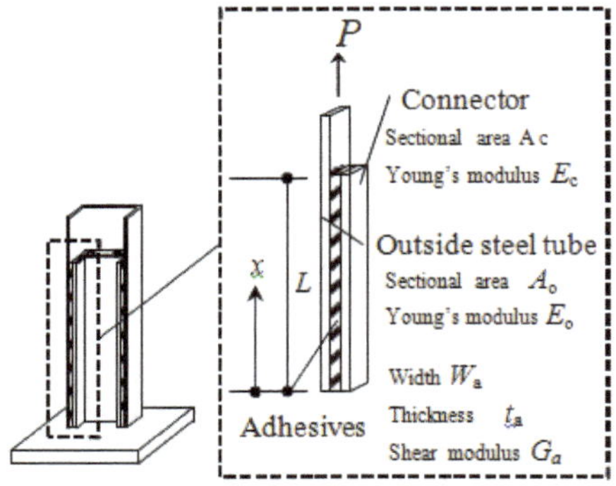


Figure 11: Dynamic model.

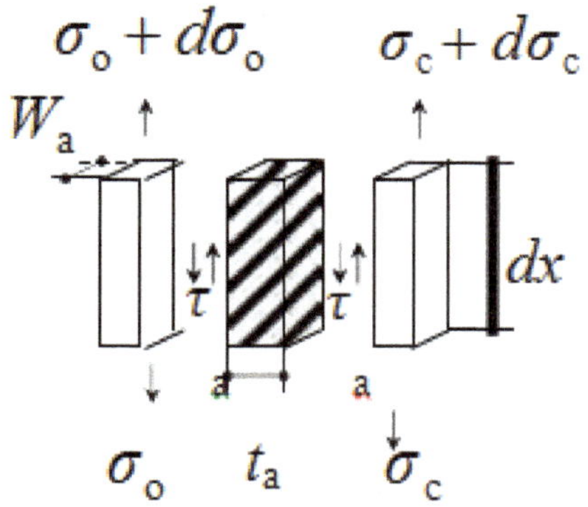


Figure 12: Small elements of members.

| Parts | A-P-m | A-np-m | nA-P-m | nA-nP-m |
|---|---|---|---|---|
| Connector | 5.59 | 6.16 | 5.31 | 5.91 |
| Outside steel tube | 3.48 | 3.08 | 1.52 | 2.51 |
| Connector + Outside | 9.06 | - | 6.82 | 8.43 |
| Specimen's moment | 9.25 | 6.86 | 7.88 | 6.17 |

Table 6: Test results of bending moment.

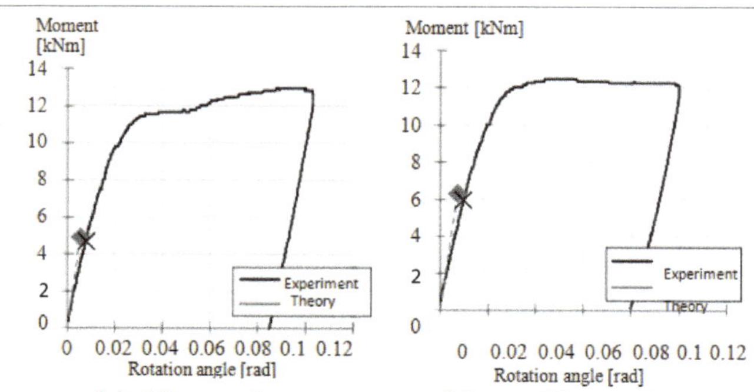


(a) Pin specimen (b) non-Pin specimen

Figure 13: Comparison between experiment and theory.

adhesives itself were not resisted bending moment because the bending strength of A-nP-m was close to one of nA-nP-m.

## Analytical Studies

### General description of analytical method

In this section, from the observation of test results, theoretical equations of bending strength and rigidity at the exfoliation of adhesives are proposed. Here in, each bending strength of the flange and the web is calculated, and the sum of them is the whole bending strength of embedded connection from principle of superposition.

### The bending strength valuation plan at exfoliation of adhesives

The Figure 11 shows the analytical model assumed from test results. When axial force P by bending moment is loaded, it is assumed that shear stress distribution by adhesives is applied at tension side as Figure 12. Thus, following equation is established by condition of forces.

$$A_o \times d\sigma_o = \tau_a \times dx \times W_a \ , A_c \times d\sigma_c = \tau_a \times dx \times W_a \quad (1)$$

Ao, σo: Sectional area and stress of outside of steel tube.

Ac, σc: Sectional area and normal stress of connector.

Wa, τa: Width and shearing stress of adhesive.

$$d\tau_a = G_a \times d\gamma_a = \frac{G_a}{Et_a} \times (\sigma_o - \sigma_c)dx \quad (2)$$

E: Young's modulus of outside steel tube and connector.

Ga: Shear modulus of adhesives.

Following equation (3) is equation (2) integrated.

$$\tau_a = \frac{KP}{W_a} \times \frac{1}{\sinh KH} \times \frac{A_O \cosh(KH - kx) + A_c \cosh kx}{A_o + A_c}$$

$$K^2 = \frac{G_a W_a}{Et_a} \times \frac{A_o + A_c}{A_o A_c} \quad (3)$$

From equation (3), the bending strength of flange Mf is expressed by equation (4).

$$M_f = \frac{\tau_a \times W_a \times (A_o + A_c) \times \sinh KH}{K \times \{A_o \cosh(KH - kx) + Ac \cosh kx\}} \times B \quad (4)$$

From Figure 7, the bending moment of web of outside steel tube and connector (Mwo, Mwc) at exfoliation of adhesives is calculated from the following equation (5).

$$M_{wo} = \frac{\sigma_3 \times L}{2} \times t_o \times \frac{2L}{3} \times 2, \ M_{wc} = \frac{\sigma_1 \times I_c}{1/2} \times 2 \quad (5)$$

*to*: Plate thickness of outside steel tube

*Ic*: Geometrical moment of area of connector

According to section 6.3 about the bending moment distribution of embedded connection, connector and outside steel tube collaborate to resist the bending moment if a pin is installed. On the other hand, in case of non-pin specimen, the connector mainly resists the bending moment. Therefore, the bending strength of embedded connection is calculated by equation (6).

$$M = M_f + M_{wo} + M_{wc} \ \ (\text{Pin})$$
$$M = M_f + M_{wo} \times \beta + M_{wc} \ \ (\text{non} - \text{Pin}) \quad (6)$$

β: Apportioning rate by stress transmitting.

Table 6 shows the comparison between theoretical and experimental value. The bending strength at exfoliation of adhesives is good agreement.

### The Bending rigidity of embedded connection at exfoliation of adhesives

The displacement of embedded connection at exfoliation of adhesives is calculated by elastic curve equation.

Part of connector $$\frac{d^2y_1}{dx^2} = \frac{P\times(L-x)}{E\times\{I_o+(1-x/H)\times I_c\}} \quad (7)$$

Part of outside steel tube $$\frac{d^2y_2}{dx^2} = \frac{P\times(L-x)}{E\times I_o} \quad (8)$$

Figure 13 shows that the comparison between experimental results and theoretical results. The theoretical equation shows good agreement from the comparison of Figure 13.

## Conclusions

In this paper, loading tests of embedded connection for lightweight steel square tube filled with adhesives were conducted to clarify the resistant mechanism and inelastic behavior as parameters with the existence of adhesives and pin. The main conclusions of this paper are:

1) If adhesives were used as filler, the stress from connector to outside steel tube was increased by the filling effect. As a result, in case of adhesive filling, the bending strength and rigidity became high.

2) The bending rigidity showed almost same in both case of pin and non-pin, however, the maximum bending strength was enhanced during ultimate state by bearing pressure in case of pin.

3) The connector and the outside steel tube collaborated to resist the bending moment when a pin was used. On the other hand, in case of non-pin, the connector mainly resisted the bending moment.

4) Theoretical equations of the bending strength and rigidity at the exfoliation of adhesives are proposed, and they are good agreement.

### References

1. Suita K (2009) Recent experiments for development of new steel joint system. Annual Meeting of Architectural of Japan Tohoku. 4-15.
2. Kuwamura H, Ito T, Tomioka Y (2005) Study on steel cone-to-cylinder socket connections. J of Struct Eng 598: 155-162.

# Non- Linear Models of Light-Duty Vehicle Emission Factors Considering Pavement Roughness

**Qing Li[1], Fengxiang Qiao[1*] and Lei Yu[1]**

*[1]Innovative Transportation Research Institute, Texas Southern University, USA*

**Abstract**

Emission factors are very important measures for developing an emission inventory, making decisions, designing control strategies, mitigating climate change, and even improving public health, in terms of respiratory system diseases. The emission factors could be either measured from field tests or estimated by an emission model. Existing models seldom consider the impacts of some special factors such as pavement roughness. As the impacts of the pavement roughness on emissions are very complicated, a linear model or physical model may not depict the mappings from affecting factors to resulted emission factors. In this paper, two non-linear models, including K-Nearest Neighbor (K-NN) and Neural Network (NN) were built to estimate vehicle emission factors using roughness involved input data. A best fitted model was identified to illustrate the emission pattern along a wide range of pavement roughness. Multiple field tests were conducted in five regions of the State of Texas, United States, with a total of 1,609 km test length. One dedicated test vehicle was employed throughout the test. Pavement roughness was tested using a smartphone based application. All tested data were separated into four groups, each representing a different range of roughness, while the modeling was conducted within each group. The predictive performance of each model was evaluated by (1) correlation coefficient; (2) relative errors; and (3) two tailed unequal variance t-test. Results suggest that, K-NN can be better than NN to model the emission factors for the Texas highway system, and driving on a smoother and rougher pavement result in higher vehicle emissions.

**Keywords:** Modeling; Emission factors; Pavement roughness; K-Nearest neighbor; Neural network

## Introduction

The U.S. Environmental Protection Agency (EPA) recommends building the emission inventory as parts of a State Implementation Plan (SIP) [1]. Emissions factors are important measures in the development of national, regional, state, and local emissions inventories for decision-making and control strategies. Users of emission factors include agencies in federal, state, and tribal levels, as well as consultants and industries [2,3]. Proper estimation of emission factors (EF) could also help in developing countermeasures in not only the environmental protections and congestion mitigation [4], but also the public health improvements [5,6]. The emission factors are also used to report to national greenhouse gas inventories under the United Nations Framework Convention on Climate Change (UNFCCC) [7].

EF could be measured directly from on-road measurement equipment and in-lab testing devices [8-13] or estimated using a suitable model such as the EPA models MOBILE 6.2 [14] and MOVES [3]. Field and in-lab tests are limited to the availabilities of equipment and testing environment/scenarios, while the model estimation may not consider all real conditions and might induce more or less errors [15]. Many studies found that vehicle emissions are very subject to many factors, such as driving behaviors [16], vehicle information [17], pavement materials [18], route's slope conditions [19], traffic control system [20], and traffic situations, such as the situation at a work zone and a signalized intersection [21-23]. However, most of the emission models seldom incorporate the impacts of pavement roughness [24-26] into the independent variables. It is hypothesized that the vehicle emissions are nonlinearly correlated to pavement roughness.

The objective of this paper is to identify a best fitted nonlinear model to illustrate the impacts of pavement roughness on vehicle emission pattern, based on real world measurements by a test vehicle in five regions of the State of Texas.

## Methodology and Field Test

### The non-linear problem to estimate emission factors

Many factors could affect emission factors, including (1) engine information (in-take air temperature IAT, manifold absolute pressure MAP, revolutions per minute rpm), (2) vehicle activity (velocity and acceleration), and (3) pavement information (e.g. the calculated International Roughness Index (IRI)). A nonlinear mapping could be envisioned converting these independent variables to the needed emission factors. Figure 1 illustrates such nonlinear mapping.

Among all input variables, the pavement roughness is the one that is not typically associated with emission estimations. However, studies have demonstrated that pavement roughness would possibly affect fuel consumption, which is a typical indicator of vehicle emissions [24,27]. A recent study identified that the pavement roughness can be classified

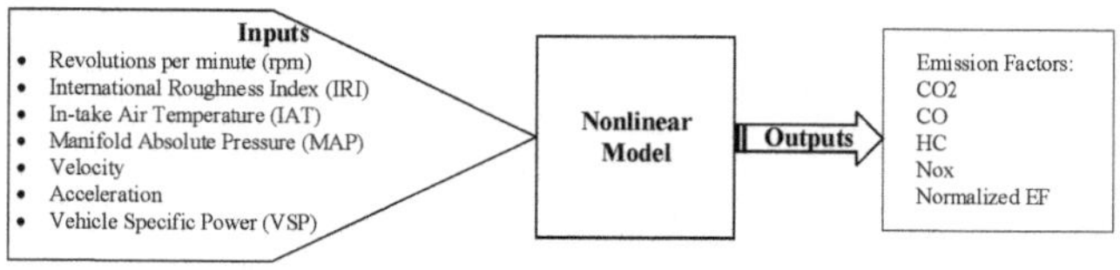


**Figure 1:** Non-linear mapping for the estimation of emission factors.

***Corresponding author:** Fengxiang Qiao, Ph.D, Professor, Innovative Transportation Research Institute Texas Southern University, 3100 Cleburne Street, Houston, Texas, 77004, USA, E-mail: qiao_fg@tsu.edu

| Category | IRI | | Average Normalized Emission Factor (ANEF) | | |
|---|---|---|---|---|---|
| | Range | Cluster Center | Avg. | Std | Evaluation |
| A | (0.00-1.99) | 1.36 | 0.051 | 0.055 | High |
| B | (1.99-3.21) | 2.54 | 0.032 | 0.017 | Low |
| C | (3.21-6.00) | 4.07 | 0.030 | 0.016 | Low |
| D | > 6.00 | 7.07 | 0.039 | 0.014 | High |

**Table 1:** Classification of pavement roughness based on Texas emission measurement (source: Li et al., 2016d).

into four groups, each group presents a specific feature in vehicle emission factors [27]. The specification of the calcination is illustrated in Table 1.

Table 1 shows higher vehicle emission factors were observed on the smoother and rougher pavement denoted by category A and D. This implies that pavement roughness is also one of determinants in air emissions. The roughness involvement may improve the accuracy of the vehicle estimation.

The nonlinear mapping in Figure 1 should be implemented through a nonlinear model, the output of which could be emission factors of major air emissions, such as: carbon dioxide ($CO_2$) in g/mi, carbon monoxide (CO) in mg/mi, hydrocarbon (HC) in mg/mi, and nitrogen oxides ($NO_x$) in mg/mi. In order to have a uniform comparison of multiple air emissions ($m$), this study adopted the normalized emission index proposed by Li et al. The Normalized Emission Factor (N) is calculated by using equation (1).

$$N_{i;j} = \frac{1}{m}\sum_{i=1}^{m}\left(\frac{x_{i;j} - Min_j(x_{i;j})}{Max_j(x_{i;j}) - \mathrm{Min}_j(x_{i;j})}\right) \quad (1)$$

where:

$N_{(i,j)}$= The i$^{th}$ normalized emission factor in the j$^{th}$ air emission,

$X_{i,j}$= The i$^{th}$ emission factor of the j$^{th}$ air emission (g/mi or mg/mi),

m = The number of studied air emissions, here is 4 for $CO_2$, CO, HC, and $NO_x$,

M in($x_{i,j}$)= The minimum emission factor of the j$^{th}$ air emission (g/mi or mg/mi), and

M $ax_j$ ($x_{i,j}$)= The maximum emission factor of the j$^{th}$ air emission (g/mi or mg/mi).

Any nonlinear models could be candidates for the required nonlinear mapping. In this paper, two typical nonlinear models were employed to estimate emission factors based on the data collected from field: (1) the K-nearest neighbors (K-NN) model, and (2) the Neural Network (NN) model. Both are machine learning based multidimensional in their respective featured spaces. The two models are also memory-based. They start with training observations, and assume that the response class of nearly observations is likely to be similar.

## The K-NN model

The K-nearest neighbors (K-NN) algorithm is based on an assumption that class probabilities are locally approximately constant. However, for most neighborhoods, it is not constant. To bring out a feasible constant class probability, distance metric needs to be improved. There are many types of distance metrics, such as Mahalanobis distance, city block metric, Minkowski metric, cosine distance, and so on [28]. Euclidean distance is a commonly used, expressed by equation (2).

$$d^2_{st} = (x_{s-}y_t)\,(x_{s-}y_t)' \quad (2)$$

where:

$d^2_{st}$ = Euclidean distance squared between vector $x_s$ and $y_t$.

$x_s = mx$, the $mx^{th}$ row vectors in an $mx$-by-$n$ data matrix X, e.g. $x_1$,$x_2$, …, $x_{mx}$ or $x_s$.

$y_t = my$, the $my^{th}$ row vectors in an $mx$-by-$n$ data matrix Y, e.g. $y_1$,$y_2$, …, $y_{mx}$ or $y_t$.

Meanwhile, the number of nearest neighbors called $k$ is essential for deliver a precise estimated result. A smaller $k$ may result in higher variance, whereas larger $k$ may lead to higher bias. The selection of $k$ quite depends on the nature data. Therefore, cross-validation is often adopted to seek for a proper nearest neighbor size with the lowest Error Log ($el$) described by equation (3).

$$el_j = \log 10\left(\sum_j^k \left(yj - y_j\right)^2\right) \quad (3)$$

where:

$k$ = The best number of neighbors,

$y_j$= The measured output at the $j^{th}$ nearest neighbor, and

$\hat{y}_j$ = The estimated output at the $j^{th}$ nearest neighbor.

The estimated output ($\hat{y}_j$) is an average of $k$ weighted nearest neighbors, described by equation (4).

$$y_j = \frac{\sum_{i=1}^{k} w_{j;i}\, y_{j;1}}{k} \quad (4)$$

where:

$y_j$ = The estimated output at the $j^{th}$ nearest neighbor,

$w_{j,i}$ = The weight of the i$^{th}$ input nearest neighbor at the $j^{th}$ measured output neighbor,

$y_{j,i}$= The measured output of the i$^{th}$ input neighbor at the $j^{th}$ measured output neighbor,

k=The best number of neighbors.

As the K-NN model excuses based on its training dataset, any noise or irrelevant features become sensitive for the model results. Meanwhile, more frequent classes may dominate the modeled result.

## The neural network model

A neural network (NN) is flexible for linear and nonlinear. For nonlinear relationship between dependent and independent variables, the neural network could provide precise estimated results. Commonly two models were executed within the network, including Multiplayer perception (MLP) and Radial Basis Function (RBF), while the RBF provides a linear combination of radial basis functions of the inputs and neuron parameters, MLP used a supervised learning technique called backpropagation for training, which allows predicting more complex relationships. MLP was chosen in this study. MLP maps sets of input data onto a set of appropriate output, which is consisted of multiple layers of nodes. Each layer is fully connected to the next one. A typical neural network structure is presented in Figure 2. [29]

In Figure 2, there are $p$ dependent variables, which are input to one or more hidden layers with q neurons. In each neuron, there is a linear activation function, which maps the weighted input variables to the out of each neuron. The main activation function is a hyperbolic tangent function in equation (5), which can be replaced by a sigmoid function in equation (6)

$$y(x_i) = \tanh(x_i) \quad (5)$$

$$y(X^i) = (1\text{-}exp(\text{-}Xi))\text{-}1 \quad (6)$$

where:

$y$ = The output variable(s),

$x$ = The input variables.

The machine learning based model results could be improved by increasing training and scoring times, which will be reflected by its structure as well. For a MLP, there would be up to two hidden layers with multiple neurons. Cross-check would be required to obtain an optimal structure, including the number of layer and neurons.

### Performance measures

The predictive performance of each modeling stage was evaluated by (1) the correlation coefficient between the observed values and the modeled values, (2) the relative errors, which were calculated from the difference between the observed and modeled emission factors divided by the observed values, and (3) the two tailed unequal variance *t*-test. The null hypothesis is that the observed and modeled population means are the same but the two population variances may differ. The null hypothesis will be accepted if the *p* value is greater than 95%. Compared with paired *t*-test, the unequal variance *t*-test is able to quantify how far apart the two means of the two population are [30].

### Emission tests and data collection

On-broad vehicle emission tests were conducted along Texas highways in various regions, including El Paso, Austin, San Antonio, College Station, Houston, and other Southeast regions, from November 2014 to June 2015 during sunny days. The tested routes include high-speed freeways, rural highways, arterial roads, and local street, covering a wide range of speed limit roadway facilities with a wide range of pavement roughness.

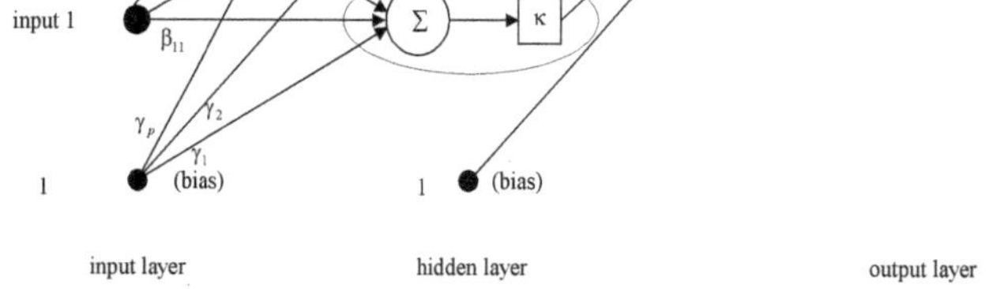


**Figure 2:** Neural network structure (source: Qiao et al., 2001).

| Category | $CO_2$ | | | CO | | | HC | | | $NO_x$ | | | Normalized Index | | |
|---|---|---|---|---|---|---|---|---|---|---|---|---|---|---|---|
| | *k* | R | *p* | *k* | R | *p* | *k* | R | *p* | *k* | R | *p* | *k* | R | *p* |
| A | 4 | 0.98 | 0.94 | 4 | 0.85 | 0.62 | 3 | 0.80 | 0.37 | 4 | 0.78 | 0.78 | 4 | 0.86 | 0.75 |
| B | 5 | 0.97 | 0.76 | 4 | 0.66 | 0.10 | 5 | 0.83 | 0.45 | 3 | 0.89 | 0.49 | 4 | 0.71 | 0.15 |
| C | 4 | 0.93 | 0.80 | 4 | 0.89 | 0.76 | 3 | 0.89 | 0.71 | 5 | 0.91 | 0.25 | 4 | 0.94 | 0.65 |
| D | 3 | 0.99 | 0.94 | 3 | 1.00 | 0.96 | 3 | 0.96 | 0.85 | 3 | 1.00 | 0.98 | 4 | 0.99 | 0.97 |

**Table 2:** Cross-validation results based on correlation relationship and significant t-test.

A Portable Emission Measurement System (PEMS) was equipped inside a dedicated test vehicle to provides second-by-second emission rate. The test vehicle is 10 years old with 10,000 starting mileages. A Global Positioning System (GPS) was paced on the top of the test vehicle to collected real-time test vehicle' geo-location information, including latitude, longitude, and altitude. Meanwhile, the engine' dynamic operation information, such as IAT, MAP, speed, acceleration and rpm were also recorded through a set of sensor arrays that were also connected to and synchronized with the PEMS.

Meanwhile, a smart phone installed with roughness measurement application (app) was mounted to the front of the windshield inside the vehicle by a phone car rack. Before each test, a simple calibration procedure was conducted. Concretely, the phone position was adjusted as straight (vertically or horizontally) as possible in order to set the phone's three dimensions (x, y and z) as close to zero as possible, which serves as a reference point for the roughness model in the app. The correlation of the calculated IRI towards laser beam measured IRI is 80% above [31]. The app provides real-time calculated International Roughness Index (IRI) for every 20-meter distance.

A total of 1,609 km (1,000 miles) highway routes were tested and about 210, 800 emission rates for $CO_2$, CO, HC, and $NO_x$, were recorded. To synchronize the IRI data, the collected emission rate data were calculated and interpolated into emission factors for every 20-meter distance. It turns out that 19,099 valid data pairs (20 meters each) were prepared. Seventy percent of the data pairs were used to train the models, while the rest were evenly separated for testing and validation.

## Results and Discussion

### Models structure identification

A total of 19,019 data pairs were prepared. Based on the four categories of pavement roughness in Table 1, the data pairs were divided into four datasets. Most of test pavement roughness fell into category A, 14,078. 3,585 data pairs were classified into category B. Only few data pairs met the category C and D with 1,304 and 52, respectively. These datasets were further randomly divided into three groups for training (70%), testing (15%), and validation (15%) in the modeling process.

**K-NN model:** Cross-validation was conducted to identify the number of *k*, the optimal number of nearest neighbors. Table 2 presents a list of *k* with the highest correlation coefficient R values in a validation stage for the four categories, which ranges from 3 to 5.

In the most cases, 4 nearest neighbors were chosen. Meanwhile, with the k values in Table 2, the validated emission factors were highly correlated to the observed values. Besides, a two tailed t-test was used to examine the variance of the observed and modeled emission factors. The null hypothesis is that two samples are equal variance. When p is greater than 0.05, the null hypothesis is accepted. All p-values in Table 2 are greater than 0.05, which means the variance of the observed and models emission factors is equal.

**Neural network:** A cross-check was conducted in the validation stage to seek for an optimal structure, with which the modeled emission factors may be highly correlated to the observed ones. There are two steps in this check. The first step was to identify the number of neurons at the first hidden layer. The second was to identify the number of hidden layer and the neural number in the second layer (if possible). Figure 3 shows the check results. In Figure 3, a blue line tells that R values increase with the increase in neurons in one hidden layer. When 10 neurons were used, the R value would be over 0.97. Thus, 10 neurons were identified for the first layer. For the possible second hidden layer, a cross-check was continued. a red line in Figure 3 demonstrates that adding second layer does not improve the R values. By the contrary, it dropped when the neuron further increases to 7 and 8. In response to this, 1 hidden layer with 10 neurons was confirmed for the structure of the NN model.

## Model testing and validation

Two models were executed three stages, including training, testing, and validation, based on the identified structure. Correlation coefficients were adopted to evaluate the level of curve fitting at the three stages. An overview of the correlation coefficients performed by the two models is listed on Table 3.

Table 3 illustrates that the R-values by the two models are mostly higher than 0.50, which indicates a good fit with the observed values. More specifically, the R values of $CO_2$ are overall higher than other air emissions in the three modeling stages. Except the R value of 0.77 and 0.64 by NN in validation for category A and D, the R values are higher than 0.89. This implies that the two models can estimate $CO_2$ emission more accurate than other air emissions. This could be attributed to their different emission patterns. Compared with $CO_2$ emission pattern, other air emission patterns are more complex. The $CO_2$ emission is proportional to the demand of power need for motion, which can be estimated by vehicle activity information, such as speed, acceleration, rpm. However, other air emissions are also subject to a number of conditions in a vehicle combustion system, such as the oxygen availability in the cylinder, sufficiency of mixture time between oxygen and fuel, and temperature. For example, Li et al. demonstrated that HC and CO emissions could be due to extremely insufficient mixture of

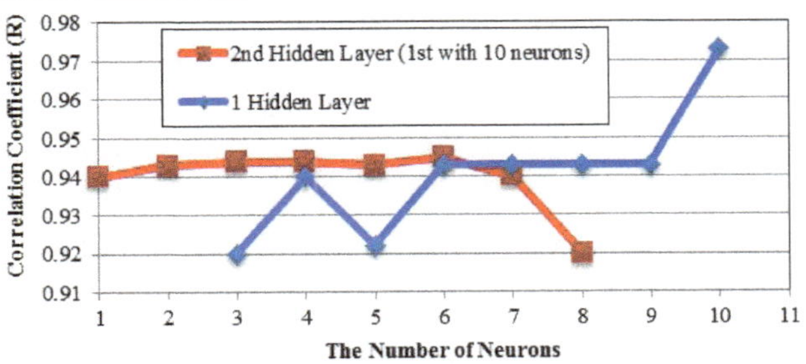


**Figure 3:** Cross-validation result of correlation coefficient for $CO_2$.

oxygen and fossil fuel in a cylinder. Inversely, excessive availability of oxygen from air results in higher $NO_x$ emissions. Moreover, the CO emission pattern is the inverse of the emission pattern for HC. The unburnt fuel can be easily escaped from an exhaust pipe as HC at higher ambient temperature.

Few R values marked in red are observed in the CO and $NO_x$ and Normalized Index (NI). In particular, the R values of CO presents lower correlative to the observed values. It is more likely that the emission pattern of CO is different from other studied air emissions here, which would be explored in next sub-section. Besides, the most lower R values in red were performed by NN model. Thus, in terms of curve fitting for these datasets, K-NN can estimate vehicle emissions more accurate than NN model.

**Fitted regression line:** To obtain an insight into the emission pattern, serval typical fitted regression lines are plotted in Figure 4. Figure 4 shows that there are obviously more data points in category A, whereas there may be insufficient data points in category D to provide a generalized picture of the CO emission pattern. Hence, the modeled CO emission factor in category D may not be reliable. Moreover, the distribution of CO emission factors in category A (Figure 4a) are apparently more dispersive than in category B and C (Figures 4b and 4c), up to 250 mg/mi. on the contrary, most CO emission factors are within 40 and 30 mg/mi in category B and C.

Similarly, the $NO_x$ emission factors in category A (Figure 4e) distribute more dispersive than in category B (Figure 4f), and the emission level is clearly higher as well. Figures 4g and 4h provided a similar view in Figures 4a and 4b.

**Emission factor:** In this study, specific emission factors were quantified for the four categories with different level of pavement roughness, based on on-board emission tests. Based on the observed results, two models were developed. The comparison of the modeled values and the observed valued is illustrated in Figure 5.

Figure 5 shows that the relationship between pavement roughness and emission factors is not linear, which is consistent with the previous study by Li et al. [6]. The smoother or the rough pavement may induce higher vehicle emissions. Besides, it seems that the developed k-NN slightly under estimate the emission factor in cross the four air emissions, particularly the estimation in category D. This could be explained by insufficient data points for such rough pavement during the on-road tests. Likewise, NN model also did not deliver a better estimate results for category D. Further, NN model slightly overestimates $CO_2$ and NOx emissions, and underestimates HC emissions.

| category | Stage | $CO_2$ | | CO | | HC | | $NO_x$ | | Normalized Index | |
|---|---|---|---|---|---|---|---|---|---|---|---|
| | | K-NN | NN | K-NN | NN | K-NN | NN | K-NN | NN | K-NN | NN |
| A | Training | 0.96 | 0.98 | 0.84 | 0.73 | 0.77 | 0.58 | 0.78 | 0.74 | 0.84 | 0.71 |
| | Testing | 0.96 | 0.97 | 0.87 | 0.41 | 0.85 | 0.72 | 0.77 | 0.42 | 0.88 | 0.36 |
| | Validation | 0.98 | 0.77 | 0.85 | 0.04 | 0.80 | 0.88 | 0.78 | 0.36 | 0.86 | 0.08 |
| B | Training | 0.96 | 0.98 | 0.85 | 0.85 | 0.78 | 0.79 | 0.87 | 0.82 | 0.86 | 0.60 |
| | Testing | 0.97 | 0.89 | 0.19 | 0.94 | 0.52 | 0.74 | 0.92 | 0.29 | 0.26 | 0.93 |
| | Validation | 0.97 | 0.89 | 0.66 | 0.33 | 0.83 | 0.80 | 0.89 | 0.43 | 0.71 | 0.27 |
| C | Training | 0.96 | 0.99 | 0.59 | 0.82 | 0.54 | 0.84 | 0.86 | 0.79 | 0.66 | 0.92 |
| | Testing | 0.99 | 0.91 | 0.69 | 0.35 | 0.98 | 0.60 | 0.95 | 0.43 | 0.81 | 0.51 |
| | Validation | 0.93 | 0.98 | 0.89 | 0.31 | 0.89 | 0.80 | 0.91 | 0.65 | 0.94 | 0.56 |
| D | Training | 0.92 | 0.89 | 0.78 | 0.69 | 0.87 | 0.97 | 0.83 | 0.85 | 0.89 | 0.89 |
| | Testing | 0.97 | 0.96 | 0.98 | -0.43 | 0.99 | 0.34 | 0.99 | 0.83 | 0.99 | 0.97 |
| | Validation | 0.99 | 0.64 | 1.00 | 0.12 | 0.96 | 0.92 | 1.00 | 0.72 | 0.99 | 0.98 |

**Table 3:** Correlation coefficients in three modeling stages.

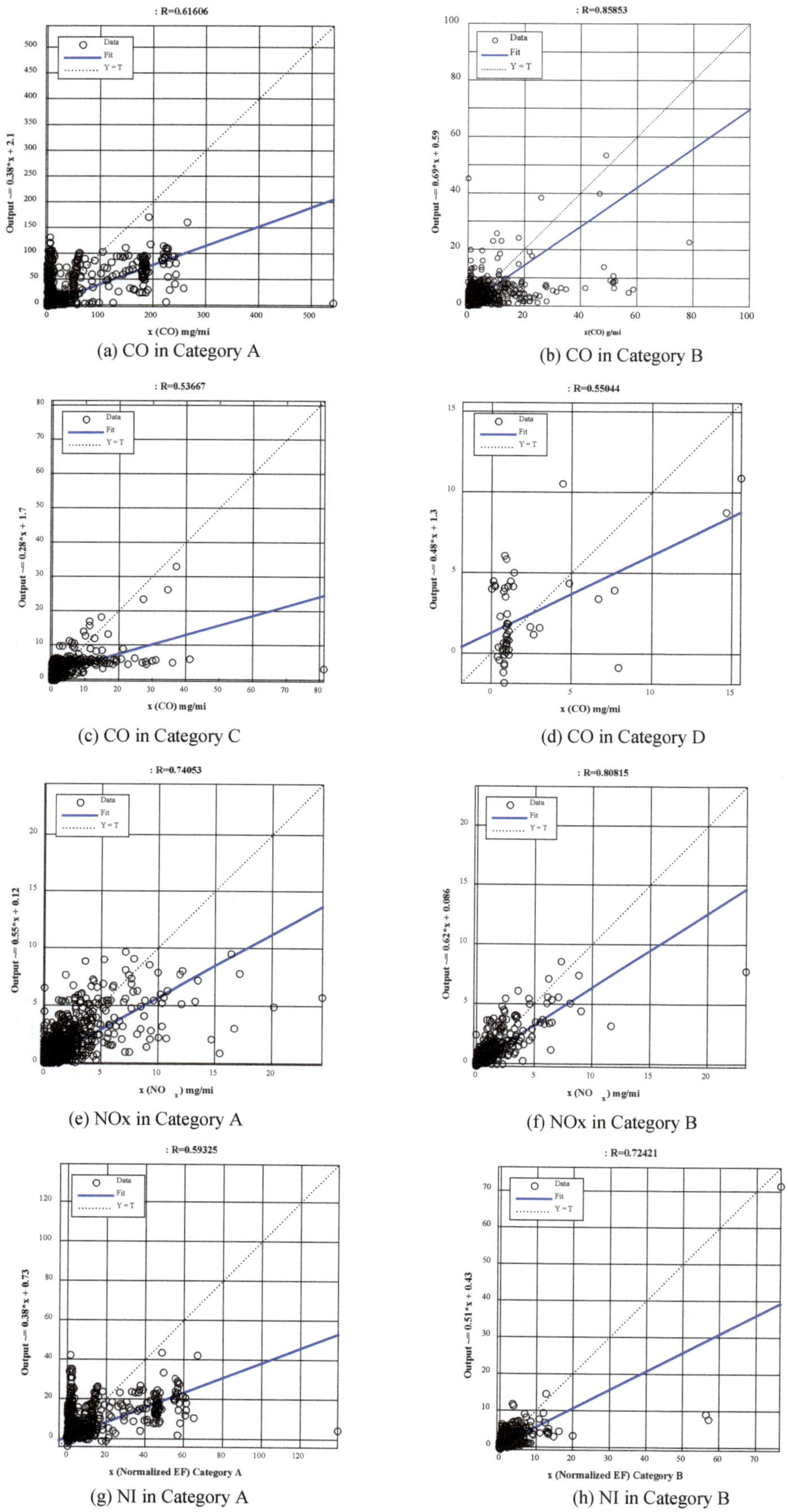


**Figure 4:** Fitted regression lines of CO in the four categories and NOx in category A and B.

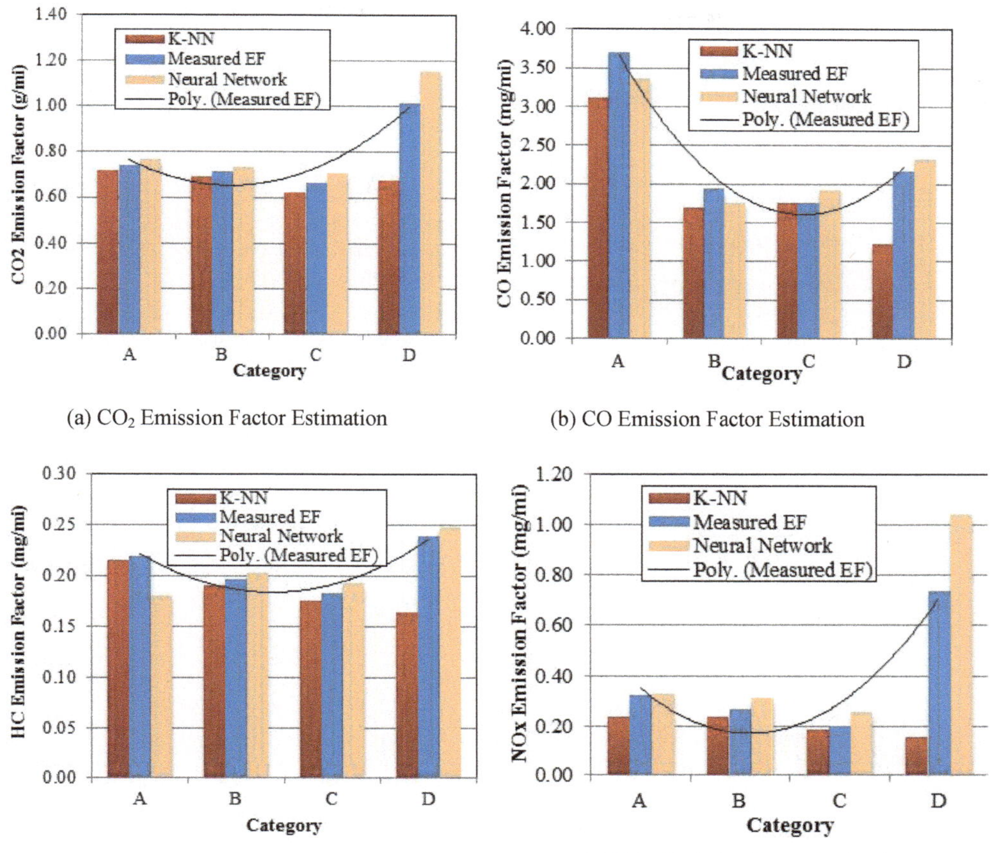


(a) $CO_2$ Emission Factor Estimation

(b) CO Emission Factor Estimation

(c) HC Emission Factor Estimation

(d) NOx Emission Factor Estimation

**Figure 5:** An overview of model estimation of emission factors under different pavement categories.

## Conclusion

In this research, the K-NN and NN models are employed to model the emission factors based on the 1,609 km on-board emission tests in the state of Texas in United States. The modeling was conducted separately based on the range of pavement roughness (categories A, B, C, and D). The input variables include vehicle operational and engine information. Results show that, the K-NN model poses more accurate on the estimate of emission factor than the NN model in the four categories of pavement roughness. Meanwhile, the nonlinear relationship between vehicle emissions and pavement roughness is further validated. Driving on a smoother and rougher pavement result in higher vehicle emissions, which is consistent with the previous study by Li et al. [27].

### Acknowledgements

Supports for this research in part by the U.S. National Tier 1 University Transportation Center (UTC) TranLIVE #DTRT12GUTC17/KLK900-SB-003, and the U.S. National Science Foundation (NSF) CREST #1137732 are gratefully acknowledged.

### References

1. Kuykendal W (2014) Emissions inventory guidance for implementation of ozone and particulate matter national ambient air quality standards (NAAQS) and regional haze regulations. DIANE Publishing, U.S. Environmental Protection Agency (US EPA).
2. US Environmental Protection Agency (US EPA) (2016a) Technology transfer network clearinghouse for inventories and emissions factors.
3. US Environmental Protection Agency (US EPA) (2016b) Moves (Motor Vehicle Emission Simulator) 2014a.
4. Qiao F, Yu L, Li L (2007) Estimating impact of nonrecurring congestion on vehicle emissions. In: Transportation Research Board 86th Annual Meeting (No. 07-0427).
5. Caiazzo F, Ashok A, Waitz IA, Yim SH, Barrett SR (2013) Air pollution and early deaths in the United States. Part I: Quantifying the impact of major sectors in 2005. Atmospheric Environment. 79: 198-208.
6. Li Q, Qiao F, Yu L (2016) Vehicle emission implications of drivers' smart advisory system for traffic operations in work zones. Journal of the Air and Waste Management Association 66(5): 446-455.
7. Intergovernmental Panel on Climate Change (IPCC) (2006) IPCC guidelines for national greenhouse gas inventories.
8. Qiao F, Yu L, Vojtisek-Lom M. (2005) On-road vehicle emission and activity data collection and evaluation in Houston, Texas. Transportation Research Record: Journal of the Transportation Research Board (1941): 60-71.
9. Frey HC, Kim K (2005) Operational evaluation of emissions and fuel use of B20 versus diesel fueled dump trucks.
10. Sjödin A, Jerksjö M (2008) Evaluation of European road transport emission models against on-road emission data as measured by optical remote sensing.
11. Franco V, Kousoulidou M, Muntean M, Ntziachristos L, Hausberger S, et al. (2013) Road vehicle emission factors development: A review. Atmospheric Environment. 70:84-97.
12. Kousoulidou M, Fontaras G, Ntziachristos L, Bonnel P, Samaras Z, et al. (2013) Use of portable emissions measurement system (PEMS) for the development and validation of passenger car emission factors. Atmospheric Environment. 64:329-38.
13. Li Q, Qiao F, Yu L (2016) Calibrating emission factors for highways considering pavement roughness information. Proceedings of the 2016 Air Quality Measurement Methods and Technology.15-7.
14. EPA U. (2003) User's guide to Mobile 6. 1 and Mobile 6. 2. Environmental Protection Agency.

15. Qiao F, Li Q, Yu L (2016) Neural network modeling of in-vehicle noises with different roadway roughness. Bridging the East and West. 27:174.

16. Munni J, Qiao F, Li Q, Yu L (2015) Driving behavior and emission analysis at yellow interval with advanced warning message under foggy weather condition: A simulator test. In: Proceedings of the 56th Annual Transportation Research Forum in Atlanta, Georgia.

17. Li Q, Qiao F, Yu L (2016) Estimating vehicle idle emissions based on on-board diagnostic II data. Proceedings of the 2016 Air Quality Measurement Methods and Technology. 15-7.

18. Li Q, Qiao F, Yu L (2015a) Implications of wireless communication system for traffic operations on vehicle emissions. Proceedings of the 107th Air & Waste Management Association (AWMA), June 22-25, 2015. Raleigh, North Carolina, USA.

19. Mansour HH, Aidin M (2012) Developing pollutant emission models for diesel public vehicles in Tehran. International Conference on Ecological, Environmental and Biological Sciences (ICEEBS'2012) Jan. 7-8, 2012 Dubai.

20. Li Q, Qiao F, Yu L (2014) Impacts of vehicles to infrastructure communication technologies on vehicle emissions. Proceedings of the International Conference on Environmental Science and Technology, American Academy of Sciences. 1: 9-13.

21. Li Q, Qiao F, Wang X, Yu L (2013) Impacts of P2V wireless communication on safety and environment in work zones through driving simulator tests (paper # 26-179). Proceedings of the 26th Annual Conference of the International Chinese Transportation Professionals Association (ICTPA). Tampa. 24-26.

22. Qiao F, Li Q, Yu L (2014) Testing impacts of work zone X2V communication system on safety and air quality in driving simulator. Publication in the proceedings of the 21st ITS World Congress. Detroit. 7-11.

23. Li Q, Qiao F (2014) How drivers' smart advisory system improves driving performance? A simulator imitation of wireless warning on traffic signal under sun glare. LAMBERT Academic Publishing.

24. Chatti K, Zaabar I (2012) Estimating the effects of pavement condition on vehicle operating costs. Transportation Research Board.

25. Li Q (2016) Impact of freeway weaving segment design on environment and public health. Ph.D. Dissertation. Environmental Toxicology. Texas Southern University.

26. Li Q, Qiao F, Yu L (2015) Will vehicle and roadside communications reduce emitted air pollution? International Journal of Science and Technology. 1:17-23.

27. Li Q, Qiao F, Yu L (2016) Clustering pavement roughness based on the impacts on vehicle emissions and public health. J Ergonomics 6:1

28. Corry L (1997) Hermann Minkowski and the postulate of relativity. Archive for History of Exact Sciences 51: 273-314

29. Qiao F, Yang H, Lam W (2001) Intelligent simulation and prediction of traffic flow dispersion. International Journal of Transportation Research.

30. GraphPad (2010) The unequal variance (Welch) t-test

31. Roadroid (2013) Road conditioning monitoring using smart phones.

# Separation of Magnesium Hydroxide and Barium Sulphate from a Barium Sulphate – Magnesium Hydroxide Mixed Sludge by Carbonation: The Effect of Temperature

**T.T. Rukuni[1]*, J.P Maree[2] and C.M Zvinowanda[1]**

*[1]Department of Environment, Water & Earth Sciences, Tshwane University of Technology, Private Bag X680, 0001, Pretoria, South Africa*
*[2]Incumbent, Rand Water Chair in Water Utilisation, Tshwane University of Technology, Private Bag X680, 0001, Pretoria, South Africa*

**Abstract**

The solids that result from mine wastewater treatment usually contain elevated levels of contaminants that were originally contained in the wastewater. These must be carefully disposed or treated to avoid shifting of the original pollutants in the waste stream to the final disposal site where they may again become free to contaminate the environment. A more reasonable approach to ultimate solids disposal is to view the sludge as a resource that can be recycled or reused. In South Africa, reverse osmosis is already being used for desalination of mine water and huge sludge volumes are also produced. The Tshwane University of Technology-Magnesium-Barium-Oxide (TUT-MBO) process and its variations is an alternative technology that offers the benefit of lower cost as magnesium hydroxide, barium hydroxide and coal are the main process raw materials. In the first stage $Mg(OH)_2$ is dosed to raise the pH of the acid mine drainage to 9 for removal of free acid, iron(II) oxidized to iron(III) and all other metals precipitated as metal hydroxides. In the second stage $Ba(OH)_2$ is dosed for magnesium and sulphate removal as $Mg(OH)_2$ and $BaSO_4$ respectively. The resultant, mixed $BaSO_4/Mg(OH)_2$ sludge is treated in a thermal stage to produce BaS and MgO. The aim of this study was to separate magnesium hydroxide from barium sulphate, produced in the second stage of the TUT-MBO Process. Magnesium hydroxide is separated from barium sulphate through the dissolution of $Mg(OH)_2$ with $CO_2$ to $Mg(HCO_3)_2$. The results showed that: (a) By adding $CO_2$ to a $BaSO_4/Mg(OH)_2$ sludge, selective dissolution of $Mg(OH)_2$ occurred due to the relatively high solubility of $Mg(HCO_3)_2$ and the low solubility of $BaSO_4$ and, (b) the solubility of $Mg(HCO_3)_2$ increased with decreasing temperatures and increasing pressures.

**Keywords:** Acid mine drainage; Precipitation; Dissolution, Reclamation, Carbon dioxide, Sludge disposal

## Introduction

South Africa currently faces both a shortage of water as well as a deterioration of the quality of the available water due to industrial activities such as mining and power generation. The gold mines in Gauteng are expected to decant 345 ML/day of acid mine drainage (AMD) when underground voids have filled up. This water will pollute surface water resources and cause serious environmental impacts [1]. South Africa was one of the first countries to implement commercial scale drinking water reclamation from mine water using reverse osmosis. The cost of reverse osmosis amounts to R10/m$^3$ and produces gypsum, ferric hydroxide rich sludge and brine that have to be disposed at a cost on sludge disposal dumps and brine ponds.

Tshwane University of Technology (TUT) has developed innovative technologies for the neutralization and desalination of acid mine drainage while avoiding gypsum crystallization. These technologies are the Magnesium-Barium-Oxide (TUT-MBO) process and its variations namely; Magnesium-Barium-Bicarbonate (TUT-MBB) process, where magnesium bicarbonate is used instead of magnesium hydroxide, and the Ammonium-Barium (TUT-NB) process [2], where ammonium hydroxide is used instead of magnesium hydroxide to avoid the formation of mixed sludges. Laboratory and pilot studies have demonstrated that magnesium hydroxide or magnesium bicarbonate can be used for removal of metals through precipitation as metal hydroxides. In the processes, magnesium and sulphate removal is achieved by treatment with $Ba(OH)_2$.

The TUT-MBO process and its variations offer the benefit of lower cost as magnesium hydroxide, barium hydroxide and coal are the main process raw materials [3,4]. It produces minimum amounts of sludge as process raw materials ($Mg(OH)_2$ or $Mg(HCO_3)_2$ and $Ba(OH)_2$) are recovered from the sludge. Sulphur, a valuable industrial raw material, can also be recovered from the sulphate removed from mine water.

The TUT-MBO process involves two stages. In the first stage, $Mg(OH)_2$ or $Mg(HCO_3)_2$ is dosed to raise the pH of the AMD to 9 for removal of free acid, iron(II) as iron(III), and all other metals as metal hydroxides. In the second stage $Ba(OH)_2$ is dosed for magnesium and sulphate removal as $Mg(OH)_2$ and $BaSO_4$, respectively. The resultant, mixed $BaSO_4/Mg(OH)_2$ sludge is treated in a thermal stage to produce a mixture of BaS and MgO. To avoid this mixture it would be beneficial if $BaSO_4$ and $Mg(OH)_2$ could be separated prior to thermal treatment. This would allow pure $BaSO_4$ to go to the thermal stage where it will be reduced to BaS. The Magnesium bicarbonate solution (TUT-MBB variation) can be used again in the first stage of TUT-MBO process instead of $Mg(OH)_2$. Figure 1 shows the schematic diagram of the TUT-MBB process.

The aim of this study was to explore the separation of magnesium hydroxide from barium sulphate, produced as a mixed sludge in the second stage of the TUT-MBO process, through dissolution of $Mg(OH)_2$ by converting it to $Mg(HCO_3)_2$ with $CO_2$. By adding $CO_2$ to a $BaSO_4/Mg(OH)_2$ sludge, selective dissolution of $Mg(OH)_2$ occurs according to the following reaction:

$$CO_{2(g)} + Mg(OH)_{2(s)} + H_2O_{(l)} \rightleftharpoons 2HCO^-_{3\,(aq)} + Mg^{2+}_{(aq)} \quad (1)$$

***Corresponding author:** Timothy T Rukuni, Department of Environment, Water & Earth Sciences, Tshwane University of Technology, Private Bag X680, 0001, Pretoria, South Africa, E-mail: chivovot@tut.ac.za

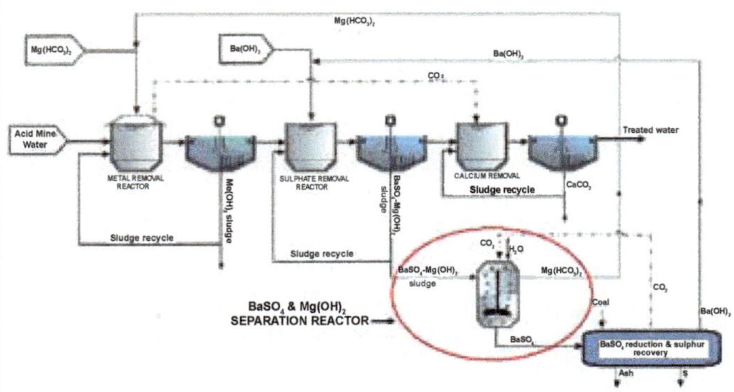


**Figure 1:** Schematic diagram of the TUT-MBB process for neutralization, metal removal and desalination of AMD.

As the system becomes enriched in $CO_2$, the dissolution rate of $Mg(OH)_2$ could be influenced by its saturation level, Ω.

$$\Omega=[Mg^{2+}][OH^-] / K_{sp} \quad (2)$$

Where $K_{sp}$ (T, S, p) is the solubility product (T is temperature, S is solubility and p is pressure) [5] with the pressure dependency adjustments by Millero et al. [6].

$$K_{sp} = [Mg^{2+}][OH^-] \quad (3)$$

**Note:** All the square-bracketed species are stoichiometric concentrations (molality or mol /kg) and disregard any complex formation or ion pairs. It is thus necessary to specify not only the pH scale used in the evaluation of the dissociation constants [7], but also their dependency on ionic strength, temperature, and pressure.

## Materials and Methods

Commercial grade $Mg(OH)_2$ (60 g) and analytical grade (14 g) $BaSO_4$ were mixed with deionized water and made up to volume (2 L). Bottled $CO_2$ was used for dissolving $Mg(OH)_2$. Batch studies were carried out using a completely-mixed pressurized reactor (Figure 2). It consisted of a 3 L reaction vessel equipped with a *BirCraft* stirrer, temperature sensor, pressure gauge, pH and conductivity control sensors. This unit was designed to handle pressures up to 10 bar and a maximum temperature of 150°C. The main body of the reactor was constructed from a Class 12 uPVC pipe with a wall thickness of 3.2 mm and an internal diameter of 560 mm. The height of the main body was 520 mm with a uPVC base and top plates that were each 65 mm thick. The dosage points were 300 mm above the effluent off-take point that was fitted at the bottom of the reactor. Calcium carbonate and barium sulphate were first fed into the reactor from the top and deionized water was pumped into the reaction vessel with a *Watson-Marlow* pump with continuous stirring. Carbon dioxide was dissolved in water, under pressure in the reactor, to lower the pH to around 6 and increase the pressure to desired levels. A pH control unit was used to control $CO_2$ dosage by monitoring the pH. At the end of the reaction the solution was filtered under pressure and the filtrate released from the pressure vessel through a valve.

The effects of the following parameters were investigated: Time (0, 10, 20, 40, 60 min (measured)), Temperature (0, 10, 25, 45°C (measured and predicted)), Pressure (predicted), $Mg(OH)_2$ concentration (60 g/L (measured and predicted)) and $BaSO_4$ concentration (0, 14 g/L (predicted)). The default conditions were 0°C, 60 g/L $Mg(OH)_2$ , 14 g/L $BaSO_4$. Filtered samples were collected over 0 - 60 min and analyzed for pH, alkalinity, calcium, barium and sulphate. The samples were filtered (0.45 μm membrane filter) to remove all solid material.

The total dissolved carbonate species were measured by titrating the sample with standard hydrochloric acid to pH 4.5. The volume of acid was used to calculate the total alkalinity of the sample. From this calculation, the carbonate and hydroxide species concentrations were calculated, and the total carbonate species concentration (as mg/L $CaCO_3$).

Magnesium hardness was determined by titration with EDTA (0.02 M) using P & R indicator and NaOH buffer. Total hardness was determined with EDTA (0.02 M) using EriochromeBlack T indicator and ammonia buffer.

Barium concentrations were determined by titrating the sample with standard 0.02 M EDTA using methylthymol blue indicator, potassium nitrate indicator and NaOH buffer (for pH 12).

The pH, conductivity and pressure in the reactor were measured directly.

## Results and Discussion

Table 1 and Figures 3-7 show the dissolution of $Mg(OH)_2$ (magnesium hydroxide) and $BaSO_4$ (barite) when contacted with $CO_2$ in water under various conditions. The Figures 3-7 contain measured values while Table 1 contains both measured and predicted values. The predictions were done using a *Visual MINTEQ* model [8,9].

### Separation of $Mg(OH)_2$ and $BaSO_4$

The mixture of $Mg(OH)_2$ and $BaSO_4$ sludge produced by the TUT-MBO process cannot be separated by solubility differences due to the low solubility of both $Mg(OH)_2$ (3.8 mg/LMg,) and $BaSO_4$ (5.6 mg/L

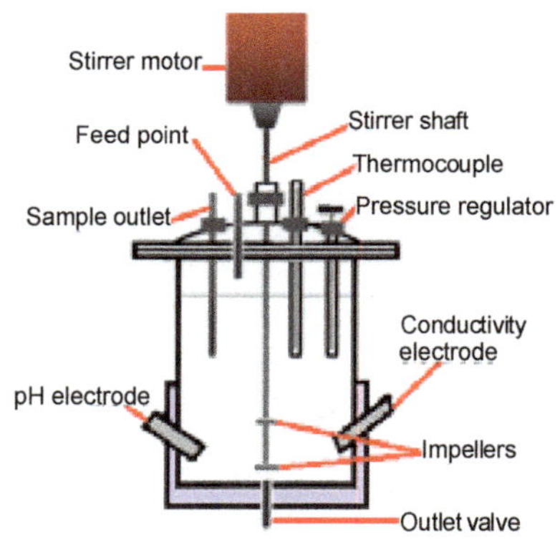


**Figure 2:** The completely-mixed pressurized reactor.

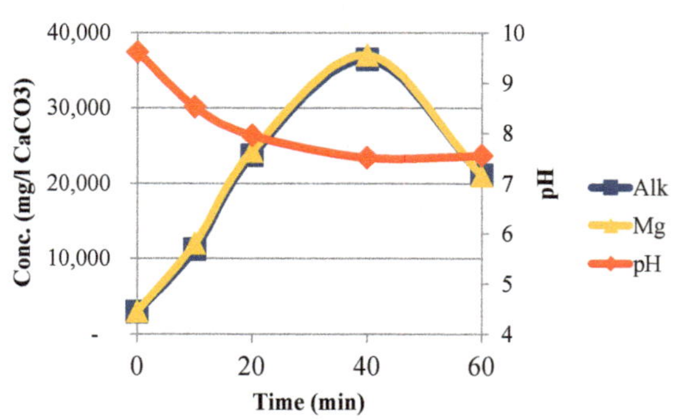


**Figure 3:** Behaviour of various parameters during contact between $Mg(OH)_2$ and $CO_2$ (Temp = 10°C; 60 g/L $Mg(OH)_2$).

| Exp No | Feedstock | | Sulphate in treated water (mg/l) | pH | | Temp | Alk (mg/l CaCO3) (aq) | | Ba in solution (mg/l) |
|---|---|---|---|---|---|---|---|---|---|
| | Mg(OH)2 (s) (g/l) | BaSO4 (s) (g/l) | | Predicted | Determined | | Predicted | Determined | Predicted |
| | DeterminedPredictedBehaviour of various parameters (see Figure 3) | | | | | | | | |
| 1.2 | 60 | 0 | 1 | | 7.55 | 10 | 60,000 | 21,100 | |
| | Solubility of Mg $(OH)_2$ and $BaSO_4$ when in contact with $CO_2$; pH and Mg measured and predicted (Figure 4) | | | | | | | | |
| 1.2 | 60 | 0 | 1 | | 7.55 | 10 | 60,000 | 21,100 | |
| 3.2 | 0 | 20 | 1 | | 6.81 | 10 | - | 50 | |
| | Effect of $BaSO_4$; pH, Mg and Ba measured and predicted (Figure 5) | | | | | | | | |
| 1.1 | 60 | 0 | 1 | | 7.44 | 0 | 60,000 | 21,000 | |
| 2.1 | 60 | 14 | 1 | 7.67 | 6.01 | 0 | 60,000 | 21,000 | 6.43 |
| 1.2 | 60 | 0 | 1 | | 7.55 | 10 | 60,000 | 21,000 | |
| 2.2 | 60 | 14 | 1 | | 7.55 | 10 | 60,000 | 19,796 | |
| | Effect of temperature on $Mg(OH)_2$ ; pH and Mg predicted and measured (see Figure 6) | | | | | | | | |
| 1.1 | 60 | 0 | 1 | | 7.44 | 0 | 60,000 | 21,000 | |
| 1.2 | 60 | 0 | 1 | | 7.55 | 10 | 60,000 | 21,000 | |
| 1.3 | 60 | 0 | 1 | | 7.45 | 25 | 60,000 | 17,600 | |
| 1.4 | 60 | 0 | 1 | | 7.38 | 45 | 60,000 | 12,300 | |
| | Effect of temperature on $Mg(OH)_2$ and $BaSO_4$; pH, Mg and Ba predicted and measured (see Figure 7) | | | | | | | | |
| 2.1 | 60 | 14 | 1 | 7.67 | 6.01 | 0 | 60,000 | 22,700 | 6.43 |
| 2.2 | 60 | 14 | 1 | | 7.55 | 10 | 60,000 | 19,796 | 9.17 |
| 2.3 | 60 | 14 | 1 | | 7.47 | 25 | 60,000 | 20,000 | 14.67 |
| 2.4 | 60 | 14 | 1 | | 7.48 | 45 | 60,000 | 8,000 | 24.29 |
| | Effect of sulphate on $Mg(OH)_2$ ; pH, Ba and $Mg(OH)_2$ predicted (Figure 8) | | | | | | | | |
| 4.1 | 60 | 0 | 0 | 7.67 | | 0 | 60,000 | | 6.43 |
| 4.2 | 60 | 0 | 9.6 | 7.67 | | 0 | 60,000 | | 2.75 |
| 4.3 | 60 | 0 | 48 | 7.67 | | 0 | 60,000 | | 0.66 |
| 4.4 | 60 | 0 | 96 | 7.67 | | 0 | 60,000 | | 0.34 |
| 4.5 | 60 | 0 | 192 | 7.67 | | 0 | 60,000 | | 0.17 |
| | Effect of sulphate on $Mg(OH)_2$ ; pH, Ba and $Mg(OH)_2$ predicted (Figure 8) | | | | | | | | |
| 5.1 | 60 | 14 | 0 | 7.76 | | 25 | 60,000 | | 14.67 |
| 5.2 | 60 | 14 | 9.6 | 7.75 | | 25 | 60,000 | | 9.75 |
| 5.3 | 60 | 14 | 48 | 7.76 | | 25 | 60,000 | | 3.32 |
| 5.4 | 60 | 14 | 96 | 7.76 | | 25 | 60,000 | | 1.72 |
| 5.5 | 60 | 14 | 192 | 7.76 | | 25 | 60 | | 0.87 |

**Table 1:** Effect of various parameters on the solubility of $Mg(OH)_2$ and $BaSO_4$ in a $CO_2$-rich solution.

Ba). As $CO_2$ is produced as a waste product in the TUT-MBO process, it was decided to investigate whether $Mg(OH)_2$ could be separated from the almost insoluble $BaSO_4$ by dissolving it as $Mg(HCO_3)_2$ (Equation1), by contacting the sludge mixture with $CO_2$. Figure 3 shows that 60 g/L $Mg(OH)_2$ dissolved partially when contacted with $CO_2$ at 1 atm. As $Mg(HCO_3)_2$ formed, Mg and alkalinity (Alk) values increased rapidly to 36 458 mg/L (as $CaCO_3$), whereafter it dropped to 21 100 mg/L (as $CaCO_3$). The maximum solubility value of 36 358 mg/L can be explained by the high solubility of $Mg(HCO_3)_2$, the intermediate product when $Mg(OH)_2$ is contacted with $CO_2$. Except for $Mg(HCO_3)_2$, there are also other Mg species present in the system, such as artinite, brucite, hydromagnesite, magnesite, magnesium hydroxide, magnesium carbonate, nesquehonite and periclase. The *Visual Minteq* [8,9] model showed that some of these compounds exceeded their solubilities as indicated by the positive saturation index values. Precipitation of the compounds that were over saturated would explain the observed drop in $Mg(HCO_3)_2$ concentration in solution from 36 358 to 21 100 mg/L $Mg(HCO_3)_2$ (as $CaCO_3$). It also showed the ionic balance between the measured calcium concentration and the total alkalinity of the system. Over the same period the pH dropped from 9.6 to 7.6. The dissolution of $Mg(OH)_2$ increased with decreased pH due to the increased formation of soluble $Mg(HCO_3)_2$. Therefore, $CO_2$ dosing lowers the pH, and magnesium hydroxide is converted to $Mg(HCO_3)_2$.

Alkalinity was used to monitor the formation of $Mg(HCO_3)_2$ and includes the parameters listed in Equation (4). As the system became enriched in $CO_2$, the extent of dissolution decreased as a function of changes in the $Mg(OH)_2$ saturation state to yield both magnesium ions and Alk (Equation 1).

$$Alk = 2[CO_3^{2-}] + [HCO_3^-] + [OH^-] + [H^+] \quad (4)$$

Figure 4 shows the contrast between the solubilities of $Mg(OH)_2$ (21100 mg/L as $CaCO_3$) and $BaSO_4$ (50 mg/L as $CaCO_3$). The results showed that, as expected, $BaSO_4$ does not dissolve when contacted with $CO_2$ and consequently there were negligible losses of $BaSO_4$ due to dissolution.

The effect of $BaSO_4$ on the rate of formation and solubility of $Mg(HCO_3)_2$ was studied. Figure 5 showed that $BaSO_4$ had no effect on

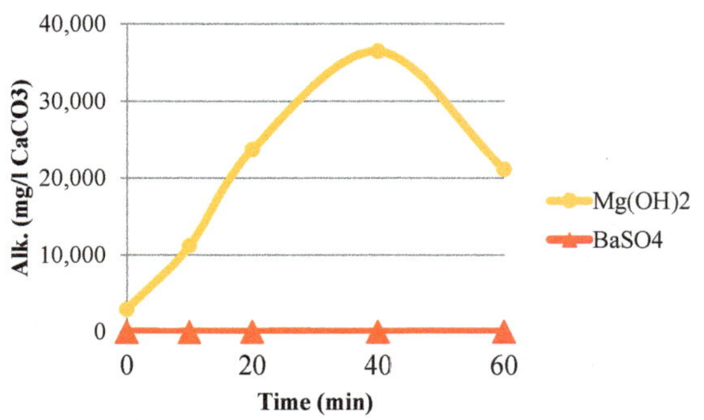


**Figure 4:** Comparison between dissolution of $Mg(OH)_2$ and $BaSO_4$ in a $CO_2$-rich solution (10°C; 60 g/L $Mg(OH)_2$; 14 g/L $BaSO_4$).

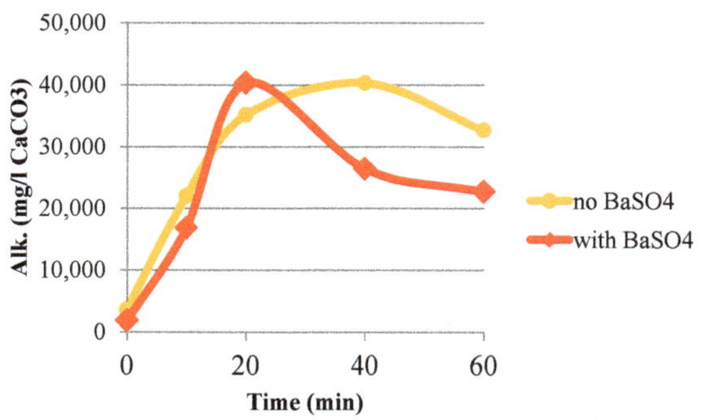


**Figure 5:** Effect of $BaSO_4$ on the rate of formation and solubility of $Mg(HCO_3)_2$ (0°C, 60 g/L $Mg(OH)_2$, no $BaSO_4$; 60 g/L $Mg(OH)_2$ and 14 g/L $BaSO_4$).

the solubility of $Mg(HCO_3)_2$. This was because $BaSO_4$ does not react with $CO_2$. The known low solubility of $BaSO_4$ was also reflected by the low predicted values for $Ba^{2+}$ in solution, when $BaSO_4$ is contacted with $CO_2$ (Table 1).

## Effect of temperature on solubility

In the previous section it was shown that $Mg(OH)_2$ can be dissolved by the formation of $Mg(HCO_3)_2$, through $CO_2$ addition without affecting the low solubility of $BaSO_4$. The sludge separation process will be more effective at higher solubility values for $Mg(HCO_3)_2$. Therefore, it was decided to determine the effect of temperature on the solubilisation of $Mg(OH)_2$ as $Mg(HCO_3)_2$. According to Henry's law (Equation 5), it was expected that the solubility should increase with decreasing temperatures. Figures 6 and 7 confirmed the validity of Henry's law. In the case of Figure 6, only $Mg(OH)_2$ was present in the slurry used, whilst in Figure 7, $BaSO_4$ was also present. As in the previous case, it was found that $BaSO_4$ had no effect on the solubilisation of $Mg(OH)_2$ when contacted with $CO_2$. These figures also showed that the concentration of $Mg^{2+}$ in solution increased to the maximum value in the first 20 minutes and stabilized for another 20 minutes and then fell to a final stable level. This implied that at a high concentration of 60 g/L, not all of the magnesium hydroxide will be carbonated to form $Mg(HCO_3)_2$.

$$P_{gas} = kC \text{ (at constant T)} \tag{5}$$

Where, P = gas partial pressure (Pa)

k = Henry's law constant (Pa m³mol)

C = concentration of the gas (mol/L)

Alaee et al. [10] showed that the air/water Henry's Law constant (K) is defined as the ratio of the concentration of a chemical in the gas phase to its concentration in the aqueous phase.

$$K = \frac{P_i}{C_w} \tag{6}$$

Where, K is in Pa.m³mol; $P_i$ is partial pressure (Pa) and $C_i$ is aqueous concentration (mol/L).

The temperature effect on the Henry constant K can be expressed as:

$$\text{Ln K} = K = \frac{P_i}{C_w} \tag{7}$$

Where T is temperature in K, A and B are constants of the Van't Hoff equation.

Ten-Hulscher et al. [11] showed that B is the ratio of the enthalpy of volatilization to the gas constant, $\Delta H_o/R$ in $K^{-1}$, and A is the ratio of the entropy of volatilization to the gas constant $\Delta S_o/R$, resulting in a dimensionless value.

## Measured versus predicted solubility values

Table 1 shows a comparison of the predicted and determined values for the effect of various parameters on the solubilisation of $Mg(OH)_2$ and $BaSO_4$ in a $CO_2$-rich solution. The predictions were done using the *Visual MINTEQ* [8,9] model. The model was designed to simulate equilibrium and speciation of inorganic solutes in natural waters.

**Temperature and sulphate:** Figures 6 (experiment 1) and 7 (experiment 2) compare the solubility of $Mg(HCO_3)_2$ at 1 atm $CO_2$ when no $BaSO_4$ and 14 g/L $BaSO_4$, respectively, were present over the temperature range 0-45°C. Both sets of results showed that the final "solubility" of $Mg(OH)_2$ increases with decreasing temperature. The measured "solubility" values for $Mg(OH)_2$ when contacted with $CO_2$ were lower than predicted by *Visual Minteq*. This can be ascribed to the high concentrations in solution which exceeded the model's operation range.

The results of experiment 2 in Table 1 show the predicted values for $Ba^{2+}$ at 0°C in the treated water. The $Ba^{2+}$ concentration increased from 6.4 to 24.3 mg/L for a temperature change of 45°C. It is preferred to have $Ba^{2+}$ concentrations of less than 2 mg/L in treated water as this is within the expected range in natural waters. This can be achieved by allowing a low residual sulphate concentration of at least 10 mg/L in the treated water. The residual sulphate ion will act as a common ion which decreases barium solubility. Figure 8 (Experiments 4 and 5) showed that, at 25°C, a $Ba^{2+}$ concentration of less than 2 mg/L at a

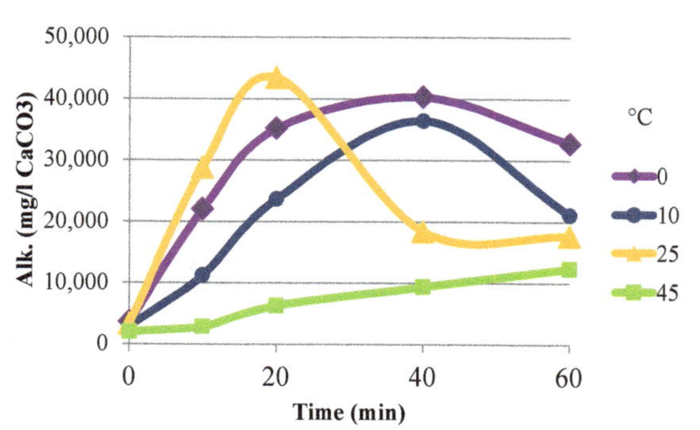


**Figure 6:** Effect of temperature on the rate of formation and solubility of $Mg(HCO_3)_2$ at low pressure (1 atm) under $BaSO_4$-free conditions (60 g/L $Mg(OH)_2$).

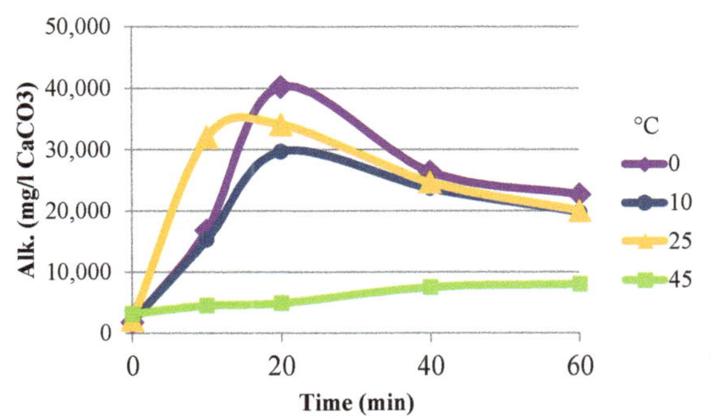


**Figure 7:** Effect of temperature on the rate of formation and solubility of $Mg(HCO_3)_2$ In the presence of $BaSO_4$ (60 g/L $Mg(OH)_2$, 14 g/L $BaSO_4$).

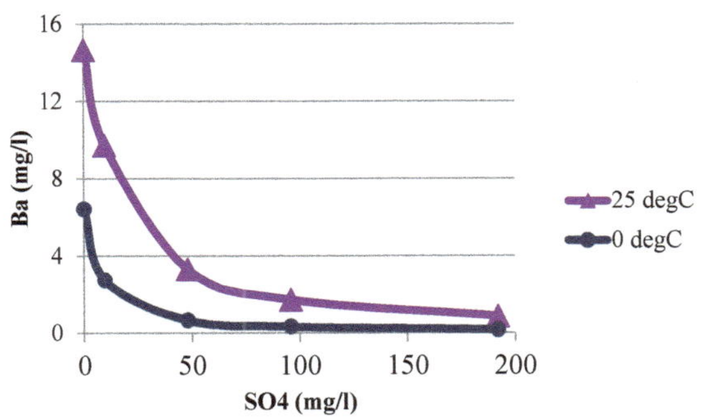


**Figure 8:** Effect of sulphate concentration in treated water on residual barium(II).

sulphate concentration of 100 mg/L. At 0°C a sulphate concentration of 25 mg/L sulphate was sufficient to keep the $Ba^{2+}$ concentration at less than 2 mg/L.

**Pressure:** Rukuni at al. [12] showed that the solubility of $Ca(HCO_3)_2$ is influenced by pressure. It was also planned to investigate the effect of pressure on $Mg(HCO_3)_2$ solubility. Due to the high solubility of $Mg(HCO_3)_2$ at ambient pressure and temperature this was seen as less practical in the context of the process being developed. The high concentration also does not allow the use of the *Visual Minteq* [8,9] model to predict the solubility at various pressures.

## Cost

The process cost for this process is low because all the process raw materials ($CO_2$ and the $Mg(OH)_2$-$BaSO_4$ sludge) are waste products of the TUT-MBO process (Figure 1). It is also foreseen that the cost can also be kept low in other applications by producing $CO_2$ on-site by burning coal and scrubbing in water than the purchase of pure $CO_2$.

## Conclusions

It was found that:

- $Mg(OH)_2$ can be separated from $BaSO_4$ and $Mg(OH)_2$ in a mixed sludge by carbonating it to $Mg(HCO_3)_2$ using $CO_2$.
- The dissolution rate of $Mg(OH)_2$ in the presence of $CO_2$ is fast in the initial 20 minutes of the reaction.
- $Mg(OH)_2$ had a high solubility of about 22 700 mg/l when in contact with $CO_2$ at 1 atm, while $BaSO_4$ is almost completely insoluble.
- The solubility of $Mg(OH)_2$ increases with decreasing temperature and increasing pressure.
- The *Visual Minteq* model was a powerful tool to predict the "solubility" of $Mg(OH)_2$ and $BaSO_4$ when contacted with $CO_2$.

Thus, the TUT-MBO process offers a sustainable method for neutralization, metal removal and desalination of AMD and recovery of saleable and reusable products from the mixed sludge produced. Because $Mg(HCO_3)_2$ has high solubility at low temperatures and high pressure, the practical optimal operation conditions for the dissolution reaction are temperatures close to 0°C and atmospheric pressure. This is because high pressures need sophisticated reactor designs and the systems are more complicated to run that those at atmospheric pressure conditions.

## Acknowledgment

The financial support from THRIP and the laboratory facilities from Tshwane University of Technology are greatly acknowledged.

## References

1. Department of Water Affairs (2010) Report to the inter-ministerial committee on acid mine drainage. Mine water management in the Witwatersrand gold fields with special emphasis on acid mine drainage.
2. Maila D, Maree JP, Zvinowanda CM, Kolesnikov AV, Louw WJ (2012) Treatment of Acid mine drainage using ammonium hydroxide. The 27th International Conference on Solid Waste Technology and Management, Philadelphia, USA. 11-14 March (2012).
3. V Bologo, JP Maree and WJ Louw (2012) Neutralization of acid mine drainage with magnesium bicarbonate, WISA Biennial Conference, Cape Town International Conference Centre, 18-22 April (2012).
4. Bologo V, Maree JP, Carlsson F (2012) Application of magnesium hydroxide and barium hydroxide for the removal of metals and sulphate from mine water. Water SA 38: 23-28.
5. Mucci A (1983) The solubility of calcite and aragonite in seawater at various salinities, temperatures and 1 atmosphere total pressure. Am J Sci 283: 780–799.
6. Millero F J (1995) Thermodynamics of the carbon dioxide system in the oceans. Geochim Cosmochim Acta 59: 661– 677.
7. Dickson AG (1984) pH scales and proton-transfer reactions in saline media such as seawater. Geochim Cosmochim Acta 48: 2299-2308.
8. Visual MINTEQ, http://www2.lwr.kth.se/English/OurSoftware/vminteq/Visual%20MINTEQ_tutorial.pdf. (2000) (Accessed: 24/11/2011).
9. Mintek, http://www2.lwr.kth.se/English/OurSoftware/vminteq/ (Accessed: 24/11/2011).
10. Alaee M, Whittal RM, Strachan WMJ (1996) The effect of water temperature and composition on Henry's law constant for various PAHs. Chemosphere 32: 1153–1164.
11. Ten-Hulscher, TEM, Van Der Velde LE, Bruggeman WA (1992) Temperature dependence of henry's law constants for selected chlorobenzenes, polychlorinated biphenyls, and polycyclic aromatic hydrocarbons. Environ Toxicol Chem 11: 1595–1603.
12. Rukuni TT, Maree JP, Zvinowanda CM (2012) Recovery of calcium carbonate from a barium sulphate-calcium carbonate mixed sludge. The 27th International Conference on Solid Waste Technology and Management, Philadelphia, USA. 11-14 March (2012).

# Permissions

All chapters in this book were first published in JCEE, by OMICS International; hereby published with permission under the Creative Commons Attribution License or equivalent. Every chapter published in this book has been scrutinized by our experts. Their significance has been extensively debated. The topics covered herein carry significant findings which will fuel the growth of the discipline. They may even be implemented as practical applications or may be referred to as a beginning point for another development.

The contributors of this book come from diverse backgrounds, making this book a truly international effort. This book will bring forth new frontiers with its revolutionizing research information and detailed analysis of the nascent developments around the world.

We would like to thank all the contributing authors for lending their expertise to make the book truly unique. They have played a crucial role in the development of this book. Without their invaluable contributions this book wouldn't have been possible. They have made vital efforts to compile up to date information on the varied aspects of this subject to make this book a valuable addition to the collection of many professionals and students.

This book was conceptualized with the vision of imparting up-to-date information and advanced data in this field. To ensure the same, a matchless editorial board was set up. Every individual on the board went through rigorous rounds of assessment to prove their worth. After which they invested a large part of their time researching and compiling the most relevant data for our readers.

The editorial board has been involved in producing this book since its inception. They have spent rigorous hours researching and exploring the diverse topics which have resulted in the successful publishing of this book. They have passed on their knowledge of decades through this book. To expedite this challenging task, the publisher supported the team at every step. A small team of assistant editors was also appointed to further simplify the editing procedure and attain best results for the readers.

Apart from the editorial board, the designing team has also invested a significant amount of their time in understanding the subject and creating the most relevant covers. They scrutinized every image to scout for the most suitable representation of the subject and create an appropriate cover for the book.

The publishing team has been an ardent support to the editorial, designing and production team. Their endless efforts to recruit the best for this project, has resulted in the accomplishment of this book. They are a veteran in the field of academics and their pool of knowledge is as vast as their experience in printing. Their expertise and guidance has proved useful at every step. Their uncompromising quality standards have made this book an exceptional effort. Their encouragement from time to time has been an inspiration for everyone.

The publisher and the editorial board hope that this book will prove to be a valuable piece of knowledge for researchers, students, practitioners and scholars across the globe.

# List of Contributors

**Jung Kwan Seo, So Young Bae, Jeom Kee Paik and Taek Soo Jang**
The Ship and Offshore Research Institute, Pusan National University, Busan, Korea

**Ghada Hasan, Mohammed Ahmed Al Kubaisy, Faten Hamed Nahhas, Ali Adil Ali, Nibras Othman and Mahir Mahmod Hason**
Civil Engineering Department, Universiti Putra Malaysia, Malaysia

**Nwachukwu MA, Umunna NA and Ntesat B**
Department of Environmental Technology, Federal University of Technology Owerri, Nigeria

**Umunna CP**
Enugu State University of Science and Technology, Park Lane, Enugu, Nigeria

**Rathod JD**
Applied Mechanics Department, Faculty of Technology and Engineering, The M.S. University of Baroda, Vadodara, India

**Owayo AA and Ou CY**
Department of Civil and Construction Engineering, National Taiwan University of Science and Technology, ROC Taipei-Taiwan

**Mohd Fadhil Md Din, Hazlini Dzinun, Ponraj Mohanadoss, Shreshivadasan Chelliapan and Zainura Zainun Noor**
Institute of Environmental Water Resources and Management (IPASA), Faculty of Civil Engineering, Universiti Teknologi Malaysia, 81310 UTM Skudai, Johor, Malaysia

**Dilshan Remaz Ossen**
Faculty of Built Environment, Universiti Teknologi Malaysia, 81310 UTM Skudai, Johor, Malaysia

**Kenzo Iwao**
Nagoya Institute of Technology, Gokiso-cho, Showa-ku, Nagoya, Aichi-Pref., 466-8555, Japan

**Tamil Selvi P, Lakshmi Narayani P and Ramya G**
Department of Civil Engineering, Dr. Sivanthi Aditanar College of Engineering, Tiruchendur-628215, Tamilnadu, India

**Ramzi Taha and Okan Sirin**
Department of Civil & Architectural Engineering, Qatar University, Doha, Qatar

**Husam Sadek**
Department of Civil & Architectural Engineering, Qatar University, Doha, Qatar
Centre for Engineering Sustainability, University of Liverpool, Liverpool, UK

**Hasan ASMZ**
Department of Civil Engineering, Rajshahi University of Engineering and Technology, Bangladesh

**Hamid R**
Department of Civil and Structural Engineering, Universiti Kebangsaan Malaysia, Malaysia

**Ariffin AK**
Department of Mechanical and Materials Engineering, Universiti Kebangsaan Malaysia, Malaysia

**Ogundipe Olumide Moses**
Civil Engineering Department, Ekiti State University, Ado-Ekiti, Nigeria

**Shaoqi Zhou**
College of Environmental Science and Engineering, South China University of Technology, Guangzhou 510640, China
State Key Laboratory of Subtropical Building Science, South China University of Technology, Guangzhou 510641, China
Key Laboratory of Environmental Protection and Eco-remediation of Guangdong Regular Higher Education Institutions, South China University of Technology, Guangzhou Higher Education Mega Center, Guangzhou 510006, PR China

**Yiming Lin and Jingjing Zhao**
College of Environmental Science and Engineering, South China University of Technology, Guangzhou 510640, China
Key Laboratory of Environmental Protection and Eco-remediation of Guangdong Regular Higher Education Institutions, South China University of Technology, Guangzhou Higher Education Mega Center, Guangzhou 510006, PR China

**Songfeng Zeng and Jianhua Zhou**
Guangdong South Solda Industry, Co Ltd, Nangang West 488, HuanPu District, Guangzhou 510760, PR China

**Saulo Rodrigues Castro and Paulo José Silva**
University Center of Formiga, UNIFOR/MG Dr. Arnaldo de Senna Av, 328, Água Vermelha 35.570-000, Formiga, MG, Brazil

**Samuel Rodrigues Castro and Liséte Celina Lange**
Federal University of Minas Gerais, Engineering School, Pampulha Campus Antônio Carlos Av, 6627, Bl; 2, 4628 31.270-901, Belo Horizonte, MG, Brazil

**Niruba S**
Civil Engineering Department, Erode Sengunthar Engineering College, Thudupathi, Erode, India

**Cheong KWD and Tham KW**
Department of Building, National University of Singapore, Singapore

**Balasubramanian R**
Division of Environmental Science and Engineering, National University of Singapore, Singapore

**Kalaiarasan M**
School of Building and Development, BCA Academy (Building and Construction Authority of Singapore), Singapore

**Dilipan Bose S**
PG Student, M.tech Structural Engineering, Bharath University, India

**Aravindan S**
Assistant Professor, Department of Civil Engineering, Bharath University, India

**Anand Kumar Varma S and Srimurali M**
Department of Civil Engineering, S.V. University, Tirupati, AP, India

**Audu HAP**
Department of Civil Engineering, Faculty of Engineering, University of Benin, Benin City, Edo State, Nigeria

**Aniekwu AN**
Department of Civil Engineering, Faculty of Engineering, University of Benin, Benin City, Edo State, Nigeria

**Oghorodje EE**
Department of Highway Engineering, Ministry of Works, Asaba, Delta State, Nigeria

**Takayuki Kinoshita and Takumi Ito**
Department of Architecture, Tokyo University of Science, Japan

**Amir Shafi Batt and Anshul Garg**
Department of Civil Engineering, Lovely Professional University, Punjab, India

**Geethakarthi A**
Department of Civil Engineering, Kumaraguru College of Technology, Coimbatore, Tamil Nadu, India

**Takumi Ito and Kenjiro Mori**
Department of Architechture, Tokyo University of Science, Japan

**Boya You, Fengxiang Qiao and Lei Yu**
Innovative Transportation Research Institute, Texas Southern University, Houston, Texas 77004, USA

**Ryotaro Arai, Hanako Satoh, Kenta Aoki, Takumi Ito and Changhoon Choi**
Department of Architecture, Tokyo University of Science, Japan

**Kento Mori, Takehiro Takahashi and Takumi Ito**
Department of Architecture, Tokyo University of Science, Japan

**Akinobu Kano, Hideaki Azu and Takumi Ito**
Department of Architechture, Tokyo University of Science, Japan

**Donghung Wu**
Wu Office, P.R. Japan

**Takashi Nagumo and Haruhiko Hirata**
Hory Corporation, Japan

**Stephen A Arhin and Asteway Ribbiso**
Department of Civil and Environmental Engineering, Howard University, Washington, DC, 20059, USA

**Maria Inmaculada García Hernández**
Uppsala University, Ångströmlab, Lägerhyddsvägen, Uppsala, Sweden

**Tomoe Onoda, Takumi Ito, Kenjiro Mori, Hiromu Fuse and Changhoon Choi**
Department of Architecture, Tokyo University of Science, Japan

**Aamir Mahawish**
Department of Highway and Transportation Engineering, University of Al-Mustansiriyah, Iraq

**Sarmed Imad Ibrahim, Ali Hussein Jawad and Ferhad Mohamed Othman**
Department of Materials Engineering, University of Technology, Baghdad, 00964, Iraq

**Marco Magnoni, Filippo Giustozzi, Emanuele Toraldo and Maurizio Crispino**
Politecnico di Milano - Department of Civil and Environmental Engineering , Transportation Infrastructures, Piazza Leonardo da Vinci 32, Milan 20133, Italy

**Itsuki Yagi, Mana Kitahonoki and Takumi Ito**
Department of Architecture, Tokyo University of Science, Japan

**Qing Li, Fengxiang Qiao and Lei Yu**
Innovative Transportation Research Institute, Texas Southern University, USA

**T.T. Rukuni and C.M Zvinowanda**
Department of Environment, Water & Earth Sciences, Tshwane University of Technology, Private Bag X680, 0001, Pretoria, South Africa

**J.P Maree**
Incumbent, Rand Water Chair in Water Utilisation, Tshwane University of Technology, Private Bag X680, 0001, Pretoria, South Africa

# Index

**R**



**S**



**T**



**U**



**W**